IET MATERIALS, CIRCUITS & DEVICES 26

High Speed Data Converters

Other volumes in this series:

Volume 2	**Analogue IC Design: The current-mode approach** C. Toumazou, F.J. Lidgey and D.G. Haigh (Editors)
Volume 3	**Analogue-Digital ASICs: Circuit techniques, design tools and applications** R.S. Soin, F. Maloberti and J. France (Editors)
Volume 4	**Algorithmic and Knowledge-based CAD for VLSI** G.E. Taylor and G. Russell (Editors)
Volume 5	**Switched Currents: An Analogue technique for digital technology** C. Toumazou, J.B.C. Hughes and N.C. Battersby (Editors)
Volume 6	**High-frequency Circuit Engineering** F. Nibler *et al.*
Volume 8	**Low-power High-frequency Microelectronics: A unified approach** G. Machado (Editor)
Volume 9	**VLSI Testing: Digital and mixed analogue/digital techniques** S.L. Hurst
Volume 10	**Distributed Feedback Semiconductor Lasers** J.E. Carroll, J.E.A. Whiteaway and R.G.S. Plumb
Volume 11	**Selected Topics in Advanced Solid State and Fibre Optic Sensors** S.M. Vaezi-Nejad (Editor)
Volume 12	**Strained Silicon Heterostructures: Materials and devices** C.K. Maiti, N.B. Chakrabarti and S.K. Ray
Volume 13	**RFIC and MMIC Design and Technology** I.D. Robertson and S. Lucyzyn (Editors)
Volume 14	**Design of High Frequency Integrated Analogue Filters** Y. Sun (Editor)
Volume 15	**Foundations of Digital Signal Processing: Theory, algorithms and hardware design** P. Gaydecki
Volume 16	**Wireless Communications Circuits and Systems** Y. Sun (Editor)
Volume 17	**The Switching Function: Analysis of power electronic circuits** C. Marouchos
Volume 18	**System on Chip: Next generation electronics** B. Al-Hashimi (Editor)
Volume 19	**Test and Diagnosis of Analogue, Mixed-signal and RF Integrated Circuits: The system on chip approach** Y. Sun (Editor)
Volume 20	**Low Power and Low Voltage Circuit Design with the FGMOS Transistor** E. Rodriguez-Villegas
Volume 21	**Technology Computer Aided Design for Si, SiGe and GaAs Integrated Circuits** C.K. Maiti and G.A. Armstrong
Volume 22	**Nanotechnologies** M. Wautelet *et al.*
Volume 23	**Understandable Electric Circuits** M. Wang
Volume 24	**Fundamentals of Electromagnetic Levitation: Engineering sustainability through efficiency** A.J. Sangster
Volume 25	**Optical MEMS for Chemical Analysis and Biomedicine** H. Jiang (Editor)
Volume 27	**Nano-Scaled Semiconductor Devices** E. A. Gutiérrez-D (Editor)
Volume 29	**Nano-CMOS and Post-CMOS Electronics: Devices and modelling** Saraju P. Mohanty and Ashok Srivastava
Volume 30	**Nano-CMOS and Post-CMOS Electronics: Circuits and design** Saraju P. Mohanty and Ashok Srivastava

High Speed Data Converters

Ahmed M.A. Ali

The Institution of Engineering and Technology

Published by The Institution of Engineering and Technology, London, United Kingdom

The Institution of Engineering and Technology is registered as a Charity in England & Wales (no. 211014) and Scotland (no. SC038698).

First published 2016

The Institution of Engineering and Technology
Michael Faraday House
Six Hills Way, Stevenage
Herts, SG1 2AY, United Kingdom

www.theiet.org

British Library Cataloguing in Publication Data
A catalogue record for this product is available from the British Library

ISBN 978-1-84919-938-4 (hardback)
ISBN 978-1-84919-939-1 (PDF)

Typeset in India by MPS Limited
Printed in the UK by CPI Group (UK) Ltd, Croydon

To Ziad, Sarah, and Rasha
To my Family

Contents

Preface

High speed data converters represent one of the most challenging, important, and exciting analog and mixed-signal systems. They are ubiquitous in our modern and highly connected world. Understanding and designing this class of converters require proficiency in analog circuit design, digital design, and signal processing. This book covers high speed data converters from the perspective of a high speed ADC designer and architect, with a strong emphasis on high speed Nyquist A/D converters. For our purposes, the term "high speed" is defined as sampling rates that are greater than 10 MS/s.

The book is intended for students and engineers who design, evaluate, or use high speed data converters. A basic foundation in circuits, devices, and signal processing is required. The book is meant to bridge the gap between analysis and design, theory and practice, circuits and systems. It covers basic analog circuits and digital signal processing algorithms. There is a healthy dose of theoretical analysis in this book, combined with discussions of practical issues and intuitive perspectives.

The book is composed of ten chapters, with problems at the end of the first nine chapters to summarize, solidify, and extend the concepts covered. Chapter 1 presents an introduction to the ideal data conversion operations from a signal processing standpoint. The concepts and analyses covered in this chapter represent some of the foundations of data conversion. Chapter 2 discusses the metrics frequently used to describe the performance of data converters. The focus is on covering both the theoretical and practical aspects of converter characterization, as well as presenting an intuitive understanding of the relations between the various performance metrics. Chapter 3 presents some of the architectures of high speed ADCs and DACs. On the ADC side, we start with the flash architecture and its variants. This is followed by multi-step ADCs, which include sub-ranging, pipelined, folding, and SAR ADCs. Finally, time-interleaved and sigma-delta ADCs are covered. On the DAC side, we cover the resistive, capacitive, and current-steering DAC architectures.

Some analog circuit challenges are discussed in Chapters 4–6. Chapter 4 covers one of the most challenging operations in high speed ADCs: the sampling of the input signal. Both buffer-less and buffered approaches are explained with the design trade-offs and optimizations of the various circuit parameters.

In Chapter 5, we discuss comparator design, which includes open loop, regenerative, and switched capacitor comparators. Design parameters covered include speed, offset, noise, and metastability. Some high speed comparator examples are presented.

Chapter 6 covers the design and characterization of analog amplifiers. These include switched capacitor amplifiers, integrators, and operational amplifiers. Important design parameters are explained, including noise, speed, and gain. Finally, some of the most common amplifier structures are presented.

Chapters 7 and 8 are focused on arguably two of the most important high speed ADC architectures, which are the pipelined and time-interleaved ADCs, respectively. The pipelined ADC is chosen as a representative of the multi-step ADC architecture. Its high speed, high performance, complexity, and popularity make it an excellent candidate for a detailed discussion. Design trade-offs and optimizations at the ADC and circuit levels are covered. These included circuit structures, non-idealities, ADC-level design decisions, and common pitfalls. Chapter 8 covers time-interleaved ADCs with a focus on theory, intuitive understanding, and implementations.

Chapter 9 discusses digitally assisted converters, a very important area in modern high speed ADCs. Some of the foreground and background calibration techniques for pipelined and time-interleaved ADCs are presented, with the pros and cons of each highlighted. Finally, Chapter 10 gives a brief overview of the evolution, trends, and future directions of the high speed ADC space.

Writing this book has been a great experience and I hope the reader will find it informative and useful. I also wish that it will help provide a glimpse of the excitement and challenges associated with this fascinating field.

Acknowledgments

Over the past two decades, I have been fortunate enough to work on very challenging and groundbreaking projects with smart, dedicated and hardworking colleagues. Every project I have led has been a team effort and has succeeded because of that effort. So, first and foremost, I would like to thank my teammates and colleagues for their hard work and dedication. It has been an absolute privilege to work with such outstanding individuals.

I would also like to express my appreciation to Analog Devices with its strong technical community; emphasis on technical excellence; and culture that encourages innovation, risk taking, teamwork, and cooperation. Special thanks go to the High Speed Converter group for entrusting me and my team with developing some of Analog Devices' most advanced, challenging, and breakthrough high speed ADCs.

I would like to express my thanks to the people who encouraged and supported me while writing this book, especially David Robertson, Stephen Lewis, Richard Schreier, and Gabriele Manganaro; and to the IET publishing staff for their help and patience throughout the process, especially Paul Deards and Jennifer Grace. My gratitude also goes to the numerous friends, colleagues, advisors, mentors, and managers who supported me throughout my career. They include Jan Van der Spiegel, Hani Ragaei, Hisham Haddara, Krishnaswamy Nagaraj, Russ Stop, David Robertson, Allen Barlow, Gabriele Manganaro, Kevin Kattmann, Huseyin Dinc, Paritosh Bhoraskar, Andy Morgan, and Tom Tice.

Last, but not least, I would like to thank my family for their continuous support and sacrifice, and for keeping up with my insane working hours.

Acknowledgements

About the Author

Ahmed M.A. Ali received the B.Sc. and M.Sc. degrees, with highest distinction, in electrical engineering from Ain Shams University, Cairo, Egypt, in 1991 and 1994, respectively. He received the Ph.D. degree in electrical engineering from the University of Pennsylvania, Philadelphia, in 1999. He is currently a Fellow at Analog Devices, where he has led the design and development of several industry firsts, and some of the fastest and highest performance converters to date. Before joining Analog Devices, he was with Texas Instruments' Mixed-Signal R&D Labs and Wireless Infrastructure Business Unit. His past industrial experience also includes Anacad/Mentor Graphics and Siemens AG. He was an Adjunct Assistant Professor at the University of Pennsylvania and is currently serving as an Associate Editor of the *IEEE Transactions on Circuits and Systems I: Regular Papers*. He is the principal author of 25 papers and holds 44 patents. His research interests include analog IC design, high linearity sampling, digitally assisted ADCs, and signal processing.

Dr. Ali received the S.J. Stein award from the University of Pennsylvania for his Ph.D. work. He is also the recipient of the George Stephenson Foundation award, the Catalyst Foundation fellowship award, the University of Pennsylvania fellowship, and multiple industry awards.

Chapter 1
Introduction

Data converters are the interface between the real analog world and the digital realm. They include Analog-to-Digital and Digital-to-Analog converters. An Analog-to-Digital converter (ADC) is a system that converts a continuous-time and continuous-amplitude (i.e. analog) signal into a discrete-time and discrete-amplitude (i.e. digital) signal. It is commonly abbreviated as A/D converter or ADC. Conversely, a Digital-to-Analog converter (or DAC) performs the opposite function, and converts a digital signal into an analog signal. Since digital signal processing and storage is more efficient and robust than analog processing, ADCs and DACs are commonly used in modern systems as interfaces between the analog signals in the real world and the digital signal processing engines.

High speed data converters are typically defined as having sampling rates higher than 10 MS/s. Their vast array of applications includes medical imaging, communications, instrumentation, automotive, and video systems. The explosive growth of wireless communications has led to an insatiable demand for high speed converters with ever increasing speeds, higher input frequencies, higher performance, and lower power. This is driven by the need to increase capacity, lower cost, improve performance, digitize larger bandwidths, and move the converter closer to the antenna in order to lower cost, enhance flexibility, and perform most of the processing in the digital domain.

High speed converters have become so ubiquitous that, on any given day, every person in the developed world will most likely depend on, and interact with, a high speed data converter. In this chapter, the basic function and operations performed by the A/D and D/A converters are discussed. Some of the signal processing concepts and foundations will also be reviewed.

1.1 Ideal data conversion

A block diagram of an ADC is shown in Figure 1.1. The analog-to-digital conversion process can be divided into two operations: *sampling* and *quantization*. The **sampler** converts the continuous-time analog signal into a discrete-time signal. It is performed by a sample-and-hold (S/H) or track-and-hold (T/H) circuit, often called a "sampler" for short. The **quantizer** then converts the resulting discrete-time signal into a discrete-time and discrete-amplitude signal, or in other words, a **digital** signal.

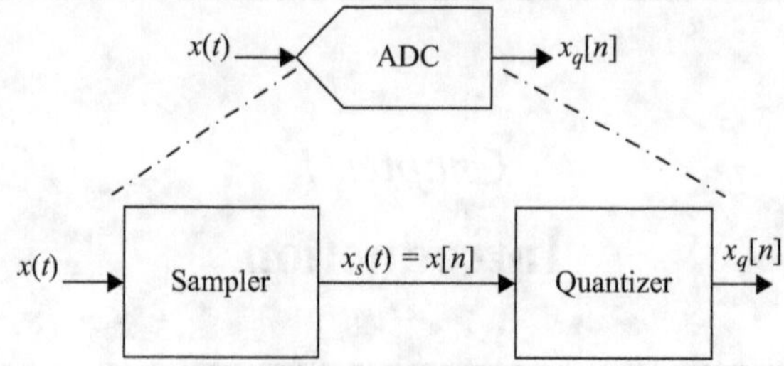

Figure 1.1 An A/D converter (ADC) is comprised of a sampler and a quantizer. The analog input signal x(t) is converted into a digital signal $x_q[n]$

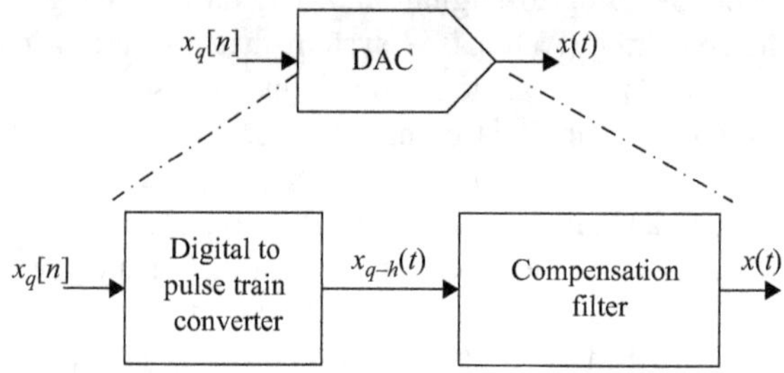

Figure 1.2 A D/A converter (DAC) is comprised of a block that converts the digital code into a train of pulses, followed by a compensation filter

The digital-to-analog conversion operation is comprised of converting the digital code into a quantized analog signal that is essentially a train of pulses representing a sampled-and-held form of the digital signal. This is then followed by a **compensation filter** to reduce the distortion in the generated analog signal, as shown in Figure 1.2.

In the following sections, we will dive into the signal processing basics of the ADC and DAC operations. Some of the concepts discussed in this chapter will be used and built upon in the rest of the book.

1.2 The sampling operation

Let's start with an arbitrary continuous-time analog signal $x(t)$, shown in Figure 1.3, whose Fourier transform is $X(f)$, as shown in Figure 1.4. The exact shape of the signal in either domain is not important at this point; what matters is that the signal is band-limited in the frequency domain, and its bandwidth is B. This signal can be sampled by a train of ideal impulses spaced apart in time by T_s, to give the discrete-time signal $x_s(t)$ whose Fourier transform is $X_s(f)$ as shown in Figure 1.4. The sampling operation is represented mathematically as follows:

$$x_s(t) = x(kT_s) = x(t) \sum_{k=-\infty}^{\infty} \delta(t - kT_s) \tag{1.1}$$

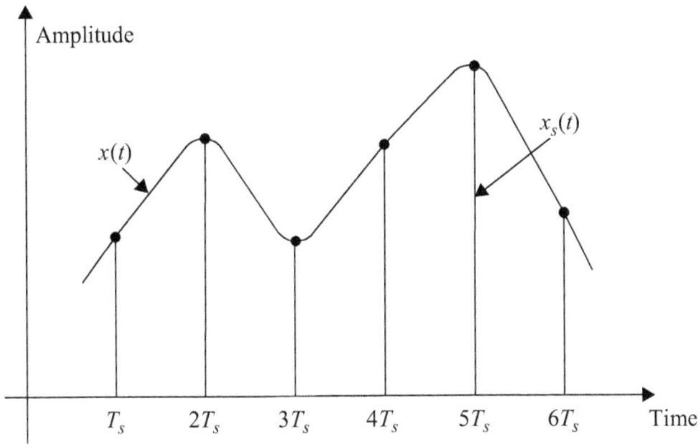

Figure 1.3 *A continuous-time signal x(t) sampled every T_s seconds to give the sampled discrete-time signal $x_s(t)$*

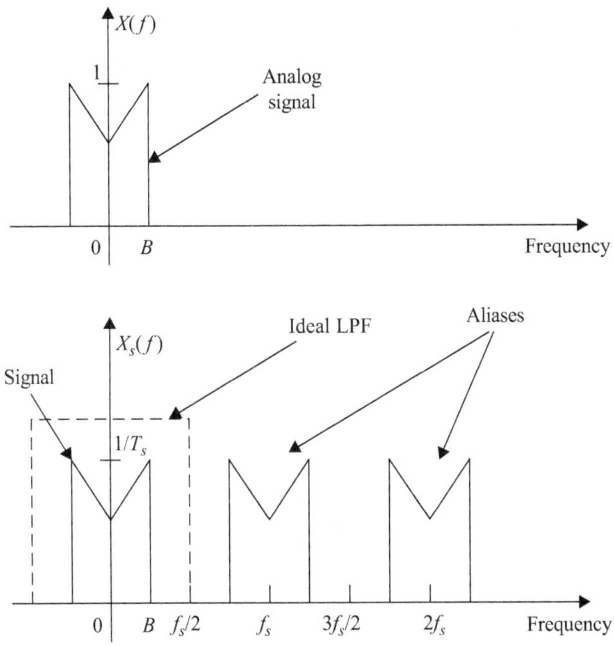

Figure 1.4 *The Fourier transform X(f) of the continuous-time signal x(t), and the Fourier transform $X_s(f)$ of the sampled discrete-time signal $x_s(t)$*

Using the "sifting property" of the impulse function [1, 2], we get

$$x_s(t) = \sum_{k=-\infty}^{\infty} x(kT_s)\delta(t - kT_s) \tag{1.2}$$

Performing the Fourier transform on both sides of (1.1), we obtain the frequency domain representation

$$X_s(f) = X(f) * F\left\{\sum_{k=-\infty}^{\infty} \delta(t - kT_s)\right\} \tag{1.3}$$

where * denotes the convolution operation. Since the Fourier transform of a train of impulses in the time domain is a train of impulses in the frequency domain,

$$X_s(f) = X(f) * \frac{1}{T_s}\sum_{k=-\infty}^{\infty} \delta(f - kf_s) \tag{1.4}$$

where T_s is the sampling period and $f_s = 1/T_s$ is the sampling frequency or sampling rate. Applying the convolution operation, we obtain the frequency representation of a sampled discrete-time signal $X_s(f)$ as

$$X_s(f) = \frac{1}{T_s}\sum_{k=-\infty}^{\infty} X(f - kf_s) \tag{1.5}$$

Equation (1.5) and Figure 1.4 show us that the Fourier transform of the sampled signal is comprised of an infinite number of copies (aliases) of the original signal's spectrum $X(f)$ spaced f_s apart. If the signal $x(t)$ is band-limited to a bandwidth B, and f_s is greater than $2B$, the aliases will not overlap, and the original signal $X(f)$ can be retrieved perfectly by digitally filtering out the aliases. This filtering is shown in Figure 1.4. On the other hand, if f_s is less than $2B$, the aliases will overlap causing distortion, and the original signal cannot be retrieved. This is commonly referred to as *aliasing*, and the analog filter used to limit the bandwidth of the analog signal to be less than half the sampling rate is called an *anti-aliasing filter*. The maximum frequency (or more accurately bandwidth) that can be sampled at a sampling rate f_s without loss of information is called the Nyquist frequency (or bandwidth) f_N and is given by

$$f_N = \frac{f_s}{2} \tag{1.6}$$

Another form of sampling is a Sample-and-Hold (S/H) operation, where the signal is sampled at an instant nT_s, then held for a duration of T_s until the next sample at $(n+1)T_s$. This operation can be represented by the convolution of an ideally sampled signal with a train of pulses [3]. The pulse, which is shown in Figure 1.5, can be represented in the time domain by the rectangle function $\Pi(t)$ as

$$p(t) = \Pi\left(\frac{t - \frac{T_s}{2}}{T_s}\right) \tag{1.7}$$

Its Fourier transform is shown in Figure 1.6 and is given by the sinc function, such that

$$P(f) = T_s e^{-\frac{j\omega T_s}{2}} \operatorname{sinc}(T_s f) \tag{1.8}$$

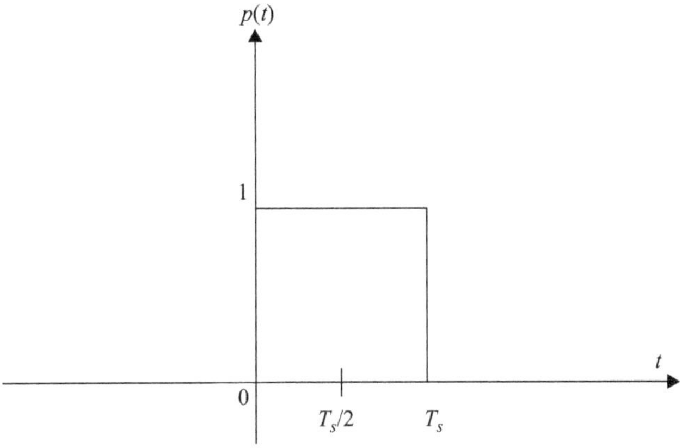

Figure 1.5 A pulse in the time-domain used for the sample-and-hold operation

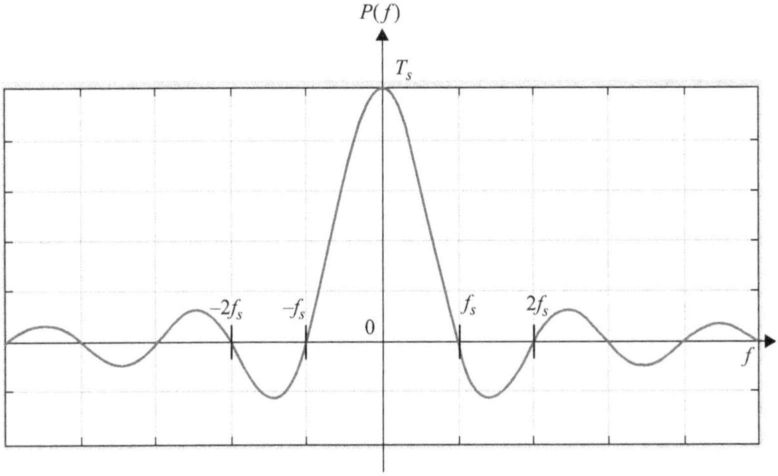

Figure 1.6 The Fourier transform of the pulse p(t) shown in Figure 1.5

where

$$\operatorname{sinc}(x) = \begin{cases} \dfrac{\sin(\pi x)}{\pi x} & x \neq 0 \\ 1 & x = 0 \end{cases} \tag{1.9}$$

The sampled-and-held signal $x_{s-h}(t)$, shown in Figure 1.7, is represented by the convolution of the ideal-sampled signal $x_s(t)$ and the pulse $p(t)$. That is

$$x_{s-h}(t) = x_s(t) * p(t) \tag{1.10}$$

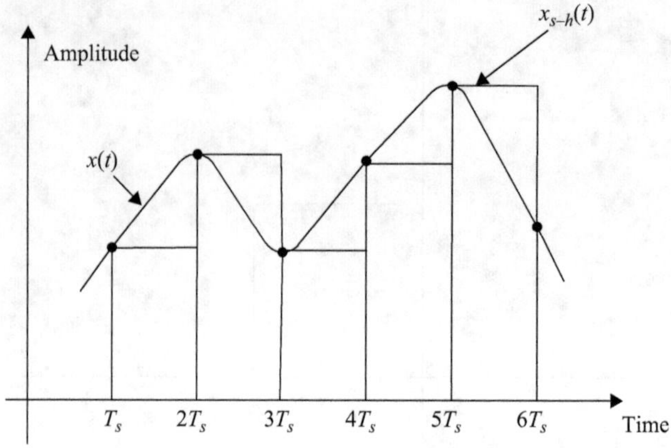

Figure 1.7 A continuous-time signal x(t) and its sampled-and-held representation $x_{s-h}(t)$

The Fourier transform of the sampled-and-held signal $X_{s-h}(f)$ is the product of $X_s(f)$ and $P(f)$, which is represented by

$$X_{s-h}(f) = e^{-j\pi f T_s}\text{sinc}(T_s f) \sum_{k=-\infty}^{\infty} X(f - kf_s) \tag{1.11}$$

This is the frequency domain representation of the sampled-and-held signal in the *analog* domain. It shows that the spectrum of the input signal will be shaped by the sinc function due to the holding process.

In a third form of sampling, a Track-and-Hold (T/H) operation, the sampler tracks the input during the tracking (or acquisition) phase, then holds the output during the hold phase. Thus, the output is the sum of a tracking signal and a held signal [3], as shown in Figure 1.8. That is, the tracked-and-held signal $x_{t-h}(t)$ is given by

$$x_{t-h}(t) = x_t(t) + x'_{s-h}(t) \tag{1.12}$$

where the tracking signal $x_t(t)$ is given by [3]

$$x_t(t) = x(t)\left[p(2t) * \sum_{k=-\infty}^{\infty} \delta(t - kT_s)\right] \tag{1.13}$$

where $p(t)$ is given by (1.7) and Figure 1.5. The factor of 2 in $p(2t)$ indicates that the pulse is half the width compared to the pulse of Figure 1.5. The sampled-and-held signal $x'_{s-h}(t)$ is given by [3]

$$x'_{s-h}(t) = \left[x(t) \sum_{k=-\infty}^{\infty} \delta(t - kT_s - T_s/2)\right] * p(2t) \tag{1.14}$$

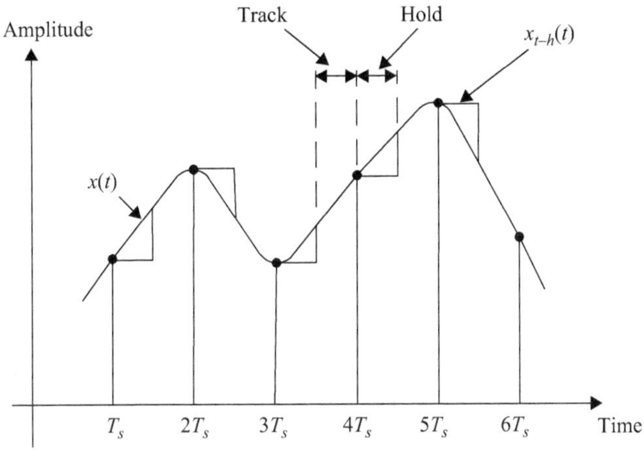

Figure 1.8 A continuous-time signal x(t) and its tracked-and-held representation $x_{t-h}(t)$

Therefore, the Fourier transform of the tracked-and-held signal $X_{t-h}(f)$ is [3]

$$X_{t-h}(f) = \sum_{k=-\infty}^{\infty} e^{-\frac{jk\pi}{2}} \frac{\sin\left(\frac{k\pi}{2}\right)}{k\pi} X(f - kf_s)$$

$$+ e^{-3j\pi f T_s/2} \frac{\sin\left(\frac{\pi f T_s}{2}\right)}{\pi f T_s} \sum_{k=-\infty}^{\infty} X(f - kf_s) \tag{1.15}$$

Like (1.11) for the sampled-and-held signal, (1.15) describes the frequency representation of the tracked-and-held signal in the *analog* domain. It is important to note that the sampling operation in ADCs happens at the sampling instant when the sampling switch is turned off and is represented at the output of the ADC in the *digital* domain as a train of sampled values (i.e. a train of impulses). Therefore, the ADC's sampling operation can indeed be represented by an ideal sampling process as described in (1.2) and (1.5). Although the analog sampled signal is usually held for some time to be processed by the following circuits, the discrete-time (i.e. digital) domain represents a snapshot of the held signal at equally spaced instants spaced T_s apart. Thus, the sampled discrete-time signal of an ADC is considered to be an impulse-sampled signal in the discrete-time domain, rather than a sampled-and-held signal. Therefore, for an ADC, the discrete-time domain signal is given by

$$x_s[n] = x_s(nT_s) = x(nT_s) \sum_{k=-\infty}^{\infty} \delta(nT_s - kT_s) \tag{1.16}$$

In the frequency domain, the Fourier transform of $x_s(t)$ is given by (1.5). That is

$$X_s(f) = \frac{1}{T_s} \sum_{k=-\infty}^{\infty} X(f - kf_s)$$

The discrete-time Fourier transform is given by

$$X_s(e^{j\omega}) = \sum_{n=-\infty}^{\infty} x[n]e^{-j\omega n} \tag{1.17}$$

and

$$X_s(e^{j\omega}) = \frac{1}{T_s}\sum_{k=-\infty}^{\infty} X\left(\frac{j\omega}{T_s} - \frac{j2\pi k}{T_s}\right) \tag{1.18}$$

where ω is the normalized frequency given by $\omega = 2\pi f T_s$ and ranges from $-\pi$ to π. $X_s(e^{j\omega})$ is the frequency-scaled version of $X_s(f)$ given in (1.5).

Because of the aliasing that results from the sampling operation, it is important to ensure that the sampled signal is band-limited. An anti-aliasing filter is usually used before the ADC in order to limit the signal bandwidth and remove any "out-of-band" signals that may alias back into the signal spectrum. This filter can be a low-pass or bandpass filter depending on the location of the input signal relative to the Nyquist frequency.

1.2.1 Sampling theorem

The Nyquist sampling theorem states that if $x(t)$ is a band-limited signal with $X(f) = 0$ for $|f| > f_N$, then $x(t)$ is uniquely represented by its samples $x[n] = x(nT_s)$ if: $f_s = 1/T_s > 2f_N$. The frequency f_N is called the Nyquist frequency and the sampling rate $f_s = 2f_N$ is called the Nyquist sampling rate [1, 2].

1.2.2 Sampling of bandpass signals

Assume that we are to sample a bandpass signal $X(f)$ with bandwidth $B = B_2 - B_1$, such that

$$X(f) = 0 \qquad \text{for: } f < B_1 \text{ and } f > B_2 \tag{1.19}$$

The center frequency f_c is given by: $f_c = (B_1 + B_2)/2$, and the upper frequency $f_c + B/2$ is a multiple integer of the bandwidth B. That is, if the upper frequency is given by

$$f_c + \frac{B}{2} = kB \tag{1.20}$$

then the signal can be recovered completely if sampled at a rate of f_s that is equal to $2B$. Therefore, the sampling rate does not have to be larger than the highest frequency content of the signal, which is $f_c + B/2$, but can be as low as $2B$ [2]. That is,

$$f_s > 2B \tag{1.21}$$

On the other hand, if the upper frequency is not an integer multiple of the bandwidth, then the sampling rate will need to be

$$f_s > \frac{2Bk'}{k} \tag{1.22}$$

where

$$k' = \frac{f_c + B/2}{B} \tag{1.23}$$

and

$$k = \text{integer}(k') \tag{1.24}$$

The function "integer(k')" gives the largest integer that is smaller than or equal to k'. Since, the upper limit of k'/k approaches two in the worst case, for a bandpass signal with bandwidth B, the minimum required sampling rate is bounded as follows [2]

$$2B \leq f_s < 4B \tag{1.25}$$

That is, for bandpass signals with bandwidth B, the minimum sampling rate required ranges between $2B$ and $4B$.

1.3 The reconstruction operation

In addition to sampling, it is imperative to understand the reconstruction process of converting a discrete-time signal into a continuous-time signal. This operation is usually used in D/A converters as shown in Figure 1.9.

From the frequency domain representation, we can see that an ideal reconstruction of the continuous-time signal from the train of impulses would be achieved using an ideal low-pass filter (LPF) of gain T_s and cutoff frequency f_c.

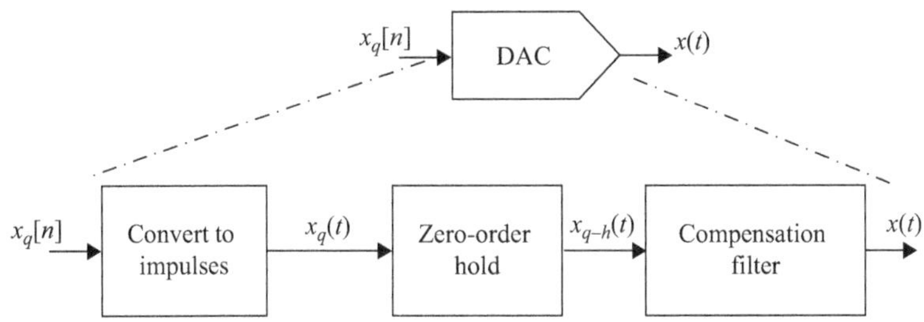

Figure 1.9 A block diagram illustrating the DAC operation

This is commonly referred to as the **reconstruction filter** whose transfer function $H_r(f)$ is given by

$$H_r(f) = \begin{cases} T_s & |f| < f_c \\ 0 & \text{otherwise} \end{cases} \tag{1.26}$$

and

$$B < f_c < f_s - B \tag{1.27}$$

Therefore,

$$H_r(f) = \begin{cases} T_s & |f| < f_s/2 \\ 0 & \text{otherwise} \end{cases} \tag{1.28}$$

This is shown in Figure 1.10. Therefore, the reconstructed waveform is given in the frequency domain by

$$X_r(f) = X_s(f) \times H_r(f) = X(f) \tag{1.29}$$

In the time domain, the reconstructed signal is given by

$$x_r(t) = \sum_{n=-\infty}^{\infty} x[n]\text{sinc}\left(\frac{t - nT_s}{T_s}\right) \tag{1.30}$$

Since $\text{sinc}(0) = 1$, the reconstructed signal is equal to the sampled signal at $t = nT_s$, as expected. In addition, the ideal LPF interpolates between samples of $x_s(t)$ to reconstruct the original continuous-time signal. If the sampled signal has been generated by sampling a band-limited signal at the Nyquist rate or higher, then the

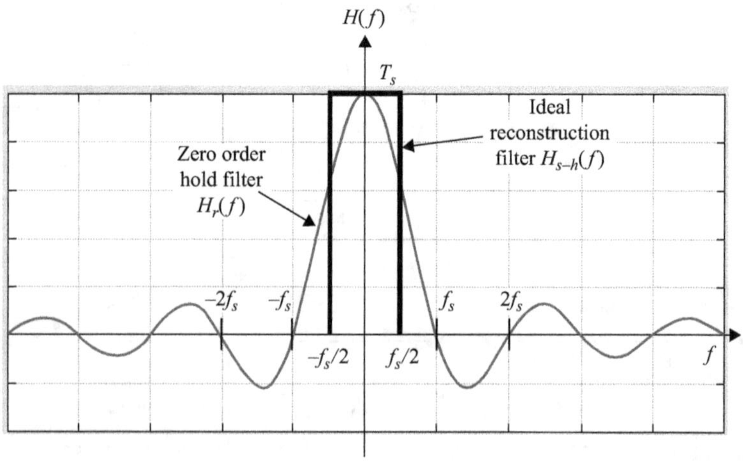

Figure 1.10 The frequency domain representation of an ideal reconstruction filter and a zero-order hold filter

ideal reconstructed signal $x_r(t)$ will be equal to the original continuous-time signal. However, since ideal LPFs are not physically realizable, real reconstruction is expected to introduce some distortion. An example is represented by the zero-order hold reconstruction filter whose response is shown in Figure 1.10 and is given by

$$H_{s-h}(f) = T_s e^{-j\pi f T_s} \mathrm{sinc}(T_s f) \tag{1.31}$$

In order to reduce the distortion due to this filter characteristic, a compensation reconstruction filter $H_c(f)$ can be used after the zero-order hold process, such that their combined effect is equal to the ideal reconstruction filter characteristic $H_r(f)$. That is,

$$H_{s-h}(f) \times H_c(f) = H_r(f) \tag{1.32}$$

Therefore,

$$H_c(f) = \frac{H_r(f)}{H_{s-h}(f)} \tag{1.33}$$

and

$$H_c(f) = \begin{cases} \dfrac{e^{j\pi f T_s}}{\mathrm{sinc}(T_s f)} & |f| < f_s/2 \\ 0 & \text{otherwise} \end{cases} \tag{1.34}$$

This ideal compensation filter consists of a time advance of $T_s/2$ to compensate for the time delay introduced by the zero-order hold filter. Such time advance cannot be realized in real-time implementations and is usually ignored, thus compensating only for the magnitude response. If the signal bandwidth is much smaller than the Nyquist frequency, the magnitude compensation may not be necessary. If that is not the case, or if the desired accuracy is high, a compensation filter is usually used. Such a filter can also be implemented in the discrete-time (i.e. digital) domain before the D/A conversion to compensate, in advance, for the zero-order hold process employed to convert the discrete-time signal into a continuous-time signal.

1.4 The quantization operation

After sampling, the other main process of digitization is the conversion of the signal's continuous amplitude into discrete levels. This is performed by the "quantizer" of the ADC. An ideal 16-level (4-bit) quantization function is shown in Figure 1.11. A conceptual block diagram of an 8-level (3-bit) quantizer is shown in Figure 1.12, where 7 comparators are used to quantize the input signal into 8 output levels. The comparator is a basic building block in almost every ADC, which outputs a logic "1" if the input is larger than a threshold value and a logic "0" if the input is smaller than the threshold. It is the interface between the input analog signal and the output discrete/digital signal.

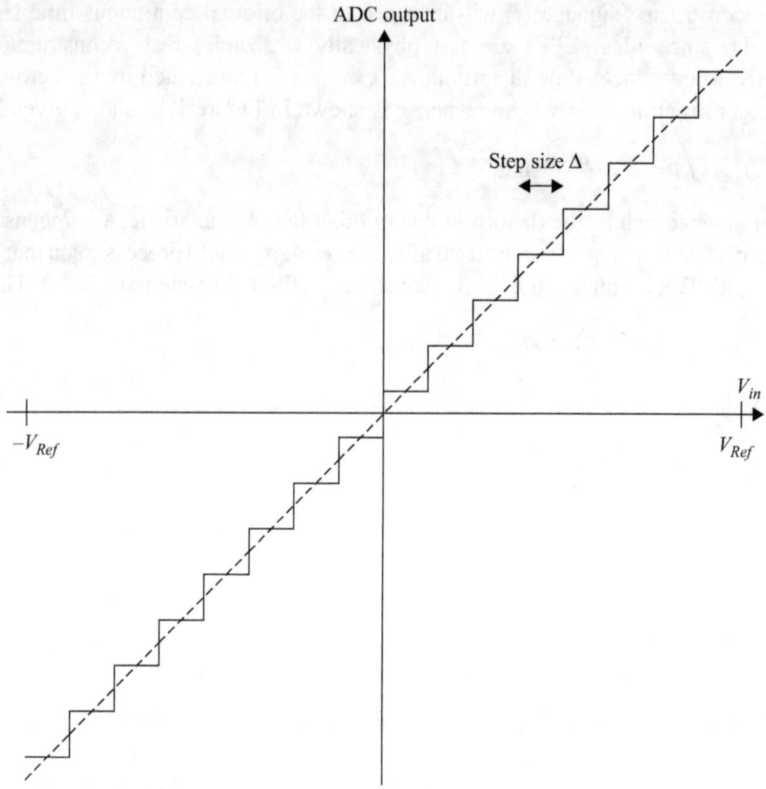

Figure 1.11 An ideal "mid-rise" quantization function with 16 levels

We should keep in mind that the "ideal" quantization is a non-linear and non-invertible function, which causes loss of information that cannot be recovered. This invariably leads to noise and distortion that are inherent to the quantization process.

For an L-level quantizer, the L-levels can be represented by N bits, such that

$$N = \log_2 L \tag{1.35}$$

If the input signal span is given by V_{FS}, then the quantization step size Δ is given by

$$\Delta = \frac{V_{FS}}{2^N} = \frac{2V_{Ref}}{2^N} \tag{1.36}$$

where V_{Ref} is the reference voltage, as shown in Figures 1.11 and 1.12. If the input exceeds the maximum full-scale range, defined by the reference voltage V_{Ref}, the output of the quantizer "clips" and the quantization error increase substantially with increasing the input amplitude.

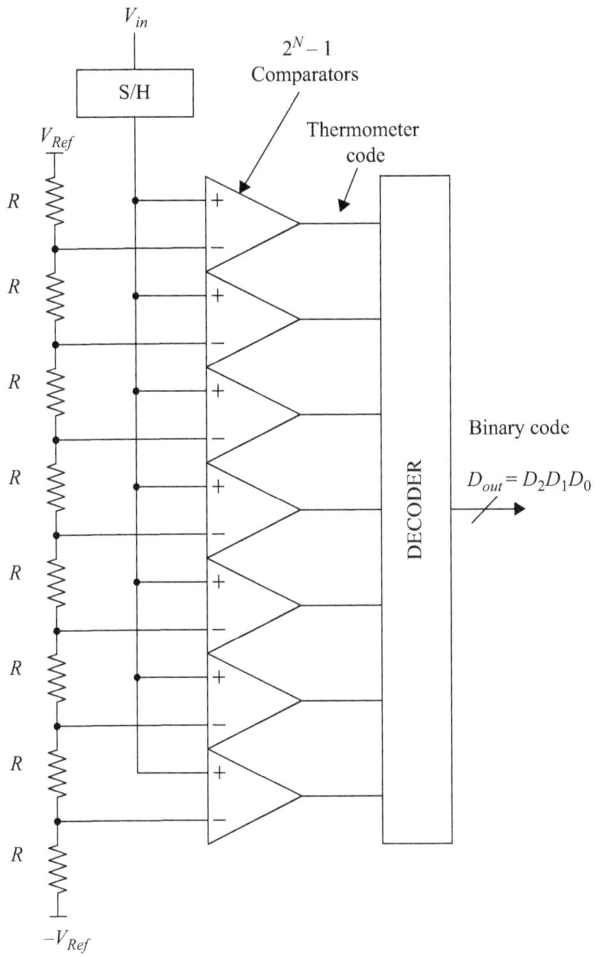

Figure 1.12 A mid-rise 3-bit flash ADC with 8 levels and 7 comparators

The quantization function shown in Figure 1.11 is called a *mid-rise* converter, where zero is an input threshold level. Alternatively, a *mid-tread* function is shown in Figure 1.13, where zero is an output code, but there is no transition, at zero. This latter implementation lacks symmetry around zero input, or it would require an odd number of levels to be symmetric. However, it is sometimes preferred in order to reduce the output "chatter," in the absence of an input signal, due to noise.

The inherent loss of information in the quantization process leads to what is commonly referred to as the "quantization error" or "quantization noise." The quantization error as a function of the input signal can be represented by a saw-tooth pattern as shown in Figure 1.14. If we assume that the quantization step size Δ is small enough such that the probability density function (pdf) of the quantized

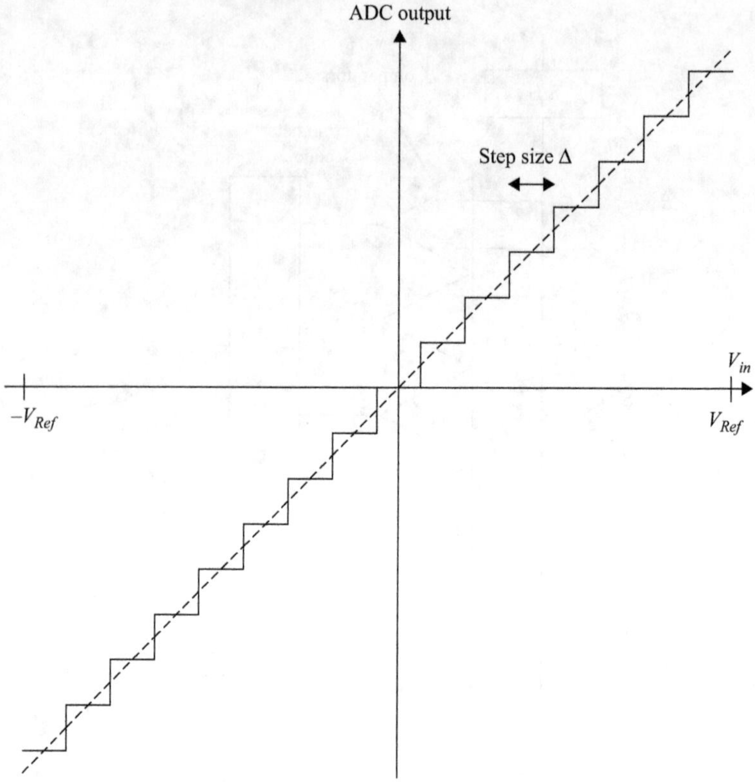

Figure 1.13 An ideal "mid-tread" quantization function with 16 levels

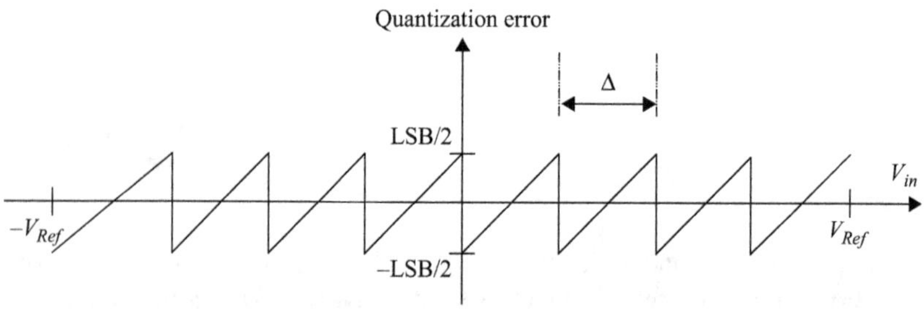

Figure 1.14 The quantization error as a function of the input voltage for a mid-rise quantizer. The quantization step Δ is equal to 1 LSB

output is practically uniform over the step size Δ (from –Δ/2 to Δ/2), then the pdf $p(e)$ is given by

$$p(e) = \begin{cases} \dfrac{1}{\Delta} & e: -\dfrac{\Delta}{2} \text{ to } \dfrac{\Delta}{2} \\ 0 & \text{otherwise} \end{cases} \tag{1.37}$$

The quantization noise power P_Q can be obtained as follows

$$P_Q = \int_{-\infty}^{\infty} e^2 p(e)de = \int_{-\Delta/2}^{\Delta/2} \frac{e^2}{\Delta} de = \frac{2\Delta^3/8}{3\Delta} = \frac{\Delta^2}{12} \tag{1.38}$$

Therefore the quantization noise power P_Q for a uniform quantizer with step size Δ is given by

$$P_Q = \frac{\Delta^2}{12} \tag{1.39}$$

For a full-scale sinusoidal input signal with an amplitude equal to $V_{FS}/2$, the input signal is given by

$$x(t) = \frac{V_{FS}}{2} \sin(\omega t) \tag{1.40}$$

The input signal power P_s is

$$P_s = \frac{V_{FS}^2}{8} \tag{1.41}$$

Therefore the signal-to-quantization noise ratio (SQNR) is given by

$$SQNR = \frac{P_s}{P_Q} \tag{1.42}$$

Substituting (1.39) and (1.41) in (1.42) gives

$$SQNR = \frac{V_{FS}^2/8}{\Delta^2/12} = \frac{V_{FS}^2/8}{V_{FS}^2/(12 \times 2^{2N})} \tag{1.43}$$

Therefore,

$$SQNR = \frac{3 \times 2^{2N}}{2} \tag{1.44}$$

where N is the number of quantization bits. When represented in dBs, the SQNR is given by

$$SQNR(\text{dB}) = 10 \log\left(\frac{P_s}{P_Q}\right) \tag{1.45}$$

Substituting (1.44) into (1.45), we get

$$SQNR(\text{dB}) = 10 \log\left(\frac{3 \times 2^{2N}}{2}\right) \tag{1.46}$$

and

$$SQNR(\text{dB}) = 2N\log(2) + 10\log\left(\frac{3}{2}\right) \tag{1.47}$$

Therefore, for a sinusoidal signal quantized by N quantization bits, the SQNR is given by

$$SQNR(\text{dB}) = 6.02 \times N + 1.76 \text{ dB} \tag{1.48}$$

If the input signal is a random signal with a Gaussian distribution of zero mean and a standard deviation equal to 1/4th the ADC full-scale, the signal power P_s will be given by [1]

$$P_s = \frac{V_{FS}^2}{16} \tag{1.49}$$

Substituting (1.39) and (1.49) into (1.45), we obtain the SQNR for a Gaussian random signal quantized by N bits:

$$SQNR(\text{dB}) = 6.02 \times N - 1.25 \text{ dB} \tag{1.50}$$

From (1.48) and (1.50), we see that every additional bit in quantization resolution improves the SQNR by 6 dB, which is a key rule of thumb for ADC specifications.

It is important to note that the quantization noise is *not* white noise. In spite of the uniform distribution assumption within the quantization step, the noise is surely not white across the whole Nyquist band. The quantization process creates input-dependent errors that result in spurs and harmonics in the output spectrum, even if the quantization function is ideal. An analysis of the expected level of those quantization spurs is described in References 4, 5. A common measure of the ADC's linearity in the frequency domain is the Spurious-Free Dynamic Range (SFDR), which is the ratio of the signal power to the power of the highest spur/harmonic, as described in more detail in Chapter 2. The SFDR of the ideal quantization process is approximately given by [4, 5]

$$SFDR(\text{dBFS}) \approx 9N - c \tag{1.51}$$

where c ranges from 0 for low resolutions to 6 for high resolutions. That is, for every additional bit of resolution, the SFDR improves by about 9 dB. The SFDR is limited by the worst harmonic, which is located around $2^N\pi$ times the fundamental. On the other hand, the low order harmonics are typically given by $9N$ dBFS.

In addition to the mathematical derivation, it is good to understand intuitively the impact of the quantization on linearity [5]. The quantization error function versus the input signal is shown in Figure 1.14. As the number of bits increases by 1 bit, the quantization error amplitude is reduced by a factor of 2. This reduces the total quantization power by a factor of 4, which results in a 6 dB improvement in SQNR. Therefore, if nothing else changes, the spur levels would be expected to drop by 6 dB too. However, an additional quantization bit leads to a doubling in the number of the saw-tooth segments, which results in doubling the number of harmonics. Since the number of harmonics is doubled, while their total power is reduced by a factor of 4 (i.e. 6 dB), each harmonic's energy is expected to be reduced by an additional factor of 2 (i.e. 3 dB). This leads to the 9 dB-per-quantization-bit rule for the maximum spur level and hence the SFDR, as shown in (1.51).

According to this intuitive understanding, the key to improving the harmonic/spur levels due to quantization, or other similar non-linearities, is to reduce the total power and/or to increase the number of segments. That is, increase the number of bits. The increase in the number of segments spreads the distortion power among more harmonics and hence reduces their peak levels. This is an important concept that is used in ADC linearization using techniques such as "dithering" as discussed in Chapter 9.

1.5 Coding

Once sampled and quantized, the ADC digital output can be coded into any format such as binary, offset binary, two's complement, Gray coding, or any other desired format. Usually, creating different digital coding formats is a straightforward digital process. In addition, some digital processing can be applied to the output to perform various functions or even improve the accuracy of the converter using digital assistance, as covered in Chapter 9.

1.6 Undersampling and oversampling

If the input signal is located between DC and $f_s/2$, which is called the first Nyquist zone, the anti-aliasing filter will need to be an LPF with a bandwidth less than the Nyquist frequency. Since an ideal LPF is not practical, the signal's bandwidth usually needs to be smaller than the Nyquist frequency (typically 2/3 of $f_s/2$) in order to allow for the roll-off of the filter. This is shown in Figure 1.15. Alternatively, the input signal can be located in one of the other Nyquist zones (such as the second Nyquist zone between $f_s/2$ and f_s). In this case, which is usually called *undersampling*, a bandpass anti-aliasing filter can be used as shown in Figure 1.16. As long as the bandwidth of the signal B is less than $f_s/2$ and satisfies the conditions discussed in Section 1.2.2, no aliasing distortion occurs, and the signal can be recovered from the sampled signal. This concept is used in IF and RF sampling applications, where the ADC samples a signal whose bandwidth is smaller than the Nyquist bandwidth, but whose center frequency may be much larger than the Nyquist frequency. This puts tough requirements on the input sampling bandwidth of the ADC, which will need to be much larger than the Nyquist bandwidth.

Conversely, the sampling rate can be chosen to be much larger than twice the signal's bandwidth ($2B$). In this case, the process is referred to as *oversampling*. If applied correctly, oversampling can simplify the anti-aliasing filter, system

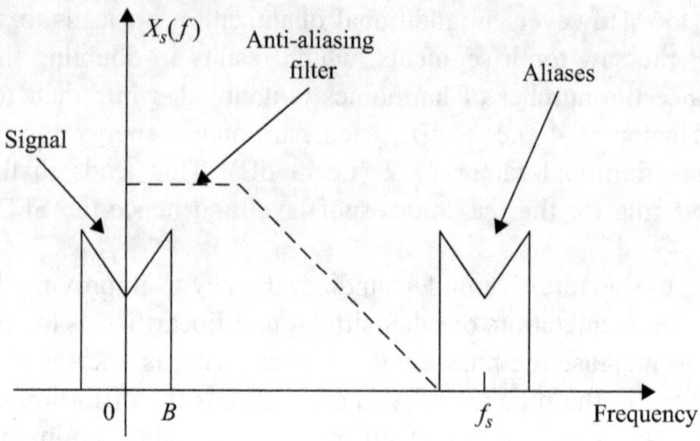

Figure 1.15 A sampled signal with an anti-aliasing filter

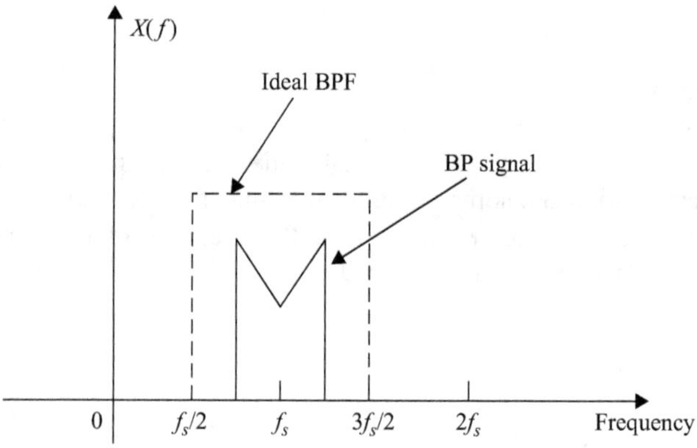

Figure 1.16 A bandpass signal and a bandpass anti-aliasing filter

design, frequency planning, cost, and performance. The oversampling ratio is given by

$$OSR = \frac{f_s}{2B} \tag{1.52}$$

One advantage of oversampling is that it enables the user to utilize digital filtering to reduce the noise of the system and to remove undesired signals and harmonics. For example, if the ADC output contains white noise uniformly distributed over the Nyquist bandwidth, oversampling with digital filtering can improve the noise by a factor commonly referred to as the *Processing Gain* (*PG*), which is given by

$$PG(\text{dB}) = 10\log(OSR) \tag{1.53}$$

The signal-to-noise ratio (SNR) within a bandwidth B that is smaller than the Nyquist bandwidth (SNR_B) is related to the SNR over the whole Nyquist zone (SNR_{Nyq}) by

$$SNR_B = SNR_{Nyq} + 10\log OSR \tag{1.54}$$

or

$$SNR_B = SNR_{Nyq} + 3\log_2 OSR \tag{1.55}$$

So, with oversampling and proper digital filtering, every doubling of the sampling rate can result in 3 dB improvement in the SNR of the ADC. In addition, some of the undesired harmonics and spurs can be filtered out digitally if they fall out of the signal band, which can improve the SFDR too.

1.7 Decimation and interpolation

Sometimes, it is necessary to change the sampling rate of a digital signal. This can be in the form of sampling rate reduction (commonly referred to as *decimation* or *downsampling*) or sampling rate increase (*interpolation* or *upsampling*). Decimation is used to *compress* the digital data, while interpolation *expands* the data.

Downsampling of a sampled signal can be done by resampling the signal at the reduced rate. Therefore, if we downsample a signal $x[n]$ by an integer factor M, we obtain a new downsampled discrete-time signal $x_d[n]$ [1], such that

$$x_d[n] = x_s[nM] = x(nMT_s) \tag{1.56}$$

This can be represented by a new sampled signal, whose sampling period $T_d = MT_s$ and sampling frequency $f_d = f_s/M$. Therefore, in the frequency domain, if the original sampled signal is given by

$$X_s(e^{j\omega}) = \frac{1}{T_s}\sum_{k=-\infty}^{\infty} X\left(\frac{j\omega}{T_s} - \frac{j2\pi k}{T_s}\right) \tag{1.57}$$

then the downsampled signal will be given by

$$X_d(e^{j\omega}) = \frac{1}{T_d}\sum_{k=-\infty}^{\infty} X\left(\frac{j\omega}{T_d} - \frac{j2\pi k}{T_d}\right) \tag{1.58}$$

which gives

$$X_d(e^{j\omega}) = \frac{1}{MT_s}\sum_{k=-\infty}^{\infty} X\left(\frac{j\omega}{MT_s} - \frac{j2\pi k}{MT_s}\right) \tag{1.59}$$

Combining (1.57) and (1.59), we obtain the Fourier transform of the downsampled signal $x_d[n]$ in terms of the sampled signal $x_s[n]$ as

$$X_d(e^{j\omega}) = \frac{1}{M}\sum_{k=0}^{M-1} X_s\left(e^{j\left(\frac{\omega}{M} - 2\pi k/M\right)}\right) \tag{1.60}$$

We can think of the downsampling process as stretching the signal representation in the frequency domain along the frequency axis by a factor of M, and scaling the amplitude by $1/M$. For aliasing distortion to be avoided $x_s[n]$ needs to be band-limited to ω_B, such that

$$X_s(e^{j\omega}) = 0 \qquad \text{for } \omega_B \leq \omega \leq \pi \tag{1.61}$$

where ω_B is given by

$$2\omega_B < 2\pi/M \tag{1.62}$$

which gives

$$2f_B < 1/MT_s \tag{1.63}$$

and

$$2f_B < f_s/M \tag{1.64}$$

An example is shown in Figure 1.17, where $M = 2$ and the signal bandwidth is less than $f_s/4 = f_d/2$. In this case, downsampling does not result in aliasing distortion. Figure 1.18 shows an example where the signal bandwidth is equal to f_d, and therefore does not satisfy (1.64). In this case, aliasing occurs and the signal cannot be recovered after downsampling. If this signal is low-pass filtered before downsampling to limit its bandwidth to be less than $f_d/2$, the filtered version of the signal would be recoverable without aliasing, after downsampling, as shown in Figure 1.19. Generally speaking, to ensure that no aliasing distortion occurs, it is desirable to digitally filter the signal $x_s[n]$ using a digital LPF of bandwidth $\omega_B = \pi/M$ before downsampling by a factor of M. The process of filtering and downsampling is called *decimation*.

On the other hand, increasing the sampling rate can be accomplished by inserting additional samples between the existing ones [1]. The new sampling period is $T_u = T_s/L$, and the new sampling frequency is $f_u = Lf_s$. To upsample by a factor L, zeroes are inserted between the existing samples such that

$$x_u[n] = \begin{cases} x_s[n/L] = x\left(\dfrac{nT_s}{L}\right) & n = 0, \pm L, \pm 2L, \ldots \\ 0 & \text{otherwise} \end{cases} \tag{1.65}$$

The signal $x_u[n]$ is the upsampled signal, as shown in Figure 1.20, which can be represented by

$$x_u[n] = \sum_{k=-\infty}^{\infty} x_s[k]\delta[n - kL] \tag{1.66}$$

In the discrete-time frequency domain, this gives

$$X_u(e^{j\omega}) = X_s(e^{j\omega L}) \tag{1.67}$$

In a manner similar to the reconstruction of a discrete-time signal into a continuous-time signal, an interpolated signal $x_i[n]$ can be obtained by low-pass

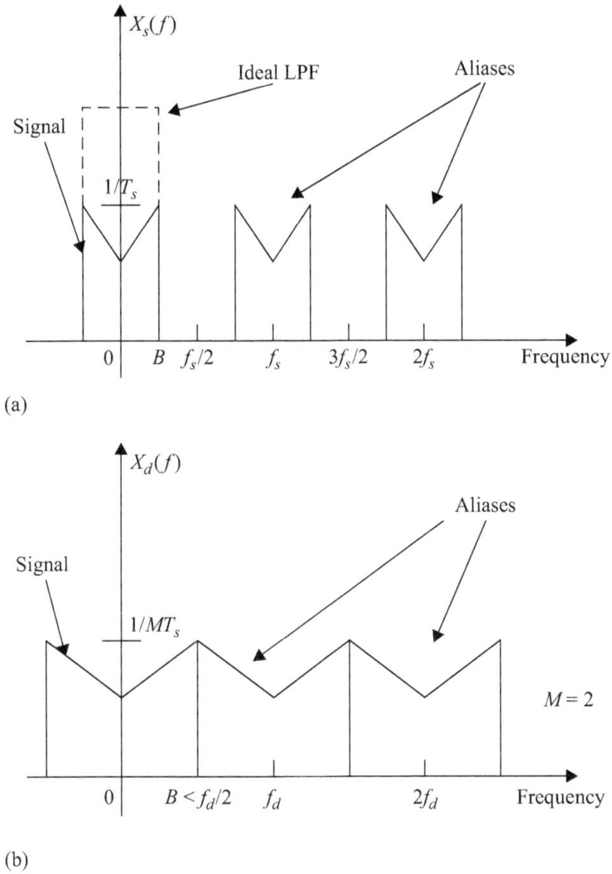

Figure 1.17 The frequency representation of (a) a sampled signal $X_s(f)$ and (b) a downsampled version of it $X_d(f)$. In this case, $2B < f_d$ and $f_d = f_s/2$ ($M = 2$)

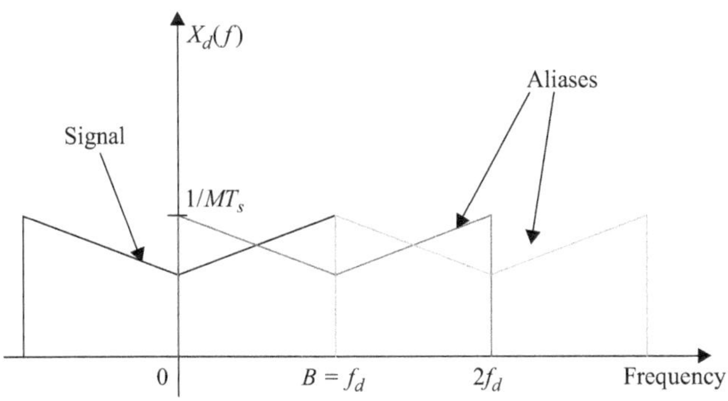

Figure 1.18 The frequency representation of a downsampled signal $X_d(f)$. In this case, $B = f_d$ and $f_d = f_s/2$ ($M = 2$)

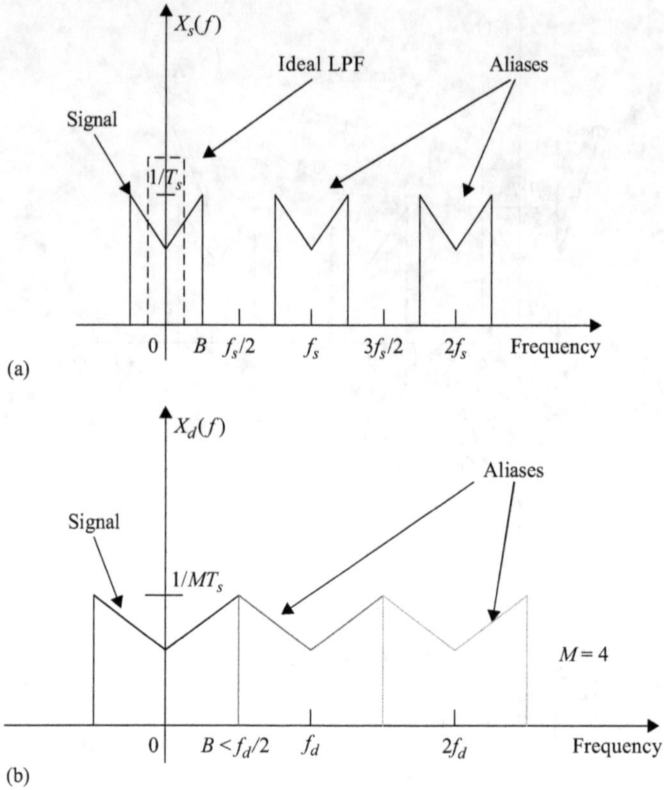

Figure 1.19 The frequency representation of (a) a sampled signal $X_s(f)$, and (b) a downsampled signal $X_d(f)$. In this case, although the original signal bandwidth is larger than $f_d/2$, it was band-limited by an ideal LPF before downsampling. Since $B < f_d/2$, no aliasing distortion occurs. In this case, $f_d = f_s/4$ ($M = 4$)

filtering the upsampled signal using an ideal LPF with gain L and cut-off frequency π/L. The filter's transfer function is given by

$$H_r\left(e^{j\omega}\right) = \begin{cases} L & |\omega| < \pi/L \\ 0 & \text{otherwise} \end{cases} \tag{1.68}$$

Therefore, we can think of the interpolated signal as being compressed in the frequency domain along the frequency axis by a factor $1/L$ and scaled up in amplitude by a factor L. In the time domain, the filter's impulse response is given by

$$h_r[n] = \text{sinc}\left(\frac{n}{L}\right) \tag{1.69}$$

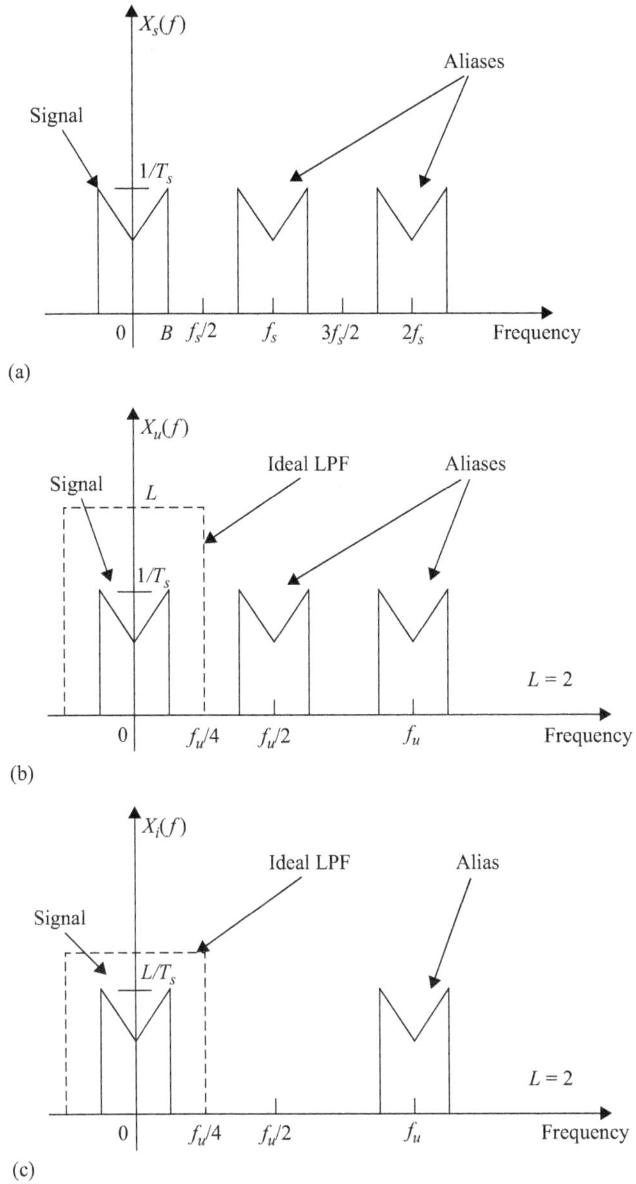

Figure 1.20 The frequency representation of (a) a sampled signal $X_s(f)$, (b) an upsampled signal $X_u(f)$ by $L = 2$, and (c) an interpolated signal $X_i(f)$ by $L = 2$. In this case, $f_u = 2f_s$ ($L = 2$)

The interpolated digital signal is given by the convolution of $x_u[n]$ with $h_r[n]$ to give [1]

$$x_i[n] = \sum_{k=-\infty}^{\infty} x_s[k] \operatorname{sinc}\left(\frac{n-kL}{L}\right) \tag{1.70}$$

Therefore, if the interpolation is performed using an ideal LPF, the interpolated signal satisfies the equation

$$x_i[n] = x_s\left[\frac{n}{L}\right] = x\left(\frac{nT_s}{L}\right) \qquad \text{for } n = 0, \pm L, \pm 2L, \ldots \tag{1.71}$$

In practice, ideal LPFs are not physically realizable, even in the digital domain, but good approximations can be implemented. In many cases, simple interpolation approaches (such as linear interpolation) can be adequate.

1.8 Conclusion

In this chapter we discussed the A/D and D/A conversion processes. The A/D conversion is comprised of sampling and quantization operations. The D/A conversion is comprised of an analog holding operation followed by a reconstruction filter. The analysis of each one of these operations was presented. In addition, the Nyquist sampling theorem was discussed, together with the concepts of undersampling and oversampling. Finally, the decimation and interpolation operations were covered.

Problems

1. For a sine wave, draw the signal at the output of the sampler and quantizer of a 4-bit ADC in the time and frequency domains.
2. What is the minimum sampling rate required to sample a single-tone sine wave of frequency 10 MHz? Discuss.
3. For a signal with a 10 MHz bandwidth, what is the minimum theoretical sampling rate required to sample the signal without distortion? What if the signal bandwidth changes to 100 MHz?
4. A signal of bandwidth 100 MHz is centered around 100 MHz. What is the minimum theoretical sampling required to sample the signal without distortion?
5. A signal of bandwidth 100 MHz is centered around 1 GHz. What is the minimum theoretical sampling required to sample the signal without distortion?
6. A signal of bandwidth 10 MHz is sampled with a 14-bit, 1 GS/s ADC.
   (a) What is the SINAD due to quantization?
   (b) If the signal is filtered after digitization using a 10 MHz digital filter, what is the SINAD of the filtered signal?

7. Using any programming language, write the code for an ideal 10-bit, 100 MS/s ADC. Plot the outputs in the time and frequency domains for the following conditions:
   (a) An input tone of 10 MHz.
   (b) Two tones of 10 MHz and 12 MHz.
   (c) Using a 4K FFT what is the level of the noise floor?
   (d) Using a 16K FFT, what is the level of the noise floor?
   (e) How does the noise floor relate to the FFT size?
8. A signal of bandwidth 10 MHz is sampled with a 14-bit, 100 MS/s ADC. If the digital output is downsampled by a factor of 4, is the signal recoverable without distortion? Repeat for downsampling by a factor of 8.

References

[1] A.V. Oppenheim and R.W. Schafer, *Discrete-Time Signal Processing*, Prentice Hall, Englewood Cliffs, NJ, 1989.

[2] J.G. Proakis and D.G. Manolakis, *Digital Signal Processing: Principles, Algorithms and Applications*, Third Edition, Prentice Hall, Upper Saddle River, NJ, 1996.

[3] B. Razavi, *Principles of Data Conversion System Design*, IEEE Press, Piscataway, NJ, 1995.

[4] H. Pan, "A 3.3-V 12-b 50-MS/s A/D Converter in 0.6-mm CMOS with over 80-dB SFDR," Ph.D. dissertation, UCLA, Dec. 1999.

[5] H. Pan and A.A. Abidi, "Spectral Spurs due to Quantization in Nyquist ADCs," *IEEE Trans. Circuits and Systems-I: Regular Papers*, 51(8), pp. 1422–1438, August 2004.

[6] F. Maloberti, *Data Converters*, Springer, Dordrecht, The Netherlands, 2010.

[7] M.J.M. Pelgrom, *Analog-to-Digital Conversion*, Second Edition, Springer, Dordrecht, The Netherlands, 2013.

[illegible] write the software [illegible] the Fourier [illegible] as the time and frequency [illegible] the following conditions:

(a) [illegible] at 20 kHz?

(b) [illegible]

(c) [illegible] the level of the noise [illegible]

(d) [illegible]

(e) [illegible]

[illegible]

References

[illegible]

Chapter 2
Performance metrics

In the previous chapter, we discussed ideal A/D and D/A conversion. Practically, there are numerous non-idealities that make the conversion process deviate from the ideal. In this chapter, we discuss the metrics used to characterize the converter's performance, with a focus on high speed ADCs. Since there are multiple dimensions of performance, the characterization of an A/D converter can be a complicated, and sometimes even confusing, process. For an ADC designer or user, it is imperative to understand the various performance metrics, and their relationship with one another.

The reconstructed digital output of an ideal ADC as a function of its input is shown in Figure 2.1, where the quantization steps (Δ) are uniform and ideal. The

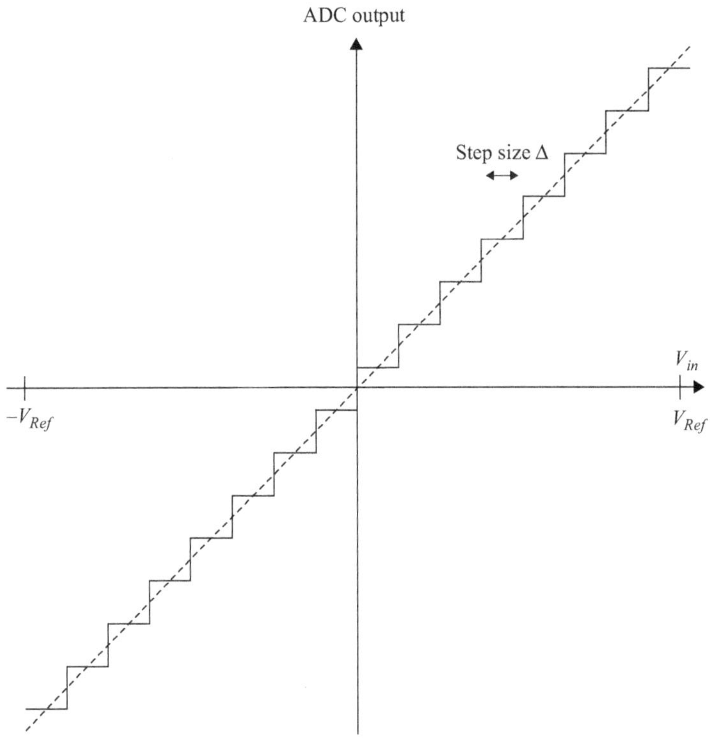

Figure 2.1 The reconstructed digital output as a function of the analog input for an ideal ADC

input full-scale is determined by the reference voltage V_{Ref}, and ranges from V_{Ref} to $-V_{Ref}$. In an ideal ADC, quantization is the only contributor to noise and distortion. In practice, there are other contributors that will be discussed in this chapter.

Some of the commonly used metrics to describe the ADC performance are:

- Resolution
- Sampling rate
- Signal-to-Noise Ratio (SNR)
- Signal-to-Noise-and-Distortion Ratio (SNDR or SINAD)
- Noise Spectral Density (NSD)
- Spurious-Free Dynamic Range (SFDR)
- Total Harmonic Distortion (THD)
- Inter-Modulation Distortion (IMD)
- Integral Non-Linearity (INL)
- Differential Non-Linearity (DNL)
- Jitter
- Offset error
- Gain error
- Bit Error Rate (BER)
- Power consumption

2.1 Resolution and sampling rate

The resolution and sampling rate are the two main performance metrics of a data converter. They are often used as identifiers to position the converter in the accuracy-speed space. The resolution of the ADC (or DAC) is the number of quantization bits (N) of the converter. Using the resolution and the input full-scale voltage (V_{FS}), we can calculate the quantization step (Δ) or the LSB size (LSB) as described in Chapter 1. This is given by

$$\Delta = \frac{V_{FS}}{2^N} = \frac{2V_{Ref}}{2^N} = LSB \tag{2.1}$$

The quantization noise power is given by P_Q such that

$$P_Q = \frac{\Delta^2}{12} = \frac{LSB^2}{12} = \frac{V_{FS}^2}{12 \times 2^{2N}} \tag{2.2}$$

The sampling rate f_s (or throughput rate) represents the number of samples that the ADC produces per second. It determines the maximum allowed signal bandwidth (f_B) that the ADC can sample according to the Nyquist sampling theorem, as discussed in Chapter 1, such that

$$f_B < f_s/2 \tag{2.3}$$

This upper limit on the bandwidth is often referred to as the quantization bandwidth or Nyquist bandwidth (B_N), to distinguish it from the input sampling circuit bandwidth of the ADC (B_{in}), which is discussed in Chapter 4. In practice, anti-aliasing filter considerations usually limit the quantization bandwidth to less than the Nyquist frequency.

Increasing the sampling rate without increasing the signal bandwidth is commonly referred to as oversampling, which is discussed in Chapter 1. Oversampling enables better performance within the signal bandwidth by applying digital filtering after the A/D conversion to remove the portion of the noise that is outside the signal band. This reduces the noise within the signal band by a factor equal to the processing gain *PG*, which is given by

$$PG(\mathrm{dB}) = 10\log(OSR) = 10\log(f_s/2f_B) = 10\log(B_N/f_B) \tag{2.4}$$

where the oversampling ratio *OSR* is given by

$$OSR = \frac{f_s}{2f_B} = \frac{B_N}{f_B} \tag{2.5}$$

In addition to improving the noise, oversampling can also improve the distortion by digitally filtering out the harmonics that fall out of the signal band.

2.2 Signal-to-noise-and-distortion ratio (SNDR or SINAD)

The SNDR/SINAD is defined as the ratio of the signal power to the noise power in dB, excluding the DC value. This is shown in Figure 2.2 and is given by

$$SNDR(\mathrm{dB}) = SINAD(\mathrm{dB}) = 10\log\left(\frac{\text{Signal Power}}{\text{Noise} + \text{Distortion Power}}\right) \tag{2.6}$$

$$SNDR(\mathrm{dB}) = SINAD(\mathrm{dB}) = 10\log\left(\frac{P_s}{P_N}\right) \tag{2.7}$$

If the only source of noise is quantization, then the *SNDR* is given by

$$SNDR(\mathrm{dB}) = SINAD(\mathrm{dB}) = SQNR(\mathrm{dB}) = 10\log\left(\frac{P_s}{P_Q}\right) \tag{2.8}$$

where P_Q is the quantization noise power. For a sinusoidal input, this gives

$$SNDR(\mathrm{dB}) = SINAD(\mathrm{dB}) = SQNR(\mathrm{dB}) = 6.02\,N + 1.76\ \mathrm{dB} \tag{2.9}$$

Typically, the SNDR/SINAD is specified using a single-tone (sinusoidal) input signal. It is usually impacted by multiple sources of noise:

- Quantization noise and distortion
- Device noise in the signal path
- Noise and distortion in the sampling clock path (jitter)
- Non-linearity in the signal path

The current state-of-the-art *SNDR* for high speed converters is in the order of 84 dBFS at 80 MS/s [1]. As the sampling rate increases, the reported *SNDR* typically decreases, where the state of the art at 250 MS/s is about 77 dBFS [2], at 1 GS/s it is about 69 dBFS [3], and at 5 GS/s it is about 63 dB [4].

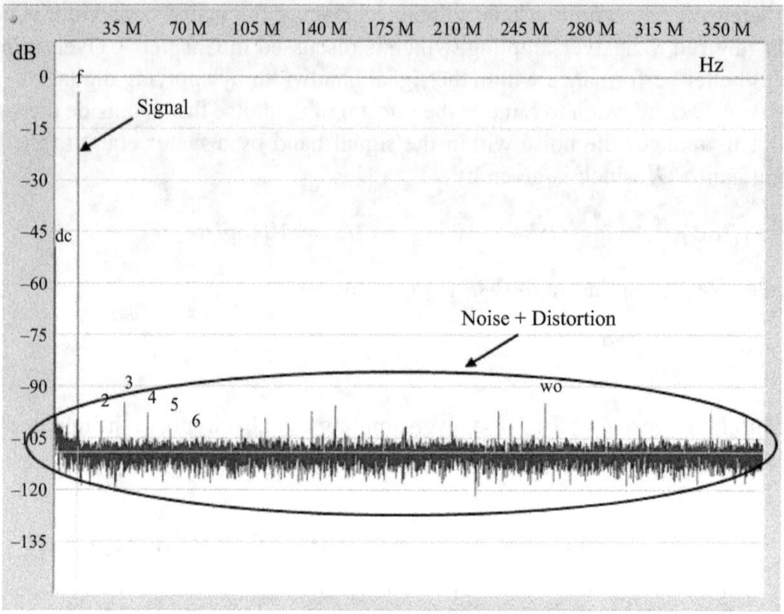

Figure 2.2 An example of the output spectrum of an ADC. The y-axis is the magnitude in dBFS and the x-axis is the frequency in Hz. The orders of the harmonics are labeled on the plot. In this graph, the term "dc" denotes the DC component, "f" denotes the fundamental frequency of the input signal, which is around 10MHz, and "wo" denotes the "worst-other" harmonic excluding the first 6 harmonics

Sometimes, it is desirable to exclude the harmonic distortion elements from the "noise" component. This is typically done by excluding the first few harmonics (such as the first 6 or 9 harmonics) to give the *SNR*, such that

$$SNR(\text{dB}) = 10 \log\left(\frac{\text{Signal Power}}{\text{Noise Power Excluding Harmonics}}\right) \tag{2.10}$$

Although the SNR, SNDR, and SINAD are ratios given in dB, in practice, they are sometimes expressed in dBc (i.e. relative to the "carrier") or dBFS (i.e. relative to the Full-Scale). That is,

$$SNDR(\text{dBc}) = 10 \log\left(\frac{\text{Signal Power}}{\text{Noise + Distortion Power}}\right) \tag{2.11}$$

and

$$SNDR(\text{dBFS}) = 10 \log\left(\frac{\text{ADC Full-Scale Power}}{\text{Noise + Distortion Power}}\right) \tag{2.12}$$

If the input signal is equal to the ADC full-scale, then the dBc and dBFS values are the same. However, when the input signal is less than the full-scale, the two values will be different. If the noise is independent of the input signal amplitude,

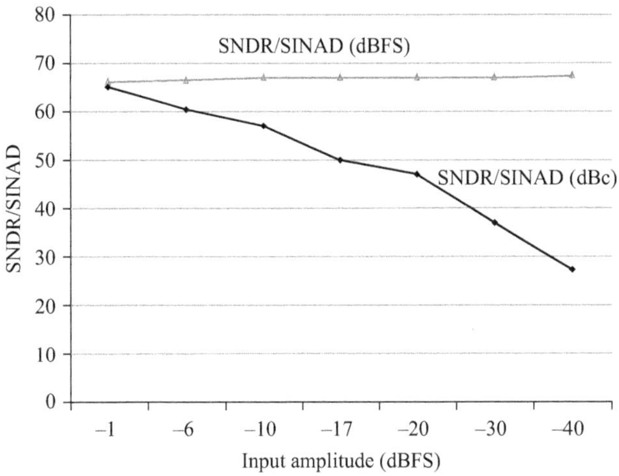

Figure 2.3 A plot of SNDR or SINAD in dBFS and dBc versus input amplitude in dBFS

we expect that as the input signal amplitude is reduced, the SNR/SNDR in dBc will go down, while the SNR/SNDR in dBFS should stay the same. However, the SNR/SNDR in dBFS typically improves with reducing the input amplitude because the noise contribution from some non-idealities (such as jitter and quantizer non-linearity) may decrease. An example is shown in Figure 2.3, where the SNDR in dBFS improves slightly with reducing the input amplitude.

As discussed in Chapter 1, oversampling improves the noise within a limited band by digitally filtering the out-of-band noise. This can be represented by the $SINAD_B$ over a bandwidth f_B, which is given as a function of the SINAD over the whole Nyquist bandwidth by

$$SINAD_B = SINAD + PG = SINAD + 10 \log OSR \tag{2.13}$$

where PG is the processing gain and OSR is the oversampling ratio.

The noise spectral density (NSD) represents the average noise per Hz over the Nyquist bandwidth. For an ADC with white noise, it is given by

$$NSD = -SINAD - 10 \log(f_s/2) \tag{2.14}$$

For example, if the SINAD is 70 dB, and the sampling rate is 1 GS/s, the noise spectral density will be −157 dB/Hz. It is important to note that the NSD depends on both the SINAD and the sampling rate. Therefore, it is influenced by both the noise performance and the speed of the ADC, which can make comparisons between the NSDs of different ADCs tricky. For some users who only use a portion of the Nyquist band, the NSD is a more useful and relevant parameter than the SINAD. The $SINAD_B$ over a specific bandwidth f_B that is less than the Nyquist bandwidth is related to the NSD in a straightforward manner, as follows

$$SINAD_B = -10 \log(f_B) - NSD \tag{2.15}$$

The SNDR is one of the most important ADC metrics. Its importance is reflected in the often quoted metric called: the Effective Number of Bits (*ENOB*), which is defined as

$$ENOB = (SNDR - 1.76\ \text{dB})/6.02 \qquad (2.16)$$

By comparing (2.16) with (2.9), we see that the *ENOB* parameter represents the hypothetical resolution of the ADC had the SNDR been limited only by the quantization noise.

Another parameter than can be used to translate the noise into an equivalent resolution is the *effective resolution* of the ADC [5], which is given by

$$\text{Effective Resolution } (N_{eff}) = \log_2\left(\frac{\text{ADC Full Scale}}{\text{Input-referred RMS Noise}}\right) \qquad (2.17)$$

The *ENOB* parameter is commonly used in sigma-delta (ΣΔ) converters to describe their effective resolution, because they employ a small number of quantization bits combined with noise-shaping to suppress the resulting quantization noise. However, for Nyquist ADCs, we should *not* think of the *ENOB* as the "real" resolution of the ADC. Although the *ENOB* value is almost always less than the resolution of a Nyquist ADC, this does not mean that the ADC's resolution is not real. For example, if an ADC has a resolution of 16 bits, and it has 16-bit linearity with a monotonic output characteristic of 2^{16} unique levels (digital codes) without missing codes, then that converter indeed has a resolution of 16 bits. These 16 bits of resolution and linearity can be useful to the user, even if the ADC's *ENOB* value is less than 16 bits.

On the other hand, if the goal is to achieve an *ENOB* of (say) 16 bits, then the resolution will need to be higher than 16 bits, otherwise the noise due to quantization alone would limit the *ENOB* to 16 bits, without taking into account yet the contributions from other noise sources, like thermal noise. This would require us to reduce the thermal noise significantly below the 16-bit level, which is a very power-expensive proposition for noise-limited ADCs. Therefore, for high performance and high speed ADCs, we often need the *ENOB* to be smaller than the resolution in order to reduce the quantization's impact on the overall noise, which decreases the power consumption.

So, in short, the resolution of the ADC and the *ENOB* are two different performance metrics. The former is a measure of the number of quantization bits and the latter is a representation of the total noise (or SNDR/SINAD) of the ADC.

An interesting exercise is to find the relation between the *ENOB* and the effective resolution (N_{eff}). From (2.16), we have

$$ENOB = \frac{SNDR}{6} - \frac{1.76}{6} = 10\frac{\log(P_s/P_N)}{6} - 0.3$$

Therefore,

$$ENOB = \log_2\left(\frac{V_{FS}/2\sqrt{2}}{n_v}\right) - 0.3 \qquad (2.18)$$

where n_v is the input-referred RMS noise. From (2.17), the effective resolution N_{eff} is given by

$$\text{Effective Resolution } (N_{eff}) = \log_2\left(\frac{V_{FS}}{n_v}\right)$$

which when combined with (2.18) gives:

$$\text{Effective Resolution } (N_{eff}) = ENOB + 1.8 \text{ bits} \tag{2.19}$$

That is, there is a 1.8-bit difference between the two measures. So, for example, an ADC with 16-bit resolution, and 14-bit *ENOB*, would have an "Effective Resolution" of 15.8 bits. This inconsistency between the two metrics is another reason why neither the *ENOB* nor the Effective Resolution (N_{eff}) should be taken literally as an indication of the "real" resolution of the ADC. They are merely representations of the ADC's noise performance, while the *real* resolution of a Nyquist ADC is determined by the number of the quantization bits (N).

2.3 Spurious-free dynamic range (SFDR)

Like the SNDR, the linearity of the converter is one of its most important performance metrics. A commonly used parameter to describe the converter's linearity is the SFDR, defined as the ratio of the signal power to the power of the largest undesired harmonic or spur, using a single-tone sinusoidal input signal. That is,

$$SFDR(\text{dBc}) = 10\log\left(\frac{\text{Signal Power}}{\text{Worst Harmonic or Spur Power}}\right) \tag{2.20}$$

Alternatively, the SFDR can be represented in dBFS as

$$SFDR(\text{dBFS}) = 10\log\left(\frac{\text{ADC Full-Scale Power}}{\text{Worst Harmonic or Spur Power}}\right) \tag{2.21}$$

Examples are shown in Figure 2.4. In Figure 2.4(a) we see a spectrum where the SFDR is about 90 dB, while in Figure 2.4(b) the SFDR is better than 110 dB, which represents the state of the art and best reported SFDR in high speed ADCs [2]. The SFDR is usually evaluated across the input amplitude. Figure 2.5 illustrates the SFDR in dBc and dBFS, and Figure 2.6 shows an SFDR plot, in dBc and dBFS, versus the input amplitude in dBFS. Figure 2.7 shows more SFDR plots in dBFS versus the input amplitude in dBFS. These examples show that the behavior of the SFDR versus input amplitude can vary widely from one ADC to another. Figure 2.6 is an example of an ADC where the SFDR in dBFS improves with reducing the input amplitude. On the other hand, Figure 2.7(a) shows a different pattern of variation, where the SFDR in dBFS degrades slightly (from 102 dBFS to 97 dBFS) with reducing the input amplitude, before significantly improving to better than 105 dBFS for input amplitudes below −20 dBFS. Figure 2.7(b) shows a

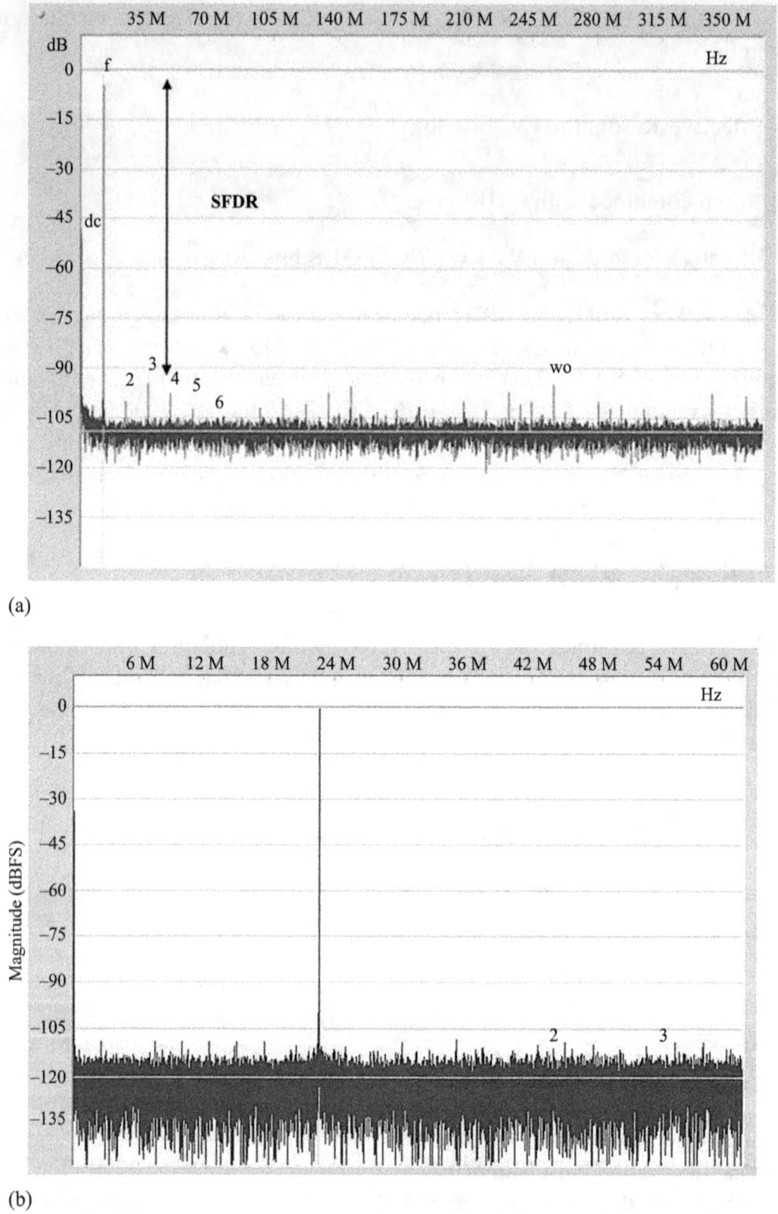

Figure 2.4 (a) The output spectrum of an ADC showing the Spurious-Free Dynamic Range (SFDR). (b) The output spectrum of an ADC showing the Spurious-Free Dynamic Range (SFDR) of better than 110 dB [2].© 2010 IEEE. Reprinted, with permission, from Reference 2

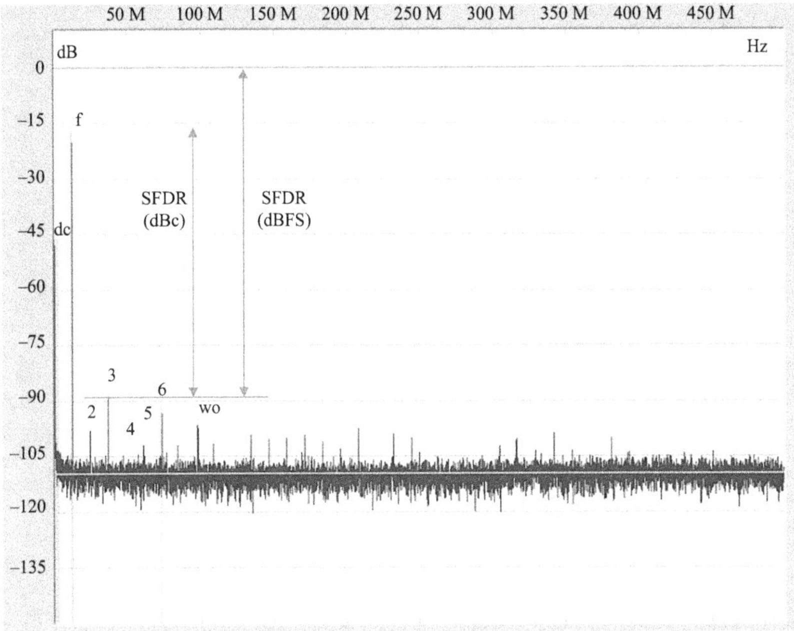

Figure 2.5 The output spectrum of an ADC showing the Spurious-Free Dynamic Range (SFDR) in dBc and dBFS

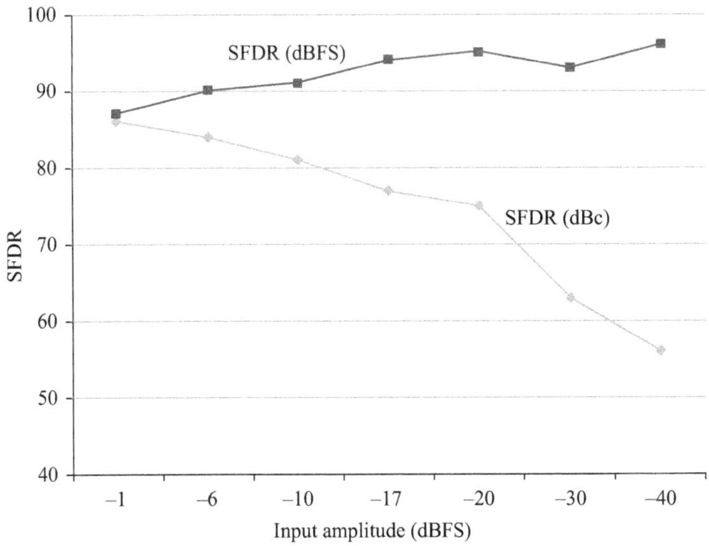

Figure 2.6 A plot of SFDR in dBc and dBFS versus input amplitude

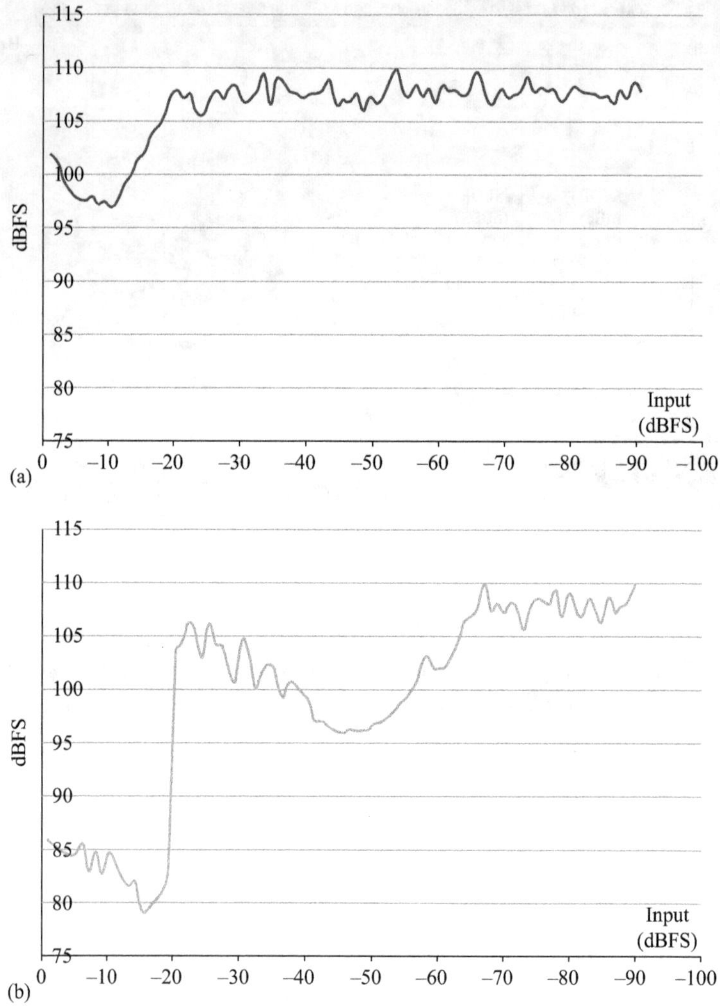

Figure 2.7 Examples of sweeps of the SFDR in dBFS versus input amplitude

case where the performance is worse and the variation is more substantial. In both examples of Figure 2.7, the SFDR's (in dBFS) pattern of variation is characteristic of pipelined ADCs with inter-stage gain errors, which are discussed in Chapters 3, 7, and 9. Some of the techniques that can be employed to improve this variation will also be covered.

Next to the SFDR, another commonly used linearity metric is the THD, which is defined as the ratio of the total harmonic power, typically the lowest 6 or 9 harmonics, to the signal power. That is

$$THD(\text{dBc}) = 10\log\left(\frac{\text{Total Harmonic Power}}{\text{Signal Power}}\right) \tag{2.22}$$

2.3.1 HD2 and HD3

To gain some insight into how the ADC's non-linearity is related mathematically to the harmonic distortion and the SFDR, assume we have a memory-less third-order non-linear system given by [6]

$$y(t) = \alpha_1 x(t) + \alpha_2 x^2(t) + \alpha_3 x^3(t) \tag{2.23}$$

where $x(t)$ is the input signal, $y(t)$ is the output; α_1, α_2, and α_3 are the first-, second-, and third-order system parameters, respectively. If we apply an input signal that is given by

$$x(t) = A\cos(\omega t) \tag{2.24}$$

the output will be given by

$$y(t) = \frac{\alpha_2 A^2}{2} + \left(\alpha_1 A + \frac{3\alpha_3 A^3}{4}\right)\cos\omega t + \frac{\alpha_2 A^2}{2}\cos 2\omega t + \frac{\alpha_3 A^3}{4}\cos 3\omega t$$

Therefore, the second-order harmonic $HD2$ and the third-order harmonic $HD3$ are given by

$$HD2 \approx \left(\frac{\alpha_2 A}{2\alpha_1}\right)^2 \quad \text{and} \quad HD3 \approx \left(\frac{\alpha_3 A^2}{4\alpha_1}\right)^2 \tag{2.25}$$

Hence, the second-order harmonic ($HD2$) in dB is given by

$$HD2(\text{dBc}) = 10\log(A^2/2) + 10\log\left(\frac{\alpha_2^2}{2\alpha_1^2}\right) = P_s(\text{in dB}) + K2 \tag{2.26}$$

The third-order harmonic ($HD3$) in dB is given by

$$HD3(\text{dBc}) = 20\log(A^2/2) + 10\log\left(\frac{\alpha_3^2}{4\alpha_1^2}\right) = 2P_s(\text{in dB}) + K3 \tag{2.27}$$

where P_s is the input signal power, $K2$ and $K3$ are the second- and third-order coefficients, respectively. Therefore, for a third-order non-linear system, every dB reduction in the input signal power leads to 1 dB improvement in $HD2$ (in dBc) and 2 dB improvement in $HD3$ (in dBc). This is an important rule of thumb that applies to third-order non-linear systems. Also, please note that although the SFDR and THD are positive quantities in dB, the $HD2$ and $HD3$ are usually expressed as negative numbers according to the definitions above.

Having said that, the astute reader would have noticed that the plots of SFDR in dBc shown in Figures 2.6 and 2.7 do not follow the trend anticipated by (2.26) and (2.27). In fact, plots of $HD2$ and $HD3$ in ADCs almost never show this trend in a consistent fashion. This is because ADCs tend to have higher-order non-linear terms that greatly influence the levels of the $HD2$ and $HD3$ at some input amplitudes. These high-order terms are caused by quantization non-idealities (errors) that manifest themselves as discontinuities in the input-output characteristic, which make it substantially different from the smooth third-order function described in (2.23). As the input amplitude is reduced, the higher-order non-linearities affect the levels of the $HD2$ and $HD3$ in a manner that may not follow (2.26) and (2.27). However, for the analog buffers and amplifiers that drive the ADC, the third-order function of (2.23) may hold.

So, if the linearity of the ADC is limited by the driver amplifier or buffer, a trend similar to that given by (2.26) and (2.27) may still be valid. However, if the linearity is limited by the ADC quantizer, as is often the case for small input amplitudes, it is unlikely that the second- and third-order non-linearity trend will hold.

2.3.2 Differential operation

In many applications, differential input signals are used in order to suppress the even-order harmonics. If the positive input is $x_p(t)$, the negative input is $x_n(t)$, the positive output is $y_p(t)$, and the negative output is $y_n(t)$, then

$$y_p(t) = \alpha_1 x_p(t) + \alpha_2 x_p^2(t) + \alpha_3 x_p^3(t) \tag{2.28}$$

and

$$y_n(t) = \alpha_1 x_n(t) + \alpha_2 x_n^2(t) + \alpha_3 x_n^3(t) \tag{2.29}$$

For input sinusoidal signals with amplitude A, it was shown in the previous section that the single-ended $HD2_{se}$ is given by

$$HD2_{se} \approx 10 \log\left(\frac{\alpha_2 A}{2\alpha_1}\right)^2 \tag{2.30}$$

The differential output $y_d(t)$ is given by

$$\begin{aligned} y_d(t) &= y_p(t) - y_n(t) \\ &= \alpha_1 x_p(t) + \alpha_2 x_p^2(t) + \alpha_3 x_p^3(t) - \left[\alpha_1 x_n(t) + \alpha_2 x_n^2(t) + \alpha_3 x_n^3(t)\right] \end{aligned} \tag{2.31}$$

If the two inputs are perfectly matched, then $x_n(t) = -x_p(t)$, which gives

$$y_d(t) = y_p(t) - y_n(t) = 2\alpha_1 x_p(t) + 2\alpha_3 x_p^3(t) \tag{2.32}$$

This shows that, with perfect matching (or balance) between the two single-ended sides, the even-order harmonics will cancel out and the odd harmonics will remain unchanged.

If there are amplitude and phase mismatches between the two single-ended sinusoidal inputs, they can be represented as

$$x_p(t) = A_p \sin \omega t \quad \text{and} \quad x_n(t) = -A_n(\sin \omega t + \phi) \tag{2.33}$$

where ϕ is the phase mismatch in radians, and $A = (A_p + A_n)/2$. The relative amplitude mismatch Δ can be defined as

$$\Delta = \frac{A_p - A_n}{A} = 2\frac{A_p - A}{A} = 2\frac{A - A_n}{A} = 2\delta \tag{2.34}$$

The impact of an amplitude mismatch alone on the second harmonic can be derived by substituting (2.33) and (2.34) into (2.31), while setting $\phi = 0$, and solving for the $HD2$, which will be given by

$$HD2_{diff} \approx HD2_{se} + 20 \log(\Delta) \tag{2.35}$$

where $HD2_{diff}$ is the differential second harmonic level in dB, $HD2_{se}$ is the single-ended second harmonic level for each side in dB, and Δ is the amplitude imbalance.

The impact of a phase mismatch alone on the second harmonic can be derived by substituting (2.32) into (2.31), while setting $\Delta = 0$, and solving for the $HD2$, which will be given by

$$HD2_{diff} = HD2_{se} + 20\log\left(\frac{\sin\phi}{\cos(\phi/2)}\right) \tag{2.36}$$

where $HD2_{diff}$ is the differential second harmonic level in dB, $HD2_{se}$ is the single-ended second harmonic level for each side in dB, and ϕ is the phase mismatch (imbalance) between the two inputs in radians. Equation (2.36) is represented in Figure 2.8 for an example where the single-ended $HD2$ ($HD2_{se}$) is equal to -50 dB. From equation (2.36), and Figure 2.8, we note the following:

- For small phase mismatches, $HD2_{diff}$ is substantially better than $HD2_{se}$.
- The differential $HD2_{diff}$ degrades as the phase mismatch increases.
- The phase mismatch, which results in no improvement in $HD2_{diff}$ over $HD2_{se}$, is 60°.
- Increasing the phase mismatch beyond 60° results in further degradation in the differential $HD2_{diff}$ where it becomes worse than the single-ended $HD2_{se}$.
- A 90° phase mismatch results in the $HD2_{diff}$ being 3 dB worse than $HD2_{se}$.
- The worst-case $HD2_{diff}$ is worse than $HD2_{se}$ by a value that asymptotically approaches 6dB, as the phase mismatch approaches 180°.

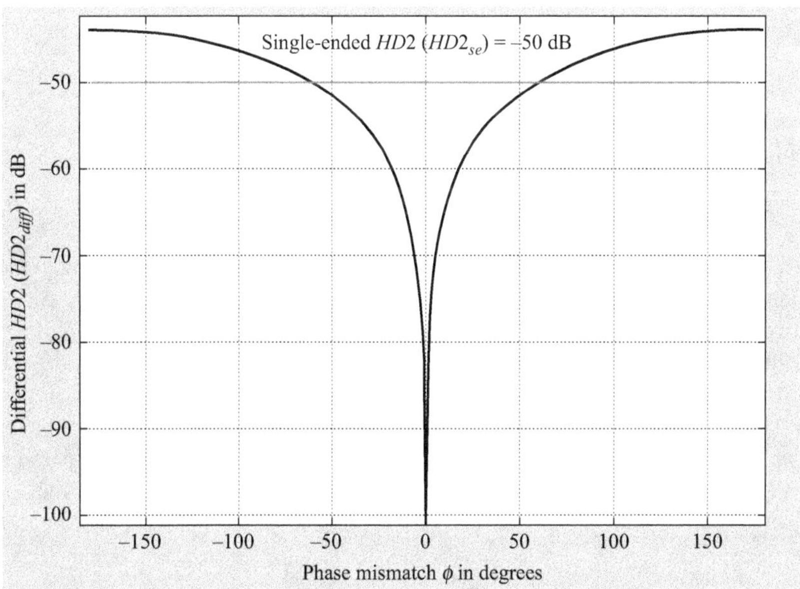

Figure 2.8 The differential $HD2_{diff}$ as a function of the phase mismatch ϕ for a single-ended $HD2_{se}$ of -50 dB

For small phase mismatches, (2.36) can be approximated by

$$HD2_{diff} \approx HD2_{se} + 20\log(\phi) \tag{2.37}$$

where ϕ is the phase mismatch in radians. It is clear from (2.35)–(2.37) that since the $HD2$ is a negative quantity, the amplitude and phase mismatches will degrade the differential second harmonic level, such that every doubling in the magnitude of the mismatch results in 6 dB degradation in the differential second harmonic amplitude.

In the presence of both amplitude and phase mismatches together, the $HD2$ can be obtained by substituting (2.33) and (2.34) into (2.31), which gives

$$HD2_{diff} \approx HD2_{se} + 10\log\left(\frac{A_p^4 + A_n^4 - 2A_n^2A_p^2\cos 2\phi}{A^2\left(A_p^2 + A_n^2 + 2A_pA_n\cos\phi\right)}\right) \tag{2.38}$$

Using (2.34) into (2.38) we obtain

$$HD2_{diff} \approx HD2_{se} + 10\log\left(\frac{1 + 6\delta^2 + \delta^4 - \left(1-\delta^2\right)^2\cos 2\phi}{1 + \delta^2 + \left(1-\delta^2\right)\cos\phi}\right) \tag{2.39}$$

which can be simplified to give

$$HD2_{diff} \approx HD2_{se} + 10\log\left(\frac{4\delta^2 + \left(1-\delta^2\right)^2\sin^2(\phi)}{\delta^2 + \left(1-\delta^2\right)\cos^2\left(\frac{\phi}{2}\right)}\right) \tag{2.40}$$

For small mismatches, this can be represented as

$$HD2_{diff} \approx HD2_{se} + 10\log\left(\Delta^2 + 4\sin^2\left(\frac{\phi}{2}\right)\right) \tag{2.41}$$

That is,

$$HD2_{diff} \approx HD2_{se} + 10\log\left(\Delta^2 + \phi^2\right) \tag{2.42}$$

where $HD2_{diff}$ is the differential second harmonic level, $HD2_{se}$ is the single-ended second harmonic level for each side, ϕ is the phase mismatch (imbalance) between the two inputs in radians, and Δ is the amplitude imbalance. It is interesting to note that the amplitude and phase mismatches are orthogonal; therefore, they are added after being squared.

If the mismatch is in the system transfer characteristics, as opposed to the input signal, the $HD2$ cancellation can be compromised too. In this case, the parameter α_2 in (2.28) and (2.29) may be mismatched between the two sides, such that

$$\alpha_{2p} = \alpha_2 + \varepsilon\alpha_2 \quad \text{and} \quad \alpha_{2n} = \alpha_2 - \varepsilon\alpha_2 \tag{2.43}$$

That is, the system amplitude mismatch is given by

$$\varepsilon = \frac{\alpha_{2p} - \alpha_{2n}}{2\alpha_2} \tag{2.44}$$

Then, in the presence of a system amplitude mismatch without input signal imbalance, the *HD*2 is shown to be

$$HD2_{diff} \approx HD2_{se} + 20\log(\varepsilon) \tag{2.45}$$

In the presence of an input amplitude mismatch as well, the effect of the mismatch in the parameter α_2 can be combined with the signal's amplitude mismatch to give

$$HD2_{diff} \approx HD2_{se} + 20\log(\Delta + \varepsilon) \tag{2.46}$$

In the presence of phase mismatch in the input and the system as well, the *HD*2 will be given by

$$HD2_{diff} \approx HD2_{se} + 10\log\left[(\Delta + \varepsilon)^2 + (\phi + \theta)^2\right] \tag{2.47}$$

where α_2, α_{2p}, and α_{2n} are complex and given by

$$\alpha_{2p} = (\alpha_2 + \varepsilon\alpha_2)e^{j\theta} \quad \text{and} \quad \alpha_{2n} = (\alpha_2 - \varepsilon\alpha_2)e^{-j\theta} \tag{2.48}$$

where θ represents the mismatch in the system's phase response, and ϕ is the phase mismatch in the input signal as before. Therefore, in (2.47), the term $(\Delta + \varepsilon)$ represents the total amplitude imbalance, and the term $(\phi + \theta)$ represents the total phase imbalance. Please note the factor of 2 difference in the definitions between the phase terms ϕ and θ, and between the amplitude mismatch terms Δ and ε. This difference is due to the squaring of the input signal by the second-order non-linearity in (2.28) and (2.29), which doubles the impact of the mismatches in the input signal compared to the mismatches in the α_2 term. Another observation is that the amplitudes mismatches are added coherently, and so are the phase mismatches. However, the amplitude and phase mismatches are combined together after squaring because of their orthogonality.

Example 1: If the single-ended *HD*2 level is −70 dB, what should the phase and gain matching on the input signal be to achieve 100 dB differential distortion, assuming the system itself is balanced?

Answer 1: To achieve $HD2_{diff} = -100$ dB from a single-ended $HD2_{se} = -70$ dB, the attenuation needs to be better than 30 dB. Using (2.35) and (2.37), we get

$$-30 \approx 20\log(\Delta) \approx 20\log(\phi)$$

That is,

$$\Delta \approx 3.16\% \; or \; \phi \approx 1.8^\circ$$

Therefore, to obtain −100 dB *HD*2, we can tolerate a gain mismatch only up to 3.16% *or* a phase mismatch only up to 1.8°.

Alternatively, if we allocate half of the mismatch budget for each type of mismatch, and use (2.42), we get

$$-30 \approx 10\log(2\Delta^2) \approx 10\log(2\phi^2)$$

Therefore,

$$\Delta \approx 2.2\% \quad \text{and} \quad \phi \approx 1.3^\circ$$

That is, to obtain -100 dB *HD*2, we can tolerate a combination of a gain mismatch up to 2.2% *and* a phase mismatch up to 1.3°. If we have to allow for some imbalance in the system itself (i.e. in α_2), then the numbers given above will need to represent the total imbalance, such that,

$$\Delta + \varepsilon \approx 2.2\% \quad \text{and} \quad \phi + \theta \approx 1.3^\circ$$

where the quantities $\Delta, \phi, \varepsilon$, and θ are defined in (2.33), (2.34), (2.44), and (2.48) above.

2.4 Inter-modulation distortion (IMD)

Unlike the single-tone SFDR, the two-tone IMD is measured by applying two input sine waves equal in magnitude and closely spaced in frequency. The IMD is the ratio of the power of largest spur to the power of one of the two tones. It can be represented in dBc or dBFS as shown in Figure 2.9. Sometimes, the IMD is a better representation of the distortion and it can capture non-linear effects that the SFDR may not capture.

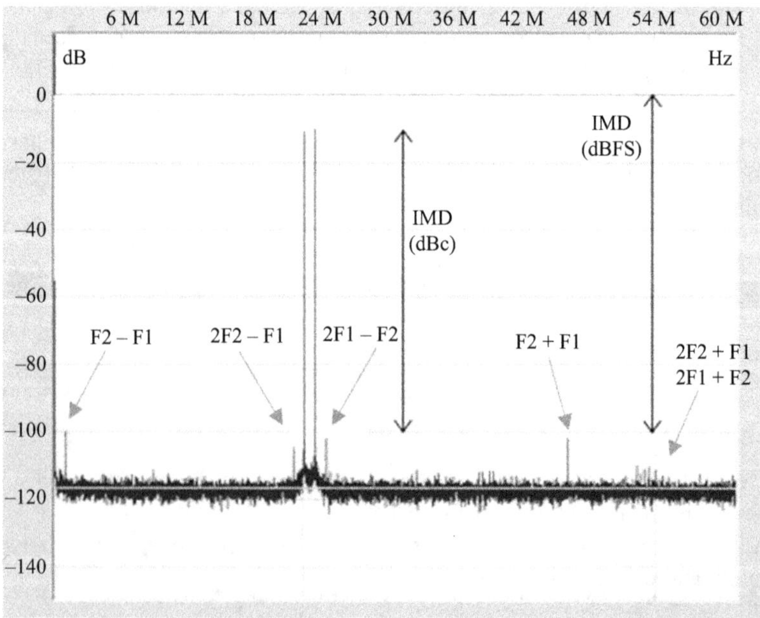

Figure 2.9 An ADC output spectrum showing the IMD performance with two-tone testing in dBc and dBFS

Assume a non-linear system given by [6]

$$y(t) = \alpha_1 x(t) + \alpha_2 x^2(t) + \alpha_3 x^3(t) \tag{2.49}$$

and we apply a two-tone signal $x(t)$, such that

$$x(t) = B_1 \cos(\omega_1 t) + B_2 \cos(\omega_2 t) \tag{2.50}$$

By substituting (2.50) into (2.49), we can prove that the fundamental $y_1(t)$ is given by

$$y_1(t) = \alpha_1 B_1 \cos(\omega_1 t) + \alpha_1 B_2 \cos(\omega_2 t) \tag{2.51}$$

The second-order components $y_{IM2}(t)$ are given by

$$y_{IM2}(t) = \alpha_2 B_1 B_2 \cos(\omega_1 + \omega_2)t + \alpha_2 B_1 B_2 \cos(\omega_1 - \omega_2)t \tag{2.52}$$

The third-order components $y_{IM3}(t)$ are given by

$$\begin{aligned} y_{IM3}(t) = \frac{3\alpha_3}{4} \Big[& B_1^2 B_2 \cos(2\omega_1 + \omega_2)t + B_1^2 B_2 \cos(2\omega_1 - \omega_2)t \\ & + B_2^2 B_1 \cos(2\omega_2 + \omega_1)t + B_2^2 B_1 \cos(2\omega_2 - \omega_1)t \Big] \end{aligned} \tag{2.53}$$

If the amplitudes of the two tones are equal, such that

$$B_1 = B_2 = B, \quad \text{and} \quad x(t) = B\cos(\omega_1 t) + B\cos(\omega_2 t) \tag{2.54}$$

then,

$$\begin{aligned} y_{IM3}(t) = \frac{3\alpha_3 B^3}{4} [& \cos(2\omega_1 + \omega_2)t + \cos(2\omega_1 - \omega_2)t + \cos(2\omega_2 + \omega_1)t \\ & + \cos(2\omega_2 - \omega_1)t] \end{aligned} \tag{2.55}$$

From (2.52), we see that the second-order inter-modulation products (*IMD*2) are ideally equal and located at frequencies $f_1 + f_2$ and $f_1 - f_2$. Equations (2.53) and (2.54) show that the third-order inter-modulation products (*IMD*3) are also ideally equal if $B_1 = B_2 = B$ and located at frequencies: $(2f_1 + f_2)$, $(2f_2 + f_1)$, $(2f_1 - f_2)$, and $(2f_2 - f_1)$. On the other hand, if the *IMD*3 component $(2f_1 - f_2)$ is different from the component $(2f_2 - f_1)$, this indicates that the input amplitudes B_1 and B_2 may not be equal.

It is important to note that if the system's bandwidth is limited, the *HD*2 and *HD*3 may be low-pass filtered and attenuated, because of their higher frequencies compared to the fundamental, therefore giving a misleading low harmonic level. On the other hand, the IMD components are comprised of sum components $(f_1 + f_2)$, $(2f_1 + f_2)$, and $(2f_2 + f_1)$ and difference components $(f_1 - f_2)$, $(2f_1 - f_2)$, and $(2f_2 - f_1)$. The sum components are located at high frequencies, near where their harmonic counterparts *HD*2 and *HD*3 are. So they would be susceptible to the low-pass filtering effect too. However, the difference component $(f_1 - f_2)$ is located near DC, and the other difference products $(2f_1 - f_2)$ and $(2f_2 - f_1)$ are located near the fundamental. These components are unlikely to be filtered out by a low-pass filter, which makes the IMD more effective in representing the non-linearity of that system. Therefore, if the difference components are much larger than the sum components, this could indicate a low-pass filtering effect.

Moreover, the presence of the IMD components in different locations, such as near DC, near the fundamental, and at locations that are twice and three-times the fundamental's frequency, enables the two-tone IMDs to capture other non-ideal effects. For example, when measuring the IMD, the response of the system to signals near DC and near the fundamental would be tested, in addition to the system's second- and third-order non-linearity. That is why it is common for the IMD measurement to expose problems that were not visible using the single-tone SFDR or the *THD* of the ADC.

Back to our analysis, from (2.52) and (2.55), we obtain

$$IMD2 \approx \left(\frac{\alpha_2 B}{\alpha_1}\right)^2 \quad \text{and} \quad IMD3 \approx \left(\frac{3\alpha_3 B^2}{4\alpha_1}\right)^2 \tag{2.56}$$

Therefore, the *IMD*2 in dBc is given by

$$IMD2(\text{dBc}) = 10\log(B^2/2) + 10\log\left(\frac{2\alpha_2^2}{\alpha_1^2}\right) = P_s(\text{in dB}) + K2 + 20\log 2 \tag{2.57}$$

The *IMD*3 is

$$IMD3(\text{dBc}) = 20\log(B^2/2) + 20\log\left(\frac{3\alpha_3}{2\alpha_1}\right) = 2P_s(\text{in dB}) + K3 + 20\log 3 \tag{2.58}$$

where *K*2 and *K*3 are the second- and third-order coefficients respectively that are identical to those defined previously in (2.26) and (2.27). Thus, for a third-order non-linear system, every dB reduction in the input signal power leads to 1 dB improvement in *IMD*2 (in dBc) and 2 dB improvement in *IMD*3 (in dBc). This is similar to the behavior of the single-tone harmonics *HD*2 and *HD*3. The *IMD*2 and *IMD*3 are also negative numbers in dB similar to their *HD*2 and *HD*3 counterparts.

Another measure of the IMD that is commonly used in communication applications is the input third-order intercept point or (*IIP*3) [6]. This is defined as the input amplitude in dBm at which the *IMD*3 level is equal to the fundamental component levels. The higher the *IIP*3, the more linear the system is. From (2.56) and (2.58), we can easily obtain the expression of the *IIP*3, which is given by

$$IIP3(\text{dBm}) = P_s(\text{dBm}) + \frac{|IMD3(\text{dBc})|}{2} \tag{2.59}$$

where *IIP*3 is the input third intercept point, P_s is the input signal power in dBm, and *IMD*3 is the third-order IMD in dBc.

2.5 Relationship between HD and IMD

From (2.26), (2.27), (2.57), and (2.58), we can readily see that for the same input amplitude (i.e. $B = A$), the *IMD* and the *HD* components are related as follows

$$IMD2(\text{dBc}) = HD2(\text{dBc}) + 20\log 2 = HD2(\text{dBc}) + 6\text{ dB} \tag{2.60}$$

and

$$IMD3(\text{dBc}) = HD3(\text{dBc}) + 20\log 3 = HD3(\text{dBc}) + 9.5\text{ dB} \tag{2.61}$$

That is, at the same input amplitude, the *IMD*2 is *worse* than the *HD*2 by 6 dB, and the *IMD*3 is *worse* than the *HD*3 by 9.5 dB. It is important to note that in this case the amplitude of *each* tone in the two-tone (IMD) test is equal to the amplitude of the single tone in the *HD* test.

Since the input tones are all at the same amplitudes, (2.60) and (2.61) above can be represented as follows

$$IMD2(\text{dBFS}) = HD2(\text{dBFS}) + 20\log 2 = HD2(\text{dBFS}) + 6\text{ dB} \tag{2.62}$$

and

$$IMD3(\text{dBFS}) = HD3(\text{dBFS}) + 20\log 3 = HD3(\text{dBFS}) + 9.5\text{ dB} \tag{2.63}$$

However, we typically measure the IMD using a smaller input amplitude for each tone, relative to the full-scale, compared to the *HD* measurement. For example, if we perform the single-tone test at −1 dBFS, we typically perform the two-tone test at −7 dBFS to avoid over-ranging the ADC. If we take the difference in amplitude into account, the *IMD*2 will be given by

$$IMD2(\text{dBc}) = HD2(\text{dBc}) + 20\log 2 - \Delta A(\text{dB}) = HD2(dBc) + 6\text{ dB} - \Delta A(\text{dB}) \tag{2.64}$$

The *IMD*3 will be given by

$$\begin{aligned} IMD3(\text{dBc}) &= HD3(\text{dBc}) + 20\log 3 - 2\Delta A(\text{dB}) \\ &= HD3(dBc) + 9.5\text{ dB} - 2\Delta A(\text{dB}) \end{aligned} \tag{2.65}$$

where $\Delta A(\text{dB})$ is the difference in amplitudes between the single-tone and the two-tone cases, which is given by

$$\Delta A(\text{dB}) = 20\log A - 20\log B \tag{2.66}$$

where A and B are the amplitudes of the cosine waves given in (2.24) and (2.54) for the single tone and the two tones, respectively. Therefore, for a difference of 6 dB between the amplitudes, we get

$$IMD2(\text{in dBc, at } -7\text{ dBFS}) = HD2(\text{in dBc, at } -1\text{ dBFS}) \tag{2.67}$$

and

$$IMD3(\text{in dBc, at } -7 \text{ dBFS}) = HD3(\text{in dBc, at } -1 \text{ dBFS}) - 2.5 \text{ dB} \quad (2.68)$$

That is, for a third-order system, the *IMD*2 in dBc (at −7 dBFS) is equal to the *HD*2 in dBc (at −1 dBFS), and the *IMD*3 in dBc (at −7 dBFS) is *better* than the *HD*3 in dBc (at −1 dBFS) by 2.5 dB.

Representing these relations in dBFS gives the following

$$IMD2(\text{in dBFS, at } -7 \text{ dBFS}) = HD2(\text{in dBFS, at } -1 \text{ dBFS}) - 6 \text{ dB} \quad (2.69)$$

and

$$IMD3(\text{in dBFS, at } -7 \text{ dBFS}) = HD3(\text{in dBFS, at } -1 \text{ dBFS}) - 8.5 \text{ dB} \quad (2.70)$$

That is, for a third-order system, the *IMD*2 in dBFS (at −7 dBFS) is *better* than the *HD*2 in dBFS (at −1 dBFS) by 6 dB, and the *IMD*3 in dBFS (at −7 dBFS) is *better* than the *HD*3 in dBFS (at −1 dBFS) by 8.5 dB.

It is important to note that (2.64)–(2.70) that relate the IMD at a certain amplitude to the HD at another amplitude rely on the assumptions that this is a third-order system and that the harmonic (and the IMD) improve with reducing the input amplitude. However, Figures 2.6 and 2.7 have shown that this relation with the input amplitude is often not valid in ADCs because of the presence of quantization errors and substantial higher-order non-linearities. Therefore, it is often *not* true that the *IMD*2 is better than the *HD*2 by 6 dB (in dBFS) or that the *IMD*3 is better than the *HD*3 by 8.5 dB (in dBFS). In fact, studying commercially available high speed ADCs, we find that this is almost never the case.

For example, consider the ADC represented by the SFDR plot in Figure 2.7(a) and assume that for that ADC the SFDR is limited by both *HD*2 and *HD*3, such that for a −1 dBFS input signal, $HD2 = -102$ dBFS and $HD3 = -102$ dBFS. If we perform a two-tone test with two tones at −7 dBFS each, according to (2.69) and (2.70), we would expect the *IMD*2 to be −108 dBFS and the *IMD*3 to be −110.5 dBF. However, that will almost certainly not be the case. If we look at the SFDR of that ADC as a function of the input amplitude, we do not see the improvement expected for a third-order non-linear system. In fact, when reducing the input amplitude by 6 dB, the SFDR *degrades* by about 5 dB from about 102 dBFS to about 97 dBFS. Since the SFDR is usually limited by the second and third harmonics at large input amplitudes, this degradation in SFDR with reducing the input amplitude indicates a possible degradation in the *HD*2 or *HD*3, or both.

If we assume that both *HD*2 and *HD*3 are around −97 dBFS at −7 dBFS, then using (2.62) and (2.63), we would expect the *IMD*2 to be *worse* than the *HD*2 at the same amplitude by about 6 dB (in dBFS) and the *IMD*3 to be worse than the *HD*3 by 9.5 dB (in dBFS). Therefore, the *IMD*2 would be equal to −91 dBFS and *IMD*3 equal to −87.5 dBFS. That is clearly much worse than the −108 dBS and −110.5 dBFS estimates obtained before for the *IMD*2 and *IMD*3 using (2.69) and (2.70), respectively. So, which estimate is correct?

Unfortunately, *neither* estimate is accurate because this is not a third-order system. The first estimate using (2.69) and (2.70) is wildly optimistic, while the second estimate using (2.62) and (2.63) is pessimistic. However, practically, experience has shown that the second estimate tends to be much closer to reality than the first one.

Therefore, we must exercise caution when trying to estimate the IMD performance from the *HD*. Although the relationship is straightforward in third-order systems, it is more complicated in ADCs. In fact, in quantizer-limited ADCs, the relationship given by (2.69) and (2.70) is almost always not valid. It is usually preferable to use the relation between the IMD and HD at the same amplitude as described by (2.62) and (2.63) for back-of-the-envelope estimation. Although this estimate will not be accurate either for quantizer-limited ADCs, it tends to practically give more realistic estimates.

Example 2: For a third-order non-linear system with $HD3 = -90$ dBc at -1 dBFS, what is the estimated $IMD3$?

Answer 2: Using (2.63): $IMD3$ (at -7 dBFS) $= HD3$ (at -7 dBFS) $+ 9.5 = (-90 - 2 \times 6) + 9.5 = -102 + 9.5 = -92.5$ dBc $= -99.5$ dBFS

Using (2.70): $IMD3$ (at -7 dBFS) $= HD3$ (at -1 dBFS) $- 2.5 = -90 - 2.5 =$ 92.5 dBc $= -99.5$ dBFS

Therefore, both equations give the same answer as expected, because this is a third-order system.

Example 3: In an ADC, *HD3* is -90 dBc at -1 dBFS and -7 dBFS. What is the estimated *IMD3* at -7 dBFS?

Answer 3: Since the $HD3$ does not follow the improvement with amplitude of a third-order system, we cannot expect accurate results using either (2.63) or (2.70).

Using (2.63) for the $IMD3$ and $HD3$ at the same amplitude gives:

$IMD3 = -90 + 9.5 = -80.5$ dBc $= -87.5$ dBFS (which is a pessimistic estimate).

If we had used (2.70), we would have got:

$IMD3 = -90 - 2.5 = -92.5$ dBc $= -99.5$ dBFS (which is a wildly optimistic estimate).

In this case using (2.63) gives a more reasonable result. It is however a pessimistic estimate because it too relies on the third-order assumption and hence assumes that the $HD3$ and $IMD3$ will get worse as the amplitude increases, which is not the case.

It is important to note that one can always simulate or measure the IMD and SFDR of any ADC to obtain their values. In fact, analysis of the IMD value for a given non-linear characteristics can be done in some cases. The purpose of the above discussion is to give some estimation tools, when the information available is not adequate. Often times, designers and users of ADCs find themselves in situations where they have to make such estimates with inadequate information. This discussion is meant to provide guidance and warn against some pitfalls in making those estimates. It is not meant to encourage using these estimates instead of rigorous simulation, measurement, or analysis.

2.6 Differential and integral non-linearity (DNL and INL)

The transfer characteristic of an ideal quantizer is given by the ideal stair-case of Figure 2.1. Practically, the ADC characteristic can look more like Figure 2.10, where we have an offset and a gain error in addition to the non-linearity represented by the unequal step sizes. Both the DNL and INL are performance metrics that describe the static non-linearity by measuring the deviation of the transfer characteristic from the ideal stair-case. Traditionally, they are measured using a very-low-frequency input or DC, and hence the "static" connotation. The DNL describes the small-signal non-linearity, while the INL tends to correlate with the large-signal non-linearity.

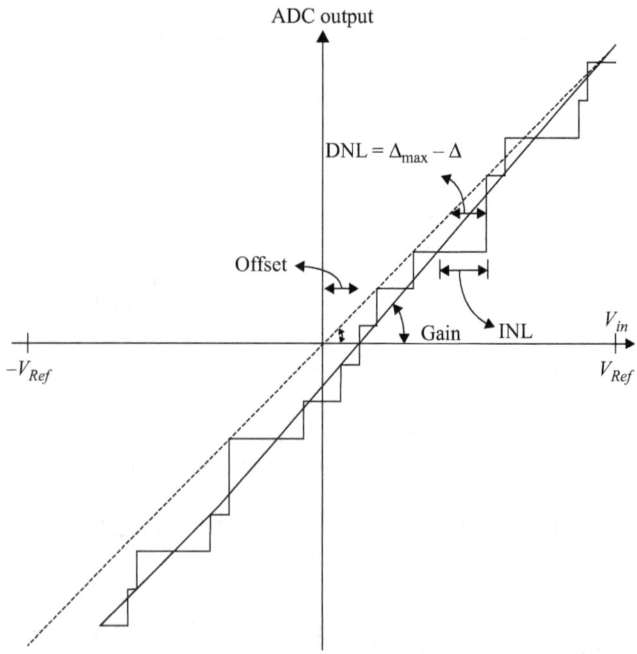

Figure 2.10 A non-ideal input–output characteristic of an ADC showing the DNL and INL

The DNL of the ADC is defined as the maximum difference between the real and ideal input-referred quantization step sizes, as shown in Figure 2.10. It can be measured by applying a ramp or a sinusoidal input. We need to note that the difference in *input* step sizes is represented on the x-axis, but in an ADC we only have access to the output codes. To measure the input step size, the histogram method is usually used, where a large number of samples are used and a histogram is built for every code. If the input is uniformly distributed, large steps will have more "hits" than average, and small steps will have fewer hits. Therefore, the number of times each code appears at the ADC output is an accurate representation of the step size.

In Figure 2.10, the DNL at code k is given by

$$DNL(k) = \frac{V_x(k) - V_x(k-1)}{LSB} - 1 \tag{2.71}$$

where $V_x(k)$ is the input threshold voltage corresponding to code k, and LSB is the ideal step size. Using the histogram method, the DNL at code k is given by

$$DNL(k) = \frac{H(k)}{\text{mean}[H]} - 1 \tag{2.72}$$

where $H(k)$ is the number of hits of code k, and mean$[H]$ is the average number of hits over all codes. The DNL value of the ADC is given by the maximum DNL over all the codes. That is,

$$DNL = \max_k[DNL(k)] \tag{2.73}$$

A positive DNL indicates a code bin that is wider than average, while a negative DNL indicates a narrower code bin than average. A DNL of -1 indicates a missing code and a DNL that is larger than $+1$ indicates non-monotonicity of the transfer characteristic. Overall, the DNL plot shows the DNL at each code versus the output code, and it describes the small-signal static non-linearity of the ADC. Examples are shown in Figure 2.11.

Since the histogram method assumes a uniformly distributed input, it can be applied directly when using a uniformly distributed ramp input. However, if a sinusoidal input signal is used, we have to compensate for its non-uniform distribution. The probability density function of a normalized sinusoidal signal is given by

$$p(V) = \frac{1}{\pi\sqrt{1 - V^2}} \tag{2.74}$$

The transition levels when using a sinusoidal input are obtained by

$$H(k) = -\cos\left(\frac{\pi \cdot \text{cumsum}(k)}{\text{sum}(H)}\right) - \left[-\cos\left(\frac{\pi \cdot \text{cumsum}(k-1)}{\text{sum}(H)}\right)\right] \tag{2.75}$$

where cumsum(k) is the cumulative sum of all the code hits up to k, and sum(H) is the sum of all the code hits. Substituting (2.75) into (2.72) gives the DNL of each code.

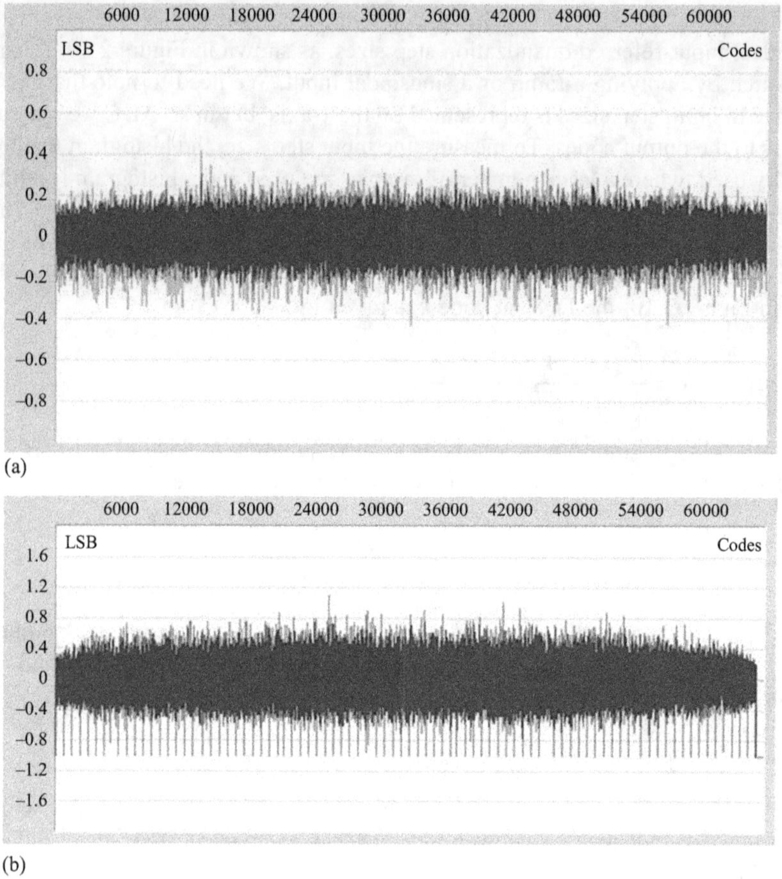

Figure 2.11 Examples of DNL plots for a 16-bit ADC. The y-axis is the DNL in LSBs, and the x-axis is the output codes. (a) The DNL is less than 0.5 LSB. (b) The DNL plot shows missing codes

When using a sinusoidal signal, it is important to avoid input frequencies that are sub-harmonics (or multiples) of the sampling rate. In addition, a large histogram size is needed in order to improve the accuracy. For a 16-bit ADC, histogram sizes of 1–2 million samples are often used.

The INL is defined as the maximum difference between the ADC input–output characteristic and the ideal one after correcting for offset and gain errors. This is shown in Figure 2.10. Practically, the INL is obtained by integrating the DNL. The INL curve is one of the most important tools used to understand the internal behavior of the ADC. Since it represents the deviation of the ADC transfer characteristic from the ideal one, it is the closest thing we have to an ADC transfer characteristic. The INL plot represents the error in the ADC output as a function of the output code. It also describes the large signal non-linearity, which is of great importance in ADC design.

The INL at code k is defined as

$$INL(k) = \frac{V_x(k) - V_x(k)|_{ideal}}{LSB} \tag{2.76}$$

where $V_x(k)$ is the input threshold value corresponding to code k, and $V_x(k)|_{ideal}$ represents the ideal threshold value. Alternatively, the INL at code k can be given by

$$INL(k) = \sum_{i=0}^{k} DNL(i) \tag{2.77}$$

The INL value of the ADC is defined as the maximum INL over all the codes. That is,

$$INL = \max_k[INL(k)] \tag{2.78}$$

The INL plot is a plot of the INL at each code versus the output code. An intuitive way to think of what the INL means is that if we start with the input-output characteristics, such as the one in Figure 2.10, subtract the ideal characteristic from the real one, and reverse the x- and y-axes, we get the INL. Examples of different ADC characteristics and INL plots are shown in Figures 2.12–2.16.

The ADC transfer characteristic of Figure 2.12 indicates a third-order non-linear behavior, as evidenced by the "S" shape. The INL of Figure 2.13(a) shows the "S" shape of a third-order non-linearity, while that of Figure 2.13(b) indicates a dominant second-order behavior, as evidenced by the "bow" shape. The plots in Figures 2.12–2.13 illustrate poor large-signal linearity, and good small-signal linearity. The likely

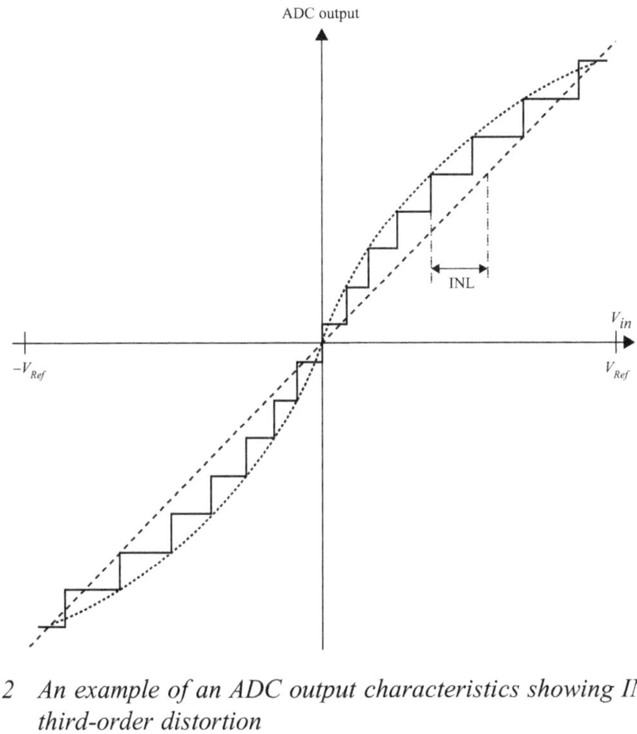

Figure 2.12 An example of an ADC output characteristics showing INL due to third-order distortion

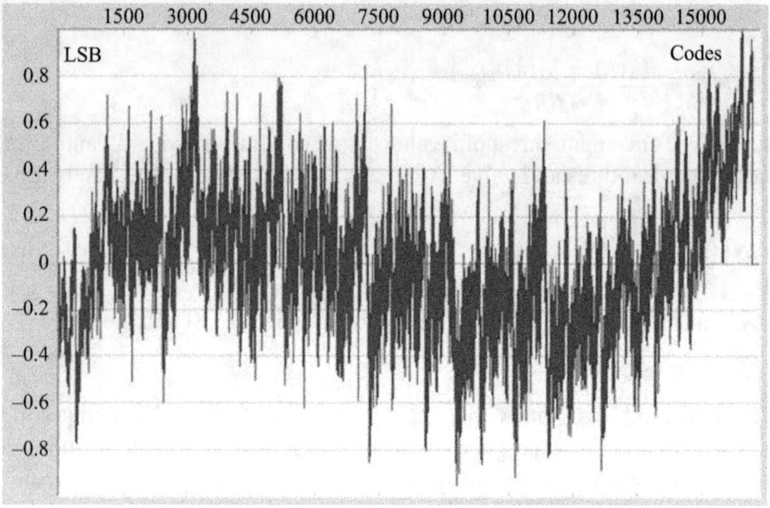

Figure 2.13(a) An example of an INL of a 14-bit ADC with third-order non-linearity

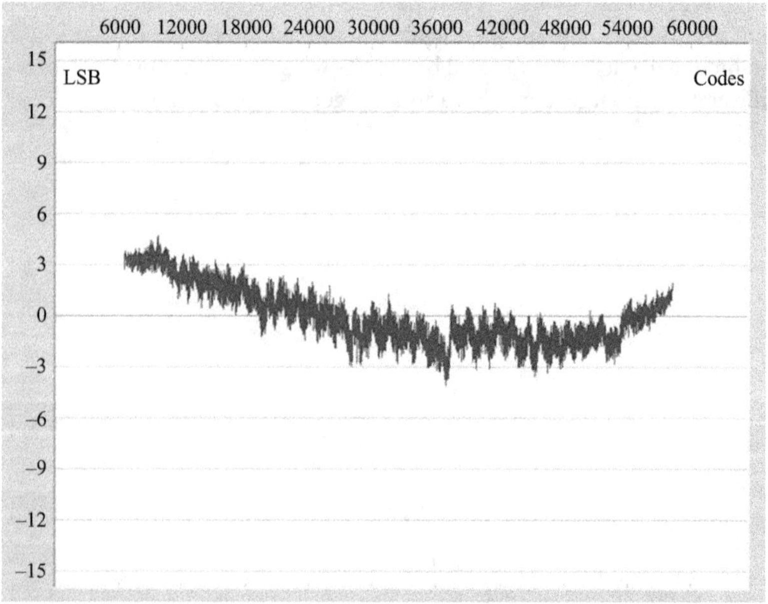

Figure 2.13(b) An example of an INL of a 16-bit ADC with second-order non-linearity

culprit of this non-linearity is not the quantizer, but the ADC's driver or sampler. The transfer characteristic of Figure 2.14 shows "jumps" or breaks in the ADC characteristic. These result in an INL with a saw-tooth pattern similar to that in Figure 2.15. A significant portion of the non-linearity in that case takes place in the quantizer,

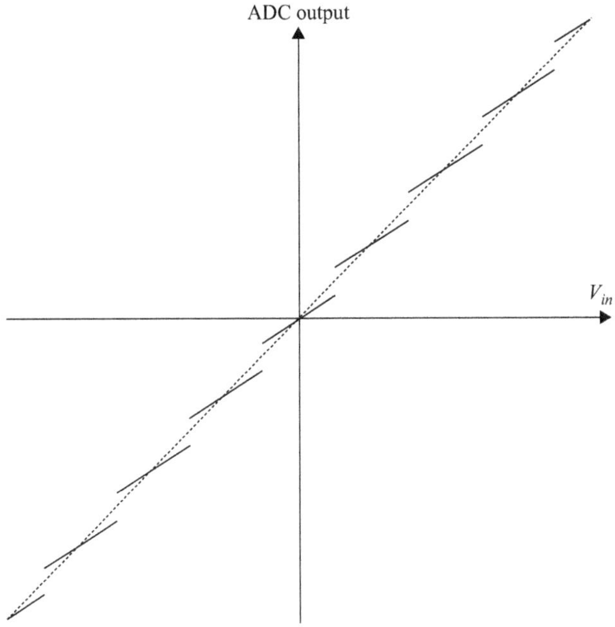

Figure 2.14 An example of ADC output characteristics with inter-stage gain errors

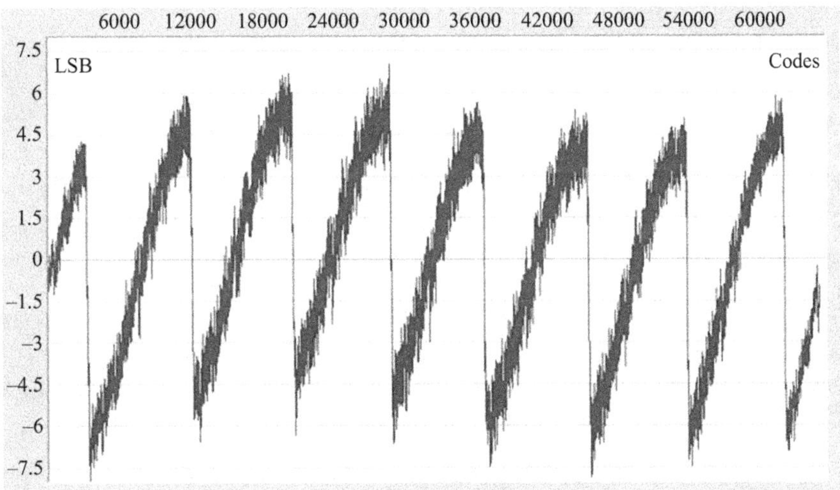

Figure 2.15 An example of INL of a 16-bit pipelined ADC with inter-stage gain errors

which indicates poor large-signal and small-signal linearity. These breaks are a hallmark of inter-stage gain errors in pipelined ADCs, as discussed in detail in chapter 7.

Although the INL is traditionally used to describe static (or DC) linearity, there is no reason not to use the INL to investigate the dynamic non-linearity too. As long

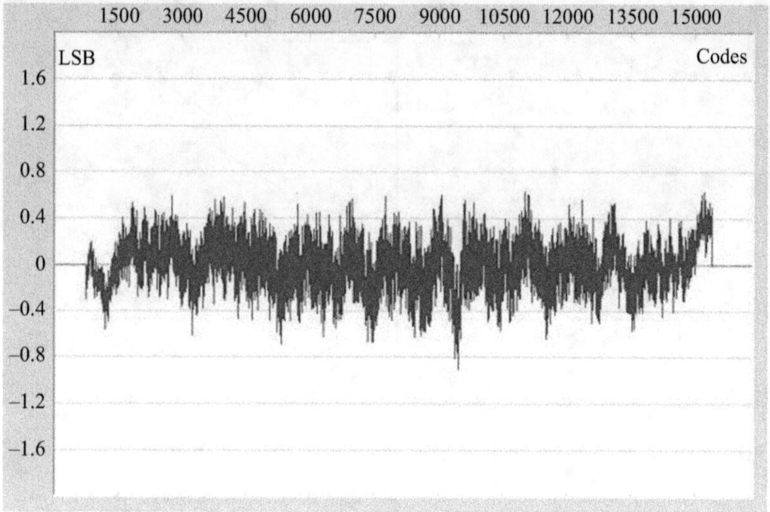

Figure 2.16 An example of a good 14-bit INL [14]

as the input sinusoidal frequency is chosen carefully to avoid sub-harmonics or multiples of the sampling rate, the INL can provide tremendous insight into the dynamic behavior of the ADC as well. Finally, Figure 2.16 shows an example of a 14-bit INL with good linearity.

2.7 Relationship between SFDR and INL

Since the INL plot represents the non-linear errors in the ADC output as a function of the output code, and hence describes the ADC's non-linearity, it is logical to expect a correlation between the INL and the harmonic distortion, and in turn the SFDR. Although the HD, IMD and SFDR tend to be the most commonly used metrics to quantify the ADC's non-linearity, the INL plot usually contains substantial amount of information about the performance and non-linearity of the converter. Therefore, it is important for the ADC designer and user to be able to effectively "read" the INL plot and deduce the behavior of the ADC from it.

2.7.1 HD2 and HD3 INL patterns

Common INL patterns tend to correspond to certain spectral contents. For example, a smooth S-shaped INL is a hallmark of a third-order non-linearity, while a symmetric "bow" represents a second-order non-linearity. These are shown in Figure 2.17. In fact, by analyzing the INL plot, one can roughly estimate the corresponding $HD2$ and $HD3$ level as follows

$$HD2(\text{dBFS}) = 20\log\left(\delta_2 \times 2^{-N}\right)\text{dB} \tag{2.79}$$

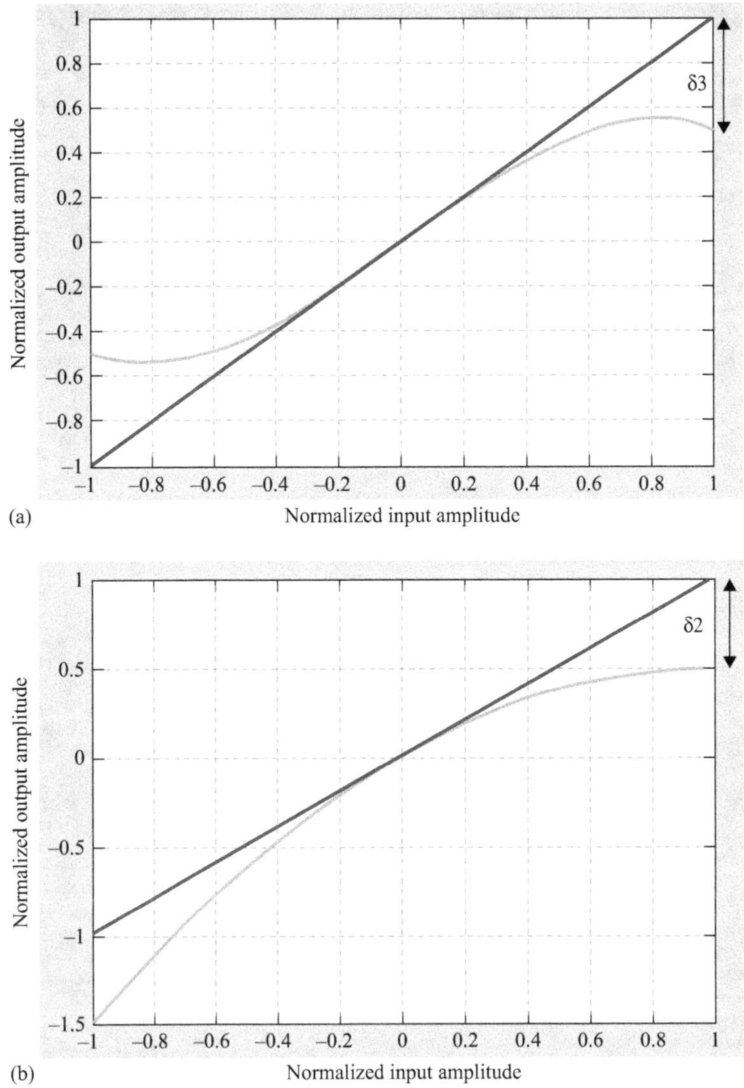

Figure 2.17 (a) An example of a third-order non-linear characteristic (b) An example of a second-order non-linearity

and

$$HD3(\text{dBFS}) = 20\log\left(\delta_3 \times 2^{-N}\right) - 6\text{ dB} \tag{2.80}$$

where δ_2 and δ_3 are the normalized INL errors in LSBs measured at the amplitude of the input signal, which may not necessarily be at the full-scale of the ADC. They are shown in Figure 2.18. These relations can be easily deduced from the

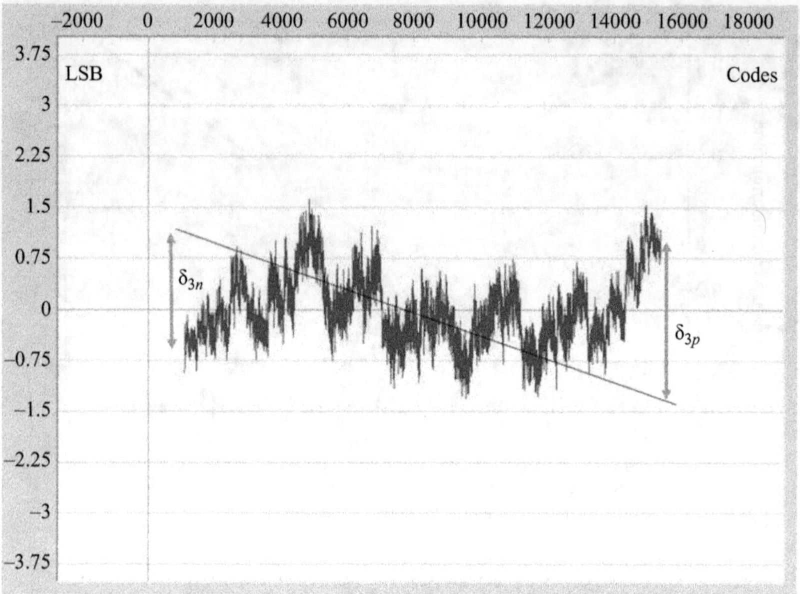

(a)

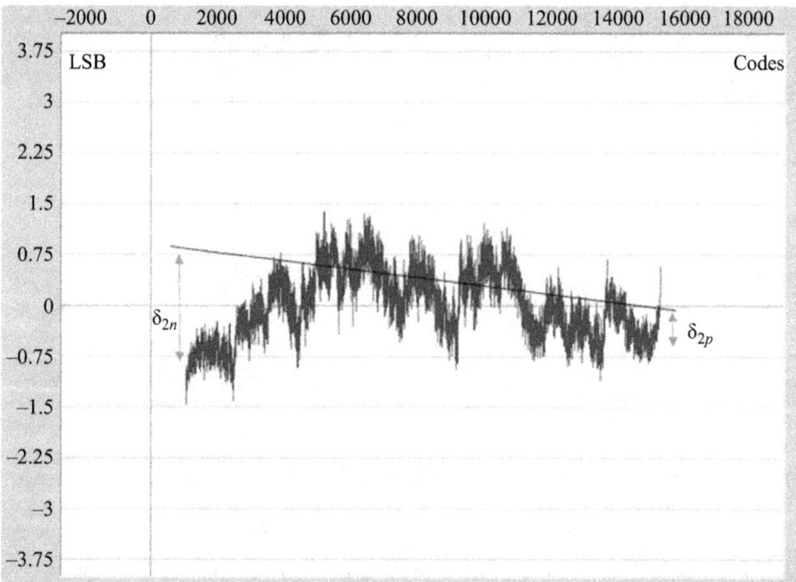

(b)

Figure 2.18 (Continued)

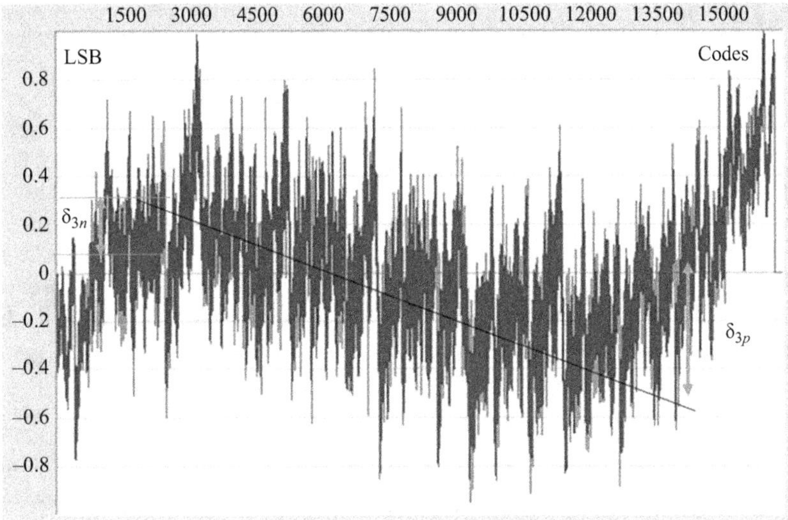

(c)

Figure 2.18 *(a) An example of an INL with third-order non-linearity. The parameter δ_3 is measured at -1 dBFS. (b) An example of an INL with second-order non-linearity. The parameter δ_2 is measured at -1 dBFS. (c) An example of an INL with third-order non-linearity. The parameter δ_3 is measured at -1 dBFS*

third-order sinusoidal response shown previously to be

$$y(t) = \frac{\alpha_2 A^2}{2} + \left(\alpha_1 A + \frac{3\alpha_3 A^3}{4}\right)\cos\omega t + \frac{\alpha_2 A^2}{2}\cos 2\omega t + \frac{\alpha_3 A^3}{4}\cos 3\omega t$$

The second-order harmonic $HD2$ and the third-order harmonic $HD3$ were given by

$$HD2 \approx \left(\frac{\alpha_2 A^2}{2\alpha_1}\right)^2 \quad \text{and} \quad HD3 \approx \left(\frac{\alpha_3 A^2}{4\alpha_1}\right)^2$$

It is clear that δ_2 and δ_3 are related to α_2 and α_3 with the proper normalization. That is,

$$HD2 \approx \left(\delta_2 \times 2^{-N}\right)^2 \quad \text{and} \quad HD3 \approx \left(\frac{\delta_3 \times 2^{-N}}{2}\right)^2$$

which leads to

$$HD2(\text{dBc}) = 20\log\left(\delta_2 \times 2^{-N}\right) - A_{out}(\text{dBFS}) \tag{2.81}$$

and

$$HD3(\text{dBc}) = 20\log\left(\delta_3 \times 2^{-N}\right) - 6\text{ dB} - A_{out}(\text{dBFS}) \tag{2.82}$$

where A_{out} is the output fundamental amplitude in dBFS, and is approximately equal to

$$A_{out}(\text{dBFS}) \approx 20\log\left(\frac{\alpha_1 A}{V_{FS}/2}\right)$$

We can see in Figure 2.18 that estimating the values of δ_2 and δ_3 may not be straightforward. Therefore, we need to exercise caution when applying these formulas, and use them only as rough estimates. The real INL may not be symmetric, and may have a combination of second-, third-, and high-order harmonics that make it difficult to estimate these values accurately. For example, in the absence of symmetry, we may have to average the δ_2 and δ_3 values on both sides to have reasonable estimates. That is,

$$\delta_2 \approx (\delta_{2p} + \delta_{2n})/2 \quad \text{and} \quad \delta_3 \approx (\delta_{3p} + \delta_{3n})/2 \tag{2.83}$$

It is interesting to note that in the presence of both *HD*2 and *HD*3 simultaneously, the difference between the individual values for a certain harmonic can be used as rough estimates for the other harmonic. That is,

$$\delta_2 \approx \frac{|(\delta_{3p} - \delta_{3n})|}{2} \quad \text{and} \quad \delta_3 \approx \frac{|(\delta_{2p} - \delta_{2n})|}{2} \tag{2.84}$$

Example 4: If δ_{3p} is 2.0 LSB and δ_{3n} is 1.5 LSB. Both are at the 14-bit level and at −1 dBFS, as shown in Figure 2.18(a). What is the estimated *HD*3 level at −1 dBFS?

Answer 4: Using (2.83), δ_3 can be estimated by averaging δ_{3p} and δ_{3n}, which is approximately equal to 1.75 LSB at the 14-bit level. Therefore,

$$\delta_3 \approx 1.75 \text{ LSB}$$

Therefore, from (2.80),

$$HD3 = -79.4 - 6 \text{ dB} = -85.4 \text{ dBFS}$$

and from (2.82),

$$HD3 = -79.4 - 6 \text{ dB} + 1 = -84.4 \text{ dBc}$$

From the FFT plot shown in Figure 2.19(a), which corresponds to the INL of Figure 2.18(a), we see that the *HD*3 is equal to −85 dBc, which is very close to our estimate.

Since δ_{3p} and δ_{3n} are different, we can estimate δ_2 using (2.84) as follows

$$\delta_2 \approx (\delta_{3p} - \delta_{3n})/2 \approx (2.0 - 1.5)/2 \approx 0.25 \text{ LSB}$$

Therefore, from (2.79),

$$HD2 = -96.3 \text{ dBFS}$$

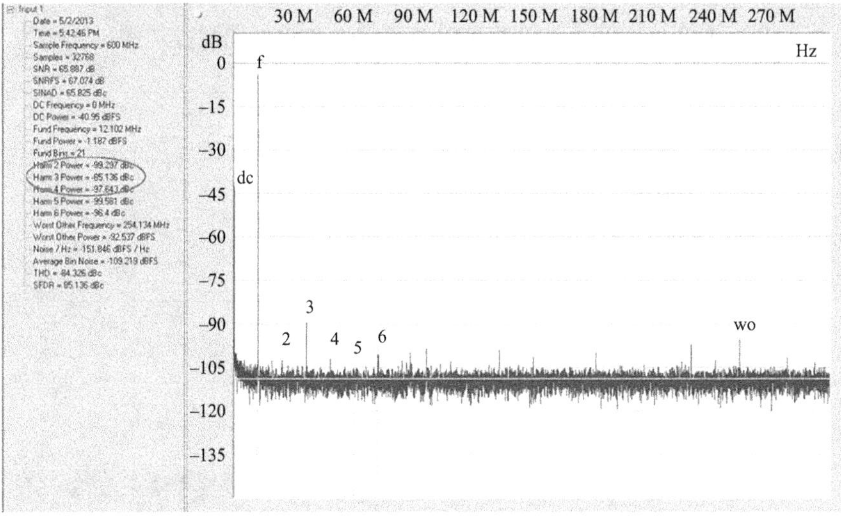

(a)

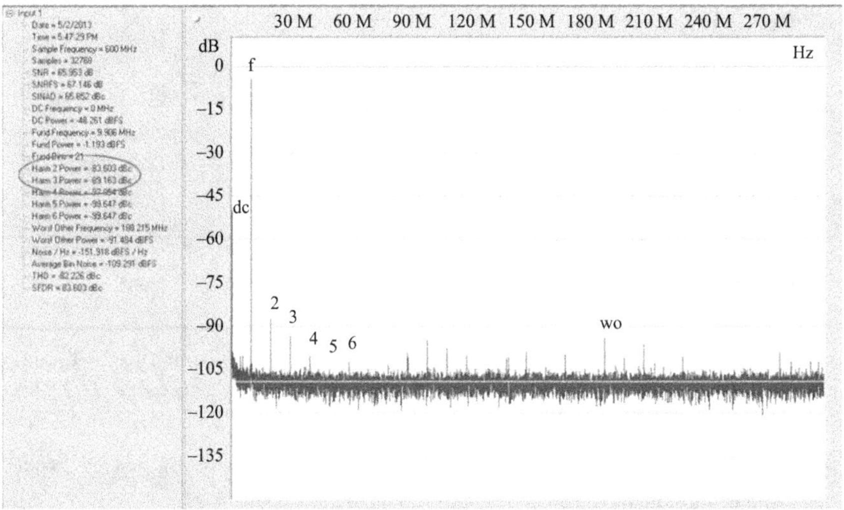

(b)

Figure 2.19 (Continued)

and from (2.81),

$$HD2 = -96.3 + 1 = -95.3 \text{ dBc}$$

From the FFT plot shown in Figure 2.19(a), which corresponds to the INL of Figure 2.18(a), we see that the $HD2$ is equal to −99 dBc, which is in the same ballpark as our estimate.

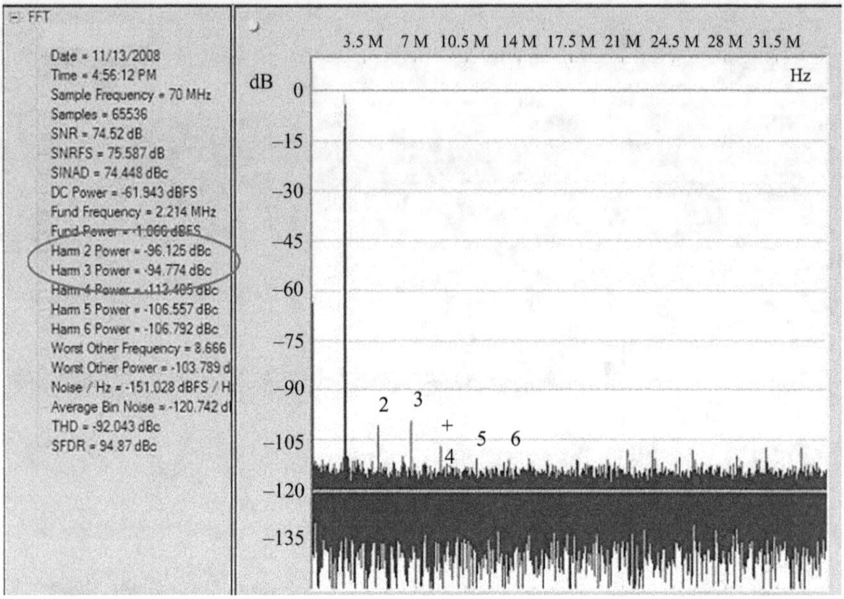

(c)

Figure 2.19 (a) The ADC output FFT corresponding to the INL of Figure 2.18(a). (b) The ADC output FFT corresponding to the INL of Figure 2.18(b). (c) The ADC output FFT corresponding to the INL of Figure 2.18(c)

Example 5: If δ_{2p} is 0.2 LSB and δ_{2n} is 1.7 LSB. Both are at the 14-bit level and at -1 dBFS, as shown in Figure 2.18(b). What is the estimated $HD2$ level at -1 dBFS?

Answer 5: Using (2.82), δ_2 can be estimated by averaging δ_{2p} and δ_{2n}, which is approximately equal to 0.95 LSB at the 14-bit level. Therefore,

$$\delta_2 \approx 0.95 \text{ LSB}$$

Therefore, from (2.27),

$$HD2 = -84.7 \text{ dBFS}$$

and from (2.81),

$$HD2 = -84.7 \text{ dB} + 1 = -83.7 \text{ dBc}$$

From the FFT plot shown in Figure 2.19(b), which corresponds to the INL of Figure 2.18(b), we see that the *HD*2 is equal to −83.6 dBc, which is very close to our estimate.

Since δ_{2p} and δ_{2n} are different, we can estimate δ_3 using (2.84) as follows

$$\delta_3 \approx (\delta_{2p} - \delta_{2n})/2 \approx (1.7 - 0.2)/2 \approx 0.75 \text{ LSB}$$

Therefore, from (2.79),

$$HD3 = -92.8 \text{ dBFS}$$

and from (2.81),

$$HD3 = -92.8 + 1 = -91.8 \text{ dBc}$$

From the FFT plot shown in Figure 2.19(b), which corresponds to the INL of Figure 2.18(b), we see that the *HD*2 is equal to −89 dBc, which is reasonably close to our estimate.

Example 6: If δ_{3p} is 0.7 LSB and δ_{3n} is 0.2 LSB. Both are at the 14-bit level and at −1 dBFS, as shown in Figure 2.18(c). What is the estimated *HD*3 level at −1 dBFS? What is the estimated *HD2* level at −1 dBFS?

Answer 6: Using (2.83), δ_3 can be estimated by averaging δ_{3p} and δ_{3n}, which is approximately equal to 0.45 LSB at the 14-bit level. Therefore,

$$\delta_3 \approx 0.45 \text{ LSB}$$

Therefore, from (2.80),

$$HD3 = -91 - 6 \text{ dB} = -97 \text{ dBFS}$$

and from (2.82),

$$HD3 = -91 - 6 \text{ dB} + 1 = -96 \text{ dBc}$$

From the FFT plot shown in Figure 2.19(c), which corresponds to the INL of Figure 2.18(c), we see that the *HD*3 is equal to −94.8 dBc, which is close to our estimate.

Since δ_{3p} and δ_{3n} are different, we can estimate δ_2 using (2.84) as follows

$$\delta_2 \approx (\delta_{3p} - \delta_{3n})/2 \approx (0.7 - 0.2)/2 \approx 0.25 \text{ LSB}$$

Therefore, from (2.79),

$$HD2 = -96.3 \text{ dBFS}$$

and from (2.81),

$$HD2 = -96.3 + 1 = -95.3 \text{ dBc}$$

From the FFT plot shown in Figure 2.19(c), which corresponds to the INL of Figure 2.18(c), we see that the *HD*2 is equal to −96 dBc, which is close to our estimate.

Some concluding remarks:

- Determining the δ_2 and δ_3 values and the straight line approximation of the INL can be difficult because of the additional artifacts in the INL plot. The methods described above are meant for rough estimation only.
- The accuracy of estimating the *HD*2 and *HD*3 using (2.83) is better than using (2.84). When we rely on subtraction to estimate the δ values from the INL, the difference is small by definition, and hence is susceptible to large errors.

2.7.2 Saw-tooth INL pattern

Another common INL pattern is the saw-tooth shown in Figure 2.20. The saw-tooth pattern often results from quantization errors, as discussed in Chapter 1. It can also result from inter-stage gain errors in pipelined and cyclic ADCs. In the case of regular quantization errors, the amplitude (or peak-to-peak) of the INL saw-tooth is related to the number of saw-tooth segments. This is shown as follows:

$$INL_{pp} = \Delta = V_{FS}/2^N \tag{2.85}$$

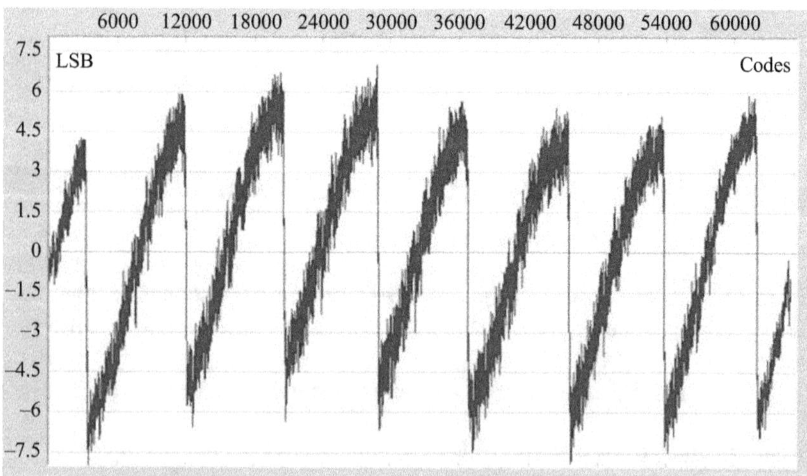

Figure 2.20 An example of an INL with a saw-tooth pattern due to inter-stage gain errors in a 16-bit pipelined ADC

where N is the number of bits, and V_{FS} is the full-scale of the ADC. The number of INL segments is given by

$$S_{INL} = 2^N \tag{2.86}$$

Therefore, the INL is related to the number of segments as follows

$$INL_{pp} = V_{FS}/S_{INL} \tag{2.87}$$

As discussed in Chapter 1, every additional bit doubles the number of segments and reduces the peak by a factor of 2, which reduces the quantization energy by 6 dB. However, since the number of segments is doubled, the number of spurs will be doubled two, which leads to a reduction in the energy of each spur by an additional 3 dB. This means that every additional quantization bit leads to an improvement of 9 dB in SFDR [7, 8]. Therefore,

$$SFDR(\text{dBFS}) \approx 9\,N - c \tag{2.88}$$

where c ranges from 0 for low resolutions to 6 for high resolutions.

However, in practice the SFDR is usually not limited by the ideal quantization noise. For example, the INL in Figure 2.20 depicts a saw-tooth pattern that consists of 8 segments in a 16-bit converter. Clearly, that is not due to quantization, which would have resulted in 2^{16} segments. In this case, the saw-tooth pattern is due to the inter-stage gain error in the first stage of a pipelined ADC. The first stage is composed of 3 bits, and hence the 8 segments of the INL. This is discussed in more detail in Chapter 7. For now, regardless of the cause of the INL pattern, it is useful to estimate the SFDR that corresponds to this INL. It is clear that the peak-to-peak value of the INL is not directly related to the number of INL segments. So, intuitively, the SFDR would approximately be given by [7, 8]

$$SFDR(\text{dBFS}) \approx 6n + 3n1\ \text{dBFS} \tag{2.89}$$

where n represents the accuracy of the INL, and is given by

$$n = INL_{pp}|_{\text{in bits}} = \log_2\left(\frac{2^N}{INL_{pp}|_{in\ LSBs}}\right) \tag{2.90}$$

and $n1$ is the number of bits corresponding to the number of INL segments, and is given by

$$n1 = \log_2 S_{INL} \tag{2.91}$$

Therefore, the intuitive explanation for expression (2.89) is that the INL_{pp} describes the total harmonic power content due to the INL errors, while the number of segment influences the maximum harmonic level because the total power is distributed among that many harmonics.

Example 7: For the INL plot shown in Figure 2.20, what is the estimated SFDR?

Answer 7: The $INLpp = 9$ LSBs at the 16-bit level ($N = 16$). Therefore $n = 12.8$.
The number of INL segments is 8 (7 and 2 halves), so $n1 = 3$. Therefore the SFDR is expected to be around:

$$SFDR \approx 6 \times 12.8 + 3 \times 3 \approx 86 \text{ dBFS}$$

It is clear from the corresponding FFT shown in Figure 2.21 that the SFDR is indeed in the order of 86 dBFS. However, we should note again that these are only meant to be rough estimates. The exact SFDR will depend on how the spur energy is spread across the harmonics and which harmonic is dominating the spectrum. This variation can easily change the SFDR by 3–6 dB.

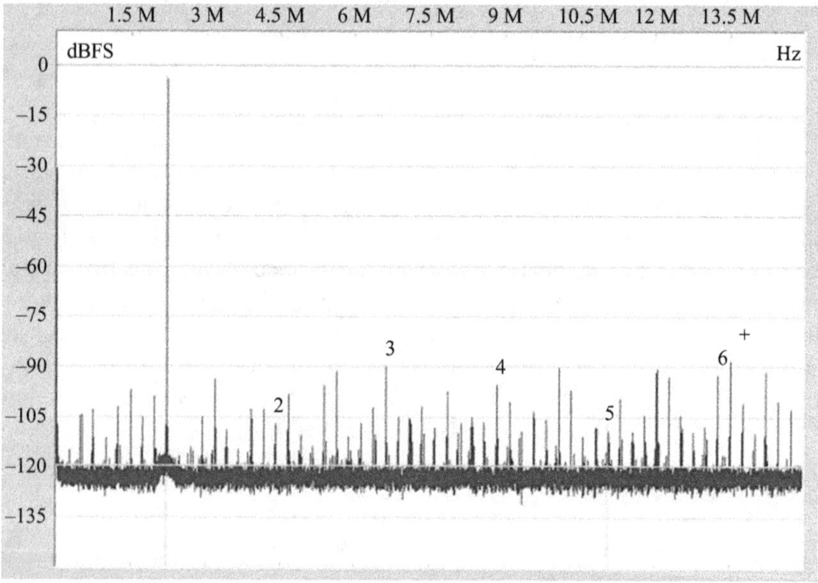

Figure 2.21 The output spectrum corresponding to the INL in Figure 2.20

Example 8: What happens to the SFDR if the input amplitude is reduced from full-scale? Can we estimate the approximate behavior of the SFDR from the INL, as we reduce the input amplitude?

Answer 8: Now it gets more interesting! As the input amplitude is reduced, the number of INL segments will decrease. Since the SFDR degrades by 3 dB for each

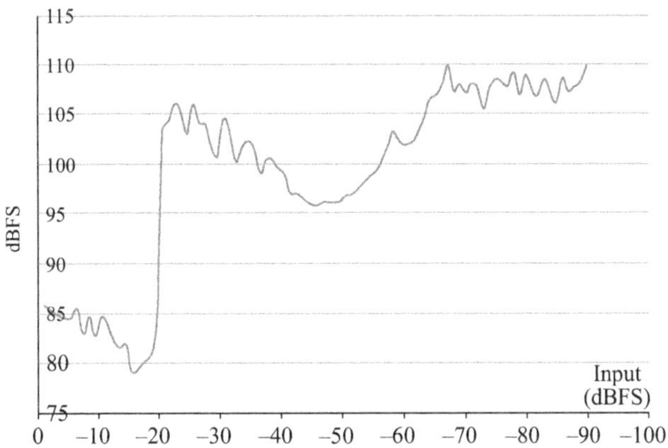

Figure 2.22 A sweep of the SFDR versus input amplitude that corresponds to the INL in Figure 2.20

2× reduction in the number of segments, we expect the SFDR curve (in dBFS) versus the input amplitude (in dBFS) to have a slope of roughly ½ until the number of segments is equal to 1, which corresponds to an input amplitude of 1/8th the full-scale, or about −18 dBFS. Therefore,

- At input amplitude = −6 dBFS, $n1 = 2$, SFDR ~ 84 dBFS.
- At input amplitude just above −18 dBFS, $n1 = 0$, SFDR ~ 77 dBFS.
- For inputs less than −18 dBFS, the INL jumps disappear, and the SFDR is expected to improve substantially.

The sweep of the SFDR versus input amplitude, corresponding to the INL of Figure 2.20, is shown in Figure 2.22, where we can see reasonable agreement between the SFDR estimates given above, and the measured SFDR for inputs from full-scale to −18 dBFS. We can also see that the slope of the SFDR curve is about −0.5 in the region between an input of 0 dBFS and −18 dBFS as expected.

In spite of being approximate estimates, this understanding of the behavior of the SFDR due to different INL patterns and at different amplitudes is very useful for ADC designers and users alike.

2.8 Offset and gain error

The offset of the ADC is defined as the input needed to give a zero output code. The gain error is defined as the deviation in the slope of the transfer characteristic from the ideal stair-case slope. Both are shown in Figure 2.10. These metrics are relatively easy to correct, and hence are usually considered benign, as long as the deviation from the ideal is not extreme.

2.9 Jitter

In Chapter 1, the ideal sampling process is discussed. A basic assumption was that the samples were equidistant in time and spaced T_s apart. In practice, there is variation in the sampling time, which is commonly referred to as *jitter*. This variation results in noise and distortion that get worse as the input frequency increases. If the jitter is random, it leads to noise degradation. If it is periodic, it leads to distortion and spurs. So, from this perspective, we can think of jitter as having two types:

- Periodic jitter: is usually due to coupling from periodic sources, and causes degradation in distortion in the form of spurs and harmonics. It leads to degradation in SFDR and SNDR.
- Random jitter: is due to noise and causes degradation in SNDR.

2.9.1 Analysis

Let's start by diving again into the sampling analysis that we covered in Chapter 1, while taking the jitter into account this time around. For an input signal $x(t)$, the sampled signal is given by

$$x_s(t) = x(t) \sum_{k=-\infty}^{\infty} \delta(t - kT_s) \tag{2.92}$$

Using the sifting property of the delta function [16], we get

$$x_s(t) = \sum_{k=-\infty}^{\infty} x(kT_s)\delta(t - kT_s) \tag{2.93}$$

If the sampling time varies by an amount ΔT, the sampled signal will be given by

$$x_s(t) = x(t) \sum_{k=-\infty}^{\infty} \delta(t - kT_s - \Delta T) \tag{2.94}$$

Using the sifting property of impulse functions again [16], we get

$$x_s(t) = \sum_{k=-\infty}^{\infty} x(kT_s + \Delta T)\delta(t - kT_s - \Delta T) \tag{2.95}$$

For a sinusoidal input signal $x(t)$, with amplitude A and input frequency $\omega_{in} = 2\pi f_{in}$, we have

$$x(t) = A\sin(\omega_{in} t) \tag{2.96}$$

Therefore, from (2.95), the sampled sinusoidal signal is given by

$$x_s(t) = \sum_{k=-\infty}^{\infty} A\sin[\omega_{in}(kT_s + \Delta T)]\delta(t - kT_s - \Delta T) \tag{2.97}$$

and

$$x_s(t) = \sum_{k=-\infty}^{\infty} A[\sin(\omega_{in}kT_s)\cos(\omega_{in}\Delta T) + \cos(\omega_{in}kT_s)\sin(\omega_{in}\Delta T)]$$
$$\times\, \delta(t - kT_s - \Delta T) \qquad (2.98)$$

If the variation in sampling time (jitter) is very small, then

$$x_s(t) \approx \sum_{k=-\infty}^{\infty} A[\sin(\omega_{in}kT_s) + \omega_{in}\Delta T\cos(\omega_{in}kT_s)]\delta(t - kT_s - \Delta T) \qquad (2.99)$$

and

$$x_s(t) \approx x_s(t)|_{ideal} + \sum_{k=-\infty}^{\infty} A\omega_{in}\Delta T\cos(\omega_{in}kT_s)\delta(t - kT_s - \Delta T) \qquad (2.100)$$

The first term on the right hand side of (2.100) represents the ideal sampled signal, while the second term represents the noise and distortion due to the sampling jitter. Therefore, the noise-plus-distortion term due to jitter is given by n_j, such that

$$n_j(t) \approx \sum_{k=-\infty}^{\infty} A\omega_{in}\Delta T\cos(\omega_{in}kT_s)\delta(t - kT_s - \Delta T) \qquad (2.101)$$

which can be represented in the continuous-time form as

$$n_j(t) \approx A\omega_{in}\Delta T \times \cos\omega_{in}t \qquad (2.102)$$

It is interesting to note the modulation effect of the sampling jitter, shown in (2.102). Unlike additive noise, jitter manifests itself in a multiplicative fashion. This is understandable because of the similarity between the sampling process and the multiplication operation. Moreover, as shown in (2.102), the effect of the jitter gets worse as the input frequency and the input amplitude increase. If ΔT is a periodic signal of frequency f_j that is represented by the Fourier series:

$$\Delta T = \sum_{k=0}^{\infty} A_{jk}\cos\left(2\pi k f_j t + \varphi_k\right) \qquad (2.103)$$

where A_{jk} and φ_k are the Fourier amplitude and phase coefficients respectively of the jitter signal. Substituting (2.103) into (2.102), we get

$$n_j(t) \approx A\omega_{in}\cos(\omega_{in}t)\sum_{k=0}^{\infty} A_{jk}\cos\left(2\pi kf_jt + \varphi_k\right) \qquad (2.104)$$

which can be represented as

$$n_j(t) \approx A\omega_{in}\sum_{k=0}^{\infty}\frac{A_{jk}}{2}\left\{\cos\left[2\pi\left(f_{in} + kf_j\right)t + \varphi_k\right] + \cos\left[2\pi\left(f_{in} - kf_j\right)t + \varphi_k\right]\right\} \qquad (2.105)$$

Thus, the jitter's frequency content appears as side bands around the signal's fundamental in a manner similar to AM modulation. If the jitter consists of a single dominant tone f_j, the jitter's "noise" signal will be given by two spurs, one on each side of the fundamental. That is

$$n_j(t) \approx A\omega_{in}\frac{A_{j1}}{2}\{\cos[2\pi(f_{in}+f_j)t+\varphi_1]+\cos[2\pi(f_{in}-f_j)t+\varphi_1]\} \quad (2.106)$$

A special case is when the input signal couples on the sampling clock, resulting in periodic sampling jitter such that $f_j = f_{in}$. Therefore, (2.106) becomes

$$n_j(t) \approx A\omega_{in}\frac{A_{j1}}{2}\{\cos[2\pi(f_{in}+f_{in})t+\varphi_1]+\cos[2\pi(f_{in}-f_{in})t+\varphi_1]\} \quad (2.107)$$

which gives a DC and a second harmonic components as follows:

$$n_j(t) \approx A\omega_{in}\frac{A_{j1}}{2}\{\cos[2\pi(2f_{in})t+\varphi_1]+1\} \quad (2.108)$$

Therefore, when the input signal couples on the sampling clock, it causes a second harmonic that increases substantially with increasing the input frequency at a rate of 6 dB/octave (20 dB/decade).

If the jitter is white noise, the RMS jitter noise power (N_j) on the sampled signal can be obtained using (2.102) and will be given by

$$N_j \approx \frac{(A\omega_{in}\Delta T_{RMS})^2}{2} \quad (2.109)$$

where ΔT_{RMS} is the RMS value of the jitter. If we call this RMS jitter value J, then

$$N_j \approx \frac{(2\pi f_{in}AJ)^2}{2} = 4\pi^2 f_{in}^2 A_{RMS}^2 J^2 \quad (2.110)$$

It is clear that the overall RMS jitter power in the sampled signal increases with increasing the input frequency and increasing the input amplitude. However, if we look at the SNDR due to the jitter, it is given by

$$SNDR = \frac{P_s}{N_j} = \frac{A^2/2}{(2\pi f_{in}AJ)^2/2} = \frac{1}{(2\pi f_{in}J)^2} \quad (2.111)$$

Therefore, the input amplitude is cancelled out, and the SNDR in dB is given by

$$SNDR(\text{dBc}) = -20\log(2\pi f_{in}J) \quad (2.112)$$

That is, the SNDR degrades with increasing the jitter magnitude and the input frequency with a slope of –6 dB/octave or –20 dB/decade.

2.9.2 *Intuitive perspective*

Now that we have analyzed the effect of jitter on the sampled signal, why not analyze it again in a different way! An alternative analysis can be done by starting with the basics of the derivative of a function, as shown in Figure 2.23. For a small

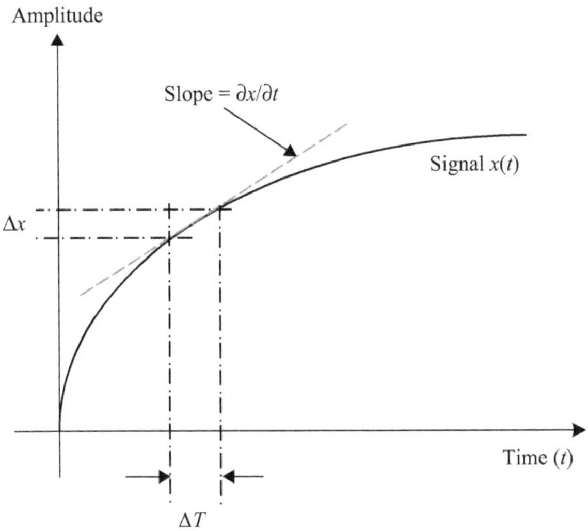

Figure 2.23 An illustration of the effect of the timing jitter on the sampled amplitude

change in time ΔT, the corresponding change in amplitude $\Delta x(t)$ can be expressed in terms of the slope $\partial x/\partial t$ as follows:

$$\Delta x(t) = \Delta T \times \frac{\partial x(t)}{\partial t} \tag{2.113}$$

This shows the multiplicative effect of the jitter. Taking the Fourier transform of both sides, the multiplication in the time domain becomes a convolution (*) in the frequency domain, and we get

$$\Delta X(f) = \Delta T(f) * F\left(\frac{\partial x(t)}{\partial t}\right) \tag{2.114}$$

which gives

$$\Delta X(f) = \Delta T(f) * j2\pi f X(f) \tag{2.115}$$

For a sinusoidal input $x(t) = A\sin\omega_{in}t$, the Fourier transform $X(f)$ is given by

$$X(f) = \frac{A}{2j}[\delta(f - f_{in}) - \delta(f + f_{in})] \tag{2.116}$$

Substituting from (2.116) into (2.115), and applying the convolution, we obtain

$$\Delta X(f) = 2\pi f_{in}\frac{A}{2}[\Delta T(f - f_{in}) - \Delta T(f + f_{in})] \tag{2.117}$$

This is a more general form of (2.105) showing the modulation effect of the clock jitter, and the resulting side bands. An example of such effect is shown in

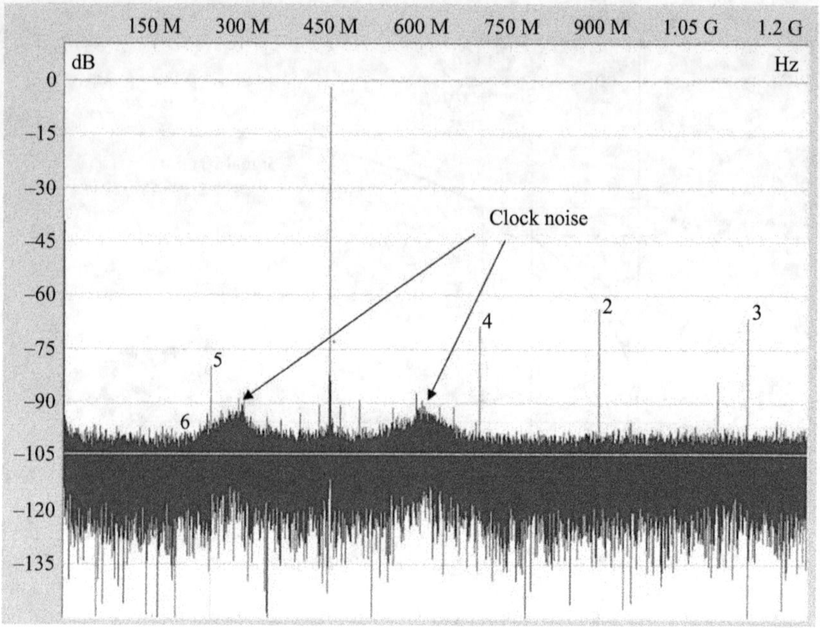

Figure 2.24 An example of an ADC output spectrum showing modulation effect of the noise on the clock source. The noise appear as two "humps" around the fundamental component

Figure 2.24, where the noise on the clock is shown to modulate the input signal manifesting itself as two noise "humps" around the fundamental frequency.

This approach can be used to work out an alternative derivation of the jitter noise and SNDR using the expression

$$\Delta x(t) = \Delta T \times \frac{\partial x(t)}{\partial t} \tag{2.118}$$

Therefore, the expectations of the squares are given by

$$E\left[\Delta x^2(t)\right] = E\left[\Delta T^2 \times \frac{\partial x^2(t)}{\partial t}\right] \tag{2.119}$$

If ΔT and $x(t)$ are independent, we get

$$E\left[\Delta x^2(t)\right] = E\left[\Delta T^2\right] \times E\left[\frac{\partial x^2(t)}{\partial t}\right] \tag{2.120}$$

Assuming zero means, we get

$$\sigma_x^2 = \sigma_T^2 \times E\left[\frac{\partial x^2(t)}{\partial t}\right] \tag{2.121}$$

where $\sigma_x{}^2$ is the variance of the sampled signal and $\sigma_T{}^2$ is the variance of the jitter. Therefore, the noise power due to jitter is given by

$$N_J = J^2 \times E\left[\frac{\partial x^2(t)}{\partial t}\right] \tag{2.122}$$

where N_J is the signal noise power due to jitter and J is the RMS value of the jitter. For a sine wave input

$$x(t) = A \sin 2\pi f_{in} t \tag{2.123}$$

The derivative is given by

$$\frac{\partial x(t)}{\partial t} = 2\pi f_{in} A \times \cos 2\pi f_{in} t \tag{2.124}$$

and

$$E\left[\frac{\partial x^2(t)}{\partial t}\right] = 4\pi^2 f_{in}{}^2 A_{RMS}^2 \tag{2.125}$$

Substituting from (2.125) into (2.122) gives the jitter noise power to be

$$N_J = 4\pi^2 f_{in}^2 A_{RMS}^2 J^2 \tag{2.126}$$

which matches (2.110). The SNDR due to jitter is given by

$$SNDR = \frac{1}{4\pi^2 f_{in}^2 J^2} \tag{2.127}$$

The SNDR in dB is given by

$$SNDR(\text{dBc}) = -20 \log(2\pi f_{in} J) \tag{2.128}$$

which matches (2.111) and (2.112).

These alternative derivations are meant to illustrate a seemingly different way to look at the impact of jitter that may hopefully help give insight, and present a different set of tools that may be useful in analyzing different situations that we may encounter.

The state of the art of the jitter performance in high speed ADCs is in the order of 40–50 fs [2, 3]. Using the expression given in (2.128), this jitter magnitude limits the SNDR to about 72 dB for a 1 GHz input frequency, and to about 60 dB for a 4 GHz input signal. This indicates that jitter can be a major contributor to noise, which must be addressed and optimized for very high speed applications.

2.9.3 Jitter measurement

One method to measure the jitter involves measuring the *SNR* at a very low input frequency (SNR_{low}) and at a high input frequency (SNR_{hi}). It is important to make sure that the degradation in the *SNR* at high frequency is due to jitter and not due to other factors. The *SNR* is given by

$$SNR_{hi} = \frac{A_{RMS}^2}{4\pi^2 f_{in}^2 J^2 A_{RMS}^2 + N_{Qth}}, \qquad SNR_{low} \cong \frac{A_{RMS}^2}{N_{Qth}} \tag{2.129}$$

where N_{Qth} represents the total noise due to quantization, thermal, and all other sources of noise, which are manifested at low input frequencies. Rearranging (2.129) gives

$$J = \frac{\sqrt{A_{RMS}^2/SNR_{hi} - A_{RMS}^2/SNR_{low}}}{2\pi f_{in} A_{RMS}} \tag{2.130}$$

Therefore, the jitter is given by

$$J = \frac{\sqrt{1/SNR_{hi} - 1/SNR_{low}}}{2\pi f_{in}} = \frac{\sqrt{\frac{N_{total} - N_{other}}{S}}}{2\pi f_{in}} \tag{2.131}$$

where N_{total} is the total noise power, N_{other} is the noise power excluding jitter, and S is the signal power.

The jitter of the ADC clock path comes from the sampling clock path inside the ADC and from the external clock applied to the ADC. So it is imperative to use a clean low-jitter clock when driving high performance ADCs for high input frequencies, such as IF and RF sampling applications. Typically, a clean sine wave clock is used, preferably filtered, to reduce its wideband noise. This can be generated using a crystal oscillator or using a signal generator. Alternatively, low-jitter clock generation circuits can be used, where a PLL is employed to lower the phase noise of the clock signal.

2.9.4 Types of random jitter

There are two dominant manifestations of random jitter [9]:

- Synchronous jitter, which occurs in driven, threshold-crossing, circuits such as digital gates and drivers, and causes phase modulation (PM) noise.
- Accumulating jitter, which occurs in autonomous circuits such as oscillators, and causes frequency modulation (FM) noise.

The synchronous jitter is usually manifested as wideband noise that can be calculated as follows:

$$J_{RMS}(t_{th}) = \frac{\sqrt{E[n^2(t_{th})]}}{dv(t_{th})/dt} \tag{2.132}$$

where $J_{RMS}(t_{th})$ is the RMS jitter value at the threshold voltage, $E[n^2(t_{th})]$ is the jitter noise power at the threshold voltage, and $dv(t_{th})/dt$ is the slope of the signal at the threshold point.

The accumulating jitter is usually manifested as close-in noise near the fundamental and is given by

$$J_{RMS} = \sqrt{aT} \tag{2.133}$$

where T is the clock period, and

$$a = L(\Delta f)\frac{\Delta f^2}{f_0^2} \tag{2.134}$$

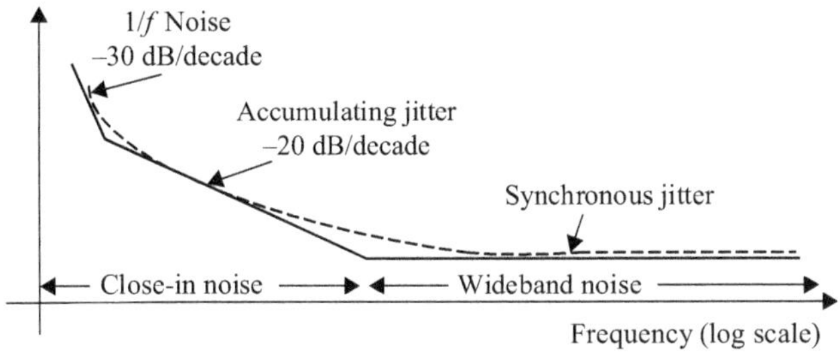

Figure 2.25 An illustration of the different regions of the clock phase noise and the resulting jitter

where $L(\Delta f)$ is the phase noise power spectrum, Δf is the offset frequency from the fundamental, and f_0 is the fundamental clock frequency.

In addition to the above two dominant sources of jitter, there is the effect of the flicker ($1/f$) noise which appears at very low frequencies. The three regions of the phase noise and jitter are illustrated in Figure 2.25.

2.9.5 Jitter and phase noise

The jitter and noise of clock sources are usually described in terms of their phase noise. For a sinusoidal clock signal [10–12], we have

$$v_{clock} = A_s \sin(2\pi f_s(t + \Delta T(t))) \tag{2.135}$$

where A_s is the amplitude of the clock signal, f_s is its frequency, and ΔT is its jitter. Therefore,

$$v_{clock} = A_s \sin(2\pi f_s t + \phi(t)) \tag{2.136}$$

such that the phase noise $\phi(t)$ is given by

$$\phi(t) = 2\pi f_s \Delta T(t) \tag{2.137}$$

In the frequency domain, the phase noise is

$$\phi(f) = 2\pi f_s \Delta T(f) \tag{2.138}$$

Therefore, the RMS jitter value can be calculated from the phase noise as follows:

$$J = \frac{1}{2\pi f_s}\sqrt{\int_{-\infty}^{\infty} \phi^2(f)df} \tag{2.139}$$

where the integration is over the entire frequency spectrum. The band of integration is not limited to the sampling frequency, the Nyquist frequency, or any multiple of them. It should be performed over the entire infinite spectrum, which is practically limited only by the bandwidth of the clock circuit.

From (2.136), if $\phi(t)$ is small, we get

$$v_{clock} = A_s \sin(2\pi f_s t) + A_s \phi(t) \cos(2\pi f_s t) \tag{2.140}$$

To represent the jitter in terms of the clock's amplitude noise, we obtain

$$J = \frac{\sqrt{2}}{2\pi f_s A_s} \sqrt{\int_{-\infty}^{\infty} \Delta v_{clock}^2(f) df} \tag{2.141}$$

If the single side band phase noise power spectrum, as measured on a spectrum analyzer, is given by $L(f)$, then

$$L(f) = 10 \log\left(\frac{\phi^2(f)}{2}\right) \tag{2.142}$$

and

$$\phi(f) = \sqrt{2 \times 10^{L(f)/10}} \tag{2.143}$$

Therefore,

$$J = \frac{1}{2\pi f_s} \sqrt{\int_{-\infty}^{\infty} \phi^2(f) df} = \frac{1}{2\pi f_s} \sqrt{2 \int_{-\infty}^{\infty} 10^{L(f)/10} df} \tag{2.144}$$

From (2.117), the representation of the sampled signal with jitter is given by

$$\Delta X(f) = 2\pi f_{in} \frac{A}{2} [\Delta T(f - f_{in}) - \Delta T(f + f_{in})] \tag{2.145}$$

Using (2.138) into (2.145), we obtain

$$\Delta X(f) = \frac{f_{in}}{f_s} \frac{A}{2} [\phi(f - f_{in}) - \phi(f + f_{in})] \tag{2.146}$$

The resulting SNR is given by

$$SNR = 10 \log\left(\frac{A^2/2}{\int \Delta X^2(f) df}\right) \tag{2.147}$$

Substituting (2.146) into (2.147), and using one side band only, we get

$$SNR = 20 \log\left(\frac{f_s}{f_{in}} \frac{1}{\sqrt{\int \phi^2(f - f_{in}) df}}\right) \tag{2.148}$$

In terms of the clock noise in dBc, the SNR is given by

$$SNR = -N_{clock}(\text{dBc}) + 20\log\left(\frac{f_s}{f_{in}}\right) \tag{2.149}$$

This indicates that for the same clock phase noise, the *SNR* due to the clock jitter degrades with increasing the input frequency and improves with increasing the clock frequency.

2.10 Bit error rate (BER)

The BER, or Sample Error Rate (SER), is an expression of the probability of error of the ADC. The error in this case is usually defined to be large enough to exclude errors due to the noise of the ADC. These errors can be due to the metastability of the comparators, as discussed in chapter 5, output data capture errors, or serializer errors. It is common for BER's to be in the order of 10^{-6} to 10^{-9} in communications applications. However, in some instrumentation applications, BER's as low as 10^{-15} may be required.

2.11 Power consumption and figure of merit

One of the most important metrics for ADCs is the power consumption. Obviously, lower power consumption is better. However, the comparison of the power efficiency between ADCs with different performance levels can be quite complicated. ADCs have multiple dimensions of performance, as shown above, which include resolution, sampling rate, SNDR, SFDR, jitter, etc. Improving any of those metrics may require additional power consumption. Moreover, the input frequency range where the performance holds can vary, and hence complicate the comparison further.

For these reasons, there were attempts to devise Figures-of-Merit (FOM) that capture the performance and power efficiency of the converter in one parameter. Unfortunately, it is practically daunting to reduce all dimensions of performance into one parameter. Nevertheless, some FOMs can still be useful in capturing some of the dimensions.

One commonly used FOM is the energy per conversion FOM defined as [13]:

$$FOM1 = \frac{\text{Power}}{2^{ENOB}ERBW}\ \text{in Joule/Conversion-step} \tag{2.150}$$

where the *ENOB* is the effective number of bits, as defined in (2.16), and the ERBW is the effective resolution bandwidth, which is defined as the bandwidth over which the *ENOB* value holds. Variations of this *FOM* include

$$FOM1 = \frac{\text{Power}}{2^{N}f_s/2} \tag{2.151}$$

and

$$FOM1 = \frac{\text{Power}}{2^{ENOB} f_s/2} \tag{2.152}$$

In spite of its popularity for low power and modest resolutions ADCs, this *FOM* does not capture the trends and fundamental trade-offs for high performance and noise-limited ADCs, where most of the power is used to achieve the required SNDR and *ENOB*. Noise-limited ADCs follow a trend where improving the SNDR by 3 dB typically requires doubling the power. An intuitive way to look at this is by assuming we have an ADC whose *ENOB* is *N* bits, and power is *P* Watts with an *ERBW* of *B*. If we use two of these ADCs in parallel and average their outputs, we would double the power consumption and improve the SNDR by 3 dB [15]. The SNDR improves by 3 dB because the signal power adds coherently, while the noise is uncorrelated.

If V_1 and V_2 represent the signals of the two ADCs, and N_1 and N_2 represent their noise powers. Then, the average of the two ADC signals is

$$V_{ave} = \frac{V_1 + V_2}{2} = V_1 = V_2 \tag{2.153}$$

The corresponding signal power is

$$P_{s\_ave} = P_{s1} = P_{s2} \tag{2.154}$$

On the other hand, the averaging process gives a noise power of

$$N_{ave} = \frac{N_1 + N_2}{4} = \frac{N_1}{2} = \frac{N_2}{2} \tag{2.155}$$

Therefore,

$$SNDR_{ave} = \frac{P_{s\_ave}}{N_{ave}} = 2 \times SNDR_1 = 2 \times SNDR_2 \tag{2.156}$$

and

$$SNDR_{ave}(\text{dB}) = SNDR_1 + 3\text{ dB} = SNDR_2 + 3\text{ dB} \tag{2.157}$$

Therefore, this process of averaging the outputs of two identical ADCs has doubled the power and improved the SNDR by 3 dB. Since the ADC core did not change, we would expect the new composite ADC to have the same *FOM* as that of each core. However, using *FOM*1, we see that

$$FOM1_{1\text{core}} = \frac{P}{2^N B} \tag{2.158}$$

and

$$FOM1_{2\text{cores}} = \frac{2P}{2^{N+0.5} B} = \frac{2P}{\sqrt{2}2^N B} = \sqrt{2} \times FOM1_{1\text{core}} \tag{2.159}$$

Therefore, the *FOM* of the composite ADC has degraded by 40%, even though we expected it to have the same *FOM* as the individual cores!

Another way to look at this is that every doubling of the sampling rate results in an increase of 3 dB in the SNDR within the same bandwidth due to the processing gain, as discussed in Chapter 1. This indicates that a factor of 2 in speed is equivalent to 3 dB in SNDR. Moreover, a factor of 2 increase in speed is equivalent to doubling the power because, in principle, we can ideally interleave 2 ADC cores to increase the speed by a factor of 2, which would increase the power by a factor of 2. Therefore, doubling the power should correspond to a 3 dB increase in SNDR, which is not what *FOM*1 predicts.

These contradictions and trends created the need for a different *FOM* that is more technically sound and more representative of noise-limited ADCs. One such *FOM* was proposed in Reference 14 and given by

$$FOM2 = \frac{\text{Power}}{2^{2\times ENOB} \times ERBW} \text{ in Joule/(Conversion-step)}^2 \tag{2.160}$$

In this case, the *FOM* is proportional to the quantization power, instead of its magnitude. Applying this *FOM* to the above example gives

$$FOM2_{\text{2cores}} = \frac{2P}{2^{2N+2\times 0.5}B} = \frac{2P}{2 \times 2^{2N}B} = \frac{P}{2^{2N}B} = FOM2_{\text{1core}} \tag{2.161}$$

Therefore, the FOM of the composite ADC is identical to that of the individual cores as expected.

This *FOM* can be represented in dB to give a third figure of merit *FOM*3, which is commonly called the Schreier *FOM*, and is given by [15]

$$FOM3(\text{dB}) = SNDR + 10\log\left(\frac{ERBW}{\text{Power}}\right) \tag{2.162}$$

Alternatively, it can be defined in terms of the sampling rate as

$$FOM3(\text{dB}) = SNDR + 10\log\left(\frac{f_s/2}{\text{Power}}\right) \tag{2.163}$$

It is interesting to note that if we combine (2.162) and (2.163) with (2.14), we can represent the Schreier *FOM* as

$$FOM3(\text{dB}) = -NSD - 10\log(\text{Power}) \tag{2.164}$$

For *FOM*1 and *FOM*2, a more efficient ADC gives a smaller *FOM* value. On the other hand, for *FOM*3, a more efficient ADC gives a larger *FOM*. Another difference is that the log scale of *FOM*3 compresses the range. So, an ADC that consumes double the power of another ADC for the same performance, would be 3 dB worse in *FOM*, which represents 100% less efficiency. The *FOM*3 is also more intuitive and easy to calculate because of its relation to the SNDR and the *NSD*.

It is important to keep in mind that all the above *FOMs* have significant limitations from a practical standpoint due to the multi-dimensional nature of the ADC performance. The power of the ADC may contain significant contributions from the digital processing, the input buffer, the bias, and the output drivers. These contributions do not necessarily follow the noise-limited trends described earlier. In addition, enabling high linearity or IF/RF sampling can result in significant power overhead, which is not necessarily captured by any of the above mentioned *FOMs*. In general, these *FOMs* tend to be more useful in comparing ADCs with similar performance. Moreover, pushing the limits of performance and speed together on a certain technology process tends to degrade the *FOM*, which is not neatly captured by a single parameter.

2.12 Conclusion

In this chapter, we covered some of the most important and common performance metrics used to evaluate and characterize data converters. The analysis, measurement, and meaning of each metric are presented. In addition, intuitive interpretations and back-of-the-envelope estimation methods are discussed to estimate and relate the different metrics. We also covered several figures of merit that are used to describe the efficiency of the converter and compare different converters. Some of these figures of merit are also useful to understand trends and trade-offs in converter design and optimization.

Problems

1. Draw the transfer characteristics of an ideal 3-bit DAC, similar to Figure 2.1 for the ADC.
2. Draw the transfer characteristics of a 3-bit DAC with offset, gain, DNL, and INL errors, similar to Figure 2.10 for the ADC. For a DAC, the errors are measured on the y-axis, which is the analog output axis. Write expressions for the DNL and INL analogous to those for the ADC.
3. A 12-bit ADC with 500 μV input referred thermal noise and 2V full-scale:
   (a) What is the expected SINAD that captures the quantization and thermal noise contributions?
   (b) What is the ENOB?
   (c) What is the effective resolution?
   (d) If the sampling rate is 100 MS/s, what is the NSD?
4. For the ADC used in Figure 2.4(a), estimate *IMD*2 and *IMD*3 at −7 dBFS? Comment on the results and the validity of the estimates.
5. For the ADC used in Figure 2.5, estimate *IMD*2 and *IMD*3 at −15 dBFS? Comment on the results and the validity of the estimates.
6. For the ADC used in Figure 2.6, if the SFDR is limited by *HD*3, estimate *IMD*3 at −6 dBFS and −10 dBFS? Comment on the results and the validity of the estimates.

7. An amplifier that is used to drive an ADC has *HD*2 and *HD*3 equal to −80 dBFS at −10 dBFS.
   (a) What are the expected values of *HD*2 and *HD*3 at 0, −1, −6, −20 dBFS?
   (b) What are the expected values of *IMD*2 and *IMD*3 at 0, −1, −6, −7, −20 dBFS?
   (c) Comment on the results and the validity of the estimates
8. Repeat Problem 7 if we use a differential input signal that is perfectly balanced.
9. Repeat Problem 7 if we use a differential input signal that has the following:
   (a) An amplitude mismatch of 0.1%.
   (b) A phase mismatch of 2°.
10. An ADC has *HD*2 and *HD*3 equal to −80 dBFS at −7 dBFS, which is limited by the integrated input buffer.
    (a) What are the expected values of *HD*2 and *HD*3 at 0, −1, −10, −20 dBFS?
    (b) What are the expected values of *IMD*2 and *IMD*3 at 0, −1, −10, −20 dBFS?
    (c) Comment on the results and the validity of the estimates
11. Derive (2.38), (2.39), and (2.41). Confirm the results with simulations.
12. Derive (2.47). Confirm the results with behavioral simulations.
13. For a non-linear system that is defined by: $y(t) = 0.9x(t) - 0.01x^2(t) - 0.05x^3(t)$, simulate the output using:
    (a) Single-tone input with unity amplitude and 100 MHz frequency.
    (b) Two-tone input with unity amplitude each and frequencies of 100 MHz and 102 MHz.
    (c) Determine *HD*2, *HD*3, *IMD*2, and *IMD*3.
14. Repeat Problem 13 with:
    (a) Changing the unity amplitudes to 2 V.
    (b) Changing the unity amplitudes to 0.5 V.
    (c) Determine *HD*2, *HD*3, *IMD*2 and *IMD*3 for each case.
15. Using any programming language, write the code for calculating the DNL and INL of an ADC using a sinusoidal input signal.
16. What are the expected *HD*2 and *HD*3 for the INL shown in Figure 2.13(b) at −1 dBFS and −6 dBFS?
17. For the ADCs represented in Figures 2.18 and 2.19, estimate *HD*2 and *HD*3 at −6 dBFS and −10 dBFS? Comment on the validity of your estimates.
18. For the ADCs represented in Figures 2.13(a) and (b), estimate *IMD*2 and *IMD*3 at −7 dBFS? Comment on the validity of your estimates.
19. For the ADCs represented in Figures 2.18 and 2.19, estimate *IMD*2 and *IMD*3 at −7 dBFS? Comment on the validity of your estimates.
20. For the INL of Figure 2.20, what is the estimated SFDR at:
    (a) −3 dBFS?
    (b) −10 dBFS?
21. For a 12-bit, 200 MS/s ADC with 500 fs of jitter:
    (a) What is the expected SNDR for an input frequency of 100 MHz?
    (b) What is the SNDR for a 10 MHz input signal?
    (c) How does the SNDR change if the ADC is 16 bits?

(d) How does the SNDR change if the ADC is 8 bits?
(e) What is the expected SNDR if the sampling rate changes to 50 MS/s, while the jitter value is the same? Explain.

22. A 16-bit 100 MS/s ADC that consumes 500 mW and has a SINAD of 80 dB.
    (a) What is the Walden FOM (*FOM*1)?
    (b) What is the Schreier FOM (*FOM*3)?
    (c) How do the FOMs compare if the ADC consumes 1 W and has a SINAD of 83 dB?
    (d) How do the FOMs compare if the ADC consumes 1 W and has a SINAD of 86 dB?
    (e) Comment on the results of (c) and (d) above.
23. A 14-bit 250 MS/s ADC with 70 dB SINAD consumes 150 mW. Using the same process, architecture, and input span, what do you expect the power consumption of the following ADCs to be:
    (a) An ADC with 70 dB SINAD at 500 MS/s?
    (b) An ADC with 73 dB SINAD at 125 MS/s?
    (c) An ADC with 73 dB SINAD at 500 MS/s?
    (d) An ADC with 76 dB SINAD at 125 MS/s?
24. Look up commercial datasheets on the web for a 12-bit, 14-bit, and 16-bit ADC at sample rates of 125 MS/s and 250 MS/s. Observe the different specifications and comment on the differences. Calculate the *FOM*1, *FOM*2, and *FOM*3 for each ADC. Examples can be found on the Analog Devices' web site at: http://www.analog.com.

References

[1] AD9446, 16 bit 80/100MS/s ADC, Analog Devices, http://www.analog.com.

[2] A.M.A. Ali, A. Morgan, C. Dillon, *et al.*, "A 16-bit 250-MS/s IF Sampling Pipelined ADC with Background Calibration," *IEEE Journal of Solid-State Circuits*, 45(12), pp. 2602–2612, Dec 2010.

[3] A.M.A. Ali, H. Dinc, P. Bhoraskar, *et al.*, "A 14b 1GS/s RF Sampling Pipelined ADC with Background Calibration," *IEEE Journal of Solid-State Circuits*, 49(12), pp. 2857–2867, Dec 2014.

[4] A.M.A. Ali, H. Dinc, P. Bhoraskar, *et al.*, "A 14-bit 2.5GS/s and 5GS/s RF Sampling ADC with Background Calibration and Dither," *IEEE VLSI Circuits Symposium*, pp. 206–207, 2016.

[5] W. Kester (Ed.), "The Data Conversion Handbook," *Analog Devices, Inc.*, Elsevier, Burlington, MA, 2005.

[6] B. Razavi, *RF Microelectronics,* Prentice Hall, Upper Saddle River, NJ, 1998.

[7] H. Pan, "A 3.3-V 12-b 50-MS/s A/D Converter in 0.6-μm CMOS with over 80-dB SFDR," Ph.D. dissertation, UCLA, Dec 1999.

[8] H. Pan and A.A. Abidi, "Spectral Spurs due to Quantization in Nyquist ADCs," *IEEE Transactions on Circuits and Systems-I: Regular Papers*, 51(8), pp. 1422–1438, Aug 2004.

[9] K. Kundert, *Predicting the Phase Noise and Jitter of PLL-Based Frequency Synthesizers*. The Designer's Guide Community.

[10] C. Azeredo-Leme, "Clock Jitter Effects on Sampling: A Tutorial," *IEEE Circuits and Systems Magazine*, 11(3), pp. 26–37, 2011.

[11] B. Brannon and A. Barlow, *Aperture Uncertainty and ADC System Performance,* Analog Devices, Inc., *Application Note AN-501*.

[12] B. Brannon, *Sampled Systems and the Effects of Clock Phase Noise and Jitter,* Analog Devices, Inc., *Applicat. Note AN-756*.

[13] R. Walden, "Analog-to-Digital Conversion in the Early Twenty-First Century," *Wiley Encyclopedia of Computer Science and Engineering*, pp. 126–138, 2008.

[14] A.M.A. Ali, C. Dillon, R. Sneed, *et al.*, "A 14-bit 125 MS/s IF/RF Sampling Pipelined ADC With 100 dB SFDR and 50 fs Jitter," *IEEE Journal of Solid-State Circuits*, 41(8), pp. 1846–1855, Aug 2006.

[15] R. Schreier and G. Temes, *Understanding Delta-Sigma Data Converters*, IEEE Press, Piscataway, NJ, 2005.

[16] A.V. Oppenheim and R.W. Schafer, *Discrete-Time Signal Processing*, Prentice Hall, Englewood Cliffs, NJ, 1989.

Chapter 3
Data converter architectures

In this chapter, we discuss some high speed ADC architectures, which include the flash, pipelined, and time-interleaved ADCs. In addition, architectures that were historically used in low speed applications, such as successive approximation (SAR) and delta-sigma converters, are covered because of their recent resurgence in the high speed space. Some DAC architectures, such as the resistive, capacitive, and current steering DACs, are also discussed.

Traditionally, ADC architectures have occupied the performance spaces shown in Figure 3.1. The *y*-axis represents a measure of performance, which can be resolution, SNDR, or ENOB. The *x*-axis represents a measure of speed, which can be sampling rate or effective Nyquist bandwidth. The highest speed space is occupied

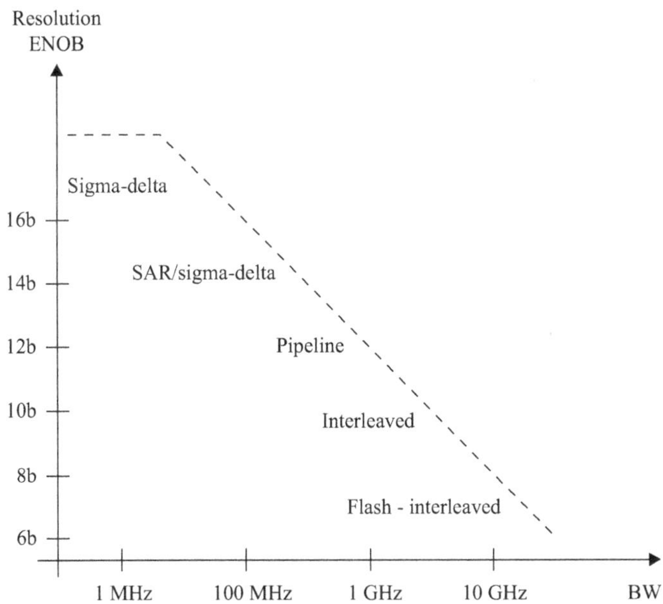

Figure 3.1 A plot of resolution (or performance) versus quantization bandwidth (or speed) showing the preferred relative spaces for various ADC architectures

by flash ADCs (with their various incarnations) and time-interleaved ADCs. The performance of both architectures is usually limited to 10 bits or less. In the middle, pipelined ADCs shine in the high speed and high resolution space with speeds reaching higher than 1GS/s and performance in the 10- to 14-bit range [1, 2]. Pipelined ADCs with resolution and linearity in the 16-bit range have been reported [1], although their ENOB is usually limited to the 12- to 14-bit range. On the lower speed side, SAR and sigma-delta ADCs occupy the relatively low speed and high resolution space, with speeds typically less than 100 MS/s and resolutions higher than 16 bits. However, recently sigma-delta ADCs have reportedly reached an effective sampling rate of up to 900 MS/s [3].

The main take-away from Figure 3.1 is the relative strength and weakness of each architecture. The boundaries between the different regions are flexible and overlapping. Moreover, with time, the boundary of the state of the art moves upward and to the right, toward higher performance and higher speeds.

3.1 Flash ADC

The flash is the fastest and most basic non-interleaved ADC architecture. Flash ADCs are used as stand-alone converters and as building blocks in other types of converters, such as in pipelined, sigma-delta, and successive-approximation ADCs. In a flash ADC, the input is sampled and compared to a set of equally spaced threshold values using comparators. Typically, an N-bit flash ADC requires a set of $(2^N - 1)$ comparators, sometimes called the "comparator bank." The threshold levels are usually set using a resistance ladder connected to the reference voltages as shown in Figure 3.2. The comparator bank generates a thermometer code word, which can be converted using a decoder into any desired format, such as two's complement or offset binary.

For ideal A/D conversion, the comparators' threshold levels need to be equally spaced. Mismatches in the resistance values result in non-uniform quantization step sizes, and hence DNL and INL errors. They can also cause offset and gain errors of the whole converter. Different resistor types have different linearity and matching characteristics. For example, salicided resistors are typically controlled better than unsalicided ones; polysilicon and metal resistors have better matching and linearity than diffusion resistors. In layout, identical units with reasonable aspect ratios are preferred for good matching. In addition, the portion of the resistor unit used by the contacts needs to be small compared to the length of the resistor. Dummy resistors can be used to improve matching near the ends of the ladder. Some of the layout matching measures are shown in Figure 3.3.

The effects of the resistance mismatches accumulate as the taps approach the middle of the ladder, with the two ends being held at ideal values. If the mismatches are random, the accumulation will be uncorrelated. However, if the mismatches of the resistance elements are correlated, such as those due to gradients, their impact will progressively increase toward the middle. For a gradient error, the value of the kth resistance in the ladder R_k can be expressed as [4, 5]

$$R_k = R + k\Delta R \tag{3.1}$$

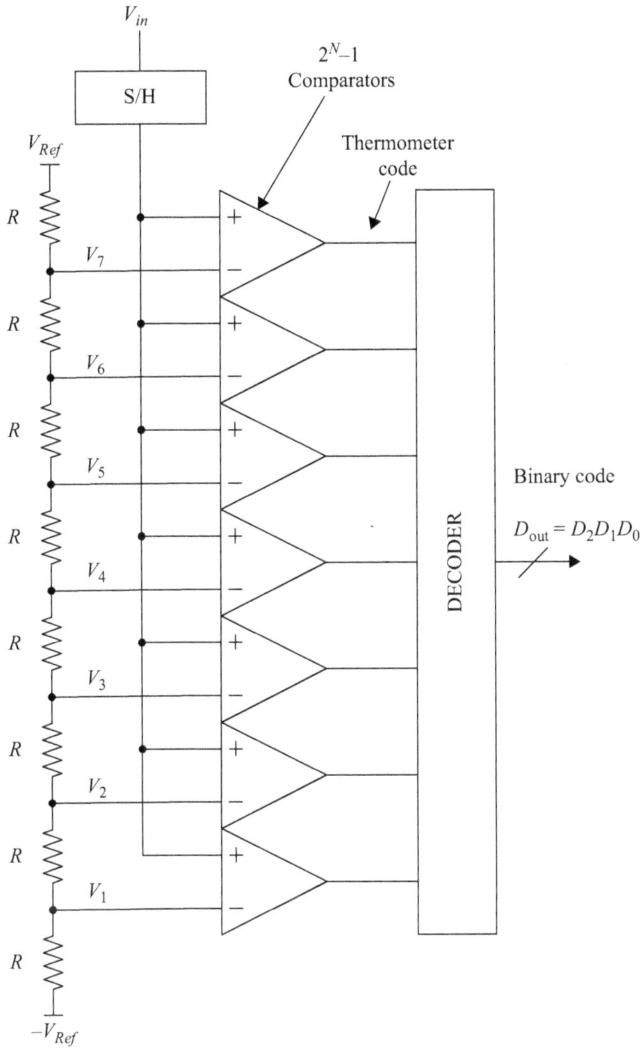

Figure 3.2 A simplified block diagram of a 3-bit flash ADC with seven comparators (N = 3)

where R is the unit resistance and ΔR is the mismatch between two neighboring resistances in the ladder. The kth output voltage is given by

$$V_k = 2V_{Ref}\left(\frac{\sum_{i=1}^{i=k}(R+i\Delta R)}{\sum_{i=1}^{i=2^N}(R+i\Delta R)} - \frac{1}{2}\right) = \frac{2V_{Ref}}{R_t}\left[\left(R+\frac{\Delta R}{2}\right)k + \frac{\Delta R}{2}k^2\right] - V_{Ref} \quad (3.2)$$

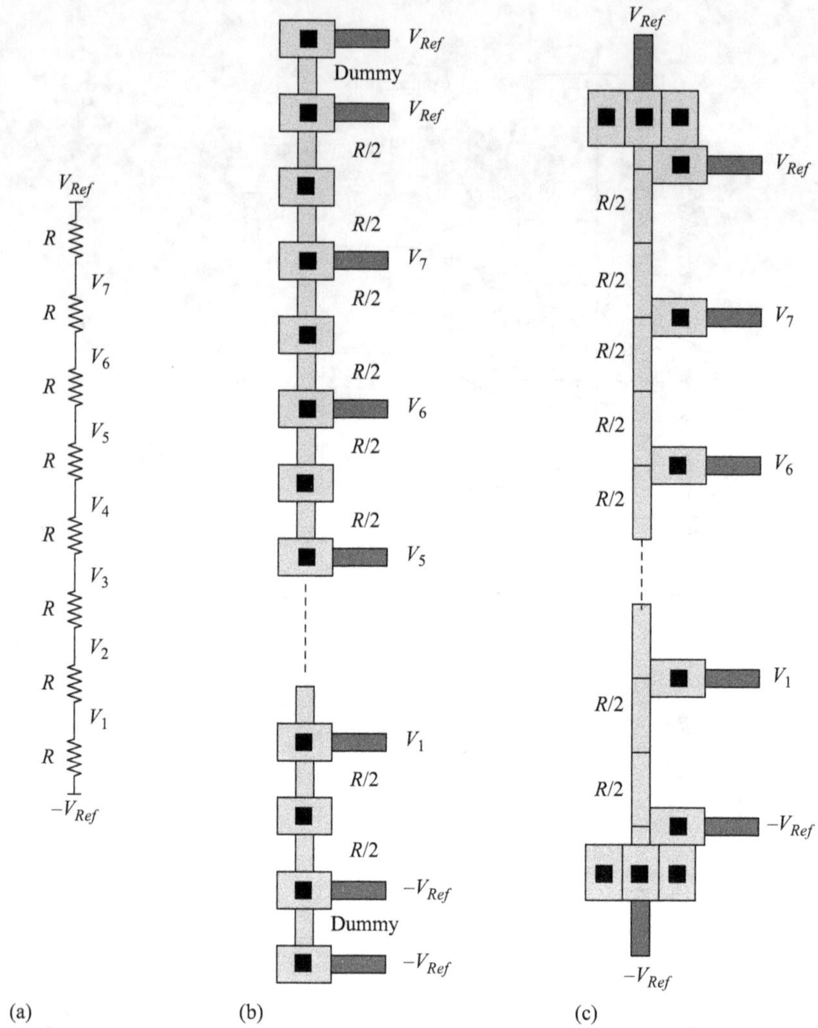

Figure 3.3 (a) A resistor ladder. (b) A layout diagram of the ladder made up of identical R/2 segments. (c) A layout of the ladder made up of a continuous resistance ladder

where V_k is the kth tap voltage, N is the number of bits, R is the unit resistance, and R_t is the total resistance of the ladder, which is given by

$$R_t = \sum_{i=1}^{i=2^N} (R + i\Delta R) = \left(R + \frac{\Delta R}{2}\right)2^N + \frac{\Delta R}{2}2^{2N} \tag{3.3}$$

Equation (3.2) indicates a parabolic relation between the tap voltages and the tap order k. The resulting INL error can be shown to be parabolic too with a

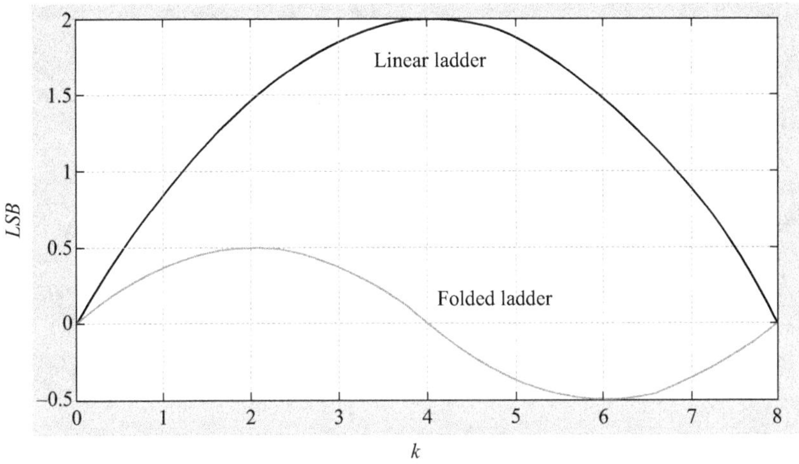

Figure 3.4 A plot of the INL versus the tap order for a linear ladder and a folded ladder

maximum at $k = 2^N/2$, as shown in Figure 3.4. As discussed in Chapter 2, the INL is given by (2.76) to be

$$INL(k) = \frac{V_k - V_{idk}}{LSB} = \frac{V_k - \left(\frac{2V_{Ref}}{2^N}k - V_{Ref}\right)}{LSB} \tag{3.4}$$

where V_{idk} is the ideal tap voltage and LSB is the size of the ADC's LSB, which is given by $2V_{Ref}/2^N$. Substituting (3.2) into (3.4), and finding the maximum by equating the derivative to zero, gives

$$\frac{\partial INL(k)}{\partial k} = \frac{\frac{2V_{Ref}}{R_t}\left[\left(R + \frac{\Delta R}{2}\right) + \frac{\Delta R}{2}2k\right] - \left(\frac{2V_{Ref}}{2^N}\right)}{LSB} = 0$$

Therefore, the maximum occurs at

$$k = \frac{2^N}{2} \tag{3.5}$$

and the magnitude of the maximum INL error is given by

$$INL = \frac{2V_{Ref} \times \Delta R \times 2^{2N}}{8R_t \times LSB} \tag{3.6}$$

Therefore, the maximum INL in Volts is

$$INL(Volts) = \frac{I \times \Delta R \times 2^{2N}}{8} \tag{3.7}$$

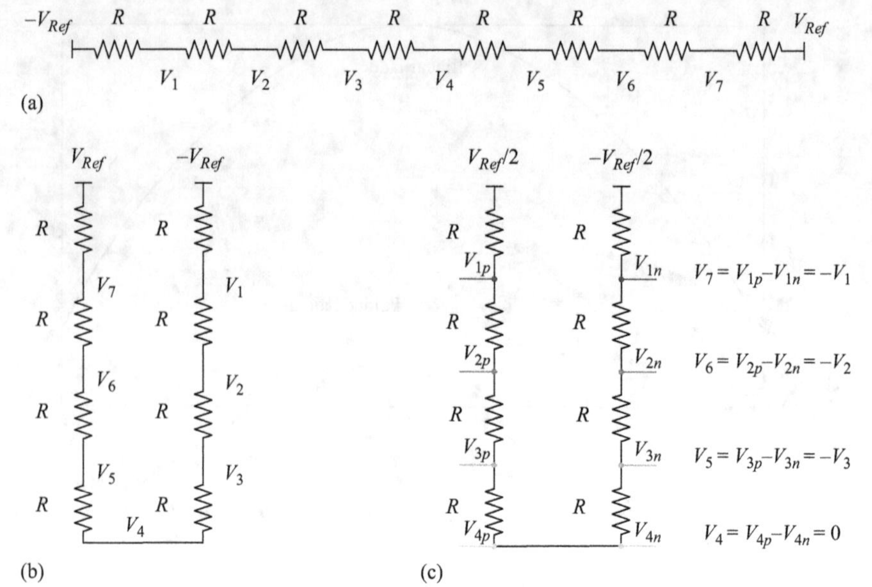

Figure 3.5 (a) A linear ladder. (b) A folded ladder. (c) A folded ladder with differential taps

where I is the current in the ladder, which is equal to $2V_{Ref}/R_t$. Since R_t is approximately equal to $R \times 2^N$, the maximum INL in LSBs is given by

$$INL(LSB) = \frac{2^{2N} \times \Delta R}{8\left(R + \dfrac{\Delta R(1 + 2^N)}{2}\right)} \cong \frac{2^{2N} \times \Delta R/R}{8} \tag{3.8}$$

Therefore, from (3.7) and (3.8), we see that the INL error is proportional to the square of the number of levels 2^N.

To reduce the INL due to the resistance gradient, a folded ladder can be used as shown in Figure 3.5. Folding minimizes the gradient error at the midpoint because the two halves are matched. The resulting INL is plotted in Figure 3.4. Since the INL is proportional to the square of the number of taps, folding the ladder leads to a $2\times$ reduction in the number of taps, and hence a $4\times$ reduction in the peak INL. Moreover, using a differential signal eliminates the first-order gradients in the ladder and hence improves the INL further. This is shown in Figure 3.5(c). In general, differential signals are preferred to reduce the even-order harmonics in all kinds of ADCs as discussed in Chapter 2. For the flash ADC, this means using differential taps on the ladder and using comparators with differential inputs, which is discussed in Chapter 5.

Typically, a comparator is composed of a cascade of one or more preamplifiers and a latch. This is shown in Figure 3.6 and is done to allow for the optimization of the different aspects of performance, as discussed in detail in Chapter 5.

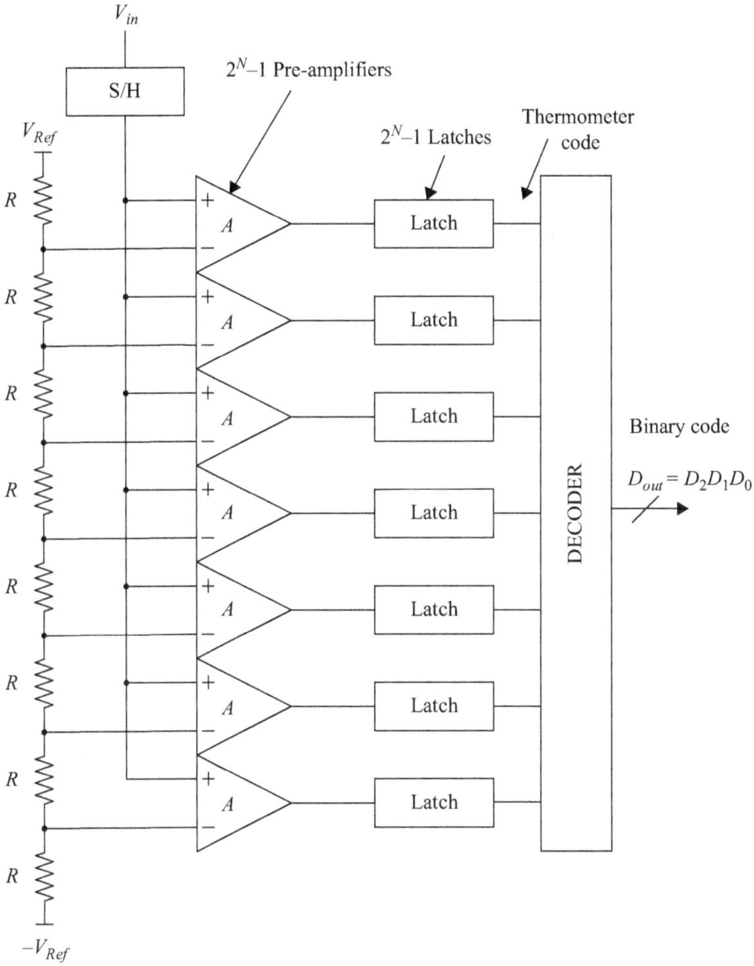

Figure 3.6 A block diagram of a flash ADC, where each comparator is made up of a preamplifier and a latch

Regenerative latches have higher speeds, better metastability and lower power than open loop comparators. The preamplifier is used to provide gain, reduce the input-referred offset of the latch, reduce the kick-back from the latch on the input, and optimize the input sampling performance of the comparator.

In addition to the gradient, random mismatches between the resistors contribute to the DNL and INL of the ADC. This mismatch is inversely proportional to the resistance area and is given by [6]

$$\frac{\sigma_R^2}{R^2} = \frac{A_R^2}{WL} \tag{3.9}$$

where σ_R^2 is the variance of the resistance value, R is the nominal resistance value, A_R is a mismatch constant, W is the resistance width, and L is its length. Matching can be improved by increasing the size of the resistance. This, however, increases the parasitics, which increase the power consumption and degrade the high-frequency performance.

Besides the resistance ladder, the DNL and INL are also affected by the comparator offsets. These are dominated by transistor mismatches inside the comparator that are expressed as

$$V_{OS}^2 = \sigma_{VT}^2 + \left(\frac{V_{gs} - V_T}{2}\right)^2 \left(\frac{\sigma_\beta^2}{\beta^2}\right)^2 \tag{3.10}$$

and

$$\sigma_{VT}^2 = \frac{A_{VT}^2}{WL} \quad \text{and} \quad \frac{\sigma_\beta^2}{\beta^2} = \frac{A_\beta^2}{WL} \tag{3.11}$$

where V_{OS} is the offset voltage, A_{VT} and A_β are mismatch constants, $\beta = \mu C_{ox}W/L$, V_T is the device threshold voltage, W is the width of the MOS device, and L is its channel length.

To reduce the offset, large devices can be used to improve their matching, as discussed in Chapter 5. This increases the power consumption and the parasitics. In addition, preamplifiers can be used to increase the gain, and hence reduce the input referred offsets. Offset cancellation techniques can also be employed to reduce the comparator offset. For example, the offset can be sampled during the sampling phase and then applied in series with the input during the comparison phase. This is shown in Figure 3.7 for the comparator's preamplifier, with the timing diagram in Figure 3.8. During $\phi 1$, the reference and the offset are sampled on the capacitor. During $\phi 2$, the input is applied in series with the capacitor. The offset is cancelled, and the net input will be $V_{in} - V_{Ref}$. Alternatively, the comparator can be taken offline for calibration and offset cancellation, then inserted back in the flash.

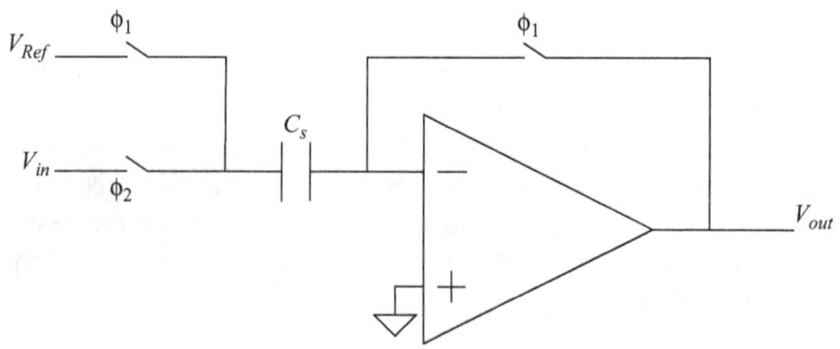

Figure 3.7 A switched capacitor preamplifier with offset cancellation

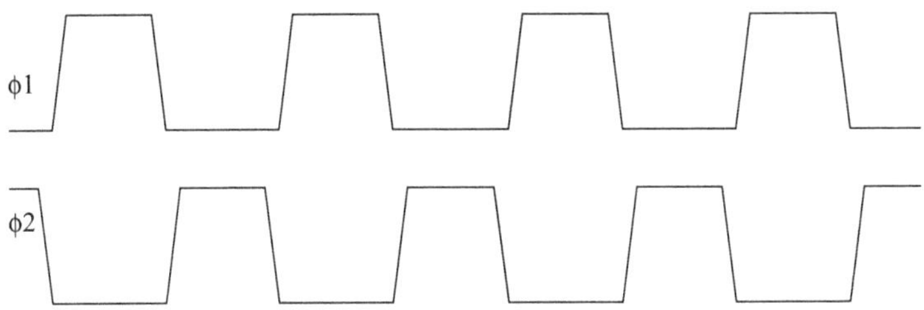

Figure 3.8 Timing diagram for the circuit in Figure 3.7

For this to work seamlessly, an additional comparator is needed to substitute for the comparator being calibrated. Offset cancellation techniques are discussed in more detail in Chapters 5 and 6.

Another design decision in the flash ADC is whether to use a dedicated sample-and-hold amplifier (SHA). This provides a "held" signal to the flash, and hence desensitizes the ADC against timing and bandwidth mismatches between the comparators. Alternatively, a SHA-less architecture can be used where the S/H circuit is integrated within the comparator sampling network. In the latter case, the sampling time and bandwidth of the various comparators need to match. In addition, the sampling bandwidth of the comparator will need to be large enough to accommodate the highest input frequency applied.

An important consideration in designing the resistance ladder is speed. Achieving high speed requires small resistance values in the ladder to reduce the settling time constant of the threshold voltages. This increases the power consumption due to the static current in the ladder, and may degrade the mismatch due to the smaller resistance values. In addition, as the number of bits increases, the number of comparators grows exponentially. This increases the capacitive load on the ladder and the input, which degrades the speed and bandwidth. It also increases the area and power significantly. These are familiar trade-off patterns, where higher speed typically leads to higher power consumption and lower accuracy. Conversely, higher accuracy leads to lower speed and higher power consumption. In the flash converter, the exponential growth in area and power with the number of bits limits the practical resolutions achievable to about 6–8 bits.

The overall "speed" of the flash converter is determined by several factors:

1. *The sampling bandwidth*: This determines the highest input frequency and sampling rate that the comparator's sampling network can handle. In the absence of a SHA, the bandwidth needs to be large enough to track the highest input frequency. In the presence of a SHA, the bandwidth should be large enough to accommodate the sampling of the held signal with the desired accuracy within the sampling phase. Assuming a first-order sampling network, the settling error ε is given by

$$\varepsilon = e^{-T_{acq}/\tau} \tag{3.12}$$

where T_{acq} is the acquisition/sampling period, which is typically equal to $1/2f_s$, where f_s is the sampling rate, and τ is the sampling time constant. Therefore,

$$\tau = \frac{T_{acq}}{\ln\left(\frac{1}{\varepsilon}\right)} \tag{3.13}$$

and the sampling bandwidth BW is given by

$$BW = \frac{\ln\left(\frac{1}{\varepsilon}\right)}{2\pi T_{acq}} \tag{3.14}$$

2. *The reference settling*: This refers to the settling of the tap voltages of the resistance ladder when sampled by the comparators. It follows a similar pattern to the input sampling described in (3.12)–(3.14). The reference settling and sampling bandwidths should be large enough to sample the reference values with the desired accuracy, within the available amount of time. Reference settling errors can limit the accuracy and/or speed of the flash ADC.
3. *The propagation delay*: This refers to the time from the sampling edge until the flash makes its decision, for relatively large inputs where metastability is not limiting the speed. This is usually dictated by the propagation delay of the latch and the gates following it.
4. *The metastability time constant*: This refers to the delay of the latch when the input is very close to the threshold. This small input causes an additional delay that can be substantial. It is represented by the bit-error-rate (BER) or the probability of error. For a comparator, the probability of error P_e is usually given by

$$P_e = \frac{V_{in\_amb}}{V_{Range}}$$

where V_{in_amb} is the input range that is ambiguous to the latch and V_{Range} is the total input range of the comparator. This can be represented in terms of the output ambiguous range V_{o_amb} and the total gain of the comparator A_{total}, as follows

$$P_e = \frac{V_{o\_amb}}{V_{Range}A_{total}}$$

The total gain is composed of the gain (or more accurately attenuation) of the sampling network A_{sample}, the gain of the preamplifier A_{pre}, and the gain of the latch A_{latch}. The regenerative nature of the latch, due to the positive feedback employed, makes its gain increase exponentially with time, and is equal to $A_{latch}e^{T/\tau_m}$, as discussed in detail in Chapter 5. Therefore,

$$P_e = \frac{V_{o\_amb}}{V_{Range}A_{sample}A_{pre}A_{latch}e^{T/\tau_m}}$$

$$BER = P_e = \frac{V_{o_{amb}}}{V_{Range}A_{sample}A_{pre}A_{latch}}e^{\frac{-T}{\tau_m}} \tag{3.15}$$

where $V_{o_{amb}}$ is the output voltage range that is ambiguous to the logic gates, V_{Range} is the voltage range of each comparator, A_{sample} is the gain/attenuation of the sampling network, A_{pre} is the gain/attenuation of the preamplifier, and A_{latch} is the gain/attenuation of the latch. T is the time available for the latch, which is dictated by the sampling rate f_s and is usually equal to $1/2f_s$, and τ_m is the metastability time constant, which is typically given by

$$\tau_m = \frac{C}{g_m} \tag{3.16}$$

where g_m is the transconductance of the latch transistors and C is the total parasitic capacitance on the latch node.

Equation (3.15) shows that the parameters outside the exponent, such as the voltage gains, have little effect on metastability and BER. The most effective parameters are the latch time constant and the available time. To reduce the metastability time constant, a larger g_m or smaller capacitance C is needed. Increasing g_m will increase the power consumption, but g_m/C is a parameter that is limited by the process technology. Fine lithography CMOS and bipolar processes can enable faster comparators due to the higher f_T of the transistors. As a data point, on the 65 nm process, flash converters can support sampling rates between 5 and 10 GS/s. A sampling rate of 3 GS/s can be supported with BER better than 10^{-10}.

In spite of their unparalleled speed, flash converters suffer from some serious limitations. The number of comparators increases exponentially with the number of bits, which leads to an exponential increase in power and area. In addition, the accuracy is usually limited due to their parallel nature and high speed. Although static errors can be trimmed or calibrated effectively, dynamic errors that are due to charge injection, settling errors, and sampling non-idealities, are more challenging to fix. Overall, these drawbacks limit the practical accuracy of flash converters to about 6 bits.

3.2 Flash ADC with interpolation

The comparators in the flash ADCs are composed of one or more preamplifiers followed by a latch. This is needed to reduce the input-referred offset, reduce the loading effect of the latch on the input driver, and improve the overall performance, as discussed in detail in Chapter 5. *Interpolation* can be employed to reduce the number of preamplifiers, and hence reduce the power and area of the flash ADC. An example is shown in Figure 3.9, where resistance dividers between the outputs of neighboring preamplifiers are used to generate the interpolated signals. This reduces the number of preamplifiers by the interpolation factor, which is equal to 2 in Figure 3.9. Larger interpolation factors can be implemented by using more intermediate taps between the preamplifiers. This can lead to a substantial power saving, area reduction, and even speed improvement. It also reduces the load on the input driver, and hence improves the sampling linearity. In addition, it improves the offsets and linearity of the flash ADC due to the offset averaging effect of the interpolation

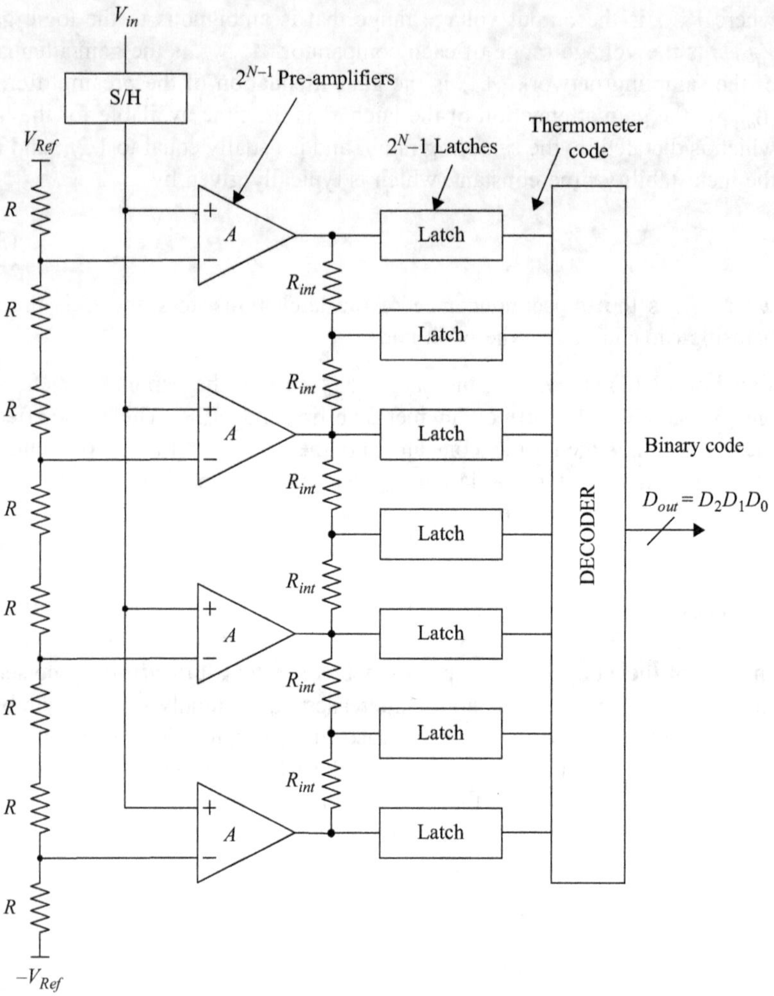

Figure 3.9 A block diagram of a 3-bit flash ADC with resistive interpolation. The interpolation factor is 2, and the number of preamplifiers is reduced by $2^N/2-1$

network [8]. Comparators with a cascade of more than one preamplifier can have multiple layers of interpolation. The number of preamplifiers in every layer of interpolation is reduced to approximately $2^N/N_F$, where N_F is the interpolation factor.

For a comparator whose threshold is far (i.e. different) from the input, the interpolation resistance ladder will source or sink current, which affects its linearity and gain. However, this is not detrimental to that comparator because of its large input difference. On the other hand, for a comparator whose threshold is near the input, the output will be near the zero crossing. Therefore, for this comparator, the ladder is practically bootstrapped, and hence does not affect the linearity or gain of

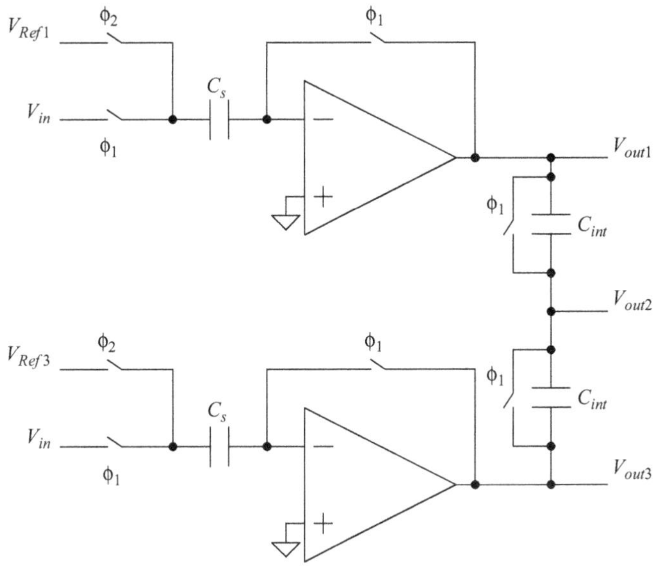

Figure 3.10 A simplified schematic of a capacitive interpolator between two preamplifiers. The three outputs are used as inputs to three latches

the comparator. In addition, the ladder has the additional effect of averaging the errors of neighboring comparators and hence improves their offsets. Another way to understand this effect, is that the ladder creates a linear superposition of the signal outputs, while the random offsets add in the root mean square. Therefore the net offset-to-signal ratio improves because of the linear interpolation.

Instead of using a resistance ladder, interpolation can be performed using a capacitance divider as shown in Figure 3.10 [4–6]. In this case, periodic charging and resetting are needed. During $\phi 1$, the input is sampled on the sampling capacitors (C_s), while the preamplifiers and interpolation capacitors are reset. During $\phi 2$, the references are applied in series, and hence the differences ($V_{in} - V_{Ref}$) are amplified at the outputs of the preamplifiers (V_{out1} and V_{out3}), with their interpolated values appearing at the midpoints (V_{out2}) of the interpolation capacitances (C_{int}). These outputs are fed to the subsequent latches to generate the comparator outputs.

3.3 Multi-step ADC

The number of comparators required by the flash architecture with its various incarnations limits its practicality to a relatively small number of bits (6–8 bits). Interpolation helps reduce the number of preamplifiers, but the number of latches can still be prohibitive. Alternatively, multi-step converters emerged as an important architecture for moderate and high-resolution ADCs. In multi-step ADCs, the conversion process is broken down into multiple steps in order to convert the exponential requirement of the number of comparators into a more linear one. For example, a 10-bit converter would require at least ($2^{10} - 1 = 1023$) comparators using the flash topology. However, if we employ a 2-step conversion process each performing 5-bit conversion, we need $2 \times (2^5 - 1) = 62$ comparators only. This is a substantial reduction that leads to a more efficient implementation, and possibly a significant improvement in accuracy. Examples of multi-step ADCs are:

- Sub-ranging ADCs
- Folding ADCs
- Pipelined ADCs
- Cyclic ADCs
- Successive Approximation (SAR) ADCs

Multi-step architectures rely on generating a *residue* signal that represents the output of one conversion step and the input to the next. The nature of the residue signal, its relative timing, and how it is generated can be substantially different in the various multi-step architectures, and hence their distinctive characteristics, strengths, and weaknesses.

In spite of their differences, multi-step converters have similarities in terms of dependencies and error patterns. Once the details, trends, patterns, and sensitivities of one of the architectures are well understood, the understanding can be extrapolated to the other architectures if the differences are taken into account. We will cover the multi-step architectures in the following sections, while dedicating Chapter 7 to the pipelined ADC. We chose the pipeline architecture because it is one of the most general and illustrative of the multi-step ADCs. It is also a very effective and attractive architecture in the high speed and high resolution ADC space.

3.4 Sub-ranging ADC

In the sub-ranging architecture, the conversion operation is divided into two steps: a coarse and a fine steps. The coarse step generates the most significant bits (MSBs) of the converter and selects the reference range that is used by the fine ADC in the next step. Alternatively, instead of using the coarse bits to select the reference range of the second stage, the first stage sometimes subtracts the digitized signal from the input signal to create a residue that is applied to the second stage. This is represented conceptually in Figure 3.11, and a block diagram of a sub-ranging ADC is shown in Figure 3.12.

If no amplification exists between the first and second stages, the fine ADC will need to have an accuracy equal to the required accuracy of the whole ADC. In addition, its noise contribution will be similar to that of the first stage. Alternatively, an amplifier can be employed between the first and second stage, which relaxes the noise and non-linearity contribution of the second ADC at the expense of the additional power in the inter-stage amplifier.

If inter-stage amplification is employed, it is common to use a gain that would ideally give a residue smaller than the full-scale range of the following stage. This is called *redundancy*. Typically, only half the full-scale is used, in order to accommodate possible errors in the first stage's ADC without over-ranging the

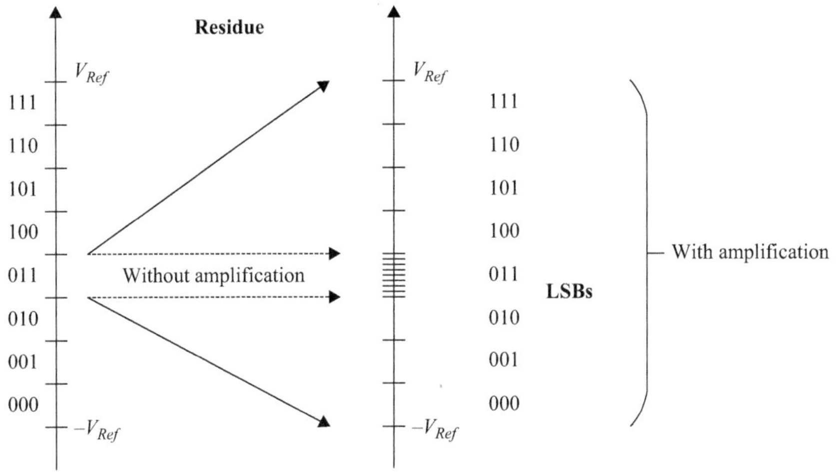

Figure 3.11 A representation of the sub-ranging concept with and without amplification. Each quantization step (sub-range) of the first stage is divided into multiple steps in the back end. In this case, the front end handles the 3 MSBs (8 steps) and the back end handles the 3 LSBs (8 steps). The total resolution is 6 bits

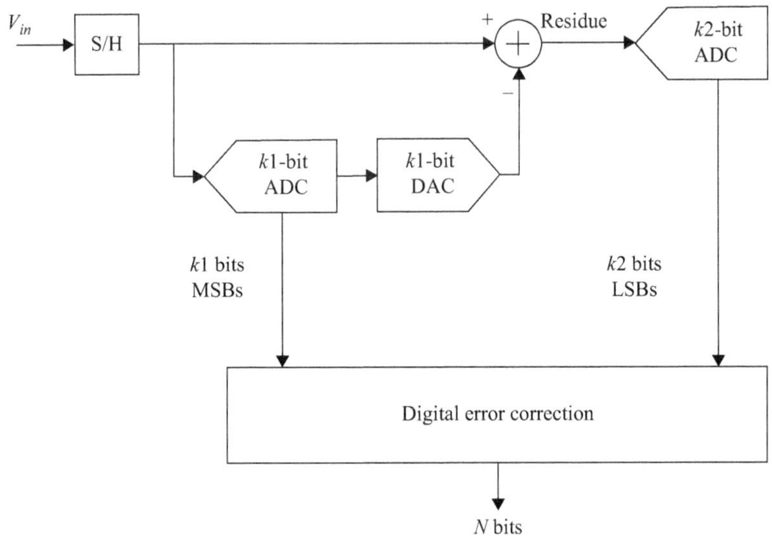

Figure 3.12 A block diagram of a sub-ranging ADC with two stages. Please note the similarity with the pipelined ADC architecture shown in Figure 3.20

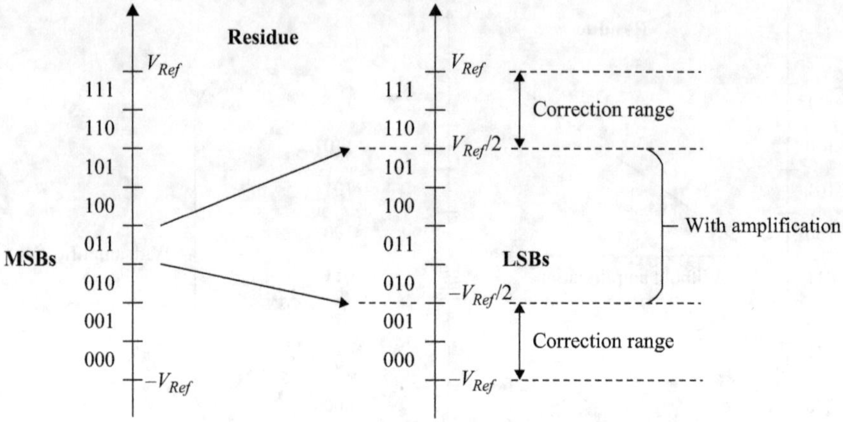

Figure 3.13 A representation of the sub-ranging concept with amplification and redundancy. Each quantization step (sub-range) of the first stage is divided into multiple steps in the back end. In this case, the front end handles the 3 MSBs (8 steps) and the back end handles the 3 LSBs (8 steps). There is a 1-bit overlap for error correction using redundancy, so the total resolution is 5 bits

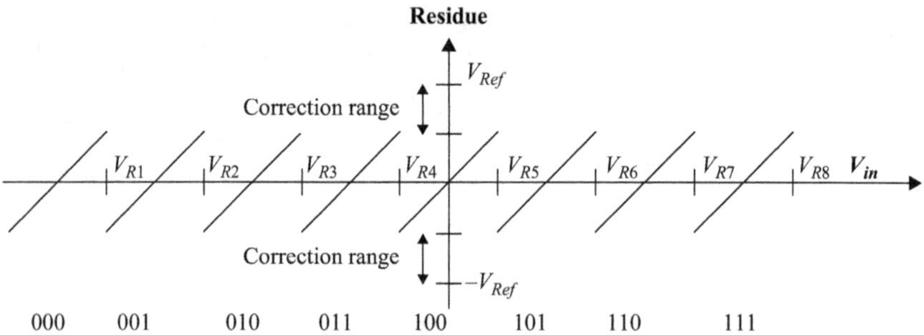

Figure 3.14 A plot of a stage's output residue as a function of its input. In this case, 1-bit redundancy is used, so the residue occupies only half the dynamic range of the stage

second stage. For example, if the first stage has $k1$ bits, the inter-stage gain will be 2^{k1-1}. A flash error in the first stage will be corrected by the following stage, as long as the residue stays within the *correction range*. This *digital error correction* by employing redundancy is an important concept in all multi-step converters. This is shown conceptually in Figure 3.13, with the resulting residue in Figure 3.14. The 1-bit overlap reduces the total resolution of the 2-step ADC by 1 bit.

Sometimes, both conversion steps are performed in one clock cycle, although multiple cycles can also be used to improve the throughput and hence the speed of the ADC. If the two stages operate in different clock phases, with inter-stage amplification, the resulting architecture will be similar to a two-stage pipelined ADC.

3.5 Folding ADC

In a folding architecture, the A/D conversion is broken down into two steps. The first step performs a coarse conversion and generates a "folded residue signal" that is passed along to a fine conversion step. This is shown in Figure 3.15. The residue signal is generated by folding the input signal into segments that depend on the number of bits in the coarse conversion step. For example, for a 1-bit coarse ADC, a 2-segment folded signal is generated; for a 2-bit coarse ADC, a 4-segment folding; for 3 bits, an 8-segment folding; and so on. The folded segments alternate in their slope between positive and negative slopes, and their peak-to-peak amplitude is limited to $1/2^{k1}$ of the input signal full-scale, where $k1$ is the number of bits in the coarse conversion step. An ideal folded residue signal for a 3-bit coarse ADC is shown in Figure 3.16.

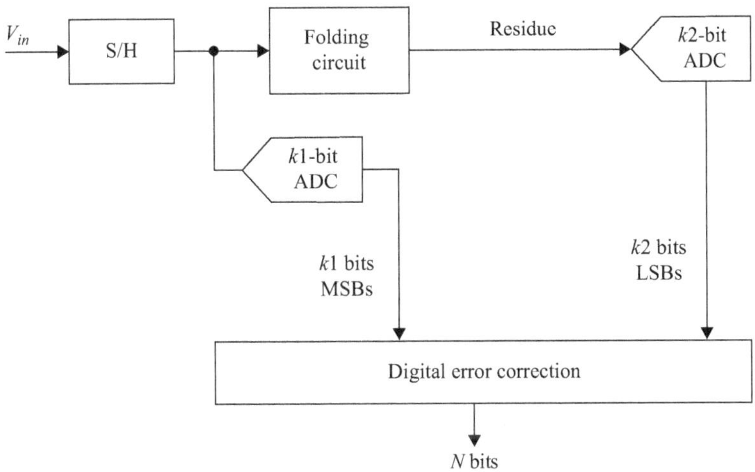

Figure 3.15 A block diagram of a folding ADC.

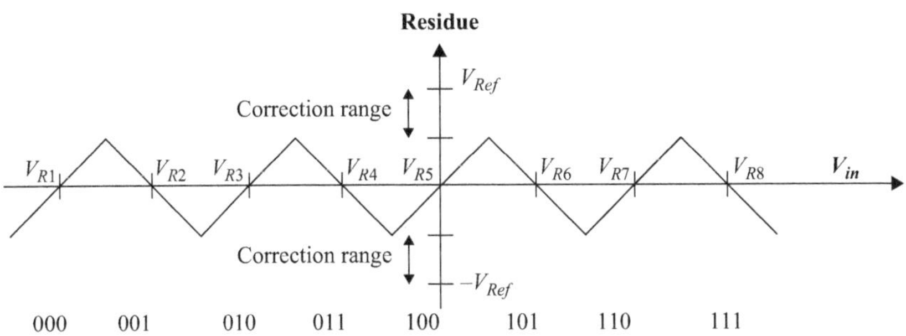

Figure 3.16 A plot of a stage's ideal folded output residue as a function of its input. In this case, 1-bit redundancy is used, so the residue occupies only have the dynamic range of the stage

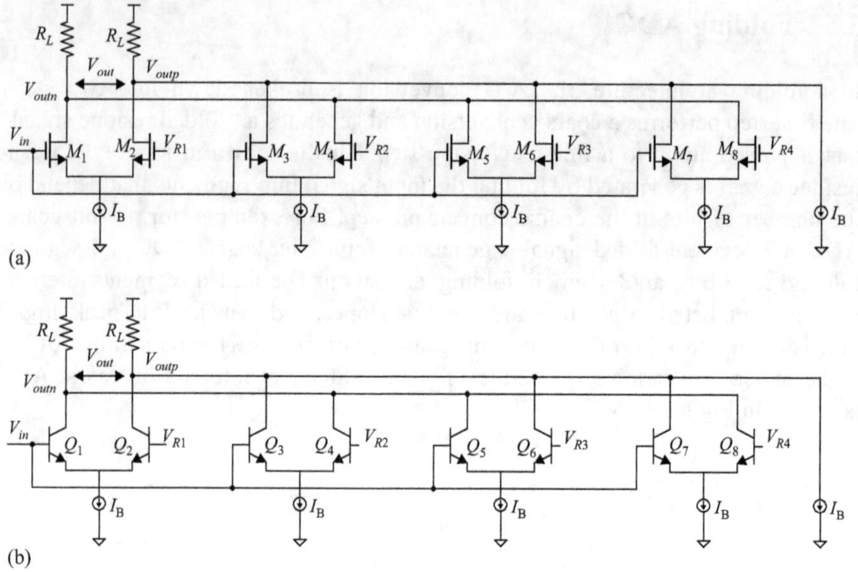

Figure 3.17 Examples of folding circuits using (a) NMOS folders. (b) BJT folders

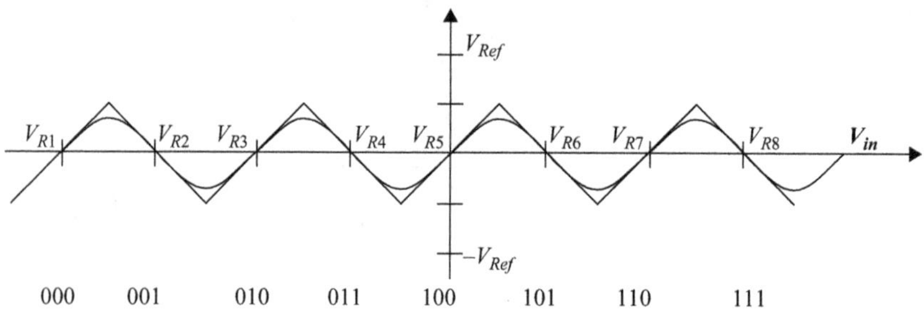

Figure 3.18 A plot of a stage's non-linear output folded residue as a function of its input. The difference between the real and ideal residues is a non-linear error that cannot be corrected by the redundancy

Typically, the folding operation is performed by current steering differential circuits as shown in Figure 3.17. Bipolar transistors were traditionally used, but MOS folders have also been used lately. The linearity of these folders degrades as the amplitude of the output signal increases, causing rounding near the peaks of the folded residue signal, as shown in Figure 3.18. This degrades the linearity of the reconstructed signal accordingly. Using additional folders shifted by half the segment size reduces the non-linear effects by ensuring that one of the folders will be near its zero crossing. In addition, interpolation can be employed to generate even more zero crossings and improve the linearity further as shown in Figure 3.19.

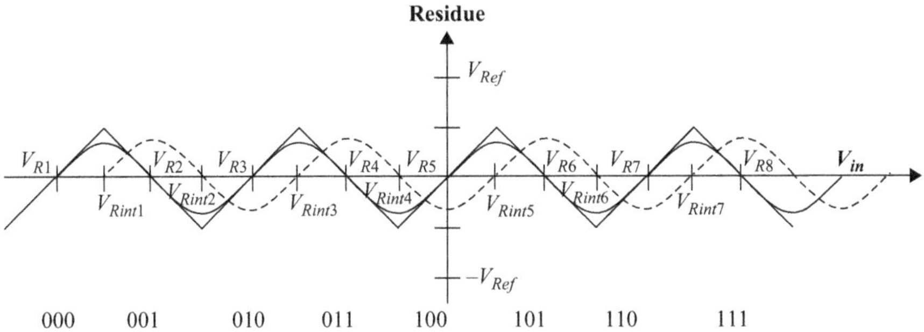

Figure 3.19 Plots of the residues when employing interpolation with folding to improve the linearity of the ADC

In fact, by increasing the interpolation factor, more intermediate points can be added, the output will respond to zero crossings instead of amplitude, and the overall linearity can be substantially improved. If enough interpolation resolution is employed to cover all the bits, the fine ADC can be removed, and the resulting *Folding-with-Interpolation* ADC will be closer to a flash ADC variant than a multi-step ADC.

Offsets in the folding circuits will also degrade the linearity. This can be improved by using more zero crossings and by employing redundancy as mentioned earlier. The concept of redundancy can be used in all multi-step architectures, where the sum of the number of bits in all stages adds up to more than the required number of bits of the converter. This overlap in the bit reconstruction ensures that an error in the front-end stages does not cause the back-end stages to over-range. This is illustrated in Figures 3.16, 3.18, and 3.19.

The fine ADC usually processes the folded signal without amplification. Alternatively, an amplifier can be employed to amplify the folded signal into a full-scale range that is suitable for the back-end ADC. The gain error of this amplifier will contribute to the non-linearity of the ADC and will need to be corrected in the digital reconstruction similar to a pipelined ADC.

Typically, in a folding ADC, the two conversion steps are performed in one clock cycle. This makes it important to have a S/H circuit in order for the ADC to "see" a held signal. Otherwise, the effect of the folding operation on the non-linearity becomes very difficult to handle. The folding transfer characteristic is non-linear and would result in generating harmonics that do not exist in the input signal. These harmonics are ideally eliminated by the reconstruction process. However, settling and gain errors in the folding operation can cause these harmonics to persist and lead to non-linearities in the output transfer function. This puts an upper limit on the typical resolution achievable by the folding architecture to about 8–10 bits. Adding a sample and hold (S/H) circuit to the front end alleviates this problem, but adds a substantial amount of power and noise to the ADC. The folding architecture could also use more than one clock cycle for the two conversion steps, in order to improve the throughput and hence the speed and bandwidth of the ADC. In this case, it will resemble a pipeline architecture, where the first-stage MDAC of the pipeline is replaced by a folding stage.

3.6 Pipelined ADC

A pipeline architecture is composed of two or more conversion stages, as shown in Figure 3.20 with the timing diagram and the progression of samples down the pipeline depicted in Figure 3.21. Each stage is composed of a sub-ADC, a DAC, a subtractor, and an inter-stage amplifier. The input is sampled by a dedicated S/H amplifier (SHA) or by the S/H circuit of the first stage. It gets quantized by the first stage's sub-ADC to generate the first $k1$ MSBs, which are converted back to a quantized analog signal by the stage's $k1$-bit DAC. The quantized DAC output is subtracted from the input to generate the residue, which is amplified before being sampled by the second stage. This process is repeated down the pipeline, with the quantization error getting progressively smaller in every stage, and passed on to the following stage until the lowest LSBs are generated by the back-end flash.

During each clock phase, the various stages will be operating simultaneously on different samples of the input. For example, at the first clock phase in Figure 3.21, as the second stage is generating a residue signal of the second sample to the third stage, the first stage will be sampling the next input sample, which is the third sample. This "pipelining" and parallel processing increase the throughput of the ADC substantially. Since the sampling rate and Nyquist bandwidth are determined by the throughput rate, these converters are capable of very high speeds.

The inter-stage amplification progressively relaxes the accuracy and noise requirements of the back-end stages compared to a front-end stage. However, the accuracy of the residue generation is critical for the overall accuracy of the ADC. Similar to other multi-step ADCs, redundancy and digital error correction are employed to reduce the sensitivity to the sub-ADC in every stage. For example, an

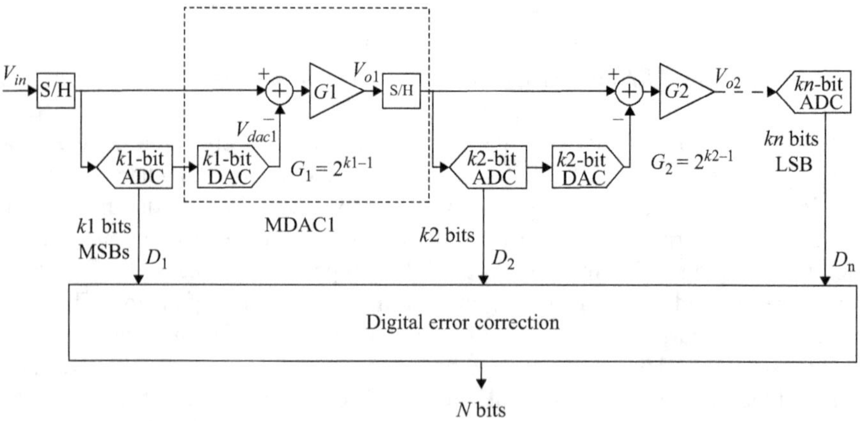

Figure 3.20 A block diagram of a pipelined ADC with n stages

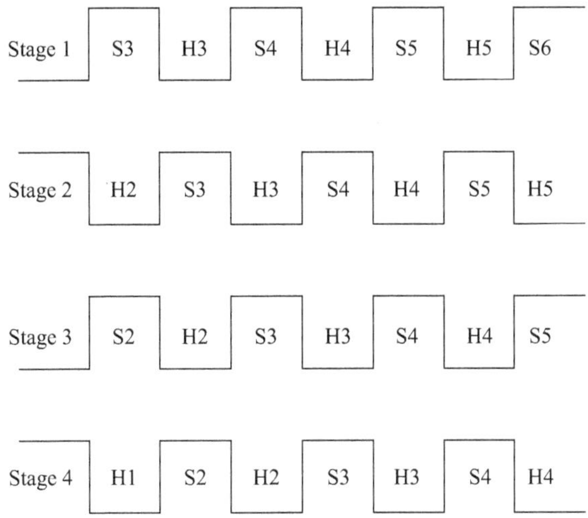

Figure 3.21 Timing diagram of four stages in a pipelined ADC. "Sn" denotes the sampling of the nth sample by the corresponding stage, and "Hn" denotes the holding of the nth sample's residue by that stage. For example, "S3" denotes the sampling of the third sample by that stage, and "H3" is the hold phase where the residue of the third sample is presented by that stage to the following stage

error in the first-stage $k1$-bit ADC can be corrected by the following stages, as long as the residue does not over-range and is within the correction range. The gain of a $k1$-bit stage is typically set to 2^{k1-1}, which allocates half the full-scale range of the back-end ADC to the correction range. As we mentioned earlier, there is a clear similarity between the pipelined ADC and the sub-ranging ADC. In fact, the sub-ranging ADC can be considered a special case of the pipeline architecture. The residue of a 3-bit pipeline stage will also look very similar to that of the sub-ranging ADC shown in Figure 3.14.

Pipelined ADCs break down the conversion process into smaller and more manageable tasks that are performed simultaneously in an assembly-line-like fashion. Their parallel processing capability enables high speeds at the expense of latency and area. Progressively reducing the quantization noise (i.e. residue) from one stage to the next enables high performance too. However, this high performance is achieved at the expense of power consumption in the inter-stage amplifiers and complexity in the form of calibration of DAC and amplifier errors. This makes the pipeline architecture one of the most effective in the high speed and high resolution space. Sampling rates up to 2.5 GS/s and resolutions up to 16 bits have been reported using this architecture [1, 2, 14]. Due to their importance, popularity, and complexity, Chapter 7 is dedicated to discussing the pipelined ADC as a representative of multi-step ADCs.

A variation of the pipeline architecture is the Cyclic (Algorithmic) ADC, in which the conversion steps are performed by the same stage in a cyclic fashion as

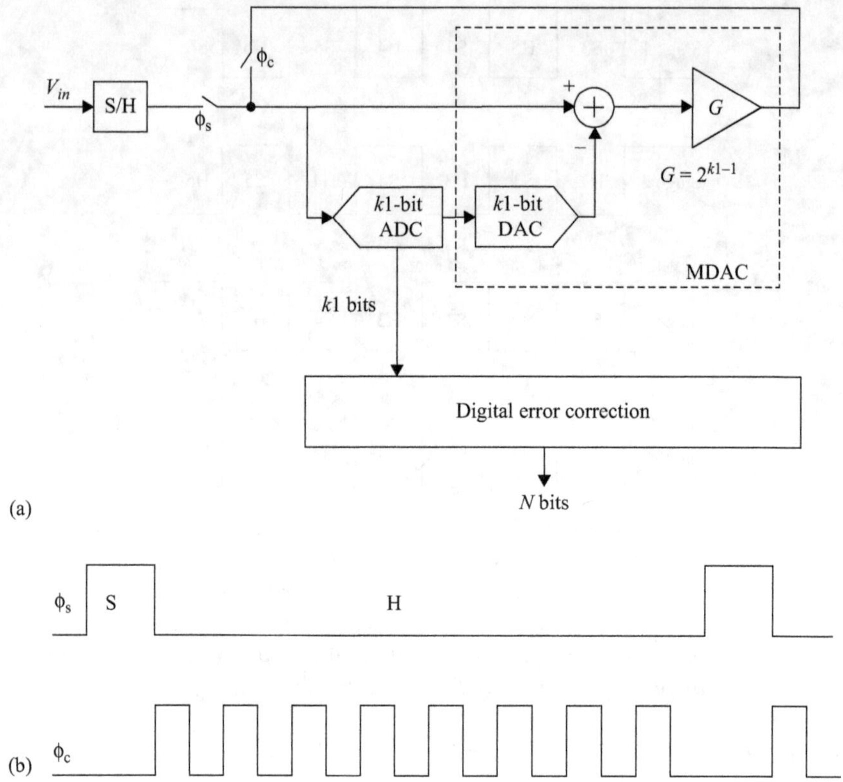

Figure 3.22 (a) A block diagram of a cyclic ADC. (b) Timing diagram for a cyclic ADC

shown in Figure 3.22 (a), with the timing diagram shown in Figure 3.22(b). The residue is generated from the first conversion step, and fed back as input to the same stage to perform the second conversion step, and so on. It is clear that this architecture will not have the high throughput that is characteristic of the pipeline architecture because the ADC cannot sample a new input until the whole conversion process is finished. This limits the application of this architecture to relatively low sampling rates. In addition, since the same stage is used to process all the bits, it is difficult in Cyclic ADCs to take advantage of the accuracy and noise relaxation in the back-end stages. Its advantage, however, is a much smaller area than the pipeline architecture.

3.7 Successive approximation (SAR) ADC

In a SAR ADC, the input is sampled and compared to the mid-scale voltage generated by an N-bit DAC, which is controlled by the successive approximation register (SAR). The digital output of the comparator is used by the SAR to select the next DAC output to be used as threshold for the comparator's next cycle. For example, if the input is larger than the DAC output, the comparator's output is "high," the MSB is set to 1, and the next bit is also set to 1, leading to a DAC output corresponding to the midpoint of the upper half of the input range. If the comparator's output is "low," indicating an input smaller than the mid-scale, the MSB is set to zero, and the next bit is set to 1, thus setting the DAC output to the midpoint of the lower half of the input range. This process is repeated N times using the binary search algorithm, in order to digitize the input signal into an N-bit digital word.

An N-bit SAR ADC requires N cycles to complete the conversion process. The next input sample can be sampled only after the whole conversion is finished. Typically, the process requires 1 cycle for sampling the input and N cycles for conversion, giving an $N+1$ total cycles. In some cases, the sampling process may require more than one clock cycle, which leads to even more cycles. The block diagram and timing diagram are shown in Figure 3.23. An example of the progression of the conversion operation and the DAC output is shown in Figure 3.24.

It is clear from the above description that SAR converters, like Cyclic ADCs, are inherently slower than other multi-step architectures. Therefore, traditionally they have been limited to low speed applications. The ADC's accuracy depends almost entirely on the accuracy of the comparator and the DAC. An offset in the comparator would affect all of the codes equally, and hence would cause an offset in the whole ADC, but does not affect its linearity. On the other hand, a decision error in one of the steps would lead to an irreversible error in the overall digital word. A comparator or DAC error affects the ADC's linearity, which puts strict requirements on the comparator's accuracy, on its code-dependent speed, and on the DAC's linearity and speed. To relax this sensitivity, redundancy can be used, such that the back-end conversion can recover from front-end errors. This is done by utilizing more conversion steps while using a radix that is less than 2, which allows the output to still reach the "right" decision in spite of intermediate errors. The additional steps, however, consume more time and hence lead to lower speeds. Moreover, the non-linearity in the DAC can also be improved by employing calibration and dynamic element matching.

There are numerous DAC implementations that can be used in the SAR ADC. One implementation is the capacitive charge redistribution DAC, where 2^N unit capacitances are used for an N-bit converter. An example is shown in Figure 3.25 for a 4-bit binary DAC. During the sampling phase ϕ_S, all the capacitances are connected to the input through their top plates, while their bottom plates are connected to ground. In the conversion phases ϕ_D, the input is disconnected and one unit capacitance is connected to ground. The other binary weighted capacitances are connected to V_{Ref} or ground depending on the step. For example, in the first MSB step, the largest capacitance is connected to the reference, and all the other capacitances are connected to ground. In the next step, the next largest capacitance is connected to the reference voltage and all the other capacitances are connected to ground, while the largest capacitance is connected to the reference or ground depending on the SAR word. Gradually, the input of the comparator will approach zero, as the output digital word approaches the correct digital word that represents the sample.

In this implementation, both the input signal and the DAC's output are sampled on the same set of capacitors and are effectively subtracted. This puts the

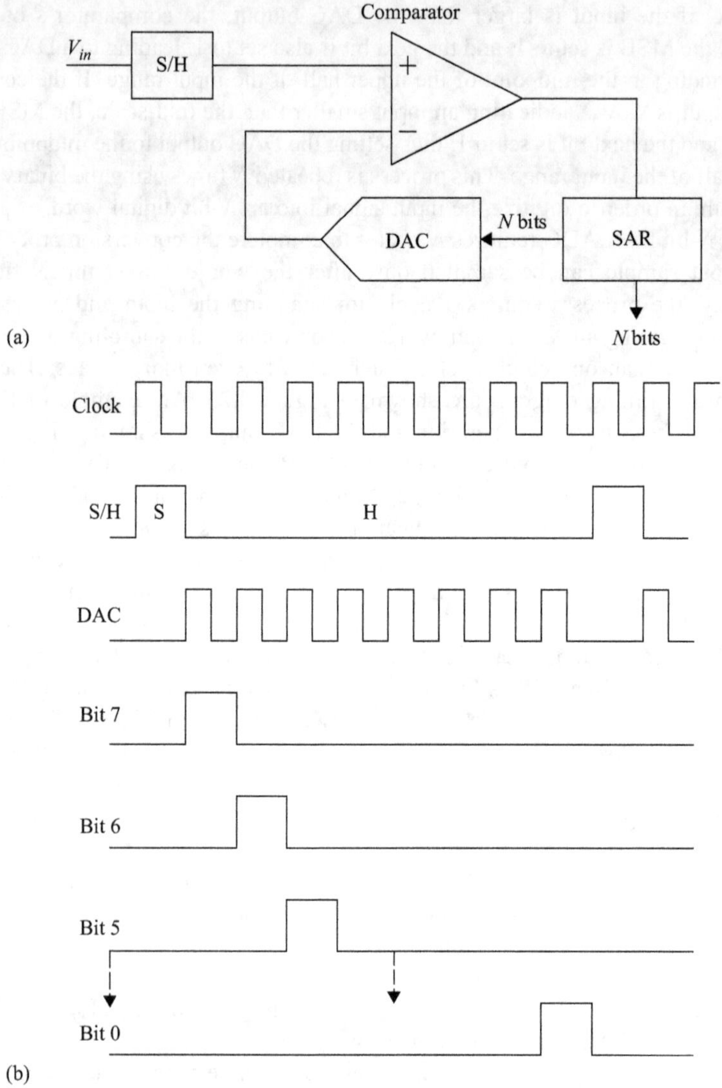

Figure 3.23 (a). A block diagram of a SAR ADC. (b) Timing diagram for an 8-bit SAR ADC

comparator's input close to zero, which is easier to handle than the large swings needed if the input is applied to one side and the DAC's output to the other. In addition, the DAC's linearity is determined by the capacitor matching, which tends to be reasonable with careful layout. The matching improves with increasing the capacitance's area and matching in the order of 10–14 bits can be achieved without calibration. The speed of the conversion step will be determined by the comparator's speed, the reference settling, and the capacitances' charge

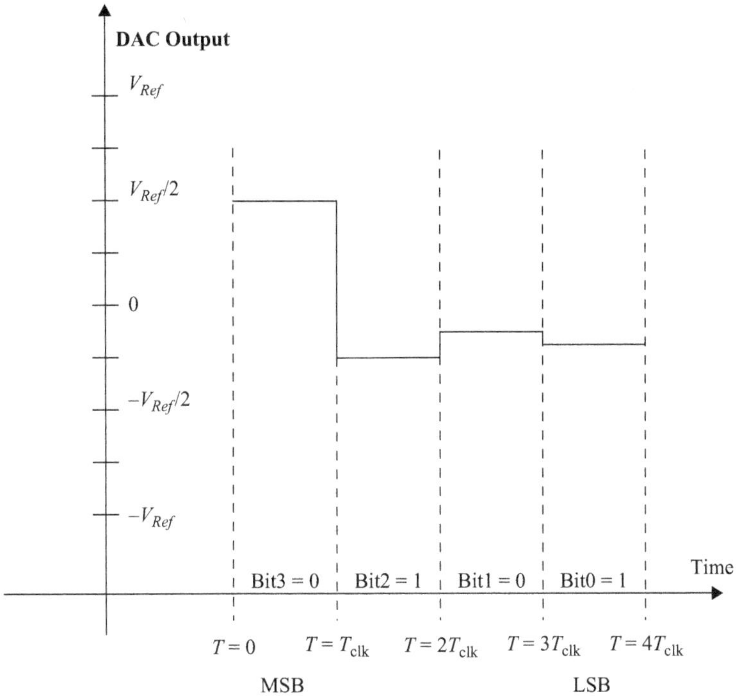

Figure 3.24 The DAC output in a 4-bit SAR ADC as a function of time. The binary search algorithm is used to reach the final digital output

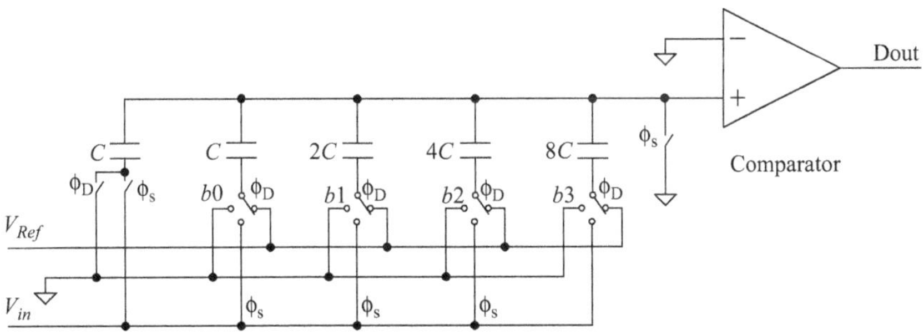

Figure 3.25 An example of the DAC implementation in a 4-bit SAR ADC. During the input sampling phase ϕ_s, the input is sampled on all the capacitors. During the DAC phase ϕ_D, one capacitor C is connected to ground, while the others are connected to either ground or V_{Ref}, depending on the DAC code

redistribution. Other possible DAC implementations will be discussed later in the DAC section of this chapter.

The noise in SAR converters is dominated by the sampling noise and the comparator noise. Their power consumption is low and can be dominated by the comparator and the power needed to charge and discharge the DAC capacitances when connected to the reference. Some creative techniques have been developed to reduce this power [12, 13, 17]. Since the main building block of a SAR converter is the comparator, the SAR ADC's power decreases with process scaling in a trend that is almost similar to digital circuits. This is different from pipelined ADCs whose reliance on inter-stage amplifiers makes them more challenging in fine lithography processes. Compared to pipelined ADCs, SAR converters are usually slower, are smaller, and consume less power. They also have smaller latency and are much simpler to implement.

In general, SAR ADCs are simple to implement and can achieve very low power consumption. This makes them good candidates for relatively low speed applications. Recently, the sampling rates of SAR converters have been pushed up by taking advantage of the advances in process technology. They have also been used in high speed and low resolution ADCs by interleaving a large number of them, or by incorporating them in a pipeline architecture. Their simple structure and low power consumption make them an attractive choice for these architectures. In addition, their short sampling clock phase, compared to the conversion clock phases, fits well with time-interleaving, since a number of channels equal to the number of conversion cycles (or number of bits) can be employed.

3.8 Pipelined and SAR ADC

We have discussed the differences between the SAR and pipeline architectures, with the SAR being slower, smaller, simpler, and lower power. However, it is also important to note the similarities, especially given that both of them are multi-step ADCs.

If we think of the SAR's comparator as a sub-ADC, we can see that the SAR architecture is similar to the cyclic architecture, which is algorithmically identical to the pipeline. However, unlike the pipeline and cyclic ADCs, the SAR ADC does not generate or amplify residues that represent the new quantization errors after every digitization step by the sub-ADC. Instead, the SAR ADC uses the same input signal and changes the DAC codes in every step to progressively reduce the quantization error. Therefore, the progressive accuracy relaxation achieved in pipelined ADC by inter-stage amplification does not exist in the traditional SAR ADC, and the accuracy of its sub-ADC and DAC should match the required ADC accuracy.

Noting these architectural similarities and differences allows us to extrapolate some of the discussions, patterns and techniques of pipelined ADCs to SAR ADCs. For example, similar to pipelined ADCs, we can employ redundancy in SAR ADCs by using more than one comparator, using sub-binary search steps, and/or increasing the number of cycles. Moreover, the error patterns tend to be similar and can be corrected using approaches that are similar to those employed in pipelined ADCs, as discussed in Chapters 7 and 9.

These similarities have also enabled the integration of the SAR and the pipeline architectures into a single ADC to reduce the power consumption [18, 19].

This is usually called a Pipelined SAR ADC. For example, the SAR ADC can be used as the sub-ADC in a pipelined ADC, with the SAR's DAC used as the pipeline stage's DAC. In other implementations, interleaved SARs are used as the sub-ADC to achieve higher speeds than what is possible using a single SAR. In addition, SAR ADCs have been used as back-end stages in pipelined ADCs.

3.9 Time-interleaved ADC

Time-interleaved ADCs employ parallel processing to increase the speed of the ADC. By interleaving M ADCs, running at f_s/M each, a net sampling rate of f_s can be achieved. An example of a two-way interleaving ADC is shown in Figure 3.26.

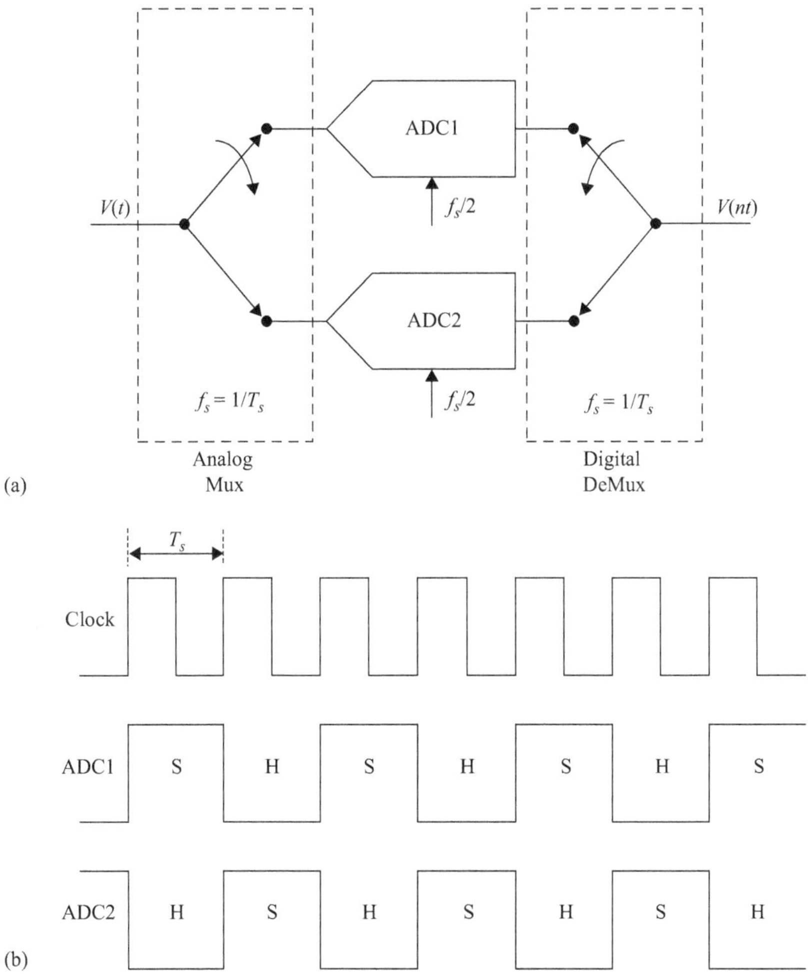

Figure 3.26 (a) A two-way interleaved ADC. (b) Timing diagram

Table 3.1 The impact of different kinds of inter-channel mismatch on the performance of an M-way interleaved ADC, with sampling rate f_s. The input frequency is f_{in}, f_{IL} is the frequency of the interleaving spurs, and the factor k is given by: 1, 2, …, M − 1

Type of mismatch	Effect on input	Spur frequency
Offset mismatch	Additive effect	$f_{IL} = \frac{k}{M} f_s$
Gain mismatch	Amplitude modulation	$f_{IL} = \pm f_{in} + \frac{k}{M} f_s$
Timing mismatch	Phase modulation	$f_{IL} = \pm f_{in} + \frac{k}{M} f_s$
Bandwidth mismatch	Frequency-dependent amplitude and phase modulation	$f_{IL} = \pm f_{in} + \frac{k}{M} f_s$

However, errors due to mismatches between the interleaved channels limit the accuracy of the resulting converter. Mismatches include offset, gain, timing, bandwidth, and non-linearity mismatches. They cause interleaving spurs that are summarized in Table 3.1 and can be quite large if the inter-channel matching is poor. The interleaving spurs need to be corrected to take advantage of the full Nyquist bandwidth. Without correction, the spur-free bandwidth is limited to the Nyquist bandwidth of each ADC, which is $f_s/2M$. However, this may be acceptable in some oversampled applications, where the interleaving spurs may fall out of the signal band and can be filtered using digital filters. In these oversampled applications, correction of the interleaving spurs may not be needed.

In spite of their limitations, time-interleaved ADCs can achieve very high speeds that are not possible otherwise, and hence represent a very important architecture for high speed ADCs. The high speed comes at the cost of area, complexity, and performance. Chapter 8 is dedicated to discussing this important class of converters.

3.10 Sigma-delta ADC

Sigma-delta converters employ oversampling and noise shaping to reduce the quantization noise and improve performance. The required oversampling makes them inherently slower than Nyquist converters. In principle, sigma-delta converters trade speed for resolution.

Sigma-delta modulators evolved in communication systems, together with delta modulators, for data compression. Since signal samples tend not to change significantly from one sample to the next, it is more efficient to transmit the difference (delta) between the signal and its estimate, which is obtained by integrating the previously transmitted samples. This is shown in Figure 3.27(a) and is known as a delta modulator. On the receiver side, a demodulator is needed to extract the original signal from the transmitted "delta" signal.

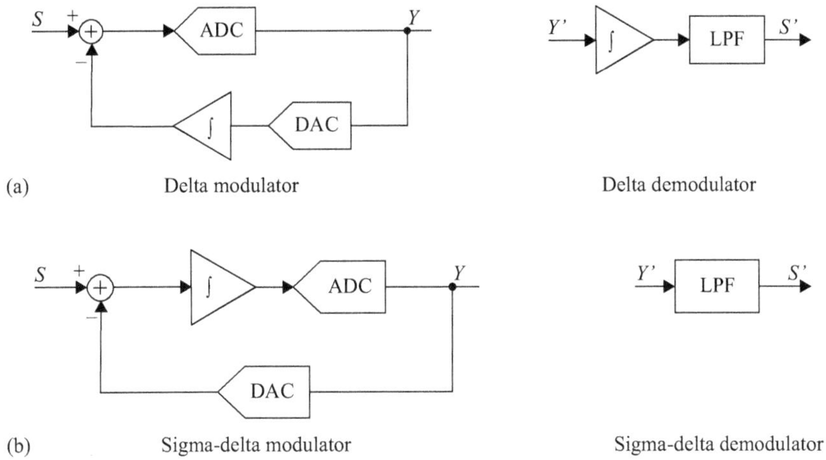

Figure 3.27 (a) A delta modulator and demodulator. (b) A sigma-delta modulator and demodulator

Alternatively, the integrator can be moved to the forward path in the delta modulator as shown in Figure 3.27(b) to give a sigma-delta (ΣΔ) modulator. This should provide similar compression and simplifies the demodulation process to a simple low-pass filtering operation. However, this equivalence between the delta and the sigma-delta modulator is not entirely accurate. It is valid if the quantization operation were a linear operation. However, since quantization is non-linear, the ordering of the operations matters. Therefore, integration after quantization, as done in a delta modulator, is not equivalent to quantization after integration, as done in a sigma-delta modulator.

Therefore, although the sigma-delta modulator may have been inspired by the delta modulator, the two operations are not equivalent. One fundamental difference is that since the integration applies to the input in the sigma-delta modulators, the response of the modulator to the input signal is a low-pass filter instead of a high-pass filter in the delta modulator. This leads to the shaping of the quantization noise relative to the input signal in sigma-delta modulators. In addition, sigma-delta modulators track the input amplitude, while delta modulators track the input slope.

As for the name, some people prefer the name "sigma-delta" because the modulation is composed of an integration (sigma) of the difference signal (delta). Others prefer the "delta-sigma" name because the signal sees a difference operation first (delta) followed by an integration (sigma). Both names are valid and commonly used.

3.10.1 Oversampling and noise shaping

The operation of a sigma-delta modulator is based on the assumption that the signal's bandwidth is much smaller than the sampling rate. This oversampling allows for the shaping of the quantization noise to attenuate it in the signal band.

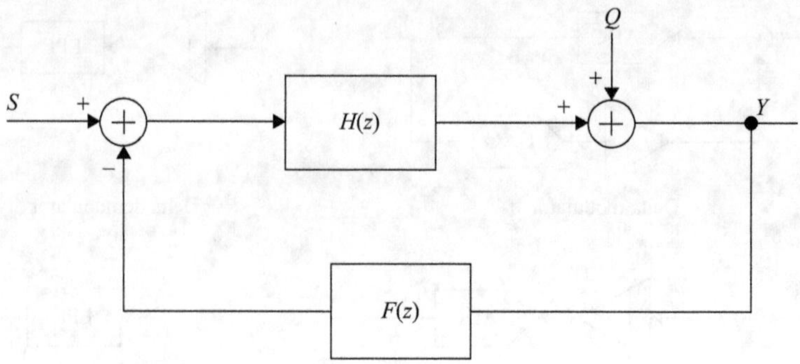

Figure 3.28 A linearized representation of a sigma-delta modulator

This is shown in Figure 3.28 for a discrete-time case, where the transfer functions are represented in the *z*-domain. The quantization noise is represented as additive noise *Q*. If we assume the feedback transfer function *F*(*z*) is unity, the noise and signal outputs are given by

$$Y_N(z) = \frac{Q(z)}{1 + H(z)} \tag{3.17}$$

and

$$Y_S(z) = \frac{S(z)H(z)}{1 + H(z)} \tag{3.18}$$

where *S*(*z*) is the *z*-domain representation of the input signal, *Q*(*z*) is the quantization error, *H*(*z*) is the transfer function of the integrator, $Y_N(z)$ is the output noise, and $Y_S(z)$ is the output signal. The noise transfer function (*NTF*) is given by

$$NTF = \frac{1}{1 + H(z)} \tag{3.19}$$

The signal transfer function (*STF*) is given by

$$STF = \frac{H(z)}{1 + H(z)} \tag{3.20}$$

The total output is *Y*(*z*), such that

$$Y(z) = S(z) \times STF + Q(z) \times NTF \tag{3.21}$$

Since *H*(z) is an integrator, it is clear from the *NTF* that the quantization noise is high-pass filtered (i.e. noise shaped) such that it is attenuated at low frequencies and increases as the frequency increases. On the other hand, the *STF* is a low-pass filter. In spite of the reduction of the quantization noise over a small bandwidth, the total quantization noise actually increases. This is shown in Figure 3.29, where the shaped quantization noise for a first-order sigma-delta modulator is plotted in comparison to the corresponding unshaped (i.e. flat) noise.

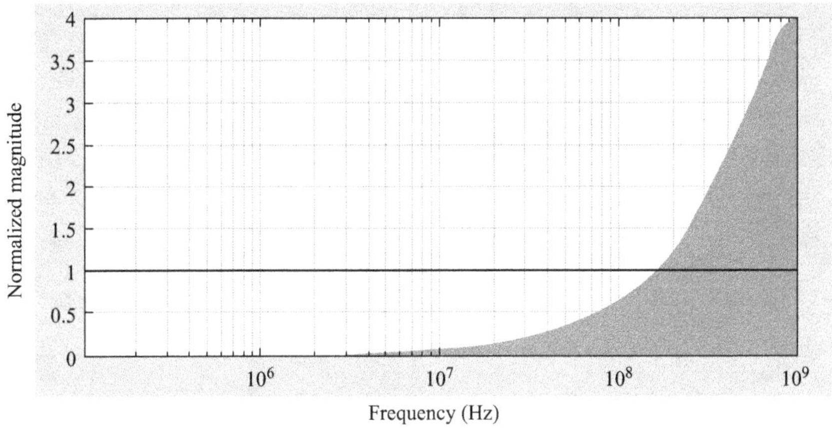

Figure 3.29 A first-order shaped noise spectrum of a sigma-delta modulator relative to a uniform noise spectrum

It is important to note that although the quantization noise is high pass filtered (i.e. "shaped"), the thermal noise of the integrator and the sampler is not shaped. The noise and non-linearity introduced in the feedback path, or in the DAC, will not be shaped either. Thus, the sigma-delta modulation relaxes the requirements on the ADC but does not relax the requirements on the linearity of the DAC. This is similar to the trend noted in multi-step ADCs with redundancy.

Another observation is that the representation of the quantization noise as an additive white noise that is uncorrelated with the signal is not entirely accurate and becomes more invalid as the number of bits is reduced, which can result in undesirable tones. In addition, there is no sample to sample correspondence in the time domain between the input and the output. That is, the output tracks the input on average, but a single output sample in isolation does not have a direct correspondence to the input amplitude. This limits the applicability of sigma-delta modulators in some time-domain applications.

3.10.2 Single-bit modulator

The stringent requirement on the DAC linearity has popularized the use of single-bit quantization. A single-bit DAC has two output values and hence is inherently linear. However, single-bit modulators suffer from large quantization noise that requires large oversampling ratios to be adequately attenuated. In addition, the quantization noise will be correlated with the input signal, which results in tones and spurs that fall in the signal band. Nevertheless, the simplicity of the single-bit sigma-delta ADC has made it a popular choice in many applications.

It is interesting to observe the output of a sigma-delta modulator with a 1-bit quantizer in the time domain. The output looks like a stream of positive and negative pulses whose width is proportional to the input amplitude, which is similar to a pulse width modulator (PWM). In addition, the average of the output pulses tracks the input amplitude, as shown in Figure 3.30.

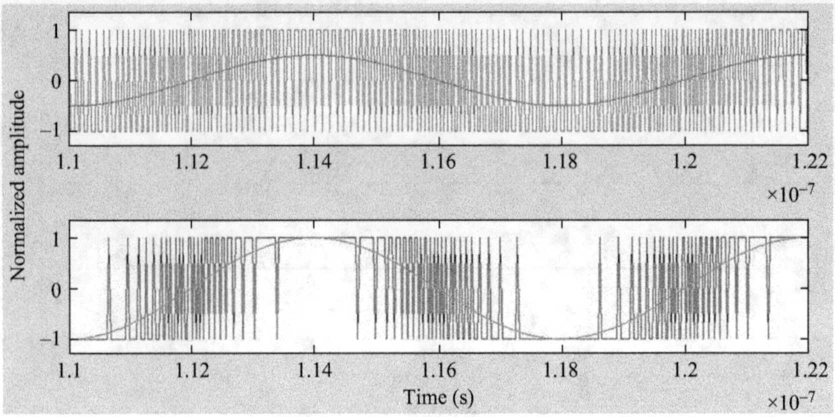

Figure 3.30 The input sine wave and the output time waveform of a sigma-delta modulator for a half-scale signal and a full-scale signal. The effect of overloading in the bottom curve is clear

3.10.3 Overloading

Overloading is a concern in sigma-delta modulators, and more so in 1-bit implementations. By examining the waveforms in Figure 3.30, one can see that the input sine wave needs to be smaller than the full-scale of the DAC, to give room for the modulator to respond to the input amplitude. As the input approaches the DAC's full-scale, the output starts clipping or oscillating, which is called "overloading." Since the period of "instability" can be significantly longer than the duration of the input overloading, it is important to prevent this from happening to avoid the long recovery time. This can be done by ensuring that the input is significantly less than the DAC's full-scale, by some factor, typically in the order of 2–4 dB.

Additionally, 1-bit quantizers produce significant tones and harmonic distortion. Although most of this energy is noise shaped to be out of band, it consumes a portion of the output dynamic range that requires even further reduction in the input signal amplitude to avoid overloading. This additional reduction can be in the order of 3 dB.

Therefore, overloading precautions can limit the practical dynamic range and SNDR of the sigma-delta ADC. Although the noise floor in the signal band may be very low, accounting for the overloading and instability budget can lead to a reduction in the practical noise floor by about 3 dB.

3.10.4 First-order modulator

An example of a discrete-time first-order modulator is shown in Figure 3.31, with the transfer function $H(z)$ given by

$$H(z) = \frac{z^{-1}}{1 - z^{-1}} \tag{3.22}$$

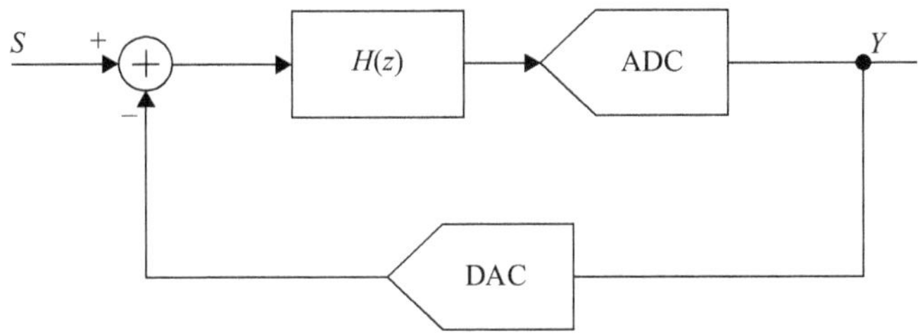

Figure 3.31 Block diagram of a sigma-delta modulator

Substituting (3.22) into (3.19) and (3.20) gives the *STF* as

$$STF = z^{-1} \tag{3.23}$$

The *NTF* is

$$NTF = 1 - z^{-1} \tag{3.24}$$

Since the output $Y(z)$ is expressed in terms of the input signal $S(z)$ and the quantization noise $Q(z)$ as

$$Y(z) = S(z) \times STF + Q(z) \times NTF \tag{3.25}$$

Substituting (3.23) and (3.24) into (3.25), we obtain

$$Y(z) = S(z) \times z^{-1} + Q(z) \times \left(1 - z^{-1}\right) \tag{3.26}$$

It is clear from (3.26) that the signal passes through the modulator delayed by one cycle, while the quantization noise gets high-pass filtered. The noise attenuation at low frequency is illustrated by replacing z^{-1} by $e^{-j\omega T_s}$. The *NTF* will be

$$NTF = 1 - e^{-j\omega T_s}$$

Therefore,

$$|NTF|^2 = 4\sin^2\left(\frac{\omega T_s}{2}\right) \tag{3.27}$$

where ω is the angular frequency and T_s is the sampling time. This indicates that the maximum and total quantization noise actually *increase* by a factor of 4 and 2, respectively, but it is shaped by a high-pass filter that significantly attenuates the low-frequency noise and amplifies the high-frequency noise. If the band of interest is defined by B, that is much smaller than the sampling rate f_s, then $\omega T_s/2$ is much less than 1, and

$$|NTF|^2 \approx 4\left(\frac{\omega T_s}{2}\right)^2 \tag{3.28}$$

The output quantization noise power will be given by

$$Y_N^2 \approx q^2 \int_0^B 4\left(\frac{2\pi f T_s}{2}\right)^2 df \approx \frac{q^2 4\pi^2 B^3 T_s^2}{3} \tag{3.29}$$

where q^2 is the quantization noise power spectral density. The quantization noise power of a quantizer with step size Δ is

$$P_Q = \frac{\Delta^2}{12}$$

where the step size is given by

$$\Delta = \frac{V_{FS}}{2^N}$$

and the quantization noise power is

$$P_Q = \frac{\Delta^2}{12} = \frac{q^2 f_s}{2}$$

Therefore, the quantization noise spectral density q^2 can be replaced in (3.29) by

$$q^2 = \frac{\Delta^2}{6f_s} = \frac{2P_Q}{f_s} \tag{3.30}$$

to give

$$Y_N^2 \approx \frac{8\pi^2 B^3}{3f_s^3} P_Q = \frac{2\pi^2 B^3}{9f_s^3} \Delta^2 \tag{3.31}$$

Since the oversampling ratio *OSR* is given by

$$OSR = \frac{f_s}{2B} \tag{3.32}$$

then using (3.32) into (3.31), we get

$$Y_N^2 \approx \frac{\pi^2}{3} \frac{P_Q}{OSR^3} \tag{3.33}$$

Therefore, the signal to quantization noise ratio is given by

$$SQNR_{1st,\Sigma\Delta}(\text{dB}) = 6.02\,N + 1.76 - 5.17 + 30 \log OSR \tag{3.34}$$

and

$$SQNR_{1st,\Sigma\Delta}(dB) = 6.02\,N + 1.76 - 5.17 + 9 \log_2 OSR \tag{3.35}$$

This shows that for first-order modulation ΣΔ, every doubling of the sampling rate or *OSR* results in 9 dB improvement in SQNR. Compared to an oversampled Nyquist ADC (in Chapter 1) where every doubling of the *OSR* results in only 3 dB improvement in SQNR, the noise shaping provides a significant advantage to oversampling in reducing the in-band quantization noise.

3.10.5 Second-order modulator

Intuitively, since the noise shaping is determined by the integrator transfer function, a higher-order filter should improve the attenuation of the quantization noise. However, using two integrators in a feedback loop can result in 180° phase shift and hence has the potential for instability. An example of a second-order modulator is shown in Figure 3.32, where the integrator is divided into two portions $H_1(z)$ and $H_2(z)$, such that

$$H_1(z) = \frac{1}{1 - z^{-1}} \tag{3.36}$$

and

$$H_2(z) = \frac{z^{-1}}{1 - z^{-1}} \tag{3.37}$$

Analyzing the loop gives the *STF* to be

$$STF = z^{-1} \tag{3.38}$$

The *NTF* is

$$NTF = \left(1 - z^{-1}\right)^2 \tag{3.39}$$

The output $Y(z)$ is given by

$$Y(z) = S(z) \times STF + Q(z) \times NTF \tag{3.40}$$

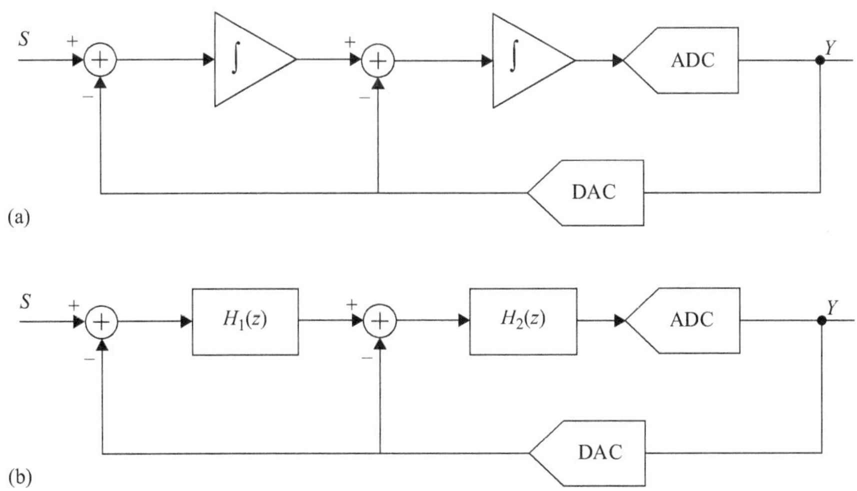

Figure 3.32 (a) A block diagram of a second-order sigma-delta modulator. (b) The integrators are represented in the z-domain

which gives

$$Y(z) = S(z) \times z^{-1} + Q(z) \times \left(1 - z^{-1}\right)^2 \tag{3.41}$$

Therefore, the quantization noise is given by

$$|NTF|^2 = 16 \sin^4\left(\frac{\omega T_s}{2}\right) \tag{3.42}$$

and

$$|NTF|^2 \approx 16\left(\frac{\omega T_s}{2}\right)^4 \tag{3.43}$$

The output quantization noise power is given by

$$Y_N^2 \approx q^2 \int_0^B 16\left(\frac{2\pi f T_s}{2}\right)^4 df \approx \frac{q^2 16\pi^4 B^5 T_s^4}{5}$$

which gives

$$Y_N^2 \approx \frac{\pi^4}{5} \frac{P_Q}{OSR^5}$$

Therefore, the signal to quantization noise ratio is given by

$$SQNR_{2\text{nd},\Sigma\Delta}(dB) = 6.02\,N + 1.76 - 12.9 + 50 \log OSR \tag{3.44}$$

and

$$SQNR_{2\text{nd},\Sigma\Delta}(dB) = 6.02\,N + 1.76 - 12.9 + 15 \log_2 OSR \tag{3.45}$$

This shows that for second-order $\Sigma\Delta$ modulation, every doubling of the sampling rate or *OSR* results in 15 dB improvement in SQNR. It is interesting to note that the second-order modulation causes a larger loss in the baseline SQNR by 12.9 dB, compared to the 5.17 dB loss for the first-order modulator. But the rate of increase in SQNR with *OSR* and sampling rate is substantially higher for the second-order modulator (15 dB/octave) compared to the first-order modulator (9 dB/octave).

3.10.6 Higher order and cascaded sigma-delta modulators

More effective noise shaping can be achieved by increasing the order of modulation beyond the second order. Extrapolating the previous analysis, it is easy to show that for an *L*th order $\Sigma\Delta$ converter, the quantization noise is given by

$$Y_N^2 \approx \left(\frac{\pi^{2L}}{2L+1}\right) \frac{P_Q}{OSR^{(2L+1)}} \tag{3.46}$$

and the *SQNR* is given by

$$SQNR_{Lth,\Sigma\Delta}(dB) = 6.02\,N + 1.76 - 10\log\left(\frac{\pi^{2L}}{2L+1}\right) + 10(2L+1)\log OSR \tag{3.47}$$

and

$$SQNR_{Lth,\Sigma\Delta}(dB) = 6.02\,N + 1.76 - 10\log\left(\frac{\pi^{2L}}{2L+1}\right) + 3(2L+1)\log_2 OSR \tag{3.48}$$

That is for every doubling in *OSR* or sampling rate, the SQNR increases by $3(2L+1)$ dB. As the modulation order increases, the noise shaping and SQNR improve. An example of the output spectrum is shown in Figure 3.33. However, in practice, keeping the loop stable for higher-order modulation can be quite challenging. Measures taken to stabilize the loop usually reduce the effectiveness of the noise shaping and degrade the SQNR below what is expected using (3.47) and (3.48).

Alternatively, multi-stage or cascaded (MASH) ΣΔ architectures can be used to achieve higher-order modulation, while relaxing the stability concerns. This is achieved by using two feedback loops, where the second loop operates on an

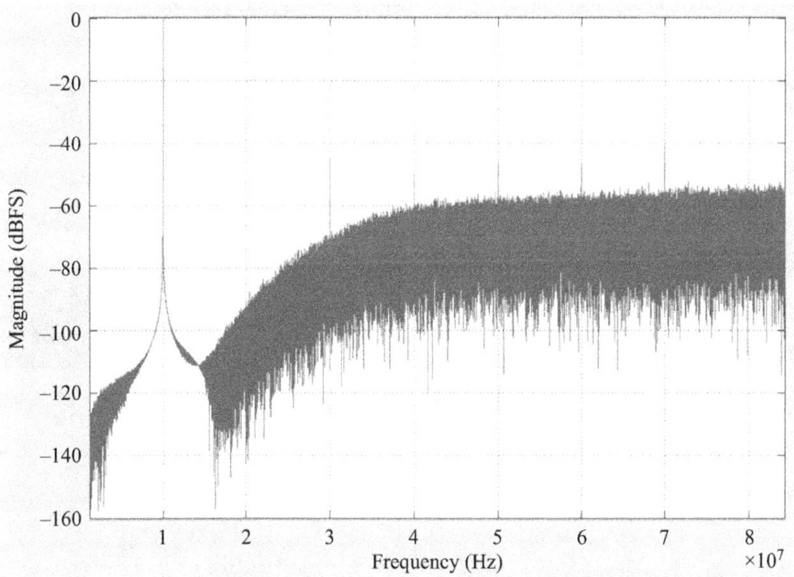

Figure 3.33 Output spectrum of a sigma-delta modulator

estimate of the quantization error generated by the first loop. This feed-forward approach makes the loops relatively independent from the stability standpoint.

The MASH ΣΔ structure is depicted in Figure 3.34. The estimate of the quantization error of the first loop is generated by subtracting the integrator output from the DAC output. This quantization error represents a "residue" signal that is further digitized and noise shaped by the second stage. The outputs of the two stages are then combined in the digital domain after processing the second loop's output with a digital filter that matches the first loop's filter. Analytically this is explained using the linearized model of Figure 3.35. The output of the first modulator $Y_1(z)$ is given by

$$Y_1(z) = S(z) \times STF_1 + Q_1(z) \times NTF_1 \quad (3.49)$$

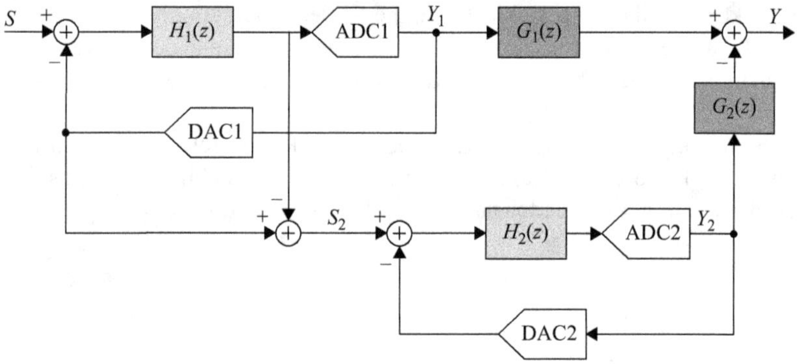

Figure 3.34 A block diagram of a cascaded (MASH) sigma-delta modulator

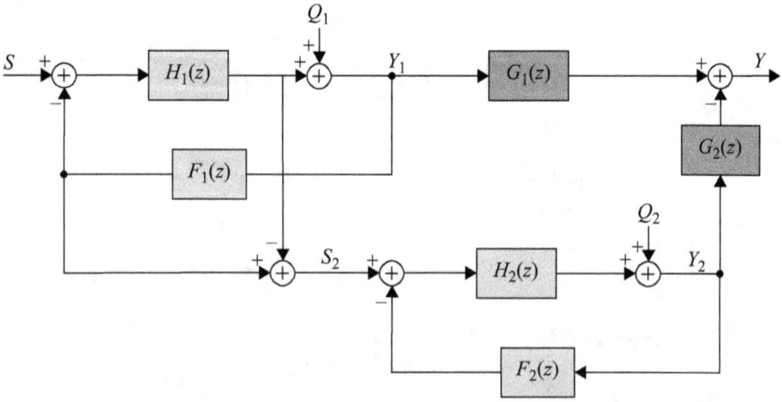

Figure 3.35 A linearized block diagram of a cascaded (MASH) sigma-delta modulator

The output of the second modulator $Y_2(z)$ is

$$Y_2(z) = S_2(z) \times STF_2 + Q_2(z) \times NTF_2 \tag{3.50}$$

since

$$S_2(z) = Q_1(z) \tag{3.51}$$

and

$$Y(z) = G_1(z) \times Y_1(z) - G_2(z) \times Y_2(z) \tag{3.52}$$

then, substituting (3.49), (3.50), and (3.51) into (3.52), gives

$$Y(z) = S(z)G_1STF_1 + Q_1(z)G_1NTF_1 - Q_1(z)G_2STF_2 - Q_2(z)G_2NTF_2 \tag{3.53}$$

if the following condition is satisfied,

$$G_1(z)NTF_1 - G_2(z)STF_2 = 0 \tag{3.54}$$

The quantization error of the first stage will cancel and the output will be given by

$$Y(z) = S(z) \times G_1 \times STF_1 - Q_2(z) \times G_2 \times NTF_2 \tag{3.55}$$

In addition, if

$$G_1(z) = k \times STF_2 \tag{3.56}$$

and

$$G_2(z) = k \times NTF_1 \tag{3.57}$$

then, the output will be given by

$$Y(z) = k \times S(z) \times STF_1 \times STF_2 - k \times Q_2(z) \times NTF_1 \times NTF_2 \tag{3.58}$$

Therefore, the total noise will be equal to the quantization noise of the second stage shaped by the cascade of the two *NTF*s. If the first stage is a second-order modulator and the second stage is a first-order modulator, the cascade will give a third-order modulation of the second-stage quantization noise, as follows

$$NTF_1 = \left(1 - z^{-1}\right)^2 \tag{3.59}$$

and

$$NTF_2 = \left(1 - z^{-1}\right) \tag{3.60}$$

Therefore,

$$NTF_1 \times NTF_2 = \left(1 - z^{-1}\right)^3 \tag{3.61}$$

In spite of the achieved third-order modulation, the stability challenge is that of the second- and first-order modulation of the first and second loops, respectively, which are more manageable.

It is interesting to note the similarity between the MASH architecture and the pipeline or sub-ranging architectures described earlier in this chapter and later in Chapter 7. The second stage of a MASH structure operates on a residue of the first stage. The accuracy of the MASH architecture depends on the matching between the digital filter and the analog filter in a manner similar to the matching required in pipelined ADCs between the digital and analog inter-stage gain factors. A mismatch between the two filters/gains results in degradation in the noise and linearity of the whole ADC. Calibration is often required to achieve acceptable performance.

3.10.7 Discrete-time and continuous-time sigma-delta modulators

So far, the discussion has assumed a discrete-time ΣΔ implementation, where the transfer functions are represented in the z-domain. This is compatible with switched-capacitor implementations, which present multiple advantages in terms of accuracy, flexibility, relative immunity to process variations, and compatibility with CMOS processes. This is discussed in Chapter 6. However, discrete-time ΣΔ ADCs require an anti-aliasing filter and are harder to drive because of the charge injection (kick-back) from the switched capacitors in their front ends. Their power consumption is determined by their amplifiers, and can be substantial at high sampling rates. This structure is shown in Figure 3.36(a).

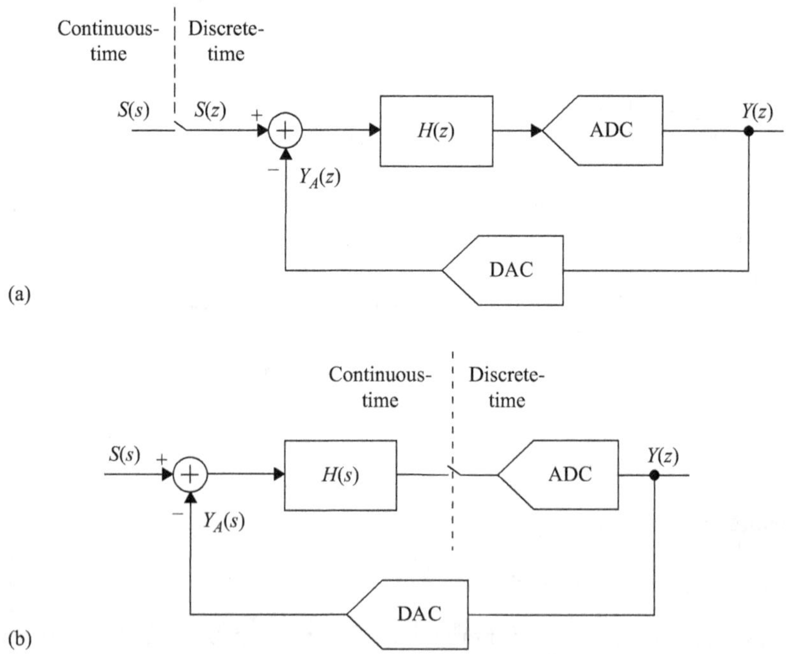

Figure 3.36 (a) Discrete-time sigma-delta modulator. (b) Continuous-time sigma-delta modulator

These disadvantages contributed to the rise in popularity of the continuous-time ΣΔ ADCs, where the sampling occurs in the quantizer, while the DAC and filter operate in the continuous-time domain. Therefore, an *s*-domain representation of the modulator is more appropriate in this case. Typically, the filter is implemented using a g_m/C structure, which tends to have lower accuracy but higher speed compared to switched capacitor implementations. This is shown in Figure 3.36(b).

Continuous-time ΣΔ modulators are easier to drive because of the relatively benign and unswitched load presented to the input. In addition, the signal is filtered by the continuous-time filter of the modulator, which performs an inherent anti-aliasing filtering that relaxes the requirements on the external anti-aliasing filtering needed in front of the ADC. Another advantage of continuous-time ΣΔ ADCs is that the sampling operation is performed in the quantizer. Therefore, the sampling non-idealities are attenuated by the noise shaping. Those include sampling switch non-linearity, sampling jitter, and charge injection. Sampling with low distortion represents one of the most challenging problems in high speed ADCs, especially for high input frequencies. Relaxing this requirement is a significant advantage of continuous-time sigma-delta ADCs.

However, the DAC implementation can be quite challenging in continuous-time ΣΔ architectures. It is typically implemented as a current steering DAC, which has higher speed but lower accuracy compared to switched-capacitor implementations. In addition, the jitter's effect on the DAC operation can be substantial, which represents one of the major limitations of the continuous-time implementation.

3.10.8 Multi-bit modulator

In spite of the simplicity and the inherent DAC linearity of the 1-bit ΣΔ ADC, it has significant limitations such as the large quantization noise and tones. The large step size also causes large signal swings at the DAC output, the subtractor, and the summing node. Moreover, in a 1-bit ADC, the gain is not well defined and overloading is a more serious problem compared to multi-bit implementations, which leads to a reduction in the input dynamic range and hence a degradation in SNDR.

On the other hand, multi-bit ΣΔ ADCs relax the OSR requirement, the gain margin for overloading, and the switching signal swings in the analog circuits. They allow for larger input amplitudes and improve the stability of the loop. However, the non-linearity of the DAC represents a significant limitation. Since it does not get attenuated by the loop, the DAC non-linearity has to match the overall desired accuracy of the ADC. Trimming, calibration, and mismatch shaping techniques are often implemented to improve the DAC linearity.

3.10.9 Bandpass sigma-delta converter

So far in our discussion, the ΣΔ discussion has assumed a low-pass filter modulator (integrator) and a high-pass filter *NTF*. Alternatively, if we use a bandpass filter modulator (i.e. a band stop filter *NTF*), we obtain a bandpass ΣΔ ADC. In this case, the signal bandwidth is a relatively narrow band that can be located at any location in the spectrum. This implementation can be used for IF and RF sampling of narrow band signals.

3.10.10 Concluding remarks

Sigma-delta ADCs represent one of the most important and attractive ADC architectures. Their popularity and dominance in the high resolution and low speed space are unparalleled. Recently, technology scaling and innovative architectures have enabled ΣΔ ADCs to achieve bandwidths that were previously restricted to Nyquist ADCs [3]. However, in spite of their fascinating progress, ΣΔ ADCs require oversampling and noise shaping that make them inherently slower than Nyquist ADCs. More detailed discussions of ΣΔ ADCs can be found in Reference 9.

3.11 DAC architectures

Digital-to-Analog converters convert a digital input signal into an analog output signal in the form of voltage or current. They are a crucial component in communication systems and many other applications. In addition, they are an important building block in some ADC architectures, such as multi-step and sigma-delta ADCs.

3.11.1 Resistive DAC

3.11.1.1 Resistive divider DAC

A 3-bit resistive divider DAC is shown conceptually in Figure 3.37. The digital code controls the switches that select which tap on the ladder is routed to the output through the buffer. The output will be an analog representation of the digital code.

The ladder voltage V_k is ideally given by

$$V_k = \frac{V_{Ref}}{2^N}\left(\frac{1}{2}+k\right) \tag{3.62}$$

where $k = 0, 1, 2, \ldots, 2^N - 1$, and N is the number of bits.

The buffer drives the DAC load with a low output impedance, while providing a high input impedance to minimize the impact of the load on the ladder. If the unit resistance of the ladder is R, then the equivalent resistance of the kth tap is given by

$$R_{eq} = \frac{\left(\frac{R}{2}+kR\right)\left[\frac{R}{2}+\left(2^N-1-k\right)R\right]}{2^N R} \tag{3.63}$$

Therefore,

$$R_{eq} = \frac{\left(\frac{2^N}{2}-\frac{1}{4}+k(2^N-1)-k^2\right)R}{2^N} \tag{3.64}$$

As shown in Figure 3.38, the tap with the highest equivalent resistance is at the middle of the ladder, and the equivalent resistance decreases as the taps approach

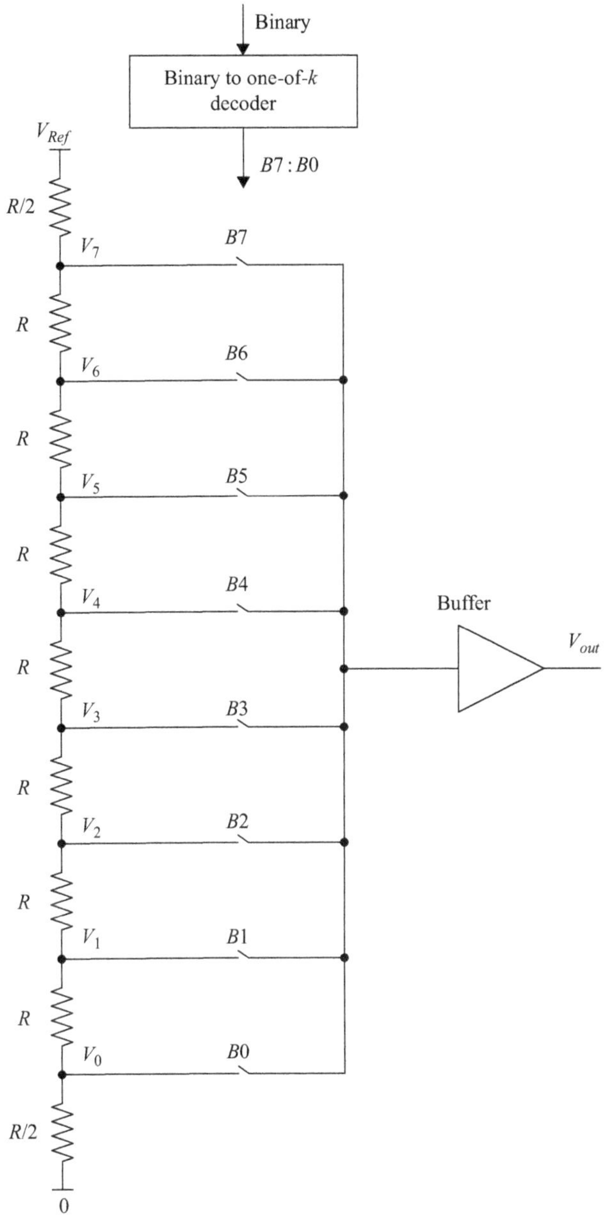

Figure 3.37 A 3-bit resistive divider DAC

the edges of the ladder. The maximum equivalent resistance is deduced by setting the derivative to zero:

$$\frac{\partial R_{eq}}{\partial k} = \frac{\left(2^N - 2k - 1\right)R}{2^N} = 0 \tag{3.65}$$

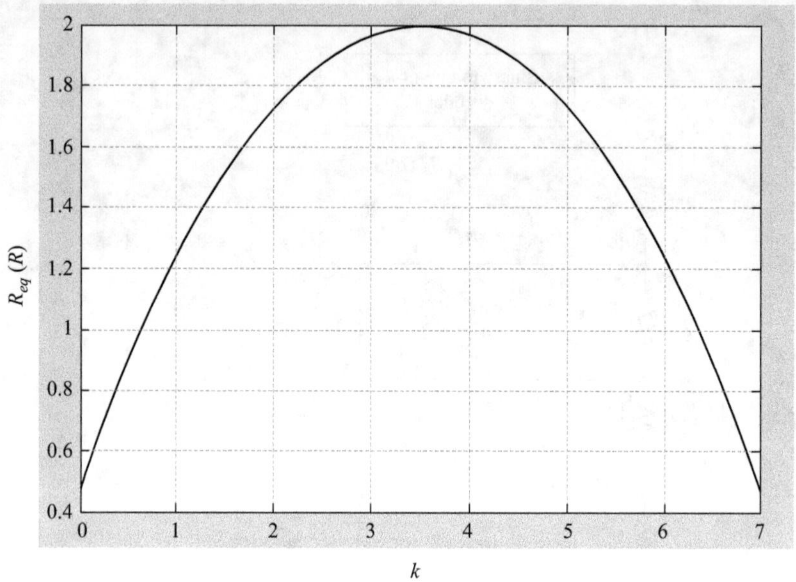

Figure 3.38 The equivalent resistance of the ladder as a function of the tap order

Therefore, the tap with the maximum resistance is given by

$$k|_{\max\_R_{eq}} = \frac{2^N - 1}{2} \tag{3.66}$$

The maximum equivalent resistance is

$$R_{eq}|_{max} = \frac{2^N}{4}R \tag{3.67}$$

That is, the maximum equivalent resistance increases exponentially with the number of bits and linearly with the unit resistance value. It is interesting to note that (3.66) gives a non-integer k value. Therefore, the maximum resistance values will be at the closest integer values of k given by

$$k|_{\max\_R_{eq}} \cong \frac{2^N}{2} \quad \text{or} \quad k|_{\max\_R_{eq}} \cong \frac{2^N}{2} - 1 \tag{3.68}$$

The maximum equivalent resistance is given by

$$R_{eq}|_{max} = \left(\frac{2^N}{4} - \frac{1}{4 \times 2^N}\right)R \cong \frac{2^N}{4}R \tag{3.69}$$

The static linearity of this DAC depends on the resistor matching and the output buffer linearity. Different types of resistors have different matching characteristics. Good layout practices can improve matching to about 8-bit accuracy.

In addition, trim and calibration can be used to improve matching further using thin film resistors or using switched resistances with fuses (or anti-fuses) to program them in or out. An example of a resistance ladder layout is shown in Figure 3.3 in the context of the flash ADC, where the resistors are made of identical units.

Similar to the resistance ladder used in flash ADCs, the mismatch's impact tends to accumulate as the taps approach the middle of the ladder, with the two ends being ideal. If the mismatch is random, the accumulation is uncorrelated. However, gradient effects lead to mismatches whose impact increases progressively toward the middle.

For a gradient error, the value of the kth resistance in the ladder R_k can be expressed as

$$R_k = R + k\Delta R \tag{3.70}$$

where $k = 1$ to $2^N - 1$, R is the unit resistance and ΔR is the mismatch between two neighboring resistances in the ladder. The kth output voltage is given by

$$V_k = V_{Ref}\left(\frac{\frac{R}{2}+\frac{\Delta R}{2}+\sum_{i=1}^{i=k}(R+i\Delta R)}{R+\left(1+2^N\right)\frac{\Delta R}{2}+\sum_{i=1}^{i=2^N-1}(R+i\Delta R)}\right) \tag{3.71}$$

which gives

$$V_k = \frac{V_{Ref}}{R_t}\left[\frac{R}{2}+\frac{\Delta R}{2}+\left(R+\frac{\Delta R}{2}\right)k+\frac{\Delta R}{2}k^2\right] \tag{3.72}$$

where V_k is the kth tap voltage, N is the number of bits, R is the unit resistance, and R_t is the total resistance of the ladder, which is given by

$$R_t = 2^N R + \frac{2^{2N}\Delta R}{2} + \frac{\Delta R}{2} \tag{3.73}$$

Equation (3.72) indicates a parabolic relation between the tap voltages and the tap order k. The resulting INL error can be shown to be parabolic too, as shown in Figure 3.4. As discussed in Chapter 2, the INL is given by (2.76) to be

$$INL(k) = \frac{V_k - V_{id_k}}{LSB} = \frac{V_k - \frac{V_{Ref}}{R_t}\left(\left(R+\Delta R 2^{N-1}\right)k + \frac{(R+\Delta R)}{2}\right)}{LSB} \tag{3.74}$$

where V_{id_k} is the ideal tap voltage, and LSB is the size of the DAC's LSB, which is given by $V_{Ref}/2^N$. Substituting (3.72) into (3.74) gives

$$INL(k) = \frac{\frac{V_{Ref}\Delta R}{2R_t}\left(\left(1-2^N\right)k + k^2\right)}{LSB} \tag{3.75}$$

Equating the derivative to zero to find the maximum gives

$$\frac{\partial INL(k)}{\partial k} = \frac{\frac{V_{Ref}\Delta R}{2R_t}((1-2^N)+2k)}{LSB} = 0$$

Therefore, the maximum INL occurs at

$$k|_{\text{max\_INL}} = \frac{2^N - 1}{2} \tag{3.76}$$

which is identical to (3.66) for the value of k that gives the maximum equivalent resistance. The magnitude of the maximum INL error is given by

$$INL = \frac{V_{Ref} \times \Delta R \times (2^N - 1)^2}{8R_t \times LSB} \tag{3.77}$$

Therefore, the maximum INL in volts is

$$INL = \frac{I \times \Delta R \times (2^N - 1)^2}{8} \cong \frac{I \times \Delta R \times 2^{2N}}{8} \tag{3.78}$$

where I is the current in the ladder, which is equal to V_{Ref}/R_t. Since the LSB is equal to $V_{Ref}/2^N$, and R_t is approximately equal to $R \times 2^N$, the maximum INL in LSBs is given by

$$INL(LSB) \cong \frac{(2^N - 1)^2 \times \Delta R/R}{8} \cong \frac{2^{2N} \times \Delta R/R}{8} \tag{3.79}$$

Therefore, from (3.78) and (3.79), we see that the INL error is proportional to the square of the number of levels 2^N.

Since the value given by (3.76) is not an integer, it can be approximated by

$$k|_{\text{max\_INL}} \cong \frac{2^N}{2} \quad \text{or} \quad k|_{\text{max\_INL}} \cong \frac{2^N}{2} - 1 \tag{3.80}$$

The resulting maximum INL in volts will still be given by

$$INL \cong \frac{I \times \Delta R \times 2^{2N}}{8} \tag{3.81}$$

The maximum INL in LSBs is still

$$INL(LSB) \cong \frac{2^{2N} \times \Delta R/R}{8} \tag{3.82}$$

To reduce the INL, a folded ladder can be used as shown in Figure 3.40. In this case, since the two halves are matched, the gradient error is minimized at the

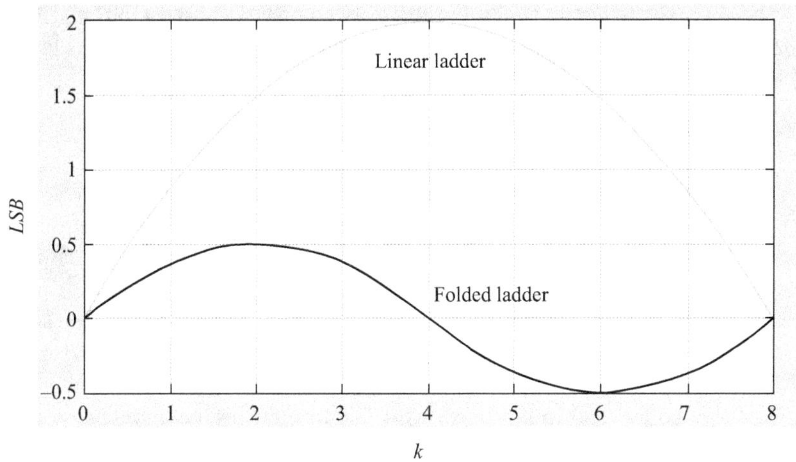

Figure 3.39 INL error as a function of the tap order for a linear ladder and a folded ladder

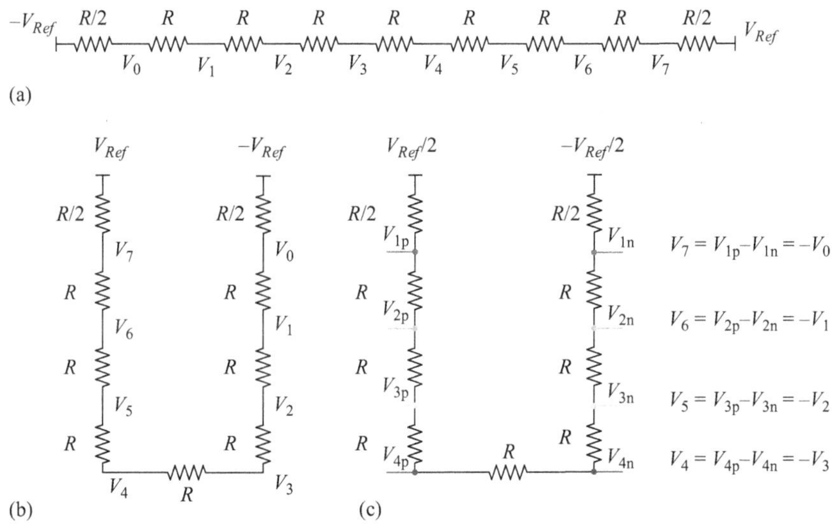

Figure 3.40 (a) A DAC linear resistor ladder. (b) A DAC folded ladder. (c) A DAC folded ladder with differential taps

midpoint, which reduces the maximum INL as shown in Figure 3.39. Since the INL is proportional to the square of the number of taps, folding the ladder leads to a 4× reduction in the peak INL because of the 2× reduction in the number of taps. Additionally, using a differential signal helps eliminate first-order gradients in the ladder and even-order harmonics.

Random mismatches between the resistors contribute to the DNL and INL of the DAC. This mismatch is inversely proportional to the resistance area as given by

$$\frac{\sigma_R^2}{R^2} = \frac{A_R^2}{WL} \tag{3.83}$$

where $\sigma_R{}^2$ is the variance of the resistance, R is the nominal resistance value, A_R is a mismatch constant, W is the resistance width, and L is its length. Matching can be improved by increasing the size of the resistance, which increases the parasitics and degrades the high-frequency performance.

The speed of the resistance ladder DAC is determined by the ladder's equivalent resistance value, the parasitic capacitances, and the switch resistance. In addition, the output buffer bandwidth can also limit the speed. The taps near the middle of the ladder tend to be the slowest due to their high equivalent resistance compared to the taps near the ends. To increase the settling speed, a smaller resistance value is needed, which increases the power consumption in the ladder. Since the maximum equivalent resistance increases exponentially with the number of bits, for a certain settling speed, the power consumption will increase exponentially with the number of bits too. Moreover, the ladder size and the number of switches increase exponentially with the number of bits (2^N), which limits the application of this architecture to low resolutions and low speeds.

3.11.1.2 R-2R Ladder DAC

In this structure, an R-2R ladder is used, as shown in Figure 3.41, to enable higher speeds compared to the resistive divider DAC. For N bits of resolution, this architecture requires only $2N$ resistors instead of 2^N resistors. Each additional bit requires an additional R-2R cell. The output can be voltage or current as shown in Figures 3.41 and 3.42, respectively. The output voltage of the R-2R DAC requires a buffer or an operational amplifier to drive the load. The current mode ladder can also be used with an operational amplifier to drive a load.

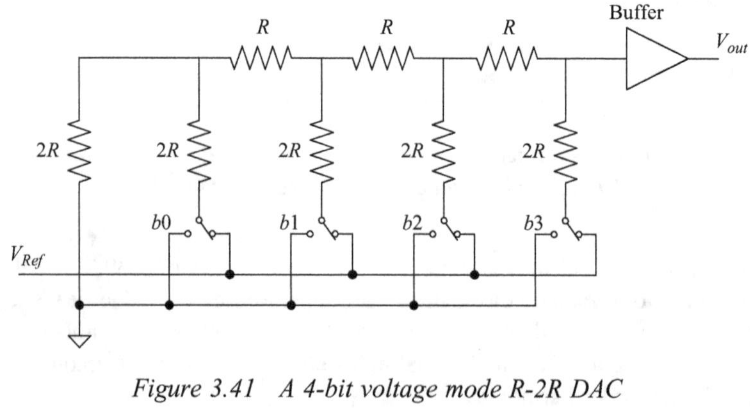

Figure 3.41 A 4-bit voltage mode R-2R DAC

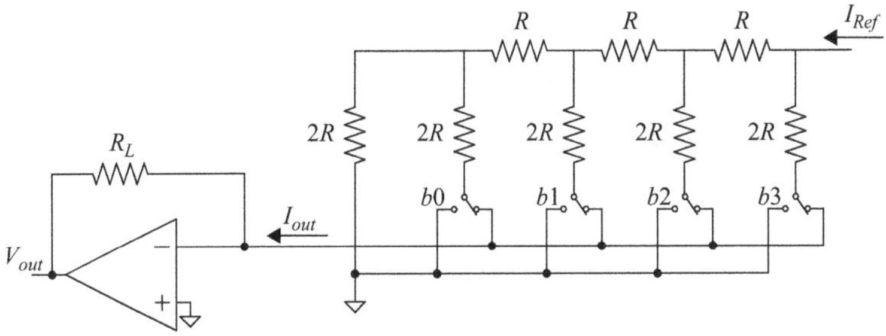

Figure 3.42 A 4-bit current mode R-2R DAC

In the voltage mode, the operation of the ladder can be illustrated using the superposition of N voltage sources (V_{Ref0}, V_{Ref1}, ..., $V_{Ref(N-1)}$). For V_{Ref0}, the Thevenin equivalent of the leftmost R-2R cell in Figure 3.41 is given by

$$R_{Th} = R, V_{Th} = \frac{V_{Ref0}}{2} \tag{3.84}$$

Applying this successively, the contribution of V_{Ref0} to the output voltage is given by

$$V_{out0} = \frac{V_{Ref0}}{2^N} \tag{3.85}$$

Similarly, the contribution of V_{Ref1} is

$$V_{out1} = \frac{V_{Ref1}}{2^{N-1}} \tag{3.86}$$

The contribution of $V_{Ref(N-1)}$ is

$$V_{out(N-1)} = \frac{V_{Ref(N-1)}}{2} \tag{3.87}$$

Therefore, the output is the binary combination of the reference voltages in all the branches. If all the reference voltages are equal to V_{Ref} and controlled by a corresponding bit to be either V_{Ref} or ground, then the output can be represented as

$$V_{out} = \sum_{i=0}^{N-1} \frac{b_i V_{Ref}}{2^{N-i}} \tag{3.88}$$

where b_i is the ith DAC bit, with b_0 the LSB and b_{N-1} the MSB.

Similarly, in current mode implementations, the output current is given by

$$I_{out} = \sum_{i=0}^{N-1} \frac{b_i I_{Ref}}{2^{N-i}} \tag{3.89}$$

Unlike the resistive divider DAC, the R-2R DAC is not inherently monotonic. The error can be worst in the middle code (binary transition). Additionally, the loading on the reference buffer depends on the digital code, which degrades its linearity. The switches can impact the DAC linearity and the speed is determined by the resistance values and the parasitics capacitances, in addition to the output buffer.

3.11.2 Capacitive DAC

Capacitive DACs use arrays of capacitors and charge sharing to convert a digital code into an analog output. An example of a capacitive DAC is depicted in Figure 3.43. During the reset phase, the capacitances are discharged by connecting them to ground. During the DAC phase, each capacitance is connected to V_{Ref} or ground depending on the DAC code. The output voltage is given by

$$V_{out} = \frac{\sum_{i=0}^{N-1} b_i C_i V_{Ref}}{\sum_{i=0}^{N} C_i} \tag{3.90}$$

where N is the number of bits, b_i is the ith DAC bit, with b_0 the LSB and b_{N-1} the MSB. C_i is the ith DAC capacitance, with $C_i = 2^i C$ for $i = 0$ to $N-1$, and $C_N = C$.

Capacitances usually have better matching than resistors and are compatible with switched capacitor circuits. Switched capacitor implementations with opamps or OTAs are usually employed for these DACs, which is discussed in more detail in Chapter 6. Typically, polysilicon, MIM, or MOM capacitances are preferred for their better matching characteristics and superior linearity. Polysilicon capacitors are formed by two polysilicon layers with a thin insulator layer in between. The insulator is preferred to be an oxide layer. With careful layout, they can achieve good matching in the order of 14- to 16-bit accuracy [1, 7] for a 125 fF unit and 750 fF total capacitance.

MIM capacitors are formed by two metal layers with a thin insulator layer in between. One of the metal layers is usually a special layer that is inserted between the regular metal layers of the process. The thin insulator is well controlled to achieve high-quality capacitances with good matching. These capacitors can achieve matching in the 10- to 12-bit range for a 50 fF unit and 200 fF total capacitance [2].

For high speed applications, capacitors with poor dielectric relaxation/absorption should be avoided, because they can lead to memory errors. Typically, oxide capacitances have better dielectric relaxation/absorption behavior than SiN capacitances [10, 11].

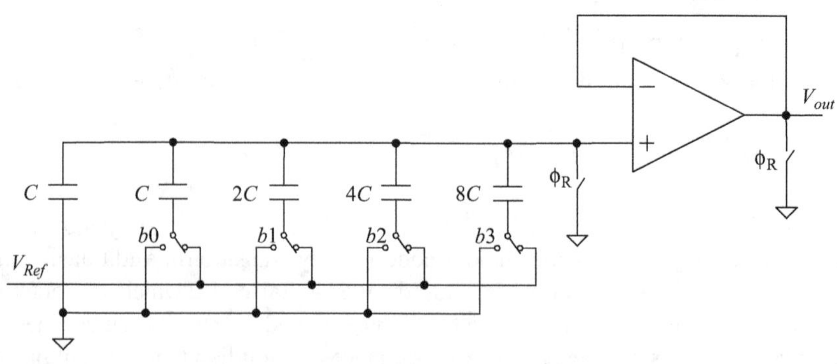

Figure 3.43 A 4-bit binary capacitive DAC

Figure 3.44 *An example of a MOM capacitance formed between nodes A and B using metal routing on the same layer. The fingers can be extended to multiple layers of metal*

MOM capacitors refer to capacitances formed using the process's field oxide as an insulator. They can be formed vertically using two regular metal layers with oxide in between. This structure, however, suffers from a thick and poorly controlled insulator, which leads to low density and poor matching. Alternatively, the MOM capacitors can be formed by inter-digitating metal "fingers" to obtain lateral capacitances on the same metal layer, as shown in Figure 3.44. By repeating the structure on multiple metal layers, high density can be achieved. This type of MOM capacitance can achieve matching that is comparable to the MIM capacitance, with a much higher density.

The random mismatch of capacitors is inversely proportional to their area and is given by

$$\frac{\sigma_C^2}{C^2} = \frac{A_c^2}{WL} \tag{3.91}$$

where $\sigma_C{}^2$ is the variance of the capacitance value, C is the nominal capacitance value, A_C is a mismatch constant, W is the capacitor's width, and L is its length. Matching can be improved by increasing the size of the capacitance, which increases the power consumption and degrades the high-frequency performance.

For the capacitive DAC shown in Figure 3.43, the linearity is determined by the capacitance mismatch and the input capacitance of the buffer, which tends to be voltage dependent. The variation of the summing node voltage with the DAC code degrades the DAC's linearity because of the buffer's variable input capacitance. This is illustrated in the expression of the output voltage, which is given by

$$V_{out} = \frac{\sum_{i=0}^{N-1} b_i C_i V_{Ref}}{\sum_{i=0}^{N} C_i + \sum_{i=0}^{N} C_{pi} + C_p} \tag{3.92}$$

where C_i is the DAC capacitances, C_{pi} is the parasitic capacitances of the DAC capacitances, and C_p is the buffer's input capacitance. The variation of C_p with the DAC code will cause non-linearity in the output voltage.

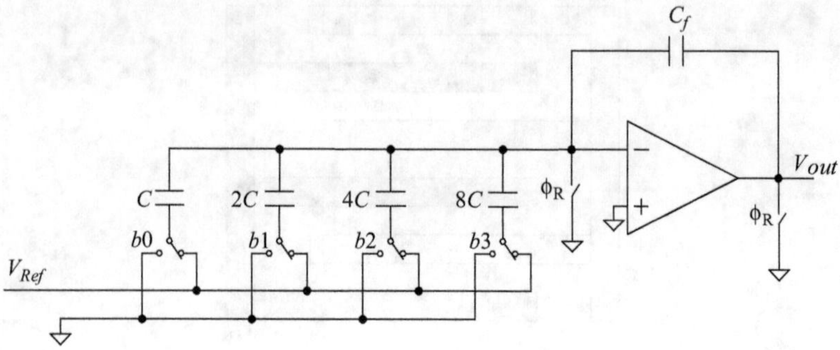

Figure 3.45 A 4-bit binary capacitive DAC with improved linearity

Alternatively, the capacitive DAC can be implemented as shown in Figure 3.45. The summing node voltage is at virtual ground, which substantially reduces the non-linear effect of the buffer's input capacitance. During the reset phase, ϕ_R is high and all the capacitors are discharged, while during the DAC phase, the capacitors are connected to the V_{Ref} or ground depending on the DAC code. The output voltage is given by

$$V_{out} = \frac{-\sum_{i=0}^{N-1} b_i C_i V_{Ref}}{C_f} \tag{3.93}$$

where C_f is the feedback capacitance. The negative sign is due to the switched capacitors being connected to the negative input of the amplifier, as discussed in more detail in Chapter 6. Alternatively, the phasing can be reversed where the DAC voltages are sampled on the capacitors when ϕ_R is high, and the charge redistribution is performed when ϕ_R is low. In this case, the DAC output is positive, and will be given by

$$V_{out} = \frac{\sum_{i=0}^{N-1} b_i C_i V_{Ref}}{C_f} \tag{3.94}$$

The binary DAC can be non-monotonic at the major code transitions because of the switching of different capacitors. Instead of binary coding, thermometer coding can be used to give *unary* DACs that consist of equal capacitors as shown in Figure 3.46. The thermometer code determines the number of capacitances connected to V_{Ref}, as opposed to ground. This DAC is inherently monotonic and smaller in size than a binary DAC. It, however, requires a larger number of control bits (2^N), as opposed to N bits for the binary DAC. The output of the unary DAC is given by

$$V_{out} = \frac{-\sum_{i=1}^{2^N} B_i C V_{Ref}}{C_f} = \frac{-\sum_{i=0}^{N-1} b_i 2^i C V_{Ref}}{C_f} \tag{3.95}$$

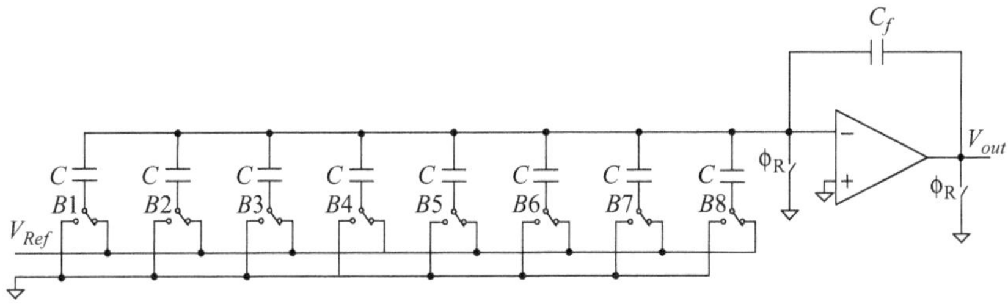

Figure 3.46 A 3-bit unary capacitive DAC with better linearity

where B_i is the thermometer code ranging from 0 to $2^N - 1$. The feedback capacitance can be set to any desired value to achieve an arbitrary gain or attenuation. If $C_f = 2^N C$, the output becomes

$$V_{out} = \frac{-\sum_{i=0}^{N-1} b_i 2^i V_{Ref}}{2^N} \tag{3.96}$$

It is important to note that using 2^N capacitors enables us to have a $(2^N + 1)$-level DAC if we utilize the code: 2^N. This will be employed in Chapter 7 in the context of the pipelined ADC's MDAC. Alternatively, we can use $2^N - 1$ capacitors, instead of 2^N, where we utilize the zero level of the thermometer code, and the output will still be given by

$$V_{out} = \frac{-\sum_{i=1}^{2^N} B_i C V_{Ref}}{C_f} \tag{3.97}$$

where B_i is the thermometer code ranging from 0 to $2^N - 1$. If the DAC voltages are applied when ϕ_R is high and the charge redistribution occurs when $\phi_R = 0$, the output polarity will be reversed, such that

$$V_{out} = \frac{\sum_{i=1}^{2^N} B_i C V_{Ref}}{C_f} = \frac{\sum_{i=0}^{N-1} b_i 2^i C V_{Ref}}{C_f} \tag{3.98}$$

A differential implementation of a unary DAC is shown in Figure 3.47, where the DAC has two levels ($V_{Ref}/2$ and $-V_{Ref}/2$). During $\phi 1$, all the capacitances are reset. During $\phi 2$, the capacitances are connected to either $V_{Ref}/2$ or $-V_{Ref}/2$, depending on the DAC thermometer code D and its inverse D'. The polarity of the output is determined by the polarity of the code (D or D') or the reference ($V_{Ref}/2$ or $-V_{Ref}/2$). The structure shown in Figure 3.47, gives a negative output. Swapping D and D', or $V_{Ref}/2$ and $-V_{Ref}/2$, will reverse the output polarity.

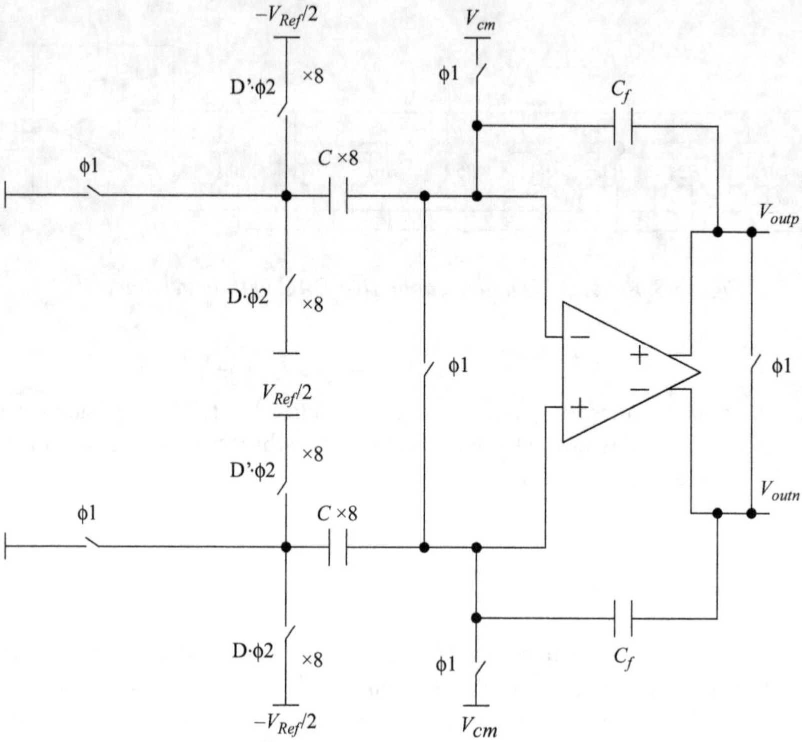

Figure 3.47 A differential implementation of the 3-bit unary capacitive DAC

It is interesting to note that the structures shown in Figures 3.45–3.47 can be used to amplify the DAC code. This is called a multiplying DAC (MDAC) and is commonly used in pipelined ADCs. The gain G of the MDAC is given by:

$$G = \frac{\sum_{i=1}^{2^N} C_i}{C_f} \tag{3.99}$$

where the capacitances C_i are the DAC capacitances whose values are nominally equal to C, and C_f is the feedback capacitance.

In the capacitive DAC structures discussed so far, the number and/or size of the DAC capacitors increases exponentially with the number of bits. For high resolutions, other architectures can be used to reduce the area and power implications. An example is shown in Figure 3.48, where an attenuation capacitance C_B is used to reduce the number and size of the DAC capacitances. For N bits, instead of requiring $2^N - 1$ unit capacitances, we can use $2 \times (2^{N/2} - 1)$ capacitances. The series combination of the attenuation capacitance C_B and the left array should be equivalent to one unit capacitance C in the right array [5, 6].

3.11.3 Current steering DAC

This represents the fastest DAC structure, where the DAC element is a current cell instead of a resistance or a capacitance. The current elements can be bipolar or MOS transistors. They rely on current sources with high output impedances that are typically implemented using cascode current sources. Current steering is controlled by the digital code as shown in Figure 3.49, such that

$$I_{out} = \sum_{i=0}^{N-1} b_i 2^i I \tag{3.100}$$

In very high speed applications, the output current can be used directly in the load as shown in Figure 3.50. In low or medium speed cases, an I-to-V converter can be used to drive the load, which is depicted in Figure 3.51.

The accuracy of this DAC structure is limited by the modulation of the current with the output voltage because of the finite output impedance of the current cells. The output impedances are depicted in Figures 3.50 and 3.51. If the DAC is driving the load directly as shown in Figure 3.50, the output voltage will be given by [5]

$$V_{out}(k) = kI\frac{R_L R_o/k}{R_L + \dfrac{R_o}{k}} \tag{3.101}$$

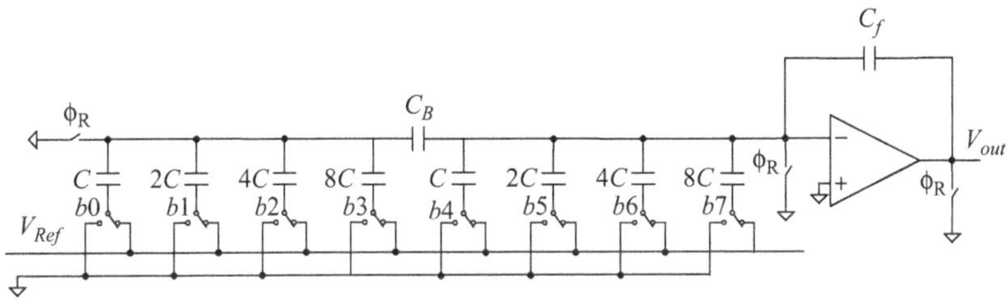

Figure 3.48 A binary capacitive DAC with attenuation capacitance C_B

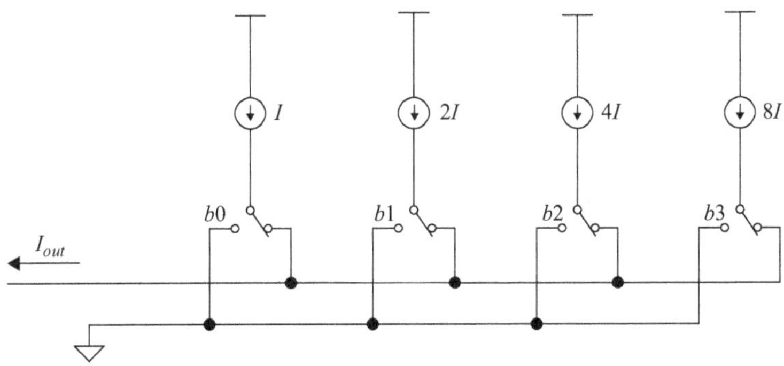

Figure 3.49 An ideal model of a 4-bit current steering DAC

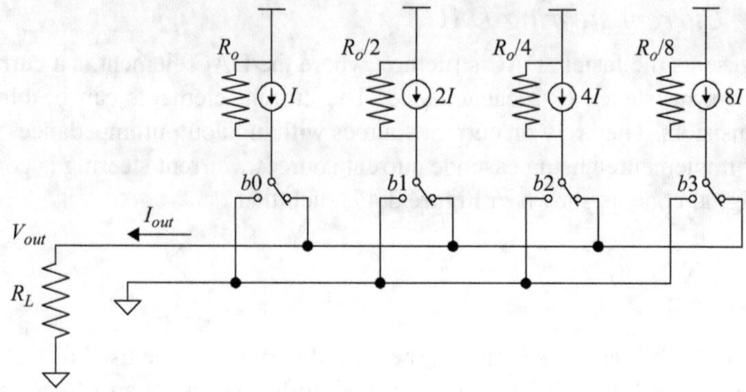

Figure 3.50 A simple model of a 4-bit current steering DAC including the output impedances of the currents sources

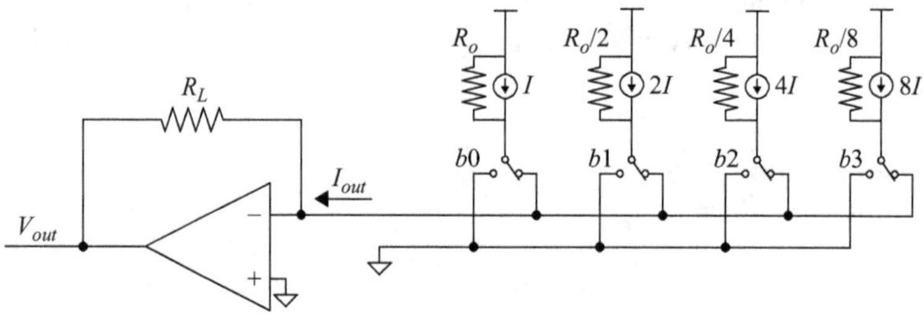

Figure 3.51 A current steering 4-bit DAC with an I-to-V buffer

where R_o is the output resistance of the unit current cell, R_L is the load resistance, and k represents the number of unit cells that are switched on, and is given by $k = 0,1,\ldots,2^N - 1$. Therefore, the output voltage will be

$$V_{out}(k) = IR_L \frac{k}{1 + \frac{kR_L}{R_o}} \tag{3.102}$$

Therefore, the finite output resistance leads to non-linearity that approaches zero as R_o becomes much larger than R_L. It is dominated by a second-order non-linearity, but higher orders exist as well to a lesser degree. Differential operation reduces the second-order distortion and the resulting $HD2$ due to the current modulation.

Another example of a current steering DAC is shown in Figure 3.52, where a unary implementation is employed. In this case, the number of current cells is proportional to 2^N and a thermometer code is used. Therefore, monotonicity is ensured and linearity is better than a binary implementation. However, it requires a large number of control bits for the switches (2^N). On the other hand, the binary coded implementation shown in Figure 3.51 is larger in size, does not guarantee monotonicity, and has poorer overall linearity. However, its control bits are reduced to N bits.

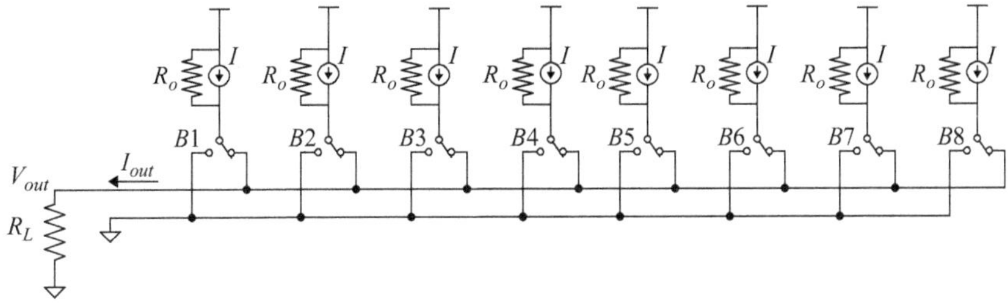

Figure 3.52 A unary current steering 3-bit DAC with a resistance load

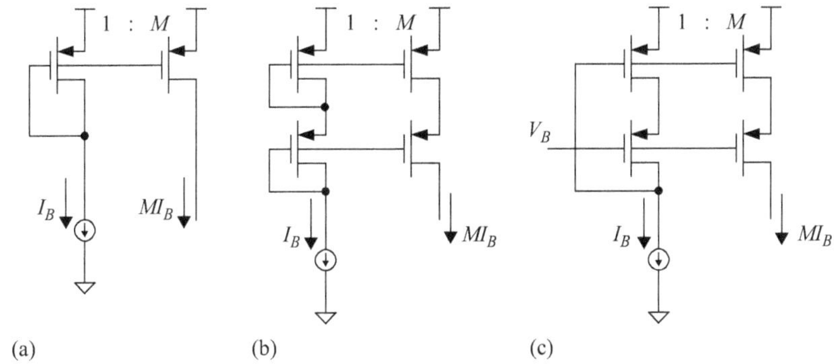

Figure 3.53 Examples of current mirrors. (a) Simple current mirror. (b) A cascoded current mirror. (c) A cascode current mirror with enhanced dynamic range

Some implementations of the current sources are shown in Figure 3.53. Figure 3.53(a) shows a simple current mirror. The output impedance is given by the output impedance of the current source device. Figure 3.53(b) is a cascoded current mirror, where the impedance of the current source is increased by about $g_m R_{ds}$ of the cascode device. The cascode device reduces the dynamic range, where the maximum voltage that keeps the devices in saturation is reduced from $V_{DD} - V_{SD}|_{\text{sat}}$ to

$$V_{out}|_{\max} = V_{DD} - \left(V_{SD}|_{\text{sat}} + V_{SG}\right) \tag{3.103}$$

Figure 3.53(c) is a cascoded current source, where the bias voltage V_B is chosen to improve the headroom, such that the maximum output voltage is given by

$$V_{out}|_{\max} = V_{DD} - 2V_{SD}|_{\text{sat}} \tag{3.104}$$

It is desirable to set the bias voltage V_B to maximize the dynamic range and minimize the over-drive used by the cascode device across process, supply, and temperature. Examples of techniques that can be employed to generate this bias voltage are described in Reference 15.

Another source of error in current steering DACs is the random mismatch between the current cells. This is typically inversely proportional to their size and is given by

$$\frac{\sigma_I^2}{I^2} = \frac{\sigma_\beta^2}{\beta^2} + \sigma_{V_T}^2 \left(\frac{g_m}{I}\right)^2 \tag{3.105}$$

where

$$\frac{\sigma_\beta^2}{\beta^2} = \frac{A_\beta^2}{WL} \tag{3.106}$$

and

$$\sigma_{V_T}^2 = \frac{A_{V_T}^2}{WL} \tag{3.107}$$

where g_m is the transconductance, $\beta = \mu C_{ox} W/L$, W is the device width, L is the length, I is the current, and the parameters A_β and A_{V_T} are mismatch constants. It is interesting to note that the effect of the threshold voltage mismatch can be reduced by reducing the g_m/I. That is by increasing $V_{gs} - V_T$, which in turn increases the overdrive of the device. This however reduces the dynamic range and can degrade linearity. In addition, increasing the area will reduce the mismatch, but it will increase the parasitics and hence reduce the speed.

An example of a current switching/steering structure is shown in Figure 3.54. The speed of this structure is determined by the output impedance of the current sources. Since current steering can be accomplished at very high speeds, this structure can typically achieve the highest speed of all the DAC structures. However, it also tends to consume the highest power.

It is important to note that the switching "off" of the current cell is achieved by steering the current away from the output path to an alternative path. The current should not be turned completely off, otherwise it will require a relatively long time to turn back on, and will degrade the speed and linearity. Therefore, the two switches are designed to overlap in their ON state. For the PMOS switches shown in Figure 3.54(a), this can be achieved using non-overlap clock generators that produce the timing diagram shown in Figure 3.54(b). The two clocks overlap in their low state, which ensures that the PMOS devices will overlap in their ON state.

3.12 Conclusion

In this chapter, we presented an overview of some of the most commonly used high speed ADC and DAC architectures. An important observation from this discussion is that speed and performance tend be fundamentally opposing parameters. Architectures and design decisions that are meant to improve performance tend to

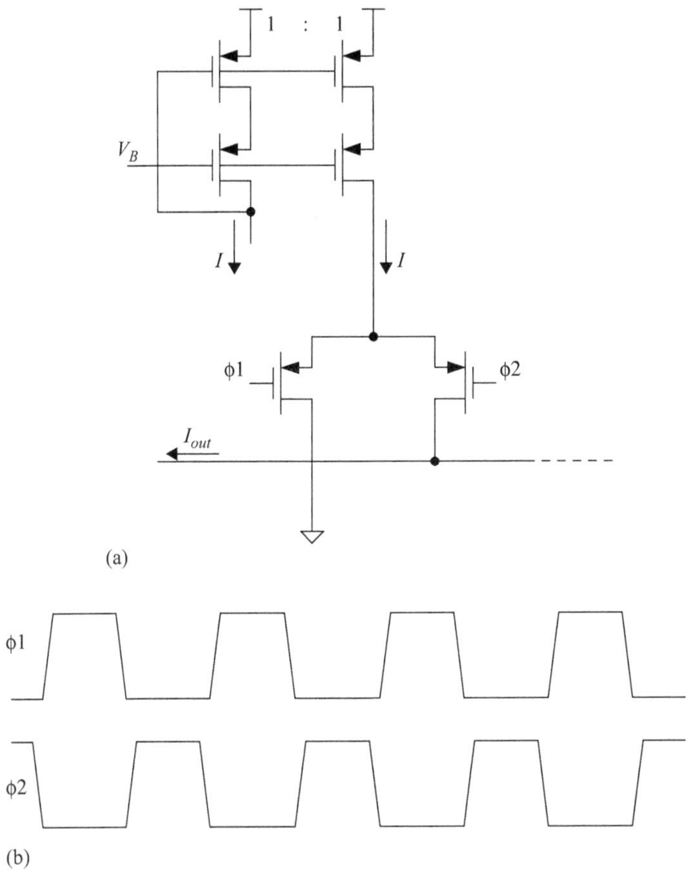

Figure 3.54 (a) A current steering circuit. (b) Timing diagram

degrade speed, and vice versa. This trend is fundamental and makes it most challenging to achieve both high performance and high speed simultaneously.

Some architectures are capable of achieving this trade-off better than others. For example, pipelined ADC have recently proven to be a formidable architecture in the high performance and high speed space. Whereas ΣΔ and SAR ADCs have achieved some recent successes, the intrinsic higher speed of pipelined ADCs give them a substantial advantage.

Another design trend is to use calibration and digital assistance to add another degree of freedom that enables higher accuracy compared to what is achievable otherwise. This applies to both ADCs and DACs. Calibration allows the analog designer to optimize for speed or power consumption on the analog side while relaxing the accuracy requirements that can be calibrated digitally. This has become more important as we migrate to fine lithography CMOS processes where the devices are inherently less analog friendly.

Problems

1. Which ADC architecture would you use for each of the following, and why?
   (a) A 16-bit, 10 MS/s ADC?
   (b) A 5-bit, 1 GS/s ADC?
   (c) A 10-bit, 100 MS/s ADC?
   (d) A 14-bit, 200 MS/s ADC?
   (e) A 6-bit, 20 GS/s ADC?
2. Consider implementing a 14-bit, 100 MS/s ADC using the following architectures:
   (a) Pipeline
   (b) SAR
   (c) Sigma-delta
   (d) Folding

   What are the pros and cons of using each architecture? Which architecture would you choose, and why?
3. Consider implementing a 6-bit, 10 GS/s ADC using the following architectures:
   (a) Flash
   (b) Flash with interpolation
   (c) Time-interleaving
   (d) Pipeline

   What are the pros and cons of using each architecture? Which architecture would you choose, and why?
4. For a 5-bit flash ADC:
   (a) Draw a simplified schematic of the flash ADC.
   (b) What is the number of comparators needed?
   (c) What is the number of resistance elements in the ladder?
5. For a 5-bit flash ADC:
   (a) For a 1% mismatch in the unit resistance value due to gradient, what is the maximum INL if a linear ladder is used?
   (b) What is the maximum INL if a folded ladder is used?
   (c) Plot the INL in both cases above.
6. For a 5-bit flash ADC with 0.5 LSB of accuracy, assume $A_R = 6$nm and $A_{VT} = 1$mV.μm. Ignoring other sources of errors, what do you expect the sizes of the ladder unit resistance and the input devices of the comparator's pre-amplifier to be in order to achieve this accuracy on a 0.18 μm CMOS process?
7. If interpolation by a factor of 4 is used for a 6-bit flash, where each comparator is composed of one pre-amplifier and one latch, draw the flash structure? How many preamplifiers are used? How many are saved compared to a straight flash?
8. For a 6-bit flash whose comparators are composed of a cascade of two pre-amplifiers and a latch, draw the block diagram. Employ two layer interpolation with a factor of 4 each, and draw the block diagram of the interpolated flash ADC. How many preamplifiers are saved?

9. A 6-bit flash employs comparators whose latches have a g_m/C of 20 Grad/s. The flash is operating at 1 GS/s. What is the estimated BER? How does the BER change if the sampling rate is increased to 10 GS/s?
10. A sub-ranging ADC consists of 2 stages, each has a resolution of 5 bits, with 1-bit redundancy. What is the resolution of the whole ADC? If there is no inter-stage amplification, what are the required accuracies of the first-stage and the second-stage sub-ADCs, if the whole ADC is to have an accuracy of 0.5 LSB?
11. A sub-ranging ADC consists of 2 stages, each has a resolution of 5 bits, with 1-bit redundancy. An inter-stage amplifier of gain 16 is used. What is the resolution of the whole ADC? What are the required accuracies of the first-stage and the second-stage sub-ADCs, if the whole ADC is to have an accuracy of 0.5 LSB? What is the required accuracy of the amplifier?
12. For the sub-ranging ADC of the previous problem, draw the residue of the first stage after amplification as a function of the input voltage for two cases: a mid-tread sub-ADC and mid-rise sub-ADC. What are the differences between the two cases? Reconstruct the final output word using the bits from the two stages in both cases.
13. For a 6-bit folding ADC with 4 bits in the first stage: draw the folded residue versus the input voltage. Draw the block diagram of the ADC and the folding circuit needed using MOS current mirrors. What is the resolution of the second stage if 1-bit redundancy is employed?
14. For a 5-bit SAR ADC:
    (a) Draw the DAC structure, timing diagram, and the DAC output as a function of time.
    (b) Express the output word as a function of the capacitance values.
    (c) Devise a method to employ redundancy.
15. For a 4-way time-interleaved ADC with a sampling rate of 1 GS/s, where will the interleaving spurs be located due to the different kinds of mismatches for an input signal of 200 MHz? Repeat the exercise if the input is composed of 2 tones at 200 MHz and 300MHz.
16. For a sigma-delta ADC, what is the expected SQNR in each of the following conditions:
    (a) 1-bit quantization, first order, $OSR = 32$?
    (b) 2-bit quantization, third order, $OSR = 4$?
    (c) 5-bit quantization, fifth order, $OSR = 8$?
    (d) How would you implement each of these? What are the pros and cons of each?
17. Using a behavioral modeling language, create a model to simulate a 6-bit SAR ADC operation. The model should include the comparator, DAC and SAR control. Introduce errors in the DAC and comparator, and comment on the impact quantitatively and qualitatively.
18. A 4-bit resistive ladder DAC with 1% gradient mismatch:
    (a) Express the analog output as a function of the DAC code.
    (b) Plot the INL versus code. What is the maximum INL?

(c) With folding the ladder, plot the INL versus code and calculate the maximum INL.
(d) Draw the DAC structure.

19. For a 5-bit R-2R ladder DAC:
    (a) Draw the DAC structure.
    (b) Express the output voltage as function of the DAC code.
    (c) Compare the implementation with a resistive ladder in terms of size, power, and linearity.
20. For a 5-bit capacitive DAC, draw the structure for the binary and unary cases. What are the pros and cons of each case?
21. A 5-bit current steering DAC composed of current elements of 100 μA each. Draw the DAC structure? What is the total power consumption if the power supply is 1.8 V?
22. For the 5-bit current steering DAC in the previous problem, what is the required output impedance if the target accuracy is 0.5 LSB, when driving a load of 100 ohms?
23. Analyze the 8-bit capacitive DAC structure of Figure 3.48 and express the output voltage as a function of the digital code. Derive the value of the capacitance C_B needed. Discuss.
24. Using a SPICE-like simulator, build a 3-bit capacitive DAC using 1 pF unit capacitance and simulate it in the time domain.
25. Using a SPICE-like simulator, build a 3-bit current-steering DAC using a 100 μA unit cell and simulate it in the time domain.

References

[1] A.M.A. Ali, A. Morgan, C. Dillon, *et al.*, "A 16-bit 250-MS/s IF Sampling Pipelined ADC with Background Calibration," *IEEE Journal of Solid-State Circuits*, 45(12), pp. 2602–2612, Dec 2010.

[2] A.M.A. Ali, H. Dinc, P. Bhoraskar, *et al.*, "A 14b 1GS/s RF Sampling Pipelined ADC with Background Calibration," *IEEE Journal of Solid-State Circuits*, 49(12), pp. 2857–2867, Dec 2014.

[3] Y. Dong, J. Zhao, W. Yang, *et al.*, "A 930mW 69-DR 465MHz-BW CT 1-2MASH ADC in 28nm CMOS," *IEEE ISSCC Digest of Technical Papers*, pp. 278–279, 2016.

[4] B. Razavi, *Principles of Data Conversion System Design*, IEEE Press, Piscataway, NJ, 1995.

[5] F. Maloberti, *Data Converters*, Springer, Dordrecht, The Netherlands, 2010.

[6] M.J.M. Pelgrom, *Analog-to-Digital Conversion*, Second Edition, Springer, Dordrecht, The Netherlands, 2013.

[7] A.M.A. Ali, C. Dillon, R. Sneed, *et al.*, "A 14-bit 125 MS/s IF/RF Sampling Pipelined ADC With 100 dB SFDR and 50 fs Jitter," *IEEE Journal of Solid-State Circuits*, 41(8), pp. 1846–1855, Aug 2006.

[8] K. Kattmann and J. Barrow, "A Technique for Reducing Differential Non-Linearity Errors in Flash A/D Converters," *IEEE ISSCC Digest of Technical Papers*, pp. 170–171, 1991.
[9] R. Schreier and G. Temes, *Understanding delta-sigma data converters*, IEEE Press, Piscataway, NJ, 2005.
[10] J.W. Fattaruso, M. De Wit, G. Warwar, *et al.*, "The Effect of Dielectric Relaxation on Charge-Redistribution A/D Converters," *IEEE Journal of Solid-State Circuits*, 25, pp. 1550–1561, Dec 1990.
[11] A. Zanchi, F. Tsay, and I. Papantonopoulos, *et al.*, "Impact of Dielectric Relaxation on a 14-bit 70-MS/s Pipeline ADC in 3-V BiCMOS," *IEEE Journal of Solid-State Circuits*, 38(12), pp. 2077–2086, Dec 2003.
[12] B.P. Ginsburg and A.P. Chandrakasan, "500-MS/s 5-bit ADC in 65-nm CMOS with Split Capacitor Array DAC," *IEEE Journal of Solid-State Circuits*, 42(4), pp. 739–747, 2007.
[13] M. van Elzakker, E. van Tuijl, P. Geraedts, *et al.*, "A 10-bit charge redistribution ADC consuming 1.9uW at 1MS/s," *IEEE Journal of Solid-State Circuits*, 45(5), pp. 1007–1015, 2010.
[14] A.M.A. Ali, H. Dinc, P. Bhoraskar, *et al.*, "A 14-bit 2.5GS/s and 5GS/s RF Sampling ADC with Background Calibration and Dither," *IEEE VLSI Circuits Symposium*, pp. 206–207, 2016.
[15] A.M.A. Ali, "Stably-biased cascode networks," US Patent 7,023,281, Apr 2006.
[16] G. Manganaro, *Advanced Data Converters,* Cambridge University Press, Cambridge, UK, 2012.
[17] R. Kapusta, J. Shen, S. Decker, *et al.*, "A 14b 80MS/s SAR ADC with 73.6dB SNDR in 65nm CMOS," *IEEE ISSCC Digest of Technical Papers*, pp. 472–473, 2013.
[18] M. Furuta and M. Nozawa, "A 10-bit, 40-MS/s, 1.21 mW pipelined SAR ADC using single-ended 1.5-bit/cycle conversion technique," *IEEE Journal of Solid State Circuits*, 46(6), pp. 1360–1370, Jun 2011.
[19] Y. Zhu, C.-H. Chan, S.-W. Sin, *et al.*, "A 50-fJ 10-b 160-MS/s pipelined-SAR ADC decoupled flip-around MDAC and self-embedded offset cancellation," *IEEE Journal of Solid State Circuits*, 47(11), pp. 2614–2626, Nov 2012.

Chapter 4
Sampling

Sampling the input signal is one of the most critical operations in the A/D conversion process. The sample-and-hold (S/H) circuit interfaces with the external analog world to convert the continuous-time signal into a discrete-time held signal. Typically, this means interacting with the package parasitics, the printed circuit board, the anti-aliasing filter, and the driver amplifier. Any distortion or noise introduced during sampling appears as is on the digital output and can be very difficult to undo.

As ADCs run faster with GS/s sampling rates, and the input frequencies rise up to the GHz range, the input sampling problem has become increasingly more challenging. The front-end sampler is tasked with sampling an input signal that is changing very rapidly in a very small amount of acquisition time with relatively high linearity and low noise added. In the literature, sampling linearity of better than 100 dB has been reached for input frequencies up to 100 MHz and sampling rates up to 125 MS/s [1]. More recently, sampling linearity of better than 80 dB up to 1 GHz input frequencies and 1 GS/s has been reported [2].

In this chapter, we discuss the challenges of input sampling in high speed ADCs. The focus will be on the design trade-offs and optimizations that enable high performance and can lead to further improvements in the future. Throughout this chapter, we present discussions and analyses of the non-linear behavior of the sampling circuits. They are meant to illustrate trends and behaviors in a simple and intuitive way, which necessitated making some assumptions and approximations, while avoiding overly complex mathematical analysis that may be more rigorous, but less insightful.

4.1 CMOS samplers

A simple S/H circuit, which is composed of a switch in series with a sampling capacitor, is shown in Figure 4.1. The switch is driven by a clock such that when the switch is on, the input signal is sampled on the capacitor for a duration that is called the *sampling*, *tracking* or *acquisition phase*. When the switch turns off, the sample is "taken" and held by the capacitor, ushering into the *hold phase*. The sampler is driven by a signal source with a finite source impedance Z_S, as shown in Figure 4.2, which plays an important role that influences all aspects of the sampling performance.

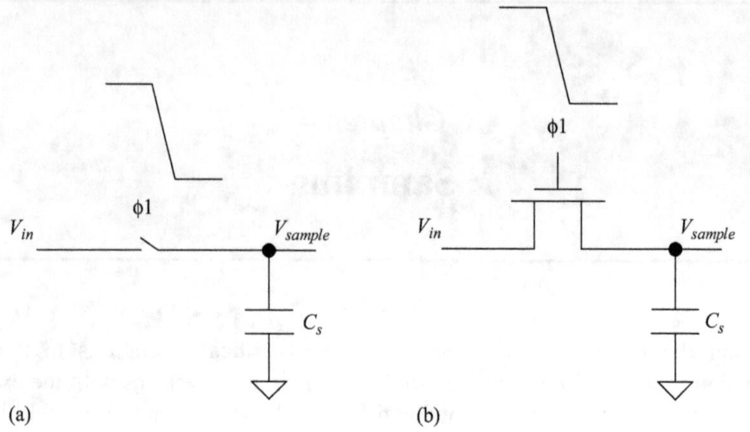

Figure 4.1 A simple sampling network with an input switch and a sampling capacitance, using (a) ideal switch, (b) NMOS switch

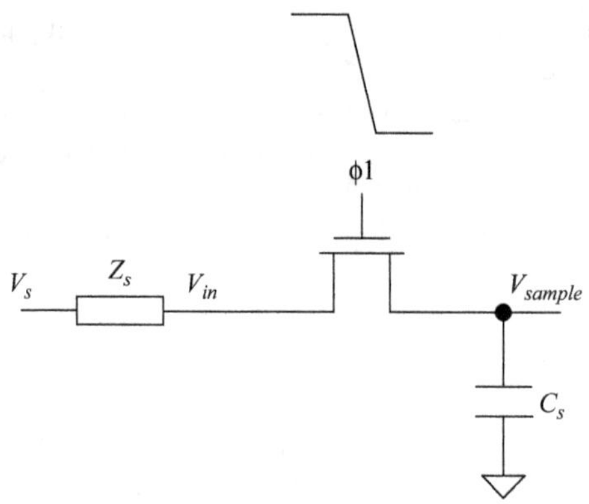

Figure 4.2 A simple sampling network with an input switch and a sampling capacitance, driven by an input source V_s with impedance Z_s

In addition to the sampling rate, two important performance metrics of the S/H circuit are:

1. noise
2. linearity

4.1.1 Sampling noise

The two main contributors to the sampling noise are the input path's noise (input noise), and the clock path's noise (jitter). The jitter noise is covered later in this chapter. The noise of the input path is due to the switch resistance's noise and other noisy components in the sampling path, including the source impedance. For the circuit shown in Figure 4.1 and ignoring the source impedance, if the resistance during the sampling (tracking) phase is R_{sw}, the noise power density S_n will be given by

$$S_n = 4kTR_{sw} \tag{4.1}$$

where k is Boltzmann's constant, which is approximately equal to 1.38×10^{-23} J/K, and T is the absolute temperature in Kelvins. For an RC network whose bandwidth is $B_{in} = 1/2\pi R_{sw}C_s$ and in which the resistor R_{sw} is the main noise contributor, the total noise power sampled on the capacitance C_s is

$$N = \int_0^\infty \frac{4kTR_{sw}}{1 + (2\pi f R_{sw}C_s)^2} df \tag{4.2}$$

Solving the integral gives

$$N = 4kTR_{sw} \times B_N = 4kTR_{sw}\left(\frac{\pi}{2}B_{in}\right) = 4kTR_{sw}\left(\frac{\pi/2}{2\pi R_{sw}C_s}\right)$$

where B_N is the effective noise bandwidth of a first-order filter of bandwidth B_{in}. Therefore, the integrated noise power is

$$N = \frac{kT}{C_s} \tag{4.3}$$

which is the famous "kT/C" formula for the sampling noise. We can see that the noise is independent of the resistance value, because the resistance R_{sw} that is the main noise contributor is itself the resistance that is limiting the bandwidth with the capacitance. If the noise generation and bandwidth limitation are decoupled from each other, this formula will not hold. For example, if there are sources of noise at the input that are dominating the total noise without impacting the bandwidth, the noise will be worse than kT/C. Conversely, if the bandwidth is limited by components that are not dominating the noise, the noise can be better than kT/C. In other words, kT/C is *not* a fundamental limit.

Having said that, the conclusion derived from the kT/C formula is practically valid and very important for high speed ADCs. That is, to improve the thermal noise we often need a larger sampling capacitance. This represents a significant limitation, as larger capacitances limit the speed, require more power, and are more difficult to drive. This indicates that noise is diametrically opposed to speed, linearity, and power consumption, which confirms the intuitive trends that are discussed in Chapter 2 in the context of ADC metrics and FOMs. In other words, it is more challenging to achieve high performance *and* high speed, than it is to achieve high performance *or* high speed alone.

4.1.2 Sampling linearity

Unlike thermal noise, the sampling linearity is impacted by many factors, some of which may be external to the ADC, and may not be properly modeled in simulations. The non-linear and multi-dimensional nature of the sampling operation make the design of a low distortion and high speed sampler as much of an art as it is a science. It is an excellent embodiment of an analog circuit design challenge.

Let's start with the simple sampler shown in Figure 4.1. The sampling linearity is degraded by two operations. The first is the input tracking, where the switch is on, and the circuit is not different from a continuous-time RC circuit. During this phase, some non-linearity exists that degrades the sampling distortion, even if the switch never turns off. The second operation is the transition from the sampling/tracking phase to the hold phase. This process of turning off the switch, and holding the signal, introduces additional possible sources of non-linearity that can degrade the performance further. One of these sources is the non-linear charge injection or kick-back, which can degrade the distortion if it is not given enough time to settle. Another source of non-linearity that occurs during the process of turning off the switch is the periodic jitter that may exist on the sampling clock. This jitter moves the sampling instant with time, modulating the sampled signal, and hence causing distortion.

Therefore, there are three contributors to the sampling non-linearity:

1. the tracking distortion,
2. the charge injection (or kick-back) distortion, and
3. the periodic jitter

We will cover the first two distortion mechanisms in this section, while the jitter will be covered separately in Section 4.4.

4.1.2.1 Tracking distortion

In Figures 4.3 and 4.4, the input switch's resistance and parasitic capacitances will vary with the input signal. This variation causes non-linear distortion during tracking, even if the switch is always on. This can be shown as follows:

$$V_{sample} = V_{in}\frac{1/sC_s}{R_{sw} + 1/sC_s} = V_{in}\frac{1}{1 + sC_sR_{sw}} \tag{4.4}$$

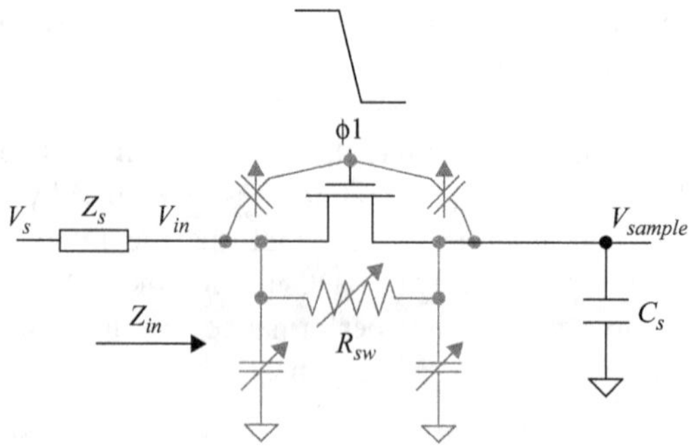

Figure 4.3 A simple sampling network with an input switch and a sampling capacitance, driven by a source with impedance Z_s. The switch parasitics are shown. The switch is clocked

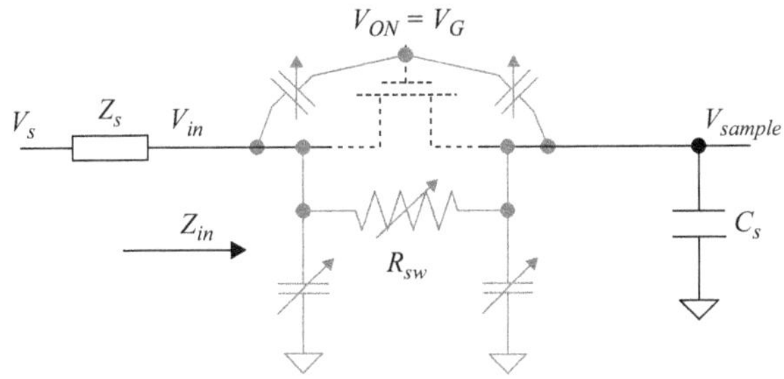

Figure 4.4 A simple sampling network during the sampling/tracking phase, with an input switch and a sampling capacitance, driven by a source with impedance Z_s. The switch parasitics are shown

where C_s is the sampling capacitance, and R_{sw} is the switch's resistance. If we ignore the source impedance and the parasitic capacitances, the input sampling bandwidth (B_{in}) will be given by

$$B_{in} = \frac{1}{2\pi R_{sw} C_s} \tag{4.5}$$

It is important to note that the input sampling bandwidth (B_{in}) is different from the ADC quantization bandwidth given by the Nyquist theorem to be half the sampling rate. Often times, in IF sampling and RF sampling applications, the input (sampling) bandwidth is much larger than the quantization bandwidth, because of the need to sample a band-limited signal in higher Nyquist zones.

In the linear region of operation, when V_{DS} is very small, the switch's resistance is approximately given by

$$R_{sw} \cong \frac{L}{\mu C_{ox} W (V_{GS} - V_T)} \cong \frac{L}{\mu C_{ox} W (V_G - V_T - V_{in})} \tag{4.6}$$

where μ is the mobility of the charge carrier, C_{ox} is the transistor oxide capacitance per unit area, V_{GS} is the gate-to-source voltage, V_T is the threshold voltage, and V_G is the clock's high voltage applied to the gate terminal. This gives

$$R_{sw} \cong \frac{L/[\mu C_{ox} W (V_G - V_T)]}{1 - \dfrac{V_{in}}{V_G - V_T}} \cong \frac{R_0}{1 - \dfrac{V_{in}}{V_G - V_T}} \tag{4.7}$$

where R_0 is the switch's resistance if its source is grounded (i.e. $V_{in} = 0$). Substituting (4.7) into (4.4) gives

$$V_{sample} = V_{in} \frac{1}{1 + sC_s \dfrac{R_0}{1 - \dfrac{V_{in}}{V_G - V_T}}}$$

which can be approximated using Taylor series expansion as

$$V_{sample} \cong V_{in} \frac{1}{1 + sC_sR_0\left(1 + \frac{V_{in}}{V_G - V_T} + \left(\frac{V_{in}}{V_G - V_T}\right)^2 + \ldots\right)}$$

By rearranging the terms, we get

$$V_{sample} \cong \frac{V_{in}}{1 + sC_sR_0} \frac{1}{1 + \frac{V_{in}sC_sR_0/(1 + sCR_0)}{V_G - V_T} + \frac{V_{in}^2 sC_sR_0/(1 + sCR_0)}{(V_G - V_T)^2} + \ldots} \tag{4.8}$$

which can be approximated further using Taylor series expansion to give

$$V_{sample} \cong \frac{V_{in}}{1 + sC_sR_0}\left[1 - \frac{\frac{V_{in}sC_sR_0}{1 + sC_sR_0}}{V_G - V_T} + \left(\frac{\frac{V_{in}sC_sR_0}{1 + sC_sR_0}}{V_G - V_T}\right)^2 - \left(\frac{\frac{V_{in}^2 sC_sR_0}{1 + sC_sR_0}}{(V_G - V_T)^2}\right) - \left(\frac{\frac{V_{in}sC_sR_0}{1 + sC_sR_0}}{V_G - V_T}\right)^3 + \ldots\right] \tag{4.9}$$

Equation (4.9) shows that the variation of the switch's resistance with the input signal leads to distortion, as evidenced by the second-, third-, and higher-order components of the input voltage V_{in}. In this analysis, we highlighted the impact of the input-dependent V_{GS}. However, the threshold voltage V_T depends on the back-gate-to-source voltage (V_{BS}) and hence can be input-dependent too. Therefore, the terms R_0 and V_T in (4.9) contribute to the distortion.

This analysis indicates a causative link between the input-dependence of the switch's characteristics and the tracking distortion. In addition to the cause of the distortion, it is also important to understand the factors that affect the *sensitivity* of the distortion to the switch's input-dependence described in (4.9). That is, are there situations where the same variation may cause more (or less) distortion than others? Equation (4.9) illustrates the sensitivity for every order of distortion individually, when the distortion is caused by a signal-dependent V_{GS}. However, deriving a similar formula that includes all causes of variation in R_{sw} can be too complicated and less insightful.

Alternatively, we can take a simpler approach if we start with (4.4), and represent the total variation in the sampled voltage in terms of the variation in the resistance value as follows

$$\left|\frac{\delta V_{sample}}{V_{sample}}\right| \cong \frac{(\omega_{in}C_sR_{sw})^2}{1 + (\omega_{in}C_sR_{sw})^2} \times \left(\frac{\delta R_{sw}}{R_{sw}}\right) \tag{4.10}$$

where δR_{sw} depicts the variation in the resistance value, ω_{in} is the input angular frequency, and δV_{sample} is the resulting change in the sampled voltage. The left hand side of (4.10) is the relative change of the sampled voltage, and hence includes the total distortion.

Equation (4.10) does not break down the distortion into the individual orders and harmonics like (4.9). It lumps together all the input-dependent variations of the switch's resistance into one parameter δR_{sw}, and the total resulting change and distortion of the sampled voltage into δV_{sample}. This simplification provides better insight into how the sensitivity of the distortion to the resistance variation is affected by the different parameters of the sampling process. This methodology can also be extended to other sources of distortion, such as input-dependent parasitic capacitances, input buffer characteristics. However, it is important to note that a portion of the resulting change in the sampled voltage δV_{sample} may be in the form of gain error, which is *not* distortion. Therefore, (4.10) and similar formulae can be useful for illustrating coarse trends qualitatively, but are not adequate for quantitative analysis or representation of the distortion.

Both (4.9) and (4.10) indicate that for the same variation in the switch's resistance δR_{sw} the resulting change in the sampled voltage (and hence the distortion) degrades with increasing the input frequency, the sampling capacitance, and/or the switch's resistance. As the sampling capacitance increases, which is needed to improve the kT/C noise, it becomes more challenging to achieve good distortion. It is also more difficult to achieve good distortion at high input frequencies. This trend is valid for input frequencies (ω_{in}) that are smaller than, or in the same order as, the bandwidth of the sampling network ($1/C_sR_{sw}$), which is usually the case. On the other hand, for a fixed input frequency and sampling capacitance, the tracking distortion can be improved by reducing the input-dependence of the switch's resistance δR_{sw}.

To reduce δR_{sw}, the input switch can be bootstrapped using a switched capacitor level shifter as shown in Figure 4.5. This reduces the variation in the V_{GS} voltage in (4.6), and hence in the switch's resistance. Similarly, to reduce the variation in the switch's characteristics due to body effects and the parasitic capacitance of the drain and source junctions, the back-gate can be bootstrapped as well, which is also shown in Figure 4.5 [3]. This bootstrapping keeps both V_{GS} and V_{BS} almost constant, and hence reduces the switch's non-linearity.

In spite of these measures, the input switch remains the main source of distortion during the input tracking. Bootstrapping imperfections, signal-dependent parasitic capacitances, the variation of the switch's V_{DS}, and the swapping of its drain and source terminals with the input signal, all contribute to the tracking distortion. For example, in (4.6), the parasitic capacitance at the gate of the switch (C_G) attenuates the signal at the gate compared to the source. This is due to voltage division between the bootstrap capacitance C_B and the gate capacitance C_G, which gives

$$R_{sw} \cong \frac{L}{\mu C_{ox} W (V_G - V_T - V_{in})} \cong \frac{L}{\mu C_{ox} W \left(\dfrac{C_B V_{in}}{C_B + C_G} - V_T - V_{in} \right)} \tag{4.11}$$

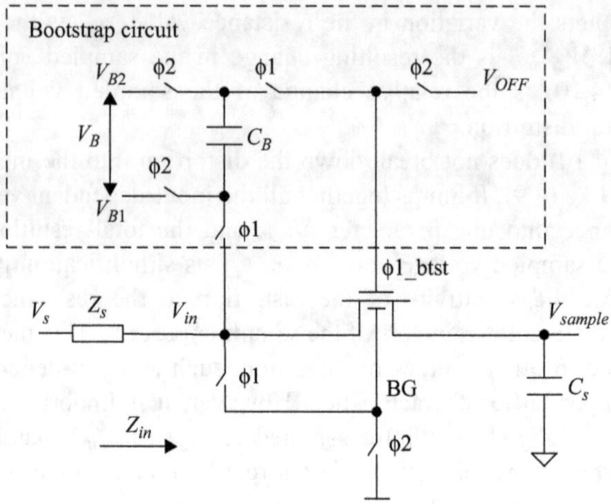

Figure 4.5 A sampling network using an NMOS bootstrapped input switch. The gate and back-gate of the input switch are bootstrapped

where C_B is the bootstrap capacitance, and C_G is the parasitic capacitance on the gate of the input switch. If C_B is much larger than C_G, the gate voltage will be almost equal to the input voltage, as desired. However, a large C_B has large parasitics, requires large switches, and can degrade distortion at very high speeds. The same limitation applies to the back-gate bootstrapping. Therefore, practically, the bootstrapping has limited accuracy, and we may have to tolerate some variation in the input switch.

On a final note, if we choose to represent (4.4) in the time domain, we get

$$V_{sample} = V_{in} - R_{sw} C_s \frac{dV_{sample}}{dt} \tag{4.12}$$

where R_{sw} is dependent on the input voltage and the V_{DS} of the switch. This is a non-linear equation that includes non-linear terms of the sampled voltage and its derivatives. Therefore, it is a non-linear system with memory, which can be analyzed using Volterra series. Unfortunately, that analysis is very complex, and would usually provide little intuitive insight [4]. Although it is important to note the presence of non-linear memory components in the sampled output, memory errors are usually dominated by the non-linear kick-back (charge injection), rather than the tracking memory.

4.1.2.2 Kick-back distortion

In addition to the tracking distortion, the charge injection (kick-back) from the input switch when the switch turns off can be non-linear, which results in additional distortion. The stored charge in the switch is approximately given by

$$Q_{ch} \approx WLC_{ox}(V_G - V_T - V_{in}) \tag{4.13}$$

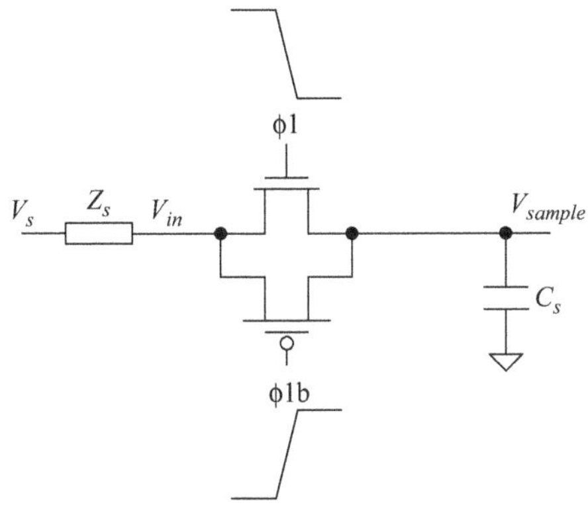

Figure 4.6 A simple sampling network using a CMOS switch

where W is the width of the transistor, L is its length, C_{ox} is the MOS oxide capacitance per unit area, V_G is the gate voltage, V_T is the threshold voltage, and V_{in} is the input voltage, which is assumed to be on the source terminal of the MOS switch. Unless V_G is equal to V_{in} and the threshold voltage is fixed, the stored charge will be input-dependent and hence potentially non-linear.

Another contributor to the charge injection is the clock feed-through, which is due to the coupling of the clock signal from the gate to the source and drain of the MOS switch. This input-dependent coupling leads to non-linear kick-back and distortion. Using a CMOS transmission gate can help with this issue if the devices are sized properly, as shown in Figure 4.6. It can also reduce the variation of the switch resistance with the input voltage. However, the improvement is relatively limited and is degraded by the added parasitics of the PMOS device on the input path. Another option is to use dummy NMOS devices on one or both sides of the NMOS input switch that are sized properly to cancel its charge injection. This is shown in Figure 4.7, and tends to be more effective in cancelling the kick-back. However, the achieved performance is usually limited to less than 8 bits of linearity.

A more effective way to reduce the non-linear charge injection is by bootstrapping the input switch, which is used to improve the tracking linearity as shown in Figure 4.5. Since bootstrapping improves the switch's linearity during tracking, it will reduce its non-linear charge injection too. This has been shown to be substantially better than using a CMOS transmission gate or dummy NMOS switches. Bootstrapping the gate and the back-gate reduces the input-dependence of the $(V_G - V_{in})$ and V_T terms in (4.13), respectively. This reduces the input-dependence of the stored charge, which improves the linearity of the kick-back substantially. However, bootstrapping imperfections and clock feed-through usually limit the performance to the 8–10 bit range.

To improve the non-linear charge injection and clock feed-through further, bottom-plate sampling can be employed, by adding a "sampling" switch on the other side of the capacitance, whose clock's falling edge (ϕ1a) is advanced relative

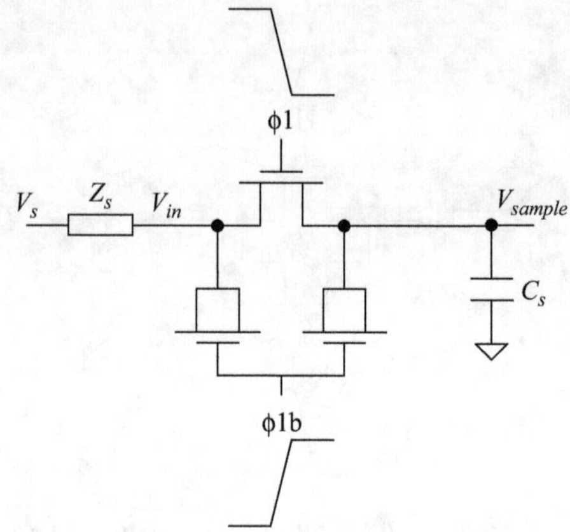

Figure 4.7 A simple sampling network using an NMOS switch with dummy devices for charge-injection cancellation

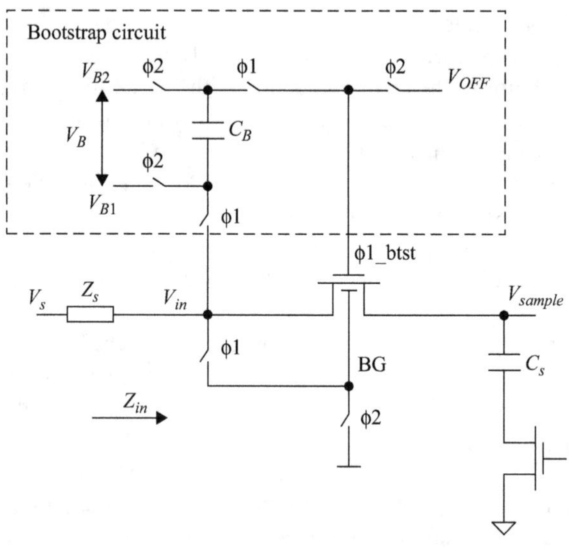

Figure 4.8 Sampling network using "bottom-plate sampling"

to the input switch's clock. This is shown in Figure 4.8 with the timing diagram in Figure 4.9. Since the voltage across the sampling switch is approximately equal to zero, with one of its terminal connected to ground, its charge injection is relatively input-independent. On the other hand, the charge injection from the input switch will ideally not affect the voltage on the sampling capacitance, because the sampling switch will be open before the input switch turns off.

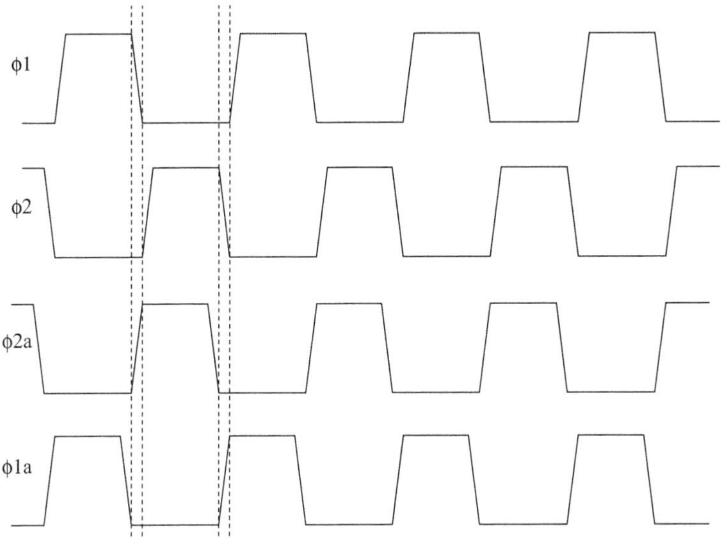

Figure 4.9 Timing diagram of the clocks used in bottom-plate sampling

In practice, the voltage across the sampling switch is not zero, and hence its charge injection can have some slight input-dependence. Moreover, the parasitic capacitances on the bottom side of the sampling capacitance can lead to some of the input switch's charge injection to appear on the sampling capacitance. However, these effects are usually very small, and the bottom-plate sampling can substantially improve the distortion due to charge injection.

One remaining issue to observe in Figure 4.8 is that the input switch is bootstrapped during the tracking phase, but not during the hold phase. The switching from a fixed gate voltage (V_{OFF}) when the switch is off to a value that tracks the input signal when the switch is on causes an input-dependent kick-back that can degrade the distortion. This can be improved further by using the "two-phase bootstrapping" shown in Figure 4.10 [5]. In this approach, the input is bootstrapped in both the on and off phases, and hence the kick-back is relatively constant and input-independent.

Another possible source of kick-back distortion is the memory errors. Some of the non-linear kick-back can be related to previous samples or a quantized version of them. This causes distortion that is a non-linear function of the previous samples, in addition to the current samples. Correcting these error requires clearing the memory in addition to the linearization discussed earlier [2].

The impact of the non-linear kick-back on distortion depends on the sampling rate and input bandwidth. Given enough time, the non-linear charge injection would dissipate and hence not degrade the sampling distortion. Therefore, lowering the sampling rate and increasing the input bandwidth will improve the distortion due to kick-back. A critical factor in determining the input bandwidth and the response of the sampler to the kick-back is the *source impedance*.

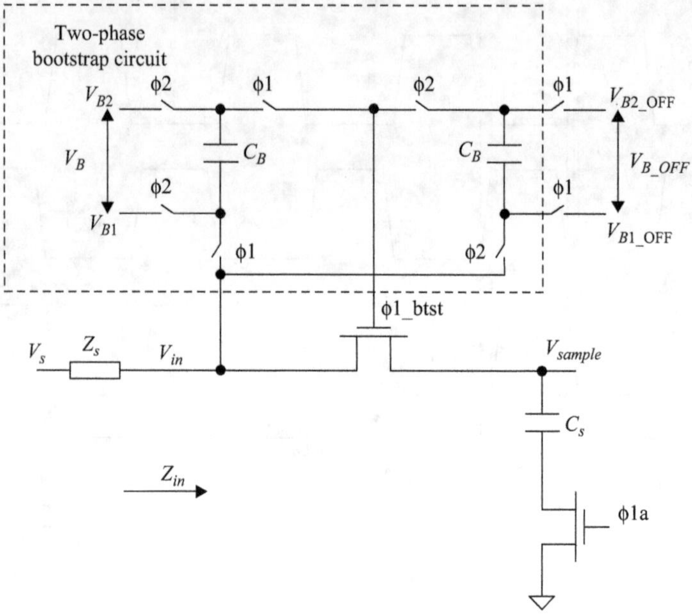

Figure 4.10 A sampling network using an NMOS bootstrapped input switch where the gate is bootstrapped in both phases [5]

4.1.2.3 Source impedance

Despite heavily influencing both the tracking and kick-back distortion, the source impedance is often overlooked. A simple way to understand its impact is by representing the ADC sampling network by its non-linear input impedance Z_{in}, which is driven by a source whose impedance is Z_s. This is shown in Figure 4.11(a). The input voltage applied to the ADC is given by

$$V_{in} = \frac{V_s Z_{in}}{Z_s + Z_{in}} \tag{4.14}$$

During tracking, the variation in the ADC's input voltage δV_{in} due to the variation in the ADC's input impedance δZ_{in} can be expressed as

$$\frac{\delta V_{in}}{V_{in}} = \frac{\delta Z_{in}/Z_{in}}{1 + Z_{in}/Z_s} \tag{4.15}$$

Therefore, to improve the tracking distortion, we need to reduce the non-linearity in the ADC's sampling network ($\delta Z_{in}/Z_{in}$), which has been the focus of our discussion so far. In addition, we also need to increase the input impedance (Z_{in}) and reduce the source impedance Z_s. For the same sampling network and sampling rate, the higher the driving impedance, the worse the distortion. An intuitive explanation of this is that the non-linear input impedance of the ADC (Z_{in}) draws a non-linear

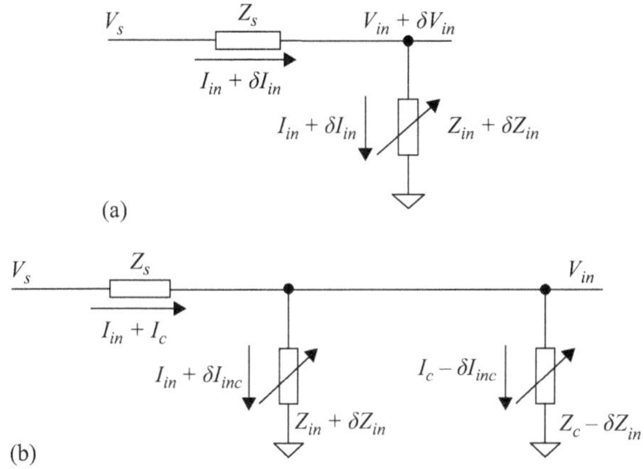

Figure 4.11 (a) A representation of the input impedance of the sampling network (Z_{in}) with the non-linear current drawn from it δI_{in}. (b) The input network with a compensation impedance (Z_c) in parallel [1, 7]

current (δI_{in}) from the source. This causes a non-linear voltage drop on the source impedance. For the same non-linear current, the higher the source impedance, the higher the non-linear drop, and hence the worse the distortion. It is interesting to note that this distortion happens to the input signal *outside* the ADC, but it is caused by the ADC's non-linearity.

Another approach to reduce the non-linear current in the source impedance is to cancel this distortion using a parallel path that provides an equal and opposite non-linear current. This is shown conceptually in Figure 4.11(b). The compensation impedance Z_c provides an equal and opposite variation that reduces the non-linear current flowing in the source impedance and hence improves the linearity in the main path [2].

A practical example was described in Reference 2 and is shown in Figure 4.12, where the distortion in the main sampling path source impedance was reduced by using a parallel compensation path formed by the flash sub-ADC sampling network. The input switch is bootstrapped and the residual non-linearity in the main path is dominated by the parasitic capacitances of the input bootstrapped switch M_{sw}. Practically, the bootstrapping is not perfect because of the attenuation of the gate voltage due to the parasitic capacitances. Therefore, the signal's amplitude on the gate is smaller than that on the source, and the switch's V_{GS} will be out of phase with the input signal. On the other hand, the flash path's distortion is dominated by the input device of the comparators' preamplifiers (M_{pre}). The device's V_{GD} signal is in phase with the input signal, which is opposite to the V_{GS} of the input switch in the main path. With proper sizing, this opposite input-dependence gives an opposite non-linearity, which reduces the non-linear current in the source impedance.

In addition to the tracking distortion, the source impedance controls the input bandwidth B_{in} and hence influences the kick-back distortion too. A smaller source

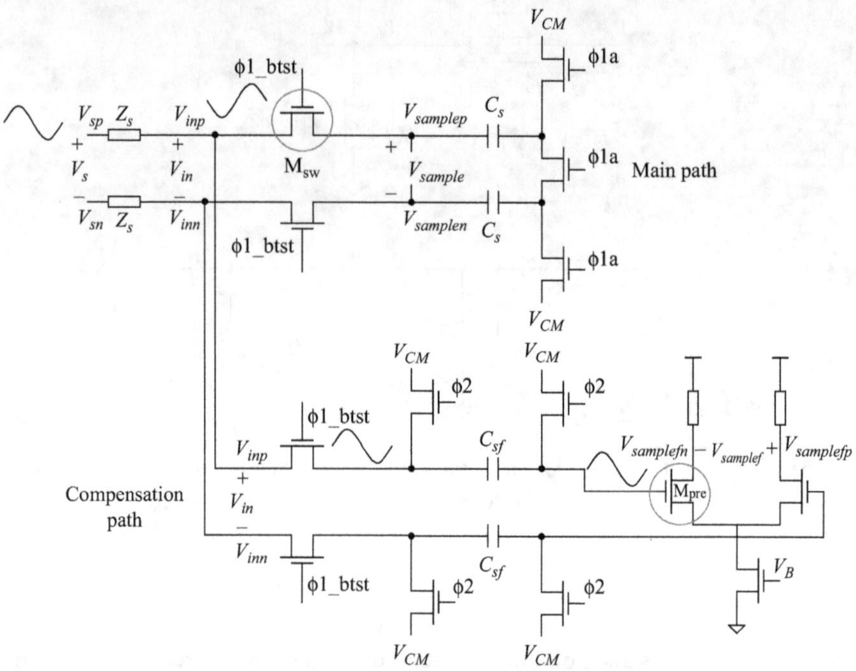

Figure 4.12 An example of non-linear compensation using the flash path to compensate for the non-linearity in the MDAC input path [1, 7]

impedance increases the bandwidth, which improves the settling of the non-linear kick-back. Alternatively, a source impedance that includes a bypass capacitance across the input terminals of the ADC, can reduce the kick-back glitches, and hence improve distortion.

It is important to note that the source impedance represents the equivalent impedance of all the elements between the signal source and the ADC. This includes the resistance termination, possibly an anti-alias filter, series resistances, bypass capacitances, parasitic inductances and capacitances of the package's bond wires, and the driver amplifier. An example is shown in Figure 4.13. The input bandwidth B_{in} is usually limited by the source impedance Z_s and the input impedance Z_{in} of the ADC. In addition, the reactive components in Z_s create a resonant network that must be optimized for the maximum bandwidth and/or best distortion, while avoiding "ringing."

4.1.2.4 Differential operation

To improve the even-order harmonic distortion, high speed ADCs usually employ differential input paths, as shown in Figure 4.14. If perfectly matched, this cancels the even-order harmonics as discussed in Chapter 2. In addition to the differential sampling switch, the single-ended sampling switches that are connected to the common-mode voltage V_{CM} are needed to attenuate the common-mode signal at the

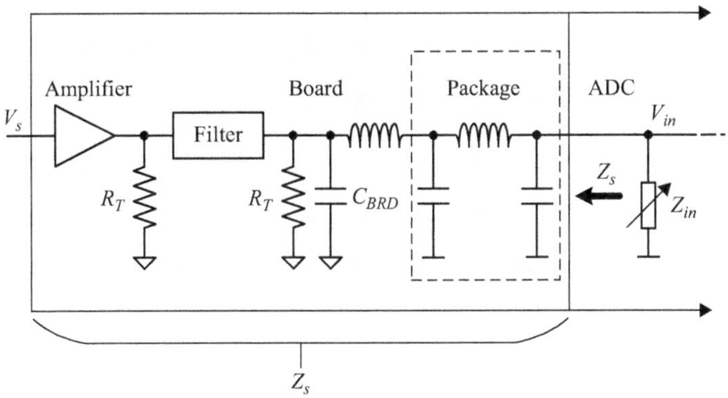

Figure 4.13 *An illustration of the source impedance driving the ADC. It includes an amplifier, filter, termination resistors R_T, board capacitance C_{BRD}, and package parasitics*

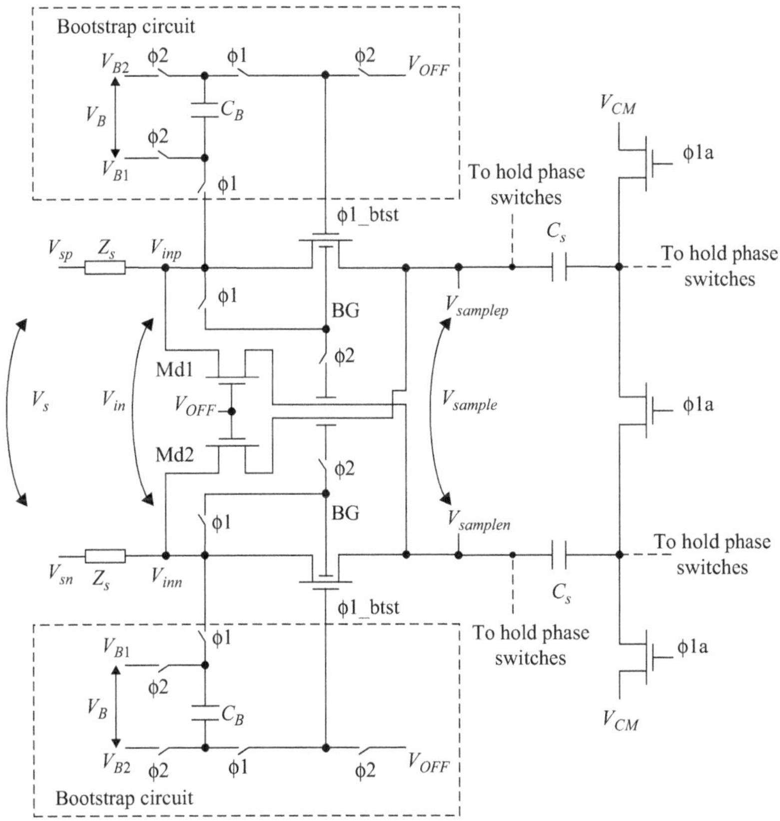

Figure 4.14 *An example of a differential sampling network with bootstrapped input switch. Devices Md1 and Md2 are used to cancel the signal feed-through when the input switches are turned off*

terminals of the differential switch, which in turn helps with the even-order harmonic cancellation and the overall distortion.

To understand the effect of the differential operation on the even-order harmonics, we represent the sampler as a second-order non-linear system, such that

$$V_{samplep}(t) = \alpha_1 V_{inp}(t) + \alpha_2 V_{inp}^2(t) \tag{4.16}$$

and

$$V_{samplen}(t) = \alpha_1 V_{inn}(t) + \alpha_2 V_{inn}^2(t) \tag{4.17}$$

The differential output is given by

$$V_{sample}(t) = \alpha_1 \left(V_{inp}(t) - V_{inn}(t)\right) + \alpha_2 \left(V_{inp}^2(t) - V_{inn}^2(t)\right) \tag{4.18}$$

If the two single-ended inputs are matched sinusoidal signals, we have

$$V_{inp}(t) = A \sin \omega t \qquad \text{and} \qquad V_{inn}(t) = -A \sin \omega t$$

and the sampled voltage will be free of even-order harmonics, and given by

$$V_{sample}(t) = 2\alpha_1 A \sin \omega t$$

If there is an amplitude mismatch between the two inputs, the two inputs will be given by

$$V_{inp}(t) = A_p \sin \omega t \qquad \text{and} \qquad V_{inn}(t) = -A_n \sin \omega t \tag{4.19}$$

where the relative amplitude mismatch is given by

$$\Delta = \frac{A_p - A_n}{A} = 2\frac{A_p - A}{A} = 2\frac{A - A_n}{A} = 2\delta$$

If there is a phase mismatch ϕ between the two inputs, the two inputs will be given by

$$V_{inp}(t) = A \sin \omega t \qquad \text{and} \qquad V_{inn}(t) = -A \sin(\omega t + \phi) \tag{4.20}$$

The impact of an amplitude mismatch alone on the second harmonic can be derived by substituting (4.19) into (4.18), and is given by

$$HD2_{diff} \approx HD2_{se} + 20\log(\Delta) \tag{4.21}$$

where $HD2_{diff}$ is the differential second harmonic level in dB, $HD2_{se}$ is the single-ended second harmonic level for each side in dB, and Δ is the amplitude imbalance.

The impact of a phase mismatch alone on the second harmonic can be derived by substituting (4.20) into (4.18), and is given by

$$HD2_{diff} = HD2_{se} + 20\log\left(\frac{\sin\phi}{\cos(\phi/2)}\right) \tag{4.22}$$

where $HD2_{diff}$ is the differential second harmonic level in dB, $HD2_{se}$ is the single-ended second harmonic level for each side in dB, ϕ is the phase mismatch (imbalance) between the two inputs in radians. From (4.22), we note the following:

- For small phase mismatches, $HD2_{diff}$ is substantially better than $HD2_{se}$.
- The differential $HD2_{diff}$ degrades as the phase mismatch increases.
- The phase mismatch which results in no improvement in $HD2_{diff}$ over $HD2_{se}$ is 60°.
- Increasing the phase mismatch beyond 60° results in further degradation in the differential $HD2_{diff}$ where it becomes worse than the single-ended $HD2_{se}$.
- A 90° phase mismatch results in the $HD2_{diff}$ being 3 dB *worse* than $HD2_{se}$.
- The worst-case $HD2_{diff}$ is worse than $HD2_{se}$ by a value that asymptotically approaches 6 dB, as the phase mismatch approaches 180°.

For small phase mismatches, (4.22) can be approximated by

$$HD2_{diff} \approx HD2_{se} + 20\log(\phi) \tag{4.23}$$

It is clear from (4.21)–(4.23) that since the *HD*2 is a negative quantity, the amplitude and phase mismatches will degrade the differential second harmonic level, such that every doubling in the magnitude of the mismatch results in 6 dB degradation in the differential second harmonic amplitude.

In the presence of both amplitude and phase mismatches together, the *HD*2 is given by

$$HD2_{diff} \approx HD2_{se} + 10\log\left(\frac{4\delta^2 + \left(1-\delta^2\right)^2\sin^2(\phi)}{\delta^2 + \left(1-\delta^2\right)\cos^2\left(\frac{\phi}{2}\right)}\right) \tag{4.24}$$

which represents the impact of both the amplitude and phase mismatches on the second harmonic. If the mismatches are small, (4.24) can be approximated as

$$HD2_{diff} \approx HD2_{se} + 10\log\left(\Delta^2 + 4\sin^2\left(\frac{\phi}{2}\right)\right) \tag{4.25}$$

That is,

$$HD2_{diff} \approx HD2_{se} + 10\log\left(\Delta^2 + \phi^2\right) \tag{4.26}$$

where $HD2_{diff}$ is the differential second harmonic level in dB, $HD2_{se}$ is the single-ended second harmonic level for each side in dB, ϕ is the phase mismatch (imbalance) between the two inputs in radians, and Δ is the amplitude imbalance.

If the mismatch is in the system transfer characteristics between the two sides, as opposed to the input signal, the *HD*2 cancellation can be compromised too. In this case, the parameter α_2 in (4.16) and (4.17) may be mismatched between the two sides, such that

$$\alpha_{2p} = (\alpha_2 + \varepsilon\alpha_2)e^{j\theta} \quad \text{and} \quad \alpha_{2n} = (\alpha_2 - \varepsilon\alpha_2)e^{-j\theta} \tag{4.27}$$

where α_2, α_{2p}, and α_{2n} are complex, θ is the mismatch in the phase response, and ε is the amplitude response mismatch, which is given by

$$\varepsilon = \frac{|\alpha_{2p}| - |\alpha_{2n}|}{2|\alpha_2|} \tag{4.28}$$

Then, the $HD2$ will be given by

$$HD2_{diff} \approx HD2_{se} + 10\log\left[(\Delta + \varepsilon)^2 + (\phi + \theta)^2\right] \tag{4.29}$$

Therefore, in (4.29), the term $(\Delta + \varepsilon)$ represents the total amplitude imbalance, and the term $(\phi + \theta)$ represents the total phase imbalance. Please note the factor of 2 difference in the definitions between the phase terms ϕ and θ, and between the amplitude mismatch terms Δ and ε. This difference is due to the squaring of the input signal by the second-order non-linearity in (4.16) and (4.17), which doubles the impact of the mismatches in the input signal compared to the mismatches in the α_2 term.

In addition to reducing the even-order harmonics, the differential input sampling provides flexibility in processing the input signal. It also allows for the cancellation of the input signal feed-through, when the input switches are off, by utilizing dummy devices that are identical to the input switches but connected across the opposite sides. These dummy devices, denoted as Md1 and Md2 in Figure 4.14, are always off and create an equal and opposite coupling from the input to the sampling capacitances, to reduce the input feed-through during the hold phase.

It is important to note that the circuit shown in Figure 4.14 does not reject the common-mode variation. Therefore, the common-mode settling needs to be acceptable for the target performance, otherwise, the cancellation of the even-order harmonics may degrade.

4.1.2.5 Summary of trade-offs

To summarize, the factors that improve the tracking linearity are:

- Small sampling capacitor.
- Small source impedance.
- Low input frequency.
- Small variation of the sampling network with the input signal.
- Using differential operation with good matching and symmetry.

We have also seen that the kick-back distortion is affected by the tracking linearity. In other words, the better the tracking linearity, the more linear the kick-back will be. However, there are other factors that affect the kick-back distortion exclusively. These include the settling time available for the charge injection to dissipate, the isolation, and the input sampling bandwidth.

Therefore, the factors that reduce the kick-back distortion are:

- Reducing the non-linear component of the kick-back.
- Large sampling bandwidth (B_{in}).
- Low sampling rate.
- Small sampling capacitor.
- Small source impedance.
- Low input frequency.
- Small variation with the input signal.
- Better isolation.

4.2 Input buffer

Based on the above conclusions, sometimes an input buffer is needed to achieve the desired sampling linearity. The input buffer provides better isolation, a low source impedance to drive the sampling capacitance, and a high input impedance. It can also reduce the non-linear current drawn from the source impedance. Therefore, it can improve both the tracking and kick-back linearity of the sampling network. However, the overall distortion may be limited by the linearity of the buffer itself. In general, high linearity input buffers are challenging to design; they degrade the thermal noise and consume significant power. However, they can enable much higher performance than what is possible otherwise, in both the tracking and kick-back distortion. The highest performing high-speed samplers in the literature include input buffers.

An example of a sampling network with an input buffer is shown in Figure 4.15, where a single buffer drives the sampling path, the gate bootstrap network, and the back-gate bootstrap network. In some cases, the loading effect of the bootstrap circuits can be detrimental to the linearity of the main sampling path. Therefore, separate buffers may be used for the different networks to optimize each independently. An example is shown in Figure 4.16 [1]. This structure reduces the negative impact of driving the bootstrap circuits on the sampling linearity, due to their non-linear loading and kick-back. This is achieved at the expense of larger area and higher power consumption.

4.2.1 Input buffer design

The ADC's input buffer is typically a source follower or an emitter follower. These structures have a low output impedance, high input impedance, and relatively low distortion. Traditional source and emitter followers are shown in Figure 4.17. In a sampler, the load impedance Z_L is dominated by the sampling capacitance. The current I is the DC bias current, and the current i_L is the AC load current flowing in the sampling capacitance. The equivalent circuits of the MOS transistor and the source follower are shown in Figure 4.18. It is important to note that the signal is usually large enough for the parameters of the model to vary, which is a major cause of distortion in the buffer.

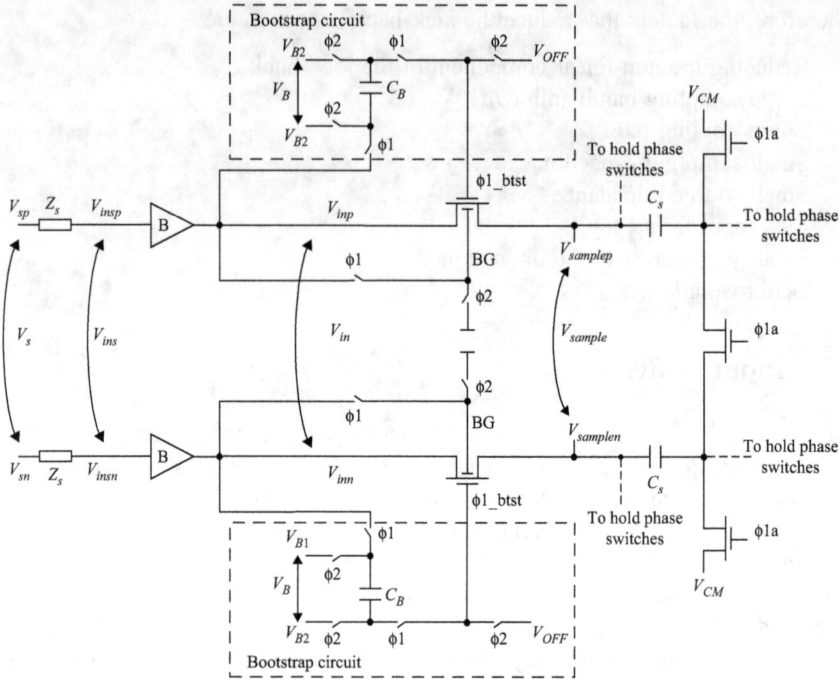

Figure 4.15 A schematic of a differential sampling network with bootstrapped input switch and input buffers. The network is driven by a source with impedance equal to Z_s

The transfer characteristic of the source follower whose source and back-gate are connected is given by

$$\frac{v_o}{v_{in}} \cong \frac{1}{1 + \dfrac{1}{(g_m + 1/Z_f)Z_L}} \tag{4.30}$$

where g_m is the transconductance of the follower device (M_f), Z_f is the input impedance of the follower device (M_f), and Z_L is the load impedance, which includes the load (i.e. the sampling capacitance) and the output impedances of the follower device and the current source. The key observation here is that the device-dependent term in the denominator of (4.30) is signal-dependent and hence contributes to the output distortion. This term is reduced by increasing the transconductance g_m and increasing the load impedance Z_L.

In the presence of source impedance Z_s, the transfer characteristic is given by

$$\frac{v_o}{v_s} \cong \frac{1}{(1 + sC_iZ_s)} \times \frac{1}{\left(1 + \dfrac{Z_{s\_eq} + Z_f}{(1 + g_mZ_f)Z_L}\right)} \tag{4.31}$$

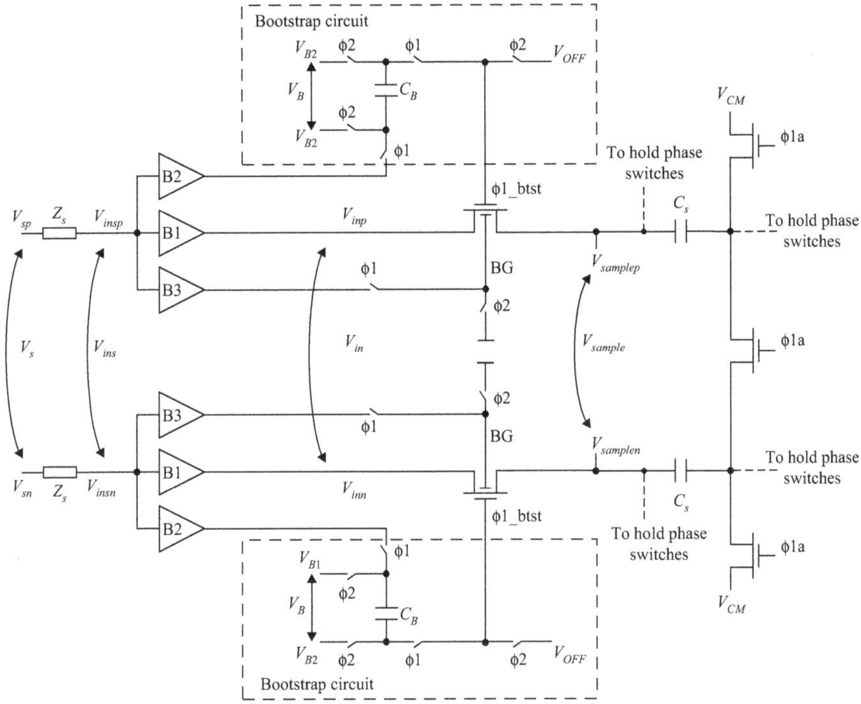

Figure 4.16 A schematic of a differential sampling network with bootstrapped input switch and input buffers. Three separate buffers drive the sampling capacitor, the switch's gate bootstrap circuit, and the switch's back-gate. © 2010 IEEE. Reprinted, with permission, from Reference 1

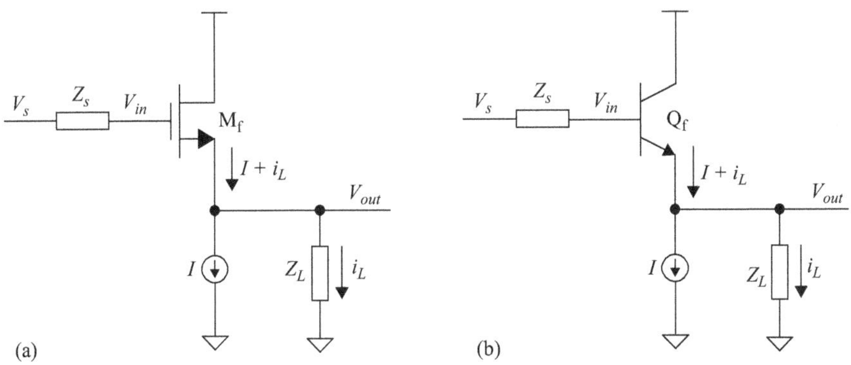

Figure 4.17 A simplified schematic of (a) source follower and (b) emitter follower

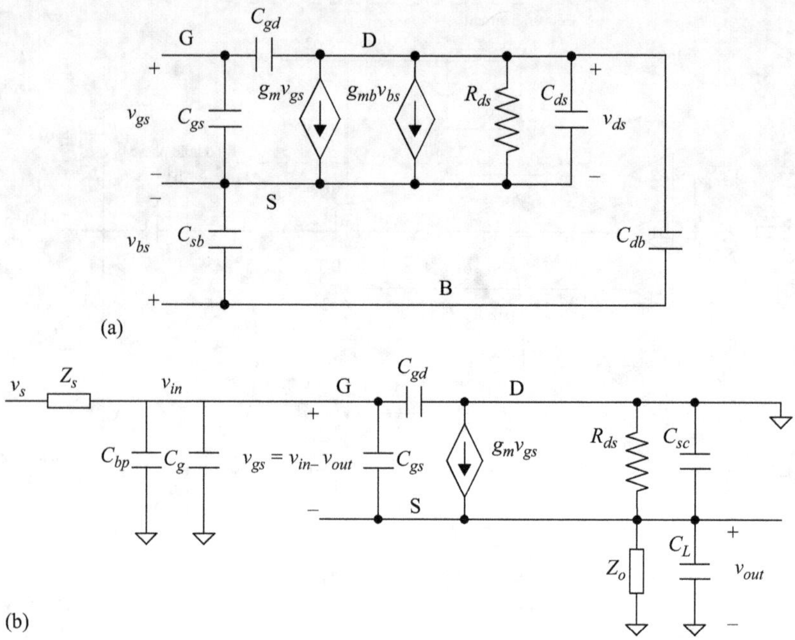

Figure 4.18 (a) A simplified model of a MOS transistor. (b) A simplified model of a source follower with the back-gate of the follower device connected to the source

where Z_{s_eq} is the parallel combination of Z_s and $1/sC_i$, which is given by

$$Z_{s\_eq} = \frac{Z_s}{1 + sC_i Z_s} \tag{4.32}$$

For the MOS transistor (M_f), the input impedance Z_f is given by

$$Z_f = \frac{1}{sC_{gs}} \tag{4.33}$$

The capacitance of the source follower's input node to ground is

$$C_i = C_{bp} + C_g + C_{gd} \tag{4.34}$$

where C_{bp} is the input bypass capacitance, C_g is the gate to ground capacitance, and C_{gd} is the gate to drain capacitance. The load impedance Z_L is given by the expression

$$\frac{1}{Z_L} = sC_L + \frac{1}{Z_o} + \frac{1}{R_{ds}} + sC_{sc} \tag{4.35}$$

where C_L is the load/sampling capacitance, Z_o is the output impedance of the current source, R_{ds} is the output impedance of the MOS transistor (M_f), and C_{sc} is the source to ground capacitance of the follower device.

The input impedance of the MOS source follower is given by

$$\frac{1}{Z_{in}} = sC_i + \frac{1}{Z_f + Z_L + g_m Z_f Z_L} \tag{4.36}$$

Therefore, to increase the input impedance, we need to:

- Increase the input impedance of the follower transistor (Z_f),
- Increase the load impedance, which includes:
  - increasing the output impedance of the follower
  - increasing the output impedance of the current source
  - increasing the load impedance by reducing the sampling capacitance
- Increase the transconductance of the follower device g_m,
- Reduce the input capacitance C_i.

On the other hand, the buffer's output impedance is given by

$$\frac{1}{Z_{out}} = \frac{1}{Z_o} + \frac{1}{R_{ds}} + sC_{sc} + \frac{1 + g_m Z_f}{Z_f + \dfrac{1}{sC_i}} \tag{4.37}$$

In the presence of a source impedance Z_s, the expression for the output impedance becomes

$$\frac{1}{Z_{out}} = \frac{1}{Z_o} + \frac{1}{R_{ds}} + sC_{sc} + \frac{1 + g_m Z_f}{Z_f + \dfrac{1}{sC_i + 1/Z_s}} \tag{4.38}$$

For large values of g_m, (4.38) can be approximated as

$$\frac{1}{Z_{out}} \approx \frac{1 + g_m Z_f}{Z_f + \dfrac{1}{sC_i + 1/Z_s}} \tag{4.39}$$

For small source impedances and large input capacitances, it can be simplified further to give

$$Z_{out} \approx \frac{1}{g_m} + \frac{Z_s}{g_m Z_f} \approx \frac{1}{g_m} \tag{4.40}$$

Therefore, a large transconductance g_m of the follower device (M_f) is key to achieving a small output impedance for the source follower. In addition, reducing the source impedance (Z_s), increasing the input impedance (Z_f) of the follower device M_f, and increasing the input capacitance C_i reduce the buffer's output impedance too.

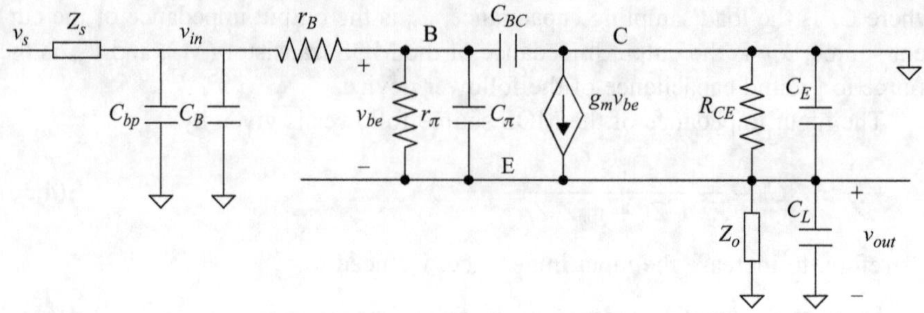

Figure 4.19 A simplified model of an emitter follower

The same analysis described earlier can be applied to emitter followers, while taking into account the finite input resistance of bipolar transistors ($r_B + r_\pi$). The equivalent circuit of the emitter follower is shown in Figure 4.19. If we ignore the base resistance r_B, the transfer characteristic will be similar to (4.30) for the source follower, which is

$$\frac{v_O}{v_{in}} \cong \frac{1}{1 + \dfrac{1}{(g_m + 1/Z_f)Z_L}} \tag{4.41}$$

where g_m is the transconductance of the follower device (Q_f), Z_f is the input impedance of the follower device (Q_f), and Z_L is the load impedance, which includes the load (i.e. the sampling capacitance) and the output impedances of the follower device and the current source. The higher output impedance of the bipolar transistor, its relatively higher bandwidth, higher transconductance, and better linearity compared to the MOS transistor enable emitter followers to achieve higher performance compared to source followers.

In the presence of source impedance Z_s, the transfer characteristic is given by

$$\frac{v_O}{v_S} \cong \frac{1}{(1 + sC_iZ_s)} \times \frac{1}{\left(1 + \dfrac{Z_{s\_eq} + Z_f}{(1 + g_mZ_f)Z_L}\right)} \tag{4.42}$$

where Z_{s_eq} is the parallel combination of Z_s and $1/sC_i$, which is given by

$$Z_{s\_eq} = \frac{Z_s}{1 + sC_iZ_s} \tag{4.43}$$

The input impedance Z_f of the bipolar transistor (Q_f) is given by

$$Z_f = \frac{1}{1/r_\pi + sC_\pi} \tag{4.44}$$

where the base resistance r_B is ignored in our analysis. The capacitance from the base node of Q_f to ground is

$$C_i = C_{bp} + C_B + C_{BC} \tag{4.45}$$

where C_{bp} is the input bypass capacitance, C_B is the base-to-ground capacitance, and C_{BC} is the base-to-collector capacitance. The load impedance Z_L is given by

$$\frac{1}{Z_L} = sC_L + \frac{1}{Z_o} + \frac{1}{R_{CE}} + sC_E \tag{4.46}$$

where C_L is the load/sampling capacitance, Z_o is the output impedance of the current source, R_{CE} is the output impedance of the bipolar transistor (Q_f), and C_E is the emitter to ground capacitance of the follower device.

The input impedance of the emitter follower, ignoring the base resistance r_B, is

$$\frac{1}{Z_{in}} = sC_i + \frac{1}{Z_f + Z_L + g_m Z_f Z_L} \tag{4.47}$$

Similar to a source follower, we can increase the input impedance of the emitter follower by:

- Increasing the input impedance of the follower transistor (Z_f),
- Increasing the load impedance,
- Increasing the transconductance of the follower device g_m,
- Reducing the input capacitance C_i.

The output impedance of the emitter follower is given by

$$\frac{1}{Z_{out}} = \frac{1}{Z_o} + \frac{1}{R_{CE}} + sC_E + \frac{1 + g_m Z_f}{Z_f + \dfrac{1}{sC_i + 1/Z_s}} \tag{4.48}$$

If g_m is large and the source impedance is small, this can be approximated by

$$Z_{out} \approx \frac{1}{g_m} + \frac{Z_s}{\beta} \approx \frac{1}{g_m} \tag{4.49}$$

where β is the short-circuit current gain (I_C/I_B) of the bipolar transistor. To reduce the output impedance of the emitter follower, we need to increase its g_m. In addition, although reducing the input capacitance is desirable to increase the input impedance, the output impedance shows an opposite trend. That is, increasing the input capacitance reduces the output impedance, which would in turn improve the tracking linearity and the kick-back linearity.

The reverse gain (isolation) of the buffer, whether it is a source follower or an emitter follower, is approximately given by

$$\frac{v_{in\_r}}{v_{out}} = \frac{1}{1 + \dfrac{Z_f}{Z_s || Z_i}} \tag{4.50}$$

where v_{in_r} is the signal appearing at the input of the buffer due to a signal v_{out} applied at its output, Z_s is the source impedance, and Z_i is the impedance from the input to ground including the bypass capacitance C_{bp}. Alternatively, the isolation can be represented by the ratio of the input voltage that results from an injected charge or current at the output i_{out}, which is given by

$$\frac{v_{in\_r}}{i_{out}} = \frac{1}{\dfrac{1}{Z_L} - sC_L + \left(g_m + \dfrac{1}{Z_L} - sC_L + \dfrac{1}{Z_f}\right)\dfrac{Z_f}{Z_s||Z_i}} \tag{4.51}$$

where i_{out} is the current injected by the load capacitance C_L (i.e. the current flowing in the capacitance C_L). It is clear from (4.50) and (4.51) that the isolation improves by increasing the input impedance of the follower transistor Z_f, and by reducing the equivalent impedance $Z_s||Z_i$. This can be accomplished by using a small source impedance and/or a small input impedance Z_i. Interestingly, a large input capacitance C_i would reduce the input impedance Z_i, and hence improve the isolation of the input buffer. That is why it is common to add an explicit capacitance on this node. Although that capacitance degrades the input bandwidth and the input impedance of the buffer, it improves the isolation, and hence improves the kick-back and its linearity, which is desirable.

4.2.2 *Input buffer non-linearity*

It is desirable for the input buffer to have a low output impedance. This improves the tracking linearity of the sampling network and reduces the sensitivity to the input switch's non-linearity. In addition, it increases the sampling bandwidth, and hence improves the settling of the non-linear kick-back at the *output* of the buffer. Reducing the output impedance requires increasing the transconductance g_m, and hence increasing the power consumption.

Another factor that affects the distortion is the input impedance Z_{in} and its non-linearity δZ_{in}. From (4.36) and (4.47), we see that the input impedance depends on signal-dependent transistor parameters, and hence are non-linear. Increasing the input impedance is desirable, but it can result in a more non-linear buffer. Therefore, in order to improve the tracking distortion at the input of the buffer, we may intentionally reduce its input impedance. This can lead to a more linear impedance, which can be better than a large but highly non-linear impedance. Moreover, as mentioned earlier, reducing the input impedance by increasing the input capacitance C_i improves the isolation of the buffer and the kick-back from the sampling network that propagates through the buffer and reaches the driver network, which can improve the kick-back linearity as well.

In addition to the above factors, which influence the buffer's non-linearity due to its interaction with its load and its driver, (4.30) and (4.41) highlight the main sources of non-linearity inside the buffer itself. The transfer characteristic of a source or emitter follower is given by

$$\frac{v_o}{v_{in}} \cong \frac{1}{1+\frac{1}{\left(g_m+\frac{1}{Z_f}\right)Z_L}} = \frac{1}{1+\frac{1}{\left(1+\frac{1}{g_m Z_f}\right)g_m Z_L}} \tag{4.52}$$

Including the source impedance, we get

$$\frac{v_o}{v_s} \cong \frac{1}{(1+sC_iZ_s)} \times \frac{1}{\left(1+\frac{Z_{s\_eq}+Z_f}{(1+g_mZ_f)Z_L}\right)}$$

If $g_mZ_f >> 1$, the transfer function can be approximated by

$$\frac{v_o}{v_{in}} \cong \frac{1}{1+\frac{1}{g_mZ_L}} \tag{4.53}$$

and

$$\frac{v_o}{v_s} \cong \frac{1}{(1+sC_iZ_s)} \times \frac{1}{\left(1+\frac{Z_{s\_eq}}{g_mZ_fZ_L}+\frac{1}{g_mZ_L}\right)} \tag{4.54}$$

The signal-dependent terms of (4.53) and (4.54), which cause the non-linearity, are reduced by increasing g_m of the follower device, increasing the load impedance Z_L, and reducing the source impedance Z_s.

If we ignore the source impedance, and assume that the non-linearity in the output voltage can be approximated as a variation in the output voltage due to the signal-dependence of g_m, we can simplify the representation of the non-linearity as a variation in the output, and using (4.53) we get

$$\frac{\delta v_o}{v_o} \cong \frac{\delta g_m/g_m}{1+g_mZ_L} \tag{4.55}$$

Therefore, the non-linear behavior is proportional to the variation in the g_m of the source/emitter follower. Moreover, a large load impedance Z_L and a large g_m are expected to reduce the distortion in the output voltage due to the variation in the g_m. Therefore, it is clear that increasing g_m is key to improving the performance of the input buffer. Since the g_m of the bipolar transistor tends to be larger and more linear than that of the MOS transistor, emitter followers usually enjoy better linearity than source followers.

The g_m of the source follower is given by

$$g_m = \sqrt{\frac{2\mu C_{ox}WI_t}{L}} \tag{4.56}$$

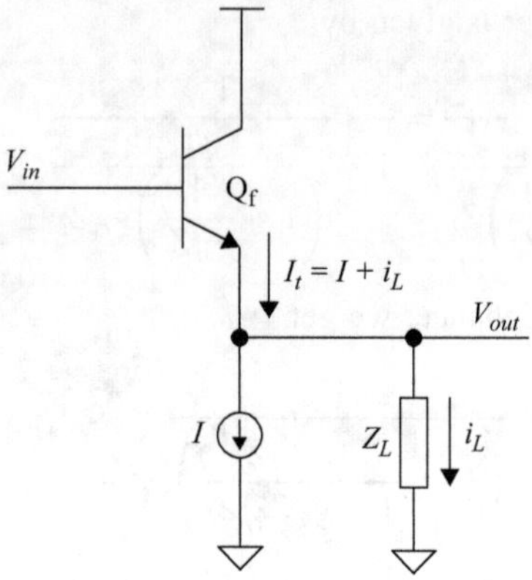

Figure 4.20 A simplified schematic of an emitter follower driving a load Z_L

The g_m of the emitter follower is

$$g_m = \frac{qI_t}{kT} \tag{4.57}$$

where I_t is the current flowing in the follower device (M_f or Q_f). Therefore, increasing the g_m requires increasing the current I_t, which increases the power consumption. As shown in Figure 4.20, the current I_t is composed of a DC component I and an AC component i_L due to the load current flowing in the load capacitance. That is,

$$I_t = I + i_L \tag{4.58}$$

The load current i_L is given by

$$i_L = \frac{v_o}{Z_L} \cong v_o \omega_{in} C_L \tag{4.59}$$

Therefore, the variable current i_L increases with increasing the load capacitance and/or the input frequency. This increases the variation in the g_m and hence degrades the distortion. Moreover, increasing the signal amplitude increases the current i_L, which degrades the distortion too.

For a MOS source follower, using (4.56), we obtain

$$\frac{\delta g_m}{g_m} = \frac{\delta I_t}{2I_t} \approx \frac{i_L}{2I} \tag{4.60}$$

Substituting (4.60) into (4.55) gives the output variation for a source follower as

$$\frac{\delta v_o}{v_o} \cong \frac{0.5\, i_L/I}{1 + g_m Z_L} \tag{4.61}$$

For a bipolar emitter follower, using (4.57), we get

$$\frac{\delta g_m}{g_m} = \frac{\delta I_t}{I_t} \approx \frac{i_L}{I} \tag{4.62}$$

Substituting (4.62) into (4.55), gives the output variation for an emitter follower as

$$\frac{\delta v_o}{v_o} \cong \frac{i_L/I}{1 + g_m Z_L} \tag{4.63}$$

Therefore, (4.61) and (4.63) indicate that to reduce the distortion we need to increase the bias current substantially, such that its variation i_L is much less than its DC bias value I. Moreover, increasing the g_m of the follower device is key to reducing its distortion. Having said that, it is important to note that increasing the current can result in reducing the dynamic range, and hence degrading the distortion. Increasing the device sizes can help with this issue, but it will degrade the parasitic capacitances, which can reduce the impedances, and hence degrade the distortion in a different way.

Before we proceed any further, we should investigate the large signal behavior of the source and emitter followers. For the source follower, V_{GS} is given by

$$V_{GS} = V_{in} - V_{out} = V_T + \sqrt{\frac{2\left(I + \dfrac{V_{out}}{Z_L}\right)}{\mu C_{ox} W/L}} \tag{4.64}$$

where V_T is the threshold voltage, μ is the mobility of the device, C_{ox} is the oxide capacitance per unit area, W is the width, and L is the length. If the source and back-gate terminals are connected, V_T will not change with V_{SB}. The main source of non-linearity will be due to the current flowing in the load V_{out}/Z_L.

For an emitter follower, V_{BE} is given by

$$V_{BE} = V_{in} - V_{out} = \frac{kT}{q} \ln\left(\frac{I + \dfrac{V_{out}}{Z_L}}{I_s}\right) \tag{4.65}$$

where I_s is the bipolar transistor's saturation current, k is the Boltzmann constant, T is the absolute temperature, and q is the electron charge. The main source of distortion is again shown to be the current variation due to the load. However, the logarithmic function of the emitter follower is much weaker than the square root relation of the source follower. That is, the variation in V_{BE} of the emitter follower due to the load current is much smaller than the variation in V_{GS} of the source follower. Another limitation of the source follower is the dependence of the threshold voltage on the V_{SB} voltage. This can be mitigated by connecting the source and the back-gate terminals. However, this adds additional parasitics on the output that can be non-linear in nature. This confirms the previous conclusion that emitter followers have better linearity than source followers.

To summarize, in order to improve the distortion *through* the buffer, we need to:

- Increase the g_m of the follower transistor.
- Reduce the variation in the buffer parameters with the input signal, especially its g_m.
- Reduce the source impedance.
- Use an emitter follower instead of a source follower if possible.
- Increase the load impedance:
  - reduce the load capacitance, which is limited by the sampling capacitance
  - reduce the input frequency, which increases the capacitive load impedance.

These last points underscore again the challenge of achieving good distortion at high input frequencies and for large sampling capacitances. Usually, the input frequency is a parameter that is not under the control of the ADC designer. The sampling capacitance is also often decided by the noise requirement of the ADC. So, this trend tells us that it is harder to achieve good distortion (SFDR) and good thermal noise (SNR) simultaneously than it is to achieve one or the other.

An alternative approach to reducing the variation in the buffer current and its g_m without substantially increasing the power is by using an alternative path to supply the load current i_L. This is shown conceptually in Figure 4.21 [1, 2, 9, 10, 12]. This reduction in the current variation in the follower device can substantially improve its linearity. An implementation of such a distortion cancellation scheme is shown in Figure 4.22 [1, 9, 10, 12]. A replica capacitance that is equal to the load capacitance is connected between the input and the cascode node of the buffer current source. Since the latter is a low impedance node, the current flowing into the replica capacitance will be approximately equal to the load current. The high output impedance of the current source device (M_s) forces most of the replica current to flow up the cascode device (Q_c) and into the load. This reduces the current variation in the follower device and hence improves the distortion.

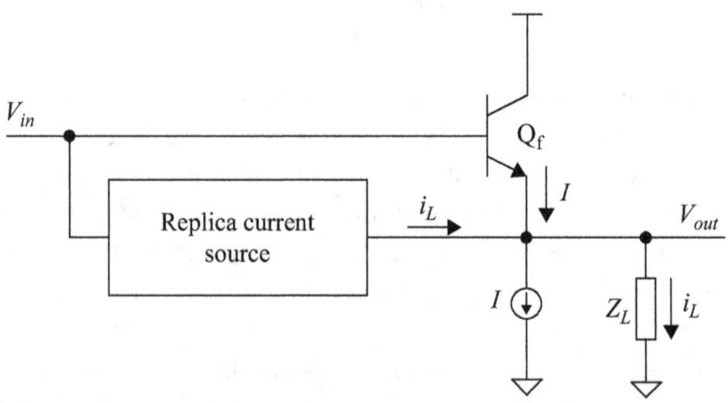

Figure 4.21 An emitter follower with distortion cancellation [1, 9, 10, 12].
© 2010 IEEE. Reprinted, with permission, from Reference 1

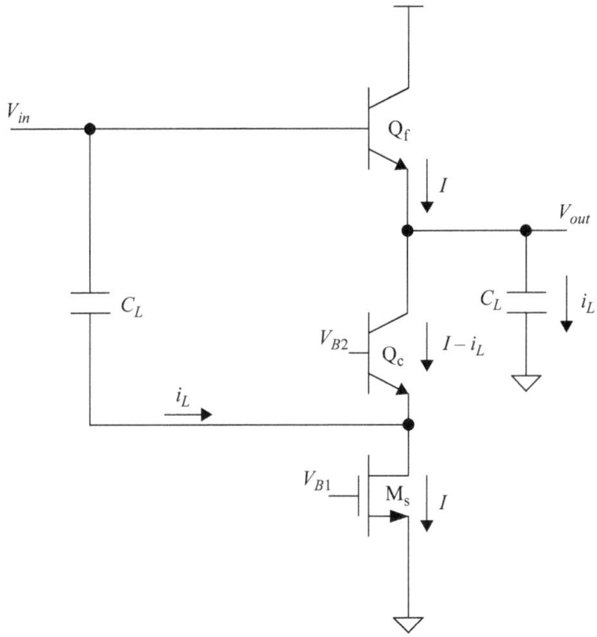

Figure 4.22 An emitter follower with distortion cancellation [1, 9, 10, 12]. © 2010 IEEE. Reprinted, with permission, from Reference 1

If the replica impedance is called Z_c, which is composed of a replica capacitance C_L in this case, g_{m2} is the transconductance of the cascode device Q_c in Figure 4.22, and Z_{f2} is its input impedance. We define the parameters ε_1 and ε_2 to be

$$\varepsilon_1 = \frac{1}{g_m Z_f} \tag{4.66}$$

and

$$\varepsilon_2 = \frac{1}{g_{m2} Z_{f2}} + \frac{1}{g_{m2} Z_c} \tag{4.67}$$

The transfer function is

$$\frac{v_o}{v_{in}} = \frac{1 + \left[\dfrac{1}{g_m Z_c} \dfrac{1}{(1+\varepsilon_1)} \dfrac{1}{(1+\varepsilon_2)}\right]}{1 + \dfrac{1}{(1+\varepsilon_1) g_m Z_L}} \tag{4.68}$$

If the compensation impedance Z_c is designed such that

$$Z_L = Z_c(1 + \varepsilon_2) \tag{4.69}$$

then, the output is given by

$$\frac{v_o}{v_{in}} \cong 1 \tag{4.70}$$

That is, if the condition in (4.69) is satisfied, this linearization technique can practically eliminate the distortion in the buffer due to the variation in its current and g_m. If g_{m2} is large, the condition in (4.69) can be approximated by

$$Z_L = Z_c(1 + \varepsilon_2) \cong Z_c \tag{4.71}$$

Equation (4.71) indicates that using a replica load should achieve the desired cancellation. Moreover, a bipolar cascode device would have a larger g_{m2} than a MOS cascode device, and hence is expected to make this distortion cancellation technique more effective.

In addition to reducing the variation in the g_m of the follower device, this distortion cancellation technique increases the input capacitance of the buffer C_i, which can be desirable to reduce the kick-back, and hence improve the kick-back linearity. However, this increase in C_i degrades the input impedance of the buffer and reduces the input bandwidth, which is the cost of this linearization scheme. If that is acceptable, using this distortion cancellation technique can lead to a 10 dB improvement in distortion and a reduction in power consumption by 70% [1].

It is important to note that, from the buffer distortion standpoint, the value of the input impedance is not as critical as its linearity. Driving a capacitance that is equal to the sampling capacitance does not mean that the buffer is useless. The sampling capacitance is harder to drive because of the presence of the input switch in series with it. That switch causes non-linear current to be drawn from the source impedance, and hence leads to tracking distortion. In addition, the kick-back from the switch is non-linear and that leads to additional distortion. On the other hand, the capacitance added for distortion cancellation draws a current that is linear during the tracking phase, and it is not switched, so there is no non-linear kick-back from it.

Having said that, the increase in the input capacitance reduces the input bandwidth, which can make it difficult for the non-linear kick-back to settle, and hence can degrade the distortion in a different way. The reduced bandwidth can also make it difficult to sample high input frequencies. This trade-off is complex and is a good example of the multi-dimensional nature of the input sampling optimization design process.

Figure 4.23 shows an emitter follower with distortion cancellation driving the whole sampling network, and Figure 4.24 shows a source follower implementation. Since the output impedance of a MOS transistor is expected to be significantly lower than that of a bipolar transistor, the source follower implementation in Figure 4.24 can suffer from a poor output impedance R_{ds} in (4.35) that can reduce the load impedance Z_L and hence degrade the distortion. This gets significantly worse in fine lithography processes due to the short-channel effects.

The output resistance of the MOS follower transistor (M_f) can be improved by using bootstrapping as shown in Figure 4.24 [2]. The input is bootstrapped to the

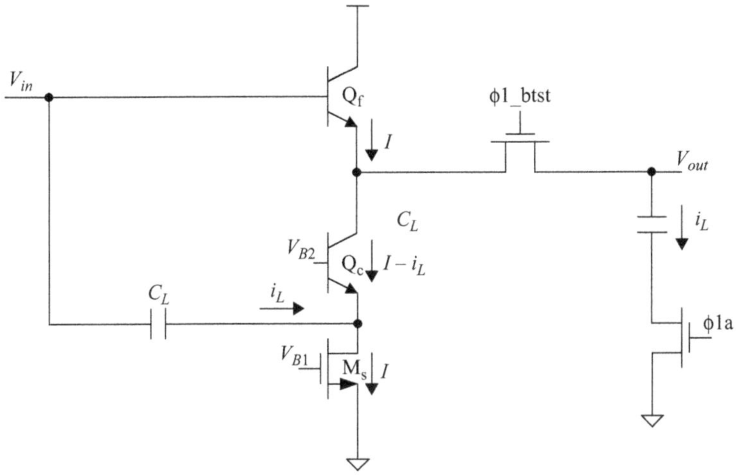

Figure 4.23 A sampling network driven by an emitter follower with distortion cancellation [1, 9, 10, 12]. © 2010 IEEE. Reprinted, with permission, from Reference 1

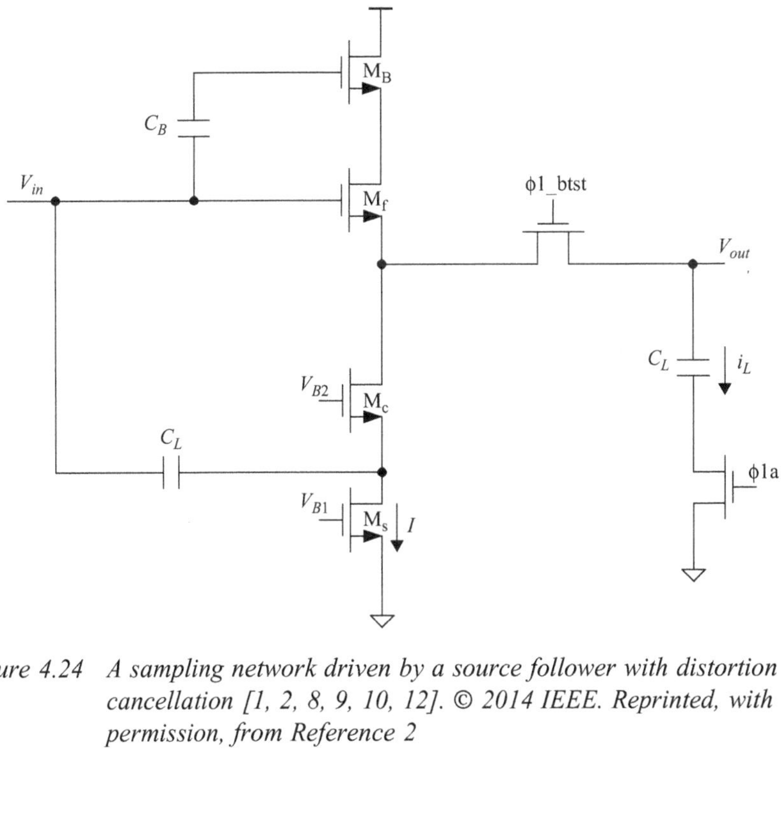

Figure 4.24 A sampling network driven by a source follower with distortion cancellation [1, 2, 8, 9, 10, 12]. © 2014 IEEE. Reprinted, with permission, from Reference 2

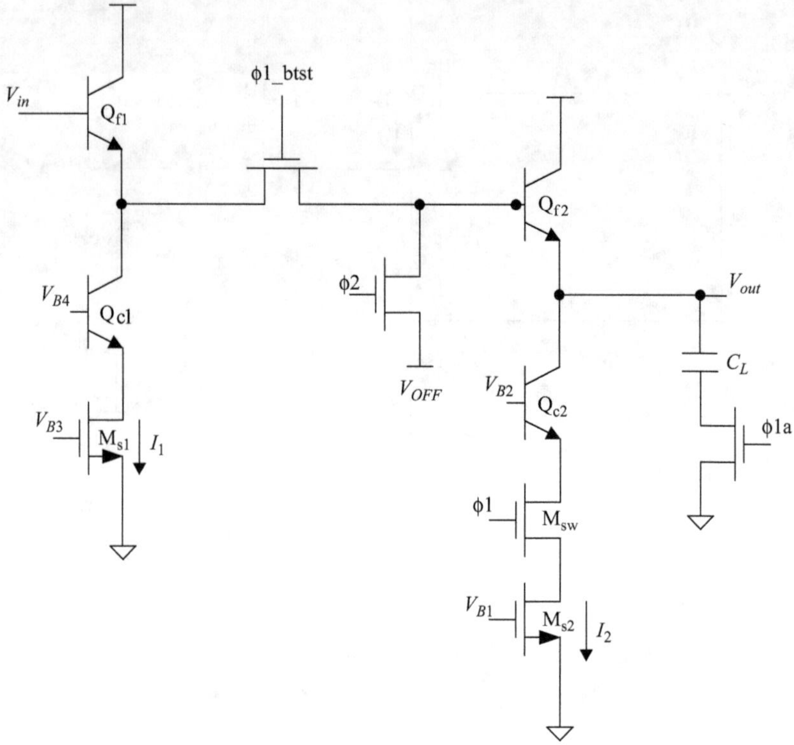

Figure 4.25 A sampling network with cascaded emitter followers for improved isolation[6, 11]. © 2006 IEEE. Reprinted, with permission, from Reference 6

drain of the follower device M_f using the auxiliary transistor M_B and the level shifting capacitance C_B. This bootstrapping reduces the variation in the drain to source voltage of the transistor M_f, and hence effectively improves its output impedance.

Another approach that was used to improve the distortion and isolation is *cascaded buffering*, as shown in Figure 4.25. This technique improves the buffer's isolation, and hence the kick-back on the ADC driver. It also reduces the requirement on the second buffer's output impedance because the input switch is moved to its input side [6, 11]. Moreover, the variable current flowing in the switch is reduced by the β of the bipolar transistor, which improves its tracking distortion substantially. However, the power consumption of this approach is usually high because of the additional buffer. The switch M_{sw} can be used to turn off the second buffer during the hold phase, and hence reduce the power consumption of that follower by 50% [11]. Distortion cancellation schemes like the ones mentioned above can also be used to lower the power consumption further [9, 10].

4.2.3 Summary of trade-offs

In order to improve the linearity of the input buffer of the ADC, we need to:

- Increase the g_m of the follower device.
- Reduce the variation in the buffer parameters, especially its g_m.
- Increase the load impedance:
  - reduce the load capacitance, which is limited by the sampling capacitance, and hence may be dictated by other factors such as noise.
  - reduce the input frequency, which increases the capacitive load impedance. This may not be in the designer's control.
  - increase the output impedance of the follower transistor and the current source.
- Reduce the variation in the input impedance of the buffer.
- Increase the input impedance of the buffer, to improve the input bandwidth and reduce its variation.
- Reduce the output impedance of the buffer.
- Use a bypass capacitance on the input node with the proper trade-off regarding the input bandwidth and kick-back settling.
- Improve the isolation of the buffer.
- Reduce the source impedance of the ADC driver.

4.3 Complementary bipolar sample and hold

In spite of the high-performance achieved by CMOS and BiCMOS samplers, complementary bipolar S/H circuits are still used especially for very high bandwidths when both NPN and PNP transistors are available with high f_t values. An example is shown in Figure 4.26, where a push-pull (class AB) design is used to reduce the power consumption [13]. The above discussion of the non-linearity of emitter followers applies to the followers in this sampler as well. The main difference is the bias point, which usually puts the followers of Figure 4.26 in a class AB operation mode, instead of the class A operation that is commonly used for emitter/source followers.

The structure in Figure 4.26 is composed of a cascade of two followers ($Q_{N1} - Q_{P4}$) and ($Q_{P1} - Q_{N4}$). The switching is performed with current steering of the current I using the devices $Q_{N2} - Q_{N3}$ and $Q_{P2} - Q_{P3}$. A clamping circuit is implemented using the devices Q_{Nc1} and Q_{Pc1} to prevent the bipolar devices from going into saturation. This S/H circuit provides very good isolation and ease of drive [13].

4.4 Clock jitter

Clock jitter can be a major contributor to the sampling noise and distortion especially at high input frequencies. It causes fluctuations in the sampling instant from sample to sample. If the jitter is random, it leads to noise degradation. If it is

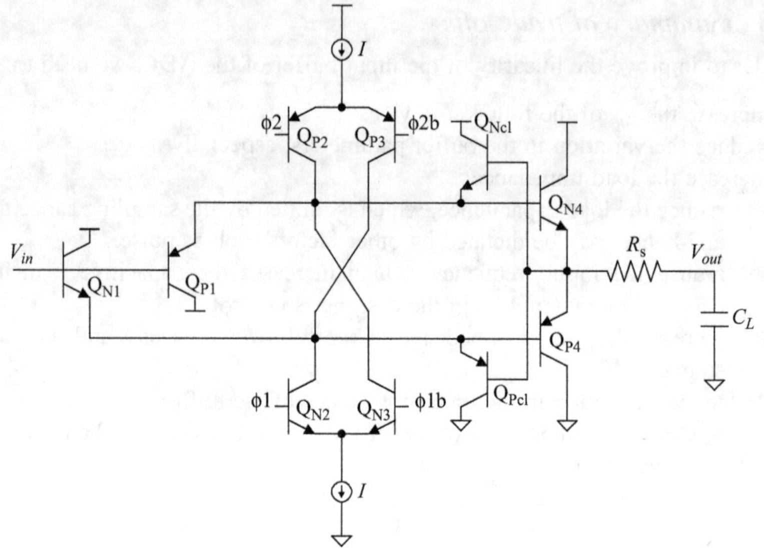

Figure 4.26 A complementary bipolar S/H circuit. © 2000 IEEE. Reprinted, with permission, from Reference 13

periodic, it leads to distortion and spurs. So, from this perspective, we can think of jitter as having two types:

- *Periodic jitter*: is usually due to coupling from periodic sources and causes distortion in the form of spurs and harmonics. It leads to degradation in SFDR and SNDR.
- *Random jitter*: is due to noise and causes degradation in SNDR.

An example of the periodic jitter is when the input signal couples on the sampling clock. As discussed in Chapter 2, the sampling process acts as a multiplier, which translates this input component on the clock into a second harmonic in the sampled signal. This second harmonic will degrade substantially with increasing the input frequency. On the other hand, examples of the random jitter are the noise sources in the clock generator and the sampling clock path inside the ADC.

A typical ADC sampling network is shown in Figure 4.27(a), and the clock path inside the ADC is shown in Figure 4.27(b). The external clock is applied differentially to an on-chip differential buffer, followed by a differential-to-single-ended converter that generates a single-ended CMOS clock. This can go to a duty-cycle stabilizer (DCS) followed by a non-overlap clock generator. The clock finally goes through drivers that generate the sampling clock with the desired amplitude, level, and rise/fall times.

Another example of the clock path is shown in Figure 4.28, where there is a DCS path and a divider path. The DCS is used to control the duty cycle of the ADC clock to the value needed by the ADC, which is typically 50%. Alternatively, a

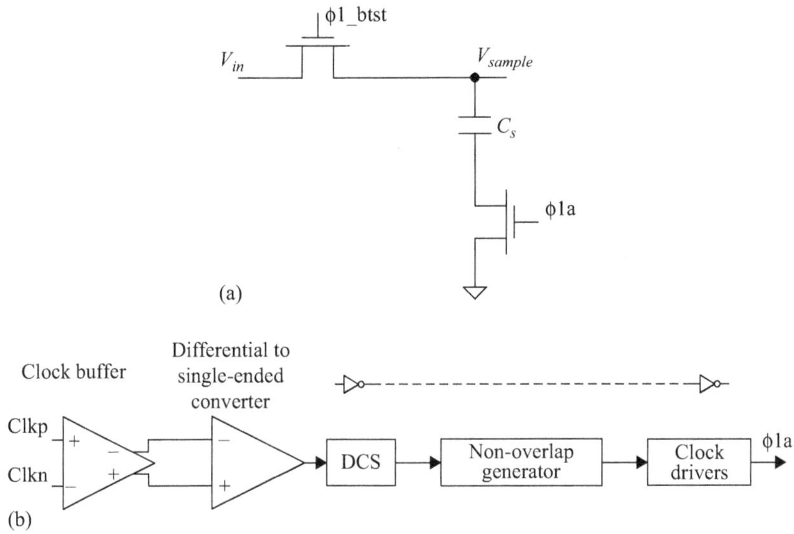

Figure 4.27 *(a) An input sampling network. (b) The clock path driving the sampling clock $\phi 1a$*

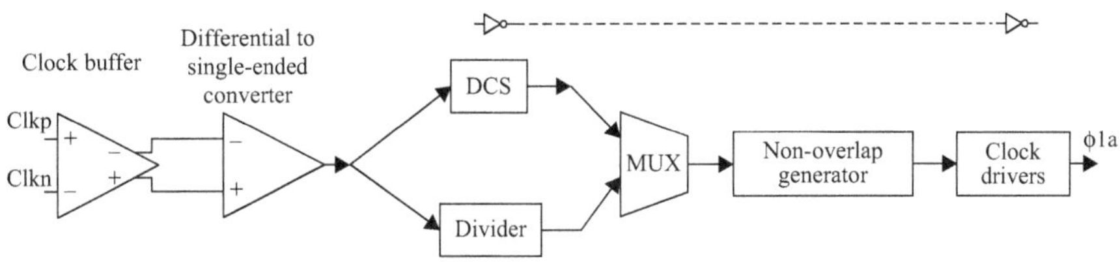

Figure 4.28 *A clock path driving the sampling clock $\phi 1a$ with either a DCS or a divider*

higher frequency clock that is 2× or 4× the ADC sampling rate can be applied to the clock buffer to improve its jitter, and a divider can be employed to divide the clock down to the sampling rate with 50% duty cycle.

Figure 4.29 shows examples of the non-overlap clock generator using NAND or NOR gates. The non-overlap time is controlled by the number of inverters in the feedback path of the non-overlap generator. With no inverters in the feedback path, we obtain a minimum non-overlap time (t_d). Inserting two inverters in the feedback path increases the delay by two gate delays to $3t_d$.

The sampling process represents a multiplication in the time domain, as discussed in Chapter 2, which is a convolution operation in the frequency domain. Ideally, the sampling clock is represented by a train of impulses in the time and frequency domains. However, the noise and spurs in the clock signal lead to impulses that contain wide-band noise, close-in noise and possibly spurs. The modulation effect of the sampling process causes those artifacts to appear around

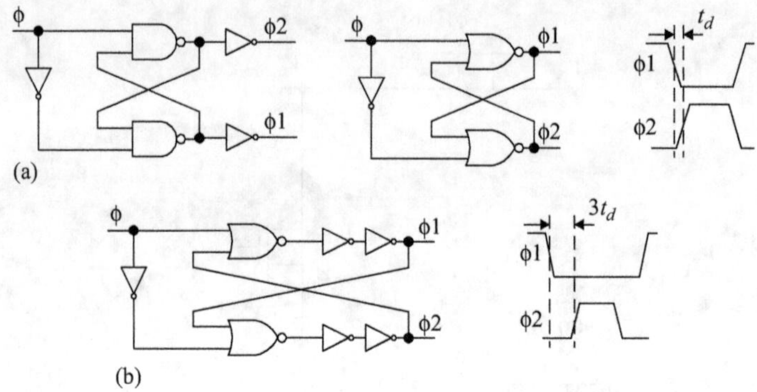

Figure 4.29 Examples of non-overlap clock generators

the fundamental of the input frequency. The modulating impact of jitter on the input signal was derived in Chapter 2 for an input sine wave and found to be

$$\Delta X(f) = \frac{2\pi f_{in} A}{2} [\Delta T(f - f_{in}) - \Delta T(f + f_{in})] \tag{4.72}$$

where $\Delta X(f)$ represents the impact of the jitter on the sampled signal, ΔT is the jitter signal, A is the input signal amplitude, and f_{in} is the input signal frequency. If the jitter is dominated by noise, the impact on the noise is given by

$$N_j = 4\pi^2 f_{in}^2 A_{RMS}^2 J^2 \tag{4.73}$$

where N_J is the noise due to jitter, J is the RMS value of the jitter, and A_{RMS} is the RMS value of the input signal. The SNDR due to jitter is

$$\text{SNDR} = \frac{1}{4\pi^2 f_{in}^2 J^2} \tag{4.74}$$

and

$$\text{SNDR(dB)} = -20\log(2\pi f_{in} J) \tag{4.75}$$

A portion of the clock jitter is due to coupling from various noise sources. These can be on the ADC chip due to supply, ground and substrate noise, or on the printed circuit board (PCB). They can cause degradation in SNDR or SFDR depending on whether the aggressor is random or periodic. They are hard to simulate, but need to be understood and minimized. Some of the measures needed to reduce coupling on the clock include:

- Sharp clock edges.
- Isolating the clock supply and grounds.
- Effective substrate isolation.
- Adequate supply decoupling.
- Effective clock isolation on the board.
- Using differential clocks.

Another contributor to the clock jitter is the noise of the devices in the ADC clock path. This random jitter degrades the ADC's SNDR. It is simulatable using the Periodic Steady State (PSS) analysis and the Periodic Noise Analysis [14], and hence can be properly optimized.

It is important to note that using traditional small-signal AC noise analysis for individual gates/stages in the clock path, then adding up their noise power can give grossly erroneous results. This is because the assumption of independence between the stages is false, and the operating point at the threshold crossing depends on the rise/fall times.

On the other hand, PSS and periodic noise analyses are suitable for cyclo-stationary noise where the ensemble averages vary periodically with time. In the PSS analysis, the simulator (such as SpectreRF [14]) calculates the large-signal steady-state periodic "operating point." Then, it linearizes the circuit around that operating point. The periodic noise analysis uses the result of the PSS analysis to modulate the noise generated by bias-dependent noise sources and to modulate the transfer function from the noise sources to the output. The linearized system is time-varying, and this method can model frequency conversion effects and solve the problem numerically. This simulation technique can be used with modulators, mixers, samplers, switched capacitor circuits, logic circuits, and oscillators. It can also be used to analyze the noise of the ADC clock path in order to simulate its total jitter.

Unlike oscillators, the jitter simulation for driven circuits and samplers requires calculating the noise at the threshold-crossing instant, not the average noise throughout the whole period. This requires using the "Strobed" Periodic Noise Analysis, which calculates the noise at a certain time point. The jitter is then calculated by integrating the noise power up to the Nyquist frequency, and dividing its square root by the slope of the clock edge obtained from the PSS analysis. Limiting the integration to the Nyquist frequency is a simulation artifact that is due to the strobing process, which samples the clock signal and aliases the wideband noise to the first Nyquist zone. An example is shown in Figure 4.30, where the jitter is given by

$$J = \frac{V_N}{Slope} = \frac{4.2 \times 10^{-3}}{84 \times 10^{9}} = 50\,\text{fs} \tag{4.76}$$

where V_N is the integrated noise voltage given as the square root of integrated noise power up to the Nyquist frequency obtained using the strobed periodic noise analysis, and "*Slope*" is the slope of the sampling clock edge at the sampling instant obtained using the PSS analysis.

This method was validated in Reference 6 and shown to consistently match the measured jitter results. It is important to note that the jitter power is *measured* by integrating the clock noise over an infinitely large bandwidth to capture all the wide band jitter noise. Limiting the integration in the simulation to the first Nyquist zone is an artifact of the aliasing effect that happens due to the sampling/strobing of the clock signal at the sampling instant during the simulation. This integration limit does not apply to other situations, especially when measuring or calculating the jitter from the phase noise as described in Chapter 2.

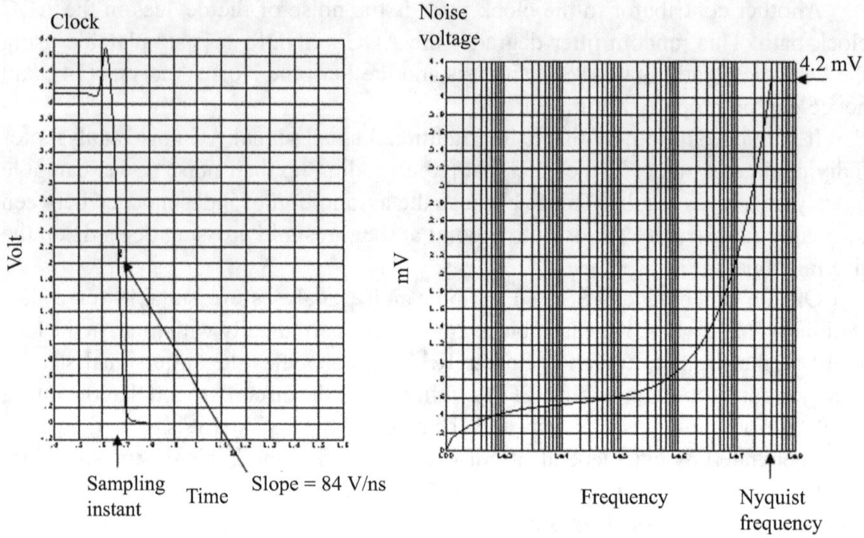

Figure 4.30 An example of the results of the PSS analysis (on the left), and the integrated noise voltage obtained using the strobed periodic noise analysis (on the right) [6]. © 2006 IEEE. Reprinted, with permission, from Reference 6

The accurate simulation of the jitter allows the designer to understand the various jitter contributors and optimize the ADC clock path design for low jitter. It also helps investigate the impact of the external clock source. However, the simulations do not account for noise coupling through the substrate, supply, and ground rails.

Typically the majority of the jitter in the ADC comes from the front-end clock buffer. This usually accounts for 60–70% of the jitter, followed by the D-to-S converter, which accounts for 20–30%. The following gates and clock distribution circuits tend to contribute minimally to the jitter due to the sharp clock edges by that point. However, an excessively large number of gates in the clock path, and/or sluggish clock edges, can change the contributions and degrade the jitter measurably. An example of a differential clock buffer is shown in Figure 4.31. The clamping diodes are used to limit the clock swing, which improves the speed and jitter of the buffer. An example of a differential-to-single-ended converter is shown in Figure 4.32.

Using a higher frequency clock at the input of the ADC, then dividing, can help the overall jitter performance [15]. This is due to the larger slope of the clock signal applied to the front-end buffer, which improves the jitter as expected by the following formula that is discussed in Chapter 2:

$$J_{RMS}(t_{th}) = \frac{\sqrt{E[n^2(t_{th})]}}{dv(t_{th})/dt} \tag{4.77}$$

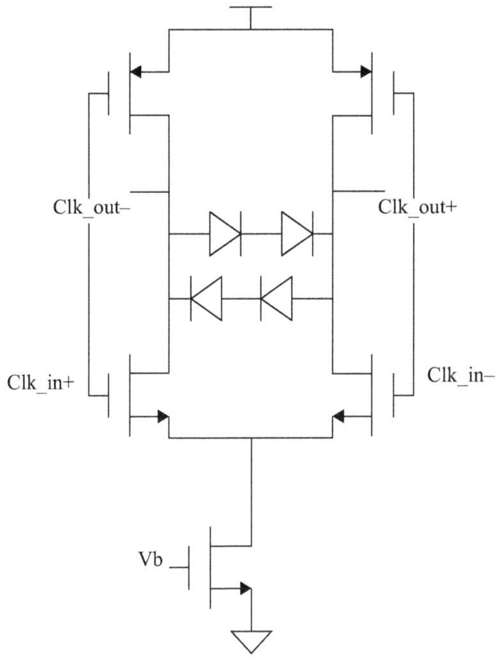

Figure 4.31 Example of a differential clock buffer

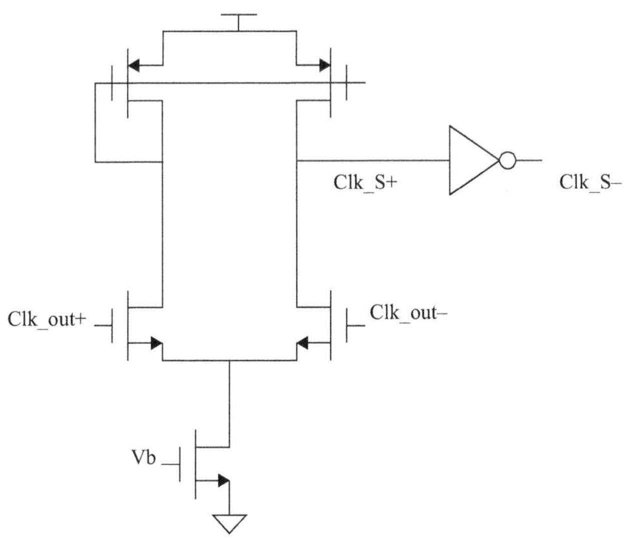

Figure 4.32 An example of a differential-to-single-ended converter

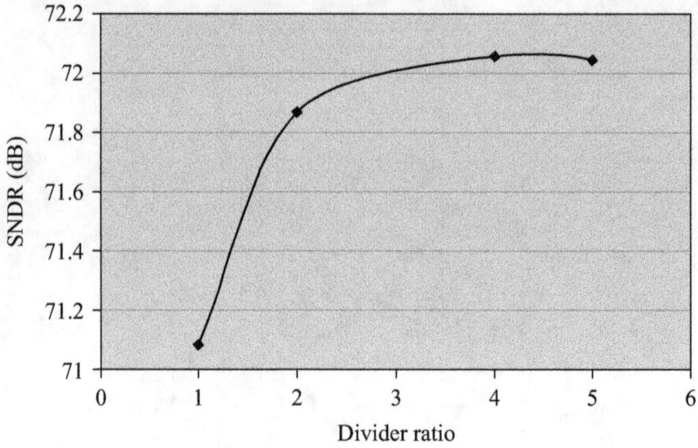

Figure 4.33 The effect of using a high-frequency clock and dividing on jitter and SNDR

where $J_{RMS}(t_{th})$ is the RMS jitter value at the threshold voltage, $E[n^2(t_{th})]$ is the jitter noise power at the threshold voltage, and $dv(t_{th})/dt$ is the slope of the signal at the threshold point. Increasing the slope of the input clock reduces the jitter contribution from the clock buffer. However, the higher frequency clock may have higher jitter. Thus, using a higher frequency clock and dividing can reduce the overall jitter if the benefit of increasing the slope outweighs the cost of the higher jitter in the input clock. An example is shown in Figure 4.33, where using a divider ratio of 2 improves the jitter substantially. Increasing the divider ratio to 4 improves the jitter further, but to a much lesser extent. Any further increase in the divider ratio does not improve the jitter any further, and can actually start to degrade it.

Since any jitter on the external clock contributes directly to the overall jitter, it is important to ensure that the external jitter is as low as possible. A low-jitter crystal oscillator is usually preferred for evaluating and testing the ADC. Alternatively, a signal generator with a sharp bandpass filter can be equally effective from the wideband noise standpoint. In both cases, the using a "clean" clock usually means using a sinusoidal clock signal.

In summary, these are some of the important points we need to note about the clock jitter:

- The clock jitter is increasingly becoming a limiting factor for the performance of high-speed and high-resolution ADCs.
- A good understanding of the clock path is necessary to optimize its design and that of the whole ADC.
- The clock jitter due to the noise in the ADC clock path is simulatable with reasonable accuracy using the PSS and the strobed periodic noise analyses.
- To reduce the clock jitter, the ADC designer needs to reduce the noise in the ADC path, the coupling on the clock path, and the noise in the external clock.
- Current state of the art for the ADC jitter is around 50 fs.

4.5 Conclusion

In this chapter we discussed the sampling process and its limitations. The sampling distortion and noise can be limited by the input path or the clock path. The sampling imperfections degrade the SNDR and SFDR. Analysis and explanation of the various contributors to the sampling non-idealities were discussed. Design trade-offs and techniques to improve the various performance parameters were covered. The optimization of the input sampling network and the input buffer were discussed in detail.

The sampling of high-frequency signals at high sampling rates is one of the most challenging analog design problems. It is a very complex and multi-dimensional problem that relies as much on intuition as it does on analysis. However, it is also one of the most enjoyable and rewarding design problems to tackle.

Problems

1. Derive (4.10), and plot the left hand side versus input frequency, sampling capacitance, and switch resistance for the same relative change in resistance value.
2. Derive (4.15) and plot the left hand side versus the input impedance and the source impedance.
3. For a differential input, how much phase and amplitude imbalance can be tolerated for the differential *HD*2 to be 90 dB if the single-ended *HD*2 is 50 dB?
4. If the sampling capacitance is 1 pF, what is the approximate bias current required for an emitter follower to have 10-bit distortion for input frequencies up to 100 MHz and 1 V amplitude? Discuss the result.
5. Repeat Problem 4 for 16-bit distortion. Discuss the result.
6. Repeat Problem 5 for 6 pF sampling capacitance. Discuss the result.
7. For a sampling capacitance of 2 pF and a source resistance of 50 Ohms, what should the switch resistance be for the input bandwidth to be at least 1 GHz? What is the sampling noise voltage? What is the SNR if the input full-scale is 1 V?
8. Repeat Problem 7 if the source resistance is 100 Ohms. Discuss.
9. For Problem 7, what is the maximum sampling rate allowed if a 10 mV kickback is to settle with 10-bit accuracy?
10. A sampling capacitance of 2 pF is driven by a source follower that is required to have an input impedance larger than 10 KOhms for input frequencies up to 200 MHz. The input capacitance of the follower device is 100 fF, what is the required g_m? What is the output impedance of the source follower?
11. If the source resistance is 200 Ohms, what is the reverse voltage gain of the buffer in Problem 10?
12. For the buffer of Problem 10, if a bypass capacitance is added at the input of the buffer equal to 2 pF, how will this change the input impedance and input bandwidth of the buffer?

13. For a bootstrapped input switch, if the parasitic capacitance at the gate of the switch is 100 fF, what should the bootstrap capacitance be for the variation in the V_{gs} of the switch to be less than 0.1%?
14. How would (4.10) change if we include the source resistance? How does the distortion change with increasing the source resistance? Discuss.

References

[1] A.M.A. Ali, A. Morgan, C. Dillon, *et al.*, "A 16-bit 250-MS/s IF Sampling Pipelined ADC with Background Calibration," *IEEE Journal of Solid-State Circuits*, 45(12), pp. 2602–2612, Dec 2010.

[2] A.M.A. Ali, H. Dinc, P. Bhoraskar, *et al.*, "A 14b 1GS/s RF Sampling Pipelined ADC with Background Calibration," *IEEE Journal of Solid-State Circuits*, 49(12), pp. 2857–2867, Dec 2014.

[3] J.M. Brunsilius, S.R. Kosic, and C.D. Peterson, "Dynamically-driven deep n-well circuit," US Patent 7,830,199, Nov 2010.

[4] P. Nikaeen and B. Murmann, "Digital Compensation of Dynamic Acquisition Errors at the Front-End of High Performance A/D Converters," *IEEE Journal of Selected Topics in Signal Processing*, 3(3), pp. 499–508, Jun. 2009.

[5] A.M.A. Ali, "Input switches in sampling circuits," US Patent 8,593,181, Nov 2013.

[6] A.M.A. Ali, C. Dillon, R. Sneed, *et al.*, "A 14-bit 125 MS/s IF/RF Sampling Pipelined ADC with 100 dB SFDR and 50 fs Jitter," *IEEE Journal of Solid-State Circuits*, 41(8), pp. 1846–1855, Aug. 2006.

[7] A.M.A. Ali and P. Bhoraskar, "Distortion cancellation in analog circuits," US Patent 8,866,541, Oct 2014.

[8] A.M.A. Ali, H. Dinc, P. Bhoraskar, *et al.*, "A 14-bit 2.5GS/s and 5GS/s RF Sampling ADC with Background Calibration and Dither," *IEEE VLSI Circuits Symposium*, pp. 206–207, 2016.

[9] A.M.A. Ali, "Buffer amplifier structures with enhanced linearity," US Patent 6,778,013, Aug 2004.

[10] A.M.A. Ali, "Signal samplers and buffers with enhanced linearity," US Patent 7,119,584, Oct 2006.

[11] A.M.A. Ali, "Buffer amplifiers with enhanced efficiency," US Patent 7,279,986, Oct 2007.

[12] A.M.A. Ali, "High performance voltage buffers with distortion cancellation," US Patent 8,339,161, Dec 2012.

[13] C. Moreland, F. Murden, M. Elliott, *et al.*, "A 14-bit 100-Msample.s subranging ADC," *IEEE Journal of Solid-State Circuits*, 35(12), pp. 1791–1798, Dec 2000.

[14] SpectreRF User Guide, Cadence Design Systems, Sep 2003.

[15] F.M. Murden and A.M.A. Ali, "Clock sources and methods with reduced clock jitter," US Patent 7,173,470, Feb 2007.

Chapter 5
Comparators

Comparators are omnipresent in A/D converters. Every ADC has at least one comparator, and the quantization operation is performed by comparators. A comparator generates a digital output bit that depends on the relation between its two inputs. A full flash converter is simply a group of comparators that quantize the input signal by comparing it to a set of equidistant thresholds.

There are numerous comparator architectures; some of which represent profoundly different trade-offs, while others tend to be largely similar in their performance. In this chapter, we discuss comparators with a focus on the architectures and design techniques that are often used in state-of-the-art high speed ADCs. The principles, design parameters, and trade-offs will be covered.

5.1 Comparator function

Ideally, a comparator performs the function shown in Figure 5.1. If the comparator's positive input (V_{inp}) is larger than its negative input (V_{inn}), the output of the comparator is "high" (V_{HI}). Otherwise, the output is "low" (V_{LO}). The inputs to the comparator can be analog or sampled-and-held signals. The outputs V_{HI} and V_{LO} represent valid digital logic levels. The transition in Figure 5.1 indicates an infinite gain. Practically, the transfer characteristic suffers from non-idealities as shown in Figure 5.2, where the comparator can have an offset (V_{OS}) and a finite DC gain A_{DC} that is given by

$$A_{DC} = \frac{V_{HI} - V_{LO}}{V_{in\_th}} \tag{5.1}$$

where V_{in_th} is the minimum input voltage difference needed to generate valid logic levels at the output. In the time domain, the comparator's output is delayed relative to its input. One measure of the comparator's speed is the propagation delay, shown in Figure 5.3 and defined as the delay between the midpoints of the input and the output transitions when the input is larger than V_{in_th}.

The comparator has multiple dimensions of performance that describe different aspects of its behavior. Some of the comparator design parameters are:

- Sampling bandwidth
- Propagation delay

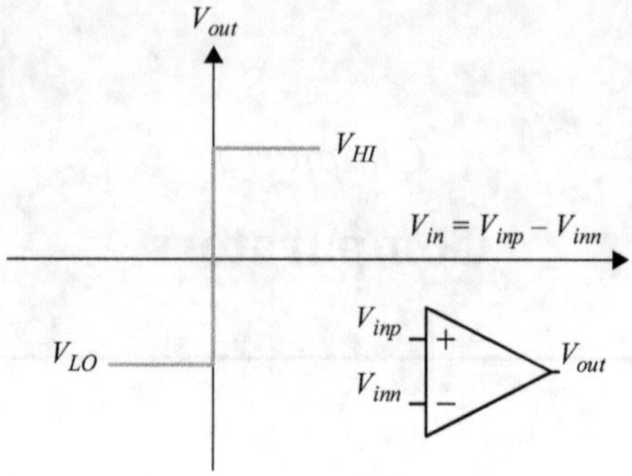

Figure 5.1 Ideal comparator characteristic

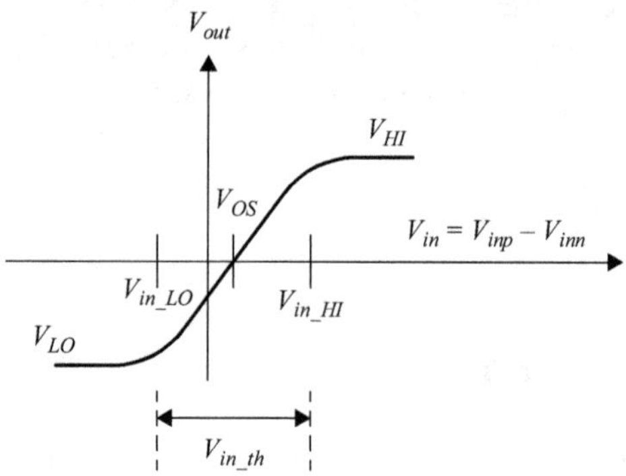

Figure 5.2 Comparator characteristic demonstrating offset (V_{OS}) and finite gain

- Metastability (BER)
- Offset
- Accuracy
- Noise
- Input common mode range
- Power consumption

The *sampling bandwidth* describes the ability of the comparator to sample wideband analog signals, or held signals at high sampling rates. It is determined by the bandwidth of its input path. In SHA-less architectures, this bandwidth dictates the upper bound on the input frequencies that the comparator, and hence the ADC, can handle.

The *propagation delay* is the time from the instant the input is sampled until the final output is a valid logic level. It is usually measured between the transition midpoints. If the output passes through a chain of gates, the propagation delay will need to include the total delay of that chain.

Metastability of the comparator happens when its input voltage difference is so small that the output is in an undefined state that is not a valid logic level.

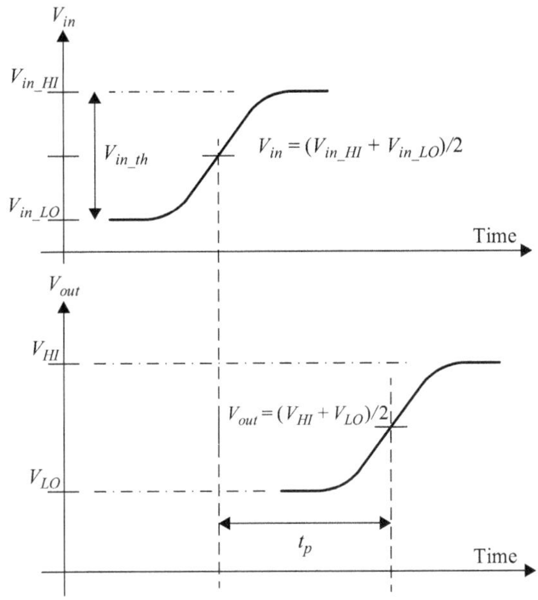

Figure 5.3 Input and output transitions of a comparator as a function of time. The propagation delay (t_p) is shown

In regenerative comparators, it is determined by the comparator's regeneration time constant. The regeneration delay is dominant when the input is very small, while the propagation delay dominates in the absence of metastability (i.e. for relatively large input values). Metastability determines the bit-error-rate (BER) and the probability of error $P(e)$ of the comparator.

The input-referred *offset* affects the accuracy of the comparator. It is often caused by mismatches between devices, which can be systematic or random. Improving matching requires increasing the area of the device to reduce the random component of the mismatch. In addition, careful layout matching techniques can be used to reduce the mismatch, such as common centroid, mirroring with half-cells, and inter-digitation [1].

Therefore, similar to many other analog blocks, the accuracy of the comparator is diametrically opposed to its speed. Measures taken to improve the accuracy of the comparator usually degrade its speed, and vice versa. For example, the offset can be improved by using large devices to reduce their mismatch, but large devices have large parasitics, which degrade the speed.

In addition to the offset, the accuracy of the comparator may be limited by dynamic and settling errors, reference errors, and noise. These non-idealities have to be addressed too, as their impact can be less predictable and more problematic than a static offset.

The noise of the comparator is the sum of the noise powers of all its building blocks, which can include a sampling network, one or more preamplifiers and a latch.

When input-referred, the noise power of any of the comparator's sub-blocks is scaled down by the square of the gain that precedes it. The noise of the sampling networks is covered in Chapter 4, and the noise of amplifiers is discussed in Chapter 6.

It is important to note that for high-speed multi-step ADCs with redundancy, the accuracy, offset, and noise of the comparator can be substantially relaxed because of the digital error correction. This helps the designer optimize the comparator for speed, sampling bandwidth, and power consumption.

The input common mode range determines the robustness of the comparator in the face of changing common mode values. The common mode range is the range of the input common mode values where the performance of the comparator does not change substantially.

5.2 Comparator structure

5.2.1 *Open loop comparator*

In its simplest form, a comparator is simply an amplifier with very high gain, as shown in Figure 5.4. In fact, simple amplifier structures like differential pairs and two-stage amplifiers can be used as comparators, especially when not compensated. An example is the differential pair shown in Figure 5.5. However, the limited gain-bandwidth product of the amplifier comparators makes them impractical for high speed applications. Instead, the open loop amplifier can be used as a gain stage driving a regenerative latch.

The input-referred offset for the differential amplifier shown in Figure 5.5, which can be used as a comparator or as a preamplifier inside a comparator, is

$$V_{OS}^2 = V_{OS_{MN1/MN2}}^2 + \left(\frac{g_{mP1}^2}{g_{mN1}^2}\right) V_{OS_{MP1/MP2}}^2 \tag{5.2}$$

where V_{OS} is the input-referred offset of the differential pair, $V_{OS_{MN1/MN2}}^2$ is the offset due to mismatches between the devices M_{N1} and M_{N2}. $V_{OS_{MP1/MP2}}^2$ is the offset due to mismatches between the devices M_{P1} and M_{P2} referred to their gates. The offsets are given by

$$V_{OS_{MN1/MN2}}^2 = \sigma_{VT\_MN1}^2 + \left(\frac{V_{gsN1} - V_{T_{MN1}}}{2}\right)^2 \left(\frac{\sigma_{\beta_{MN1}}^2}{\beta_{MN1}}\right)^2 \tag{5.3}$$

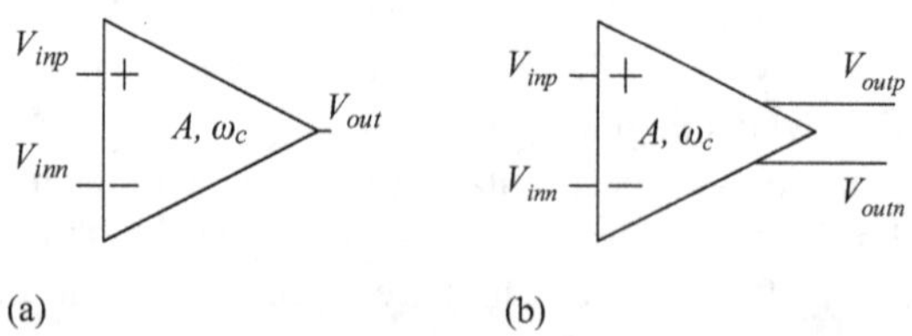

Figure 5.4 A simplified representation of (a) a single-ended comparator, and (b) a differential comparator. The gain is A and the bandwidth is ω_c

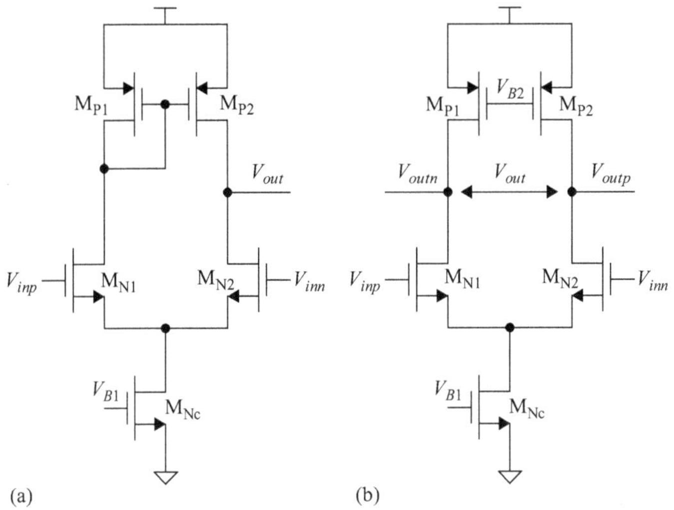

Figure 5.5 An example of a simple comparator implemented as a differential pair. (a) A single-ended comparator, and (b) a differential comparator

and

$$V^2_{OS_{MP1/MP2}} = \sigma^2_{VT\_MP1} + \left(\frac{V_{sgP1} - V_{T_{MP1}}}{2}\right)^2 \left(\frac{\sigma^2_{\beta_{MP1}}}{\beta_{MP1}}\right)^2 \tag{5.4}$$

where

$$\sigma^2_{VT} = \frac{A^2_{VT}}{WL} \quad \text{and} \quad \frac{\sigma^2_{\beta}}{\beta^2} = \frac{A^2_{\beta}}{WL} \tag{5.5}$$

such that A_{VT} and A_{β} are mismatch constants, $\beta = \mu C_{ox} W/L$, V_T is the device threshold voltage, W is the width of the MOS device, and L is its channel length. Therefore, the matching and the offset can be improved by increasing the size of the device.

If the amplifier has a DC open loop gain A_{DC}, 3-dB bandwidth of f_c and unity gain frequency of f_u, then the speed of the amplifier is determined by f_c, such that

$$f_c \approx \frac{f_u}{A_{DC}} \tag{5.6}$$

For the same unity gain frequency, the higher the gain of the comparator, the lower is its bandwidth. The small signal model of the amplifier gives [2]

$$A(s) = \frac{A_{DC}}{1 + s/\omega_c} \tag{5.7}$$

where ω_c is the 3-dB angular bandwidth, which is given by

$$\omega_c = 2\pi f_c$$

The solution to (5.7) in the time domain gives the output voltage to be

$$V_o(t) = V_{initial} + \left(V_{final} - V_{initial}\right)\left(1 - e^{-\omega_c t}\right) \tag{5.8}$$

For a step input of amplitude V_{in_th}, the DC output (i.e. V_{final}) of the comparator is given by V_{HI} and the initial voltage is given by V_{LO}. Therefore, using (5.1) in (5.8), we get

$$V_o(t) = V_{LO} + (V_{HI} - V_{LO})(1 - e^{-\omega_c t}) = V_{LO} + A_{DC}V_{in\_th}(1 - e^{-\omega_c t})$$

Therefore, the instantaneous output voltage is given by

$$V_o(t) = V_{LO} + A_{DC}(1 - e^{-\omega_c t})V_{in\_th}$$

The propagation delay is defined as the time needed to reach the midpoint between the valid high and low levels $(V_{HI} + V_{LO})/2$, and is given by [2]

$$\frac{V_{HI} + V_{LO}}{2} - V_{LO} = V_{HI} - \frac{V_{HI} + V_{LO}}{2} = \frac{V_{HI} - V_{LO}}{2} = A_{DC}(1 - e^{-\omega_c t_p})V_{in\_th} \tag{5.9}$$

Substituting (5.1) into (5.9) gives

$$\frac{V_{HI} - V_{LO}}{2} = (V_{HI} - V_{LO})(1 - e^{-\omega_c t_p})$$

Therefore, the propagation delay is

$$t_p = \ln 2/\omega_c \tag{5.10}$$

which is the well-known formula for the propagation delay in a first-order system.

For large signals, as the input amplitude increases, the delay decreases until an upper limit defined by the slew rate of the comparator. Figure 5.6 illustrates the impact of the comparator's bandwidth and over-drive on its propagation delay [2, 3]. For an input that is m times larger than V_{in_th}, (5.9) becomes

$$\frac{V_{HI} - V_{LO}}{2} = A_{DC}(1 - e^{-\omega_c t_p})m\frac{V_{HI} - V_{LO}}{A_{DC}} \tag{5.11}$$

which gives

$$\frac{1}{2m} = 1 - e^{-\omega_c t_p}$$

Thus, the propagation delay will be given by

$$t_p = \frac{\ln\left(\dfrac{2m}{2m-1}\right)}{\omega_c} \tag{5.12}$$

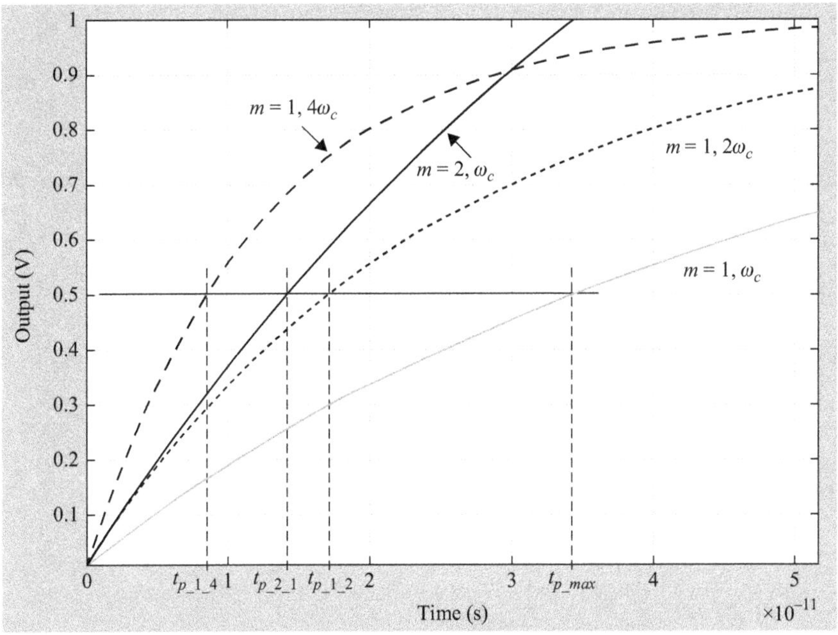

Figure 5.6 Examples of the output response of comparators for different over-drive levels ($m = 1, 2$) and different bandwidths (ω_c, $2\omega_c$, $4\omega_c$) [2, 3]

where the over-drive factor m is defined as

$$m = \frac{V_{in}}{V_{in\_th}} \tag{5.13}$$

Therefore, as m increases the propagation delay decreases. This continues until the slew rate limit is reached [2]. The slew rate defines the maximum allowed rate of change of the output voltage, and is determined by the maximum current I_{max} that is used to charge or discharge the capacitance C, which represents the total load and internal capacitance on the limiting node of the comparator. So the slew rate is given by

$$SR = \frac{\Delta V}{\Delta t} = \frac{I_{max}}{C} \tag{5.14}$$

Therefore, the slew-rate-limited propagation delay is

$$t_p = \frac{V_{HI} - V_{LO}}{2\,SR} \tag{5.15}$$

Instead of using a single amplifier, open loop comparators can be built with multi-stage amplifiers, as shown in Figure 5.7, to improve the gain-bandwidth trade-off. Using n stages of amplifiers, each with gain A_i, 3-dB angular bandwidth ω_{ci}, and unity gain angular frequency ω_{ui}, the total gain is given by [4]

$$A = \prod_{i=1}^{n} A_i \tag{5.16}$$

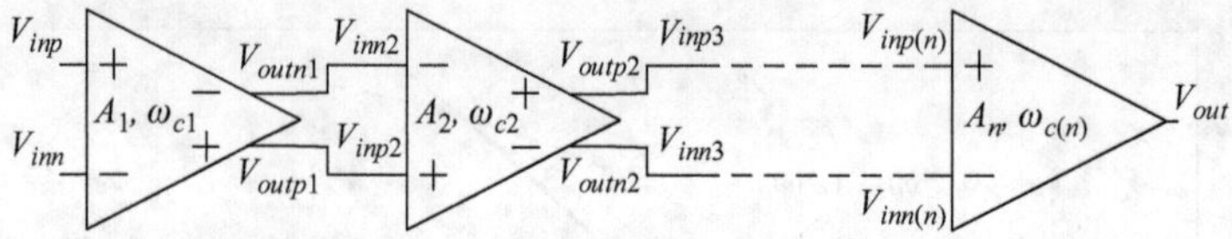

Figure 5.7 A simplified schematic showing a comparator comprised of a cascade of n gain stages [2]

If the settling time constant of each amplifier is given by τ_i, such that

$$\tau_i = \frac{1}{\omega_{ci}} \tag{5.17}$$

and the 3-dB angular bandwidth is

$$\omega_{ci} = \frac{\omega_{ui}}{A_i} \tag{5.18}$$

Then, the overall settling time constant can be approximated by

$$\tau \approx \sum_{i=1}^{n} \tau_i = \sum_{i=1}^{n} \frac{1}{\omega_{ci}} = \sum_{i=1}^{n} \frac{A_i}{\omega_{ui}} \tag{5.19}$$

If all stages have the same gain and bandwidth, then

$$\tau \approx \frac{nA}{\omega_u} \tag{5.20}$$

On the other hand, if this gain were to be achieved using a single stage, the time constant would be

$$\tau = \frac{A^n}{\omega_u} \tag{5.21}$$

Comparing (5.20) and (5.21), we see that the time constant of a single amplifier is much larger than the time constant achieved using a cascade of n stages to achieve the same gain. Therefore, using multiple stages is useful to add another degree of freedom to improve the gain with a lesser impact on the bandwidth. This approach is frequently employed in the comparator's preamplifiers as discussed later in this chapter.

5.2.2 Comparators with hysteresis

In some applications, the input noise and fluctuations can be large enough to switch the output of the comparator. This inadvertent switching is sometimes referred to as "chatter," and can lead to system problems. In these situations, it is desirable for the comparator to have *hysteresis*, as shown in Figure 5.8, where the comparator's threshold depends on the output level. If the comparator's output is high, the threshold is V_{thn}. If the output is low, the threshold is V_{thp}, where V_{thp} is larger than V_{thn}.

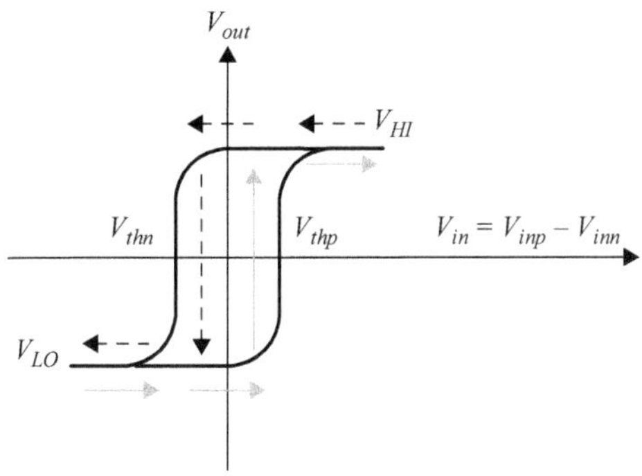

Figure 5.8 A comparator function with hysteresis centered around zero

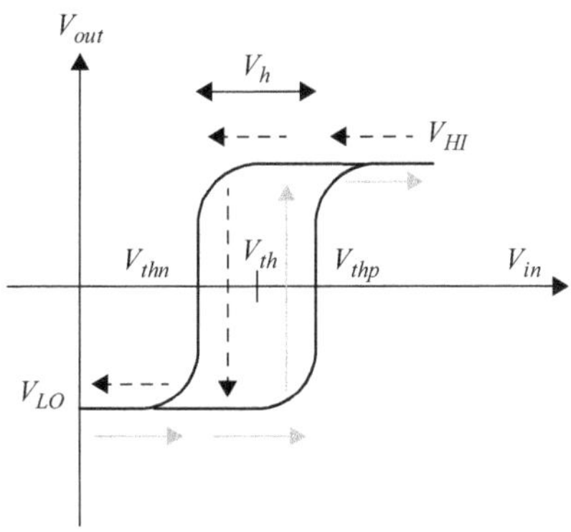

Figure 5.9 A comparator function with hysteresis centered around V_{th}

Since the threshold value depends on the *state* of the comparator, implementing hysteresis requires positive feedback.

Another example of a comparator function with hysteresis is shown in Figure 5.9, with an implementation shown in Figure 5.10. In this case, the two threshold values of the comparator are positive and given by

$$V_{thn} = \left(\frac{R_1 + R_2}{R_2}\right) V_R - \frac{R_1}{R_2} V_{HI} \tag{5.22}$$

and

$$V_{thp} = \left(\frac{R_1 + R_2}{R_2}\right) V_R - \frac{R_1}{R_2} V_{LO} \tag{5.23}$$

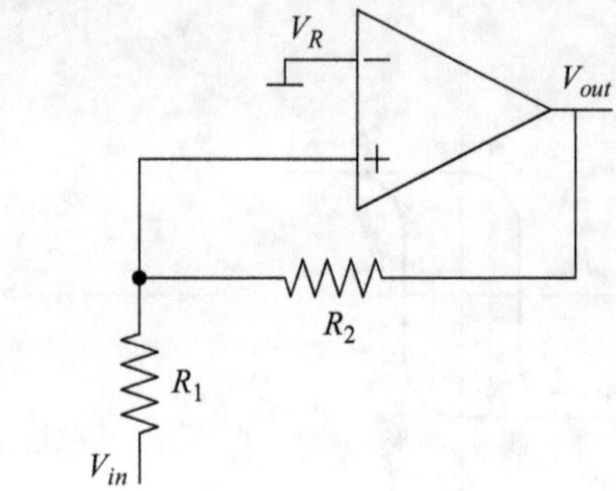

Figure 5.10 A simple implementation of the hysteresis shown in Figure 5.9

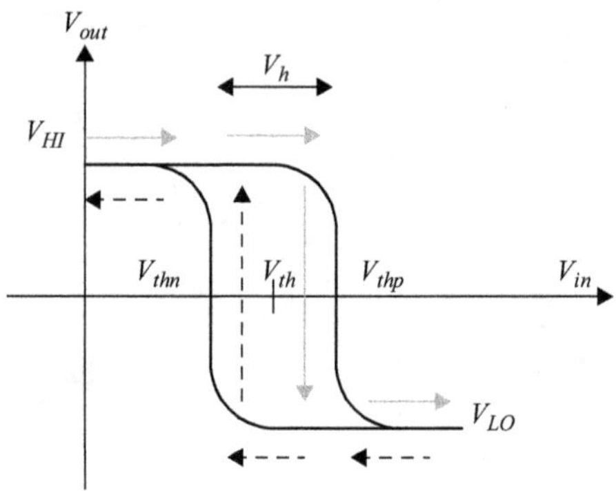

Figure 5.11 An example of inverting hysteresis

The midpoint V_{th} is given by

$$V_{th} = \left(\frac{R_1 + R_2}{R_2}\right) V_R - \frac{R_1}{R_2}\left(\frac{V_{HI} + V_{LO}}{2}\right) \tag{5.24}$$

and the magnitude or width of the hysteresis is

$$V_h = \frac{R_1}{R_2}(V_{HI} - V_{LO}) \tag{5.25}$$

Another example is shown in Figures 5.11 and 5.12, where the input is applied to the negative terminal of the comparator to give an inverting characteristic. The trip points are

$$V_{thn} = \left(\frac{R_2}{R_1 + R_2}\right) V_R + \frac{R_1}{R_1 + R_2} V_{LO} \tag{5.26}$$

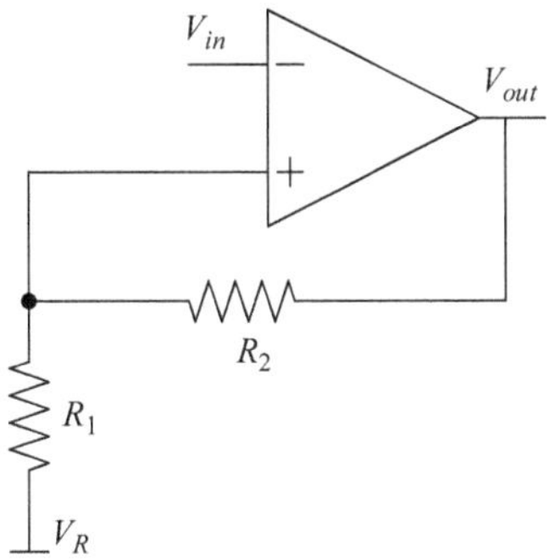

Figure 5.12 A simple implementation of the hysteresis in Figure 5.11

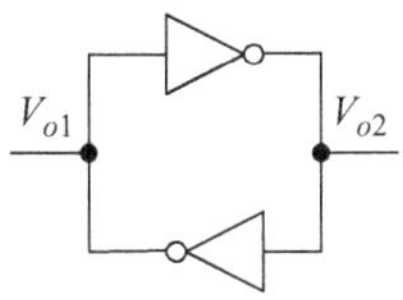

Figure 5.13 A simple representation of a regenerative latch

and

$$V_{thp} = \left(\frac{R_2}{R_1 + R_2}\right) V_R + \frac{R_1}{R_1 + R_2} V_{HI} \tag{5.27}$$

The midpoint V_{th} is given by

$$V_{th} = \left(\frac{R_2}{R_1 + R_2}\right) V_R + \frac{R_1}{R_1 + R_2} \left(\frac{V_{HI} + V_{LO}}{2}\right) \tag{5.28}$$

and the magnitude of the hysteresis is

$$V_h = \frac{R_1}{R_1 + R_2} (V_{HI} - V_{LO}) \tag{5.29}$$

5.2.3 Regenerative comparators

Regenerative comparators are based on latches (or bi-stable multi-vibrators) and are preferred for high speed applications. A latch consists of two inverters back-to-back in a positive feedback arrangement as shown in Figure 5.13. The regenerative effect of the positive feedback enables the latch to achieve better performance and speed compared to open loop amplifiers. In addition, it adds another degree of freedom that helps achieve a better overall comparator design.

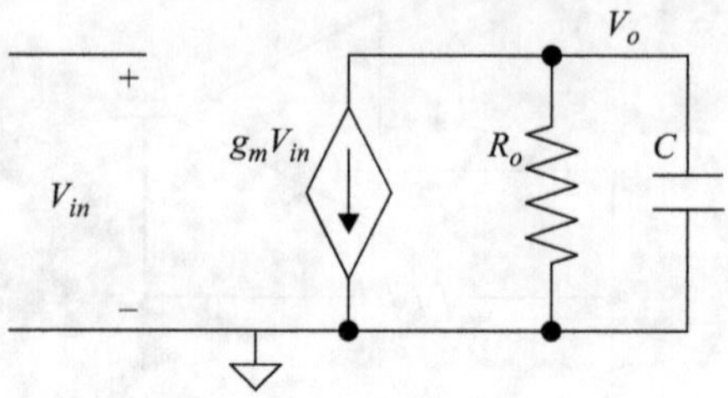

Figure 5.14 A simple model of a MOS transistor or an inverter

To analyze the latch's behavior, we represent each amplifier/inverter in Figure 5.13 with the small signal model shown in Figure 5.14. This gives [4]

$$g_m V_{o1} + \frac{V_{o2}}{R_o} = -sCV_{o2} \tag{5.30}$$

and

$$g_m V_{o2} + \frac{V_{o1}}{R_o} = -sCV_{o1} \tag{5.31}$$

Since $A = g_m R_o$ and $\tau = R_o C$, then

$$V_{o1} - V_{o2} = \frac{\tau}{A-1} s(V_{o1} - V_{o2}) \tag{5.32}$$

The solution of this differential equation depends on the initial condition at time t_0, and is given by

$$V_o(t) = V_{o1}(t) - V_{o2}(t) = V_o(t_0)e^{t(A-1)/\tau} \tag{5.33}$$

which can be represented as

$$V_o(t) \approx V_o(t_0)e^{\omega_u t} \tag{5.34}$$

where

$$\omega_u \approx \frac{g_m}{C} \approx A\omega_c \approx A/\tau \tag{5.35}$$

The latch time constant is limited by the unity gain frequency (g_m/C) of the device, which is a fundamental limit of the process. The propagation delay of the latch is given by:

$$t_p = \frac{\ln\left(\frac{\Delta V}{V_o(t_0)}\right)}{\omega_u} = \frac{\ln\left(\frac{V_{HI} - V_{LO}}{2 \times V_o(t_0)}\right)}{\omega_u} \tag{5.36}$$

Therefore, the speed of the latch is determined by its unity gain frequency, which is

$$\omega_u = \frac{g_m}{C} \tag{5.37}$$

and its gain is determined by

$$\frac{V_o(t)}{V_o(t_0)} \approx e^{\omega_u t} \tag{5.38}$$

It is clear that higher gains and speeds can be achieved using latches compared to amplifiers. Often times, comparator designers take advantage of the strengths of both amplifiers and latches by cascading one or more preamplifiers and a latch. This is shown in Figure 5.15 for a single-ended implementation, and in Figure 5.16 for a differential implementation.

For example, the open loop preamplifier can be optimized for very small input signals, while the latch is optimized for larger inputs. This is shown in the example of Figure 5.17, where we see that, for the *same time constant*, the open loop amplifier output would be faster than the latch's output for small input amplitudes,

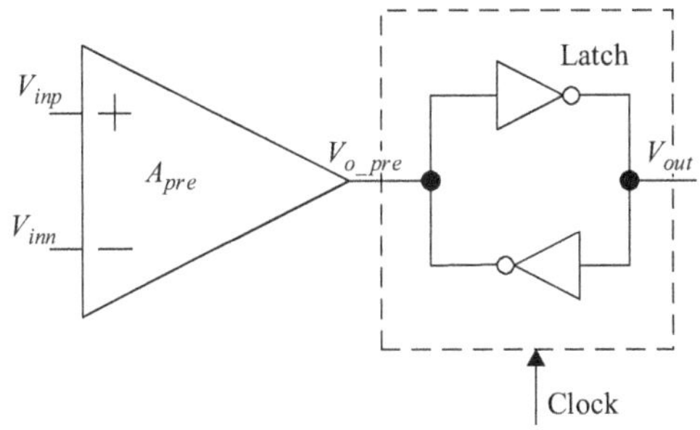

Figure 5.15 A simple single-ended comparator consisting of a preamplifier and a latch

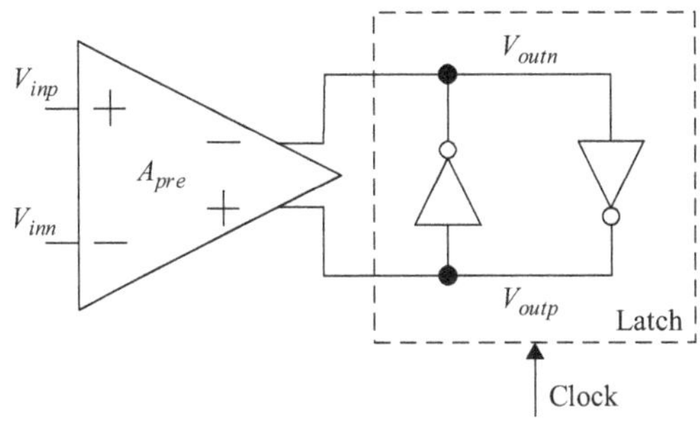

Figure 5.16 A simple differential comparator consisting of a preamplifier and a latch

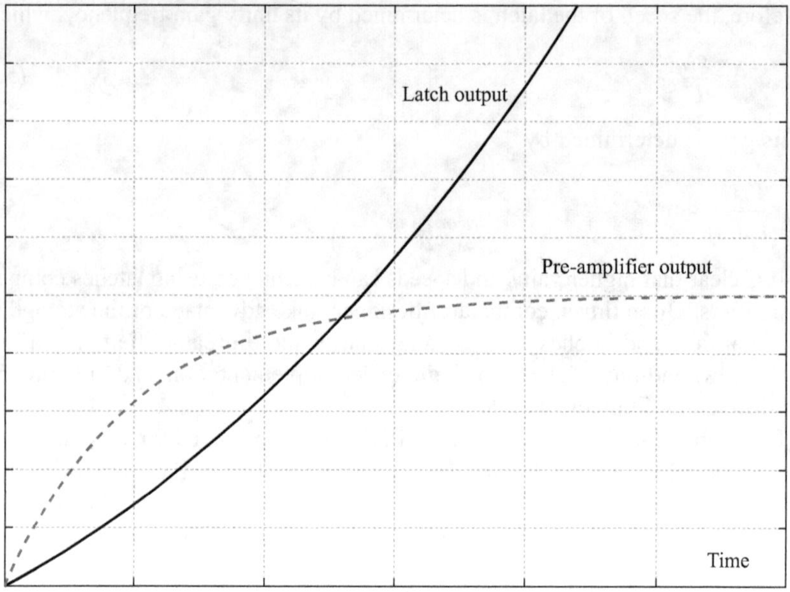

Figure 5.17 A plot showing the output versus time for an open loop amplifier and a regenerative latch. Both are assumed to have the same time constants. © 1989 IEEE. Reprinted, with permission, from Reference 3

up to a point beyond which the latch is substantially faster. In addition, the latch is usually followed by a cascade of drivers to help drive the load without overloading the latch. Therefore a comparator can be composed of a cascade of one or more preamplifiers followed by one or more latches that are followed by drivers. It has been argued that for a total desired gain G, the optimum number of preamplifiers is $\ln(G)$, each having a gain of $e = 2.7$. In addition, in some cases, a cascade of 6 low-gain preamplifiers is found to be optimum, although the speed curve tends to be nearly flat between 3 and 10 stages, which makes 3 preamplifier stages more attractive from the power and area standpoints [2, 3].

It is very important to note that the argument that the open loop preamplifier can be faster than the latch for small input amplitudes as shown in Figure 5.17 assumes the same time constant for both. However, the amplifier's time constant is determined by its 3-dB bandwidth and dominant pole frequency as shown in (5.8). If its gain A is greater than 1, its bandwidth will be A times smaller than the unity gain frequency, as shown in (5.6). On the other hand, the latch's time constant is determined by the unity gain frequency as shown in (5.34). That is, from (5.8), the output of the open loop amplifier is given by

$$V_o|_{OL} = V_{initial} + \left(V_{final} - V_{initial}\right)\left(1 - e^{-\omega_c t}\right) = A_{DC} V_{in}\left(1 - e^{-\omega_u t/A_{DC}}\right) \quad (5.39)$$

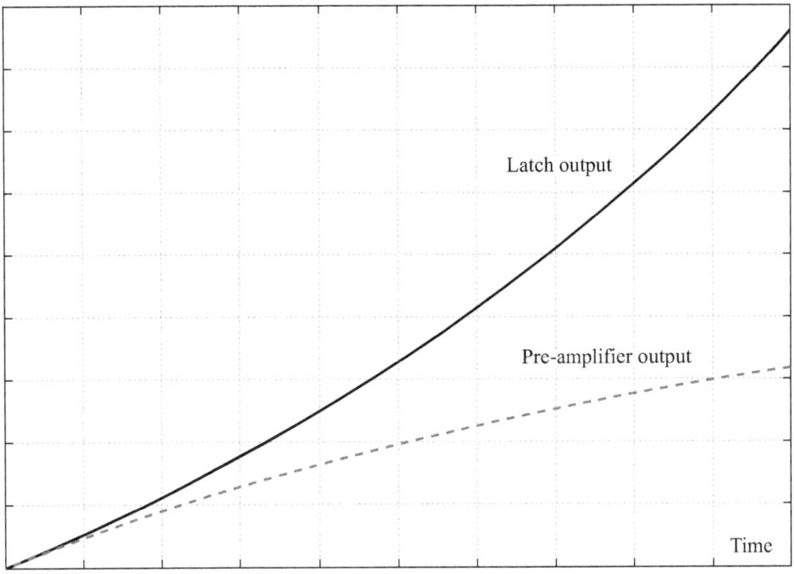

Figure 5.18 A plot showing the output versus time for an open loop amplifier and a regenerative latch. The unity gain frequency is the same for both cases

From (5.34), the output of the latch,

$$V_o|_{Latch} = V_{in}e^{\omega_u t} \tag{5.40}$$

The unity gain frequency is equal to g_m/C and is a fundamental limit of the process. If both the preamplifier and the latch are optimized for the best possible unity gain frequency, it can be shown using (5.39) and (5.40) that the latch will be faster than the preamplifier for *all* input amplitudes V_{in}. This is shown in Figure 5.18. Therefore, the main benefit of the preamplifier gain is to reduce the input-referred offset and improve the accuracy of the comparator, because the latch's offset is scaled down by the gain of preamplifier. The offset of the pre-amplifier itself can be auto-zeroed, which reduces the overall offset substantially.

However, in cases where the offset is not critical, the comparator may be reduced in its simplest form to a single latch. The preamplifier can be used to provide isolation and to help drive the latch, while the amplification and regeneration are done solely by the latch. Other reasons to use a preamplifier include providing a high input impedance for the input sampling, simplifying the resetting of the latch, and reducing the kick-back on the input. The choice of which architecture to use depends on the desired specifications in terms of offset, propagation delay, BER, and sampling bandwidth.

In regenerative comparators, two phases are usually used. In one phase, the positive feedback is disabled. In the next phase, the latch is released with the positive feedback enabled, forcing the latch into one of its two stable states, with one of its outputs high and the other output low. It is sometimes desirable to take advantage of the resetting phase for the pre-amplification, while the latching occurs in the other phase.

5.3 Metastability

Metastability refers to the period before the output of the comparator has reached a valid logic level. During the metastability period, the output is not defined from a digital standpoint, and hence it puts an upper limit on the speed of the comparator. The minimum input voltage V_{min} needed to generate a valid logic level V_{logic} for a latch is given by

$$V_{logic} \approx V_{min} e^{\omega_u T}$$

where T is the available time. The probability of error $P(e)$, or BER, is given by

$$BER \approx P(e) \approx \frac{V_{min}}{V_{Range}} \approx \frac{V_{logic}}{V_{Range}} e^{-\omega_u T}$$

where V_{Range} is the input range that the comparator covers. If the preamplifier has a net gain of A, then the BER will be given by

$$BER \approx P(e) \approx \frac{V_{logic}}{AV_{Range}} e^{-\omega_u T} \tag{5.41}$$

It is clear that the effect of the gain on metastability is small compared to the effect of its unity gain bandwidth or the available time. The BER can be made arbitrarily small by increasing the time available T and/or increasing the speed of the comparator ω_u. For example, if we ignore the terms outside the exponent, and assume a 1 GS/s ADC, where the latch phase is half the period, we get

$$T = 0.5 \text{ ns}$$

To achieve a BER of 10^{-9}, we need the unity gain frequency to be

$$\omega_u \cong \frac{\ln(10^9)}{T} = 41.4 \text{ Grad/s}$$

and

$$f_u = \frac{\omega_u}{2\pi} \cong 6.6 \text{ GHz}$$

If we want a BER of 10^{-15}, we need the unity gain frequency to be

$$f_u \cong 11 \text{ GHz}$$

On the other hand, for an open loop comparator, BER is less meaningful. The output is given by

$$V_o = A_{DC}(1 - e^{-\omega_c t})V_{in} \tag{5.42}$$

Therefore, the BER is

$$BER|_{OL} \approx P(e)|_{OL} \approx \frac{V_{min}}{V_{Range}} \approx \frac{V_{logic}}{A_{DC}V_{Range}(1 - e^{-\omega_c T})} \tag{5.43}$$

Unlike the regenerative comparator, the speed of the open loop comparator is determined by the 3-dB bandwidth (not the unity gain bandwidth). In addition, as T and ω_c are made arbitrarily large the BER is bounded by the quantity:

$$BER|_{OL} = P(e)|_{OL} < \frac{V_{logic}}{A_{DC}V_{Range}} \tag{5.44}$$

Intuitively, in the absence of positive feedback, if the input is less than the minimum needed to give a valid output logic level, the open loop comparator will stay in that undefined state indefinitely, unable to generate a valid logic level. The BER is also bounded by the quantity shown in (5.44), which is determined primarily by the gain. This is a major weakness of open loop comparators that severely limits their use in high speed ADCs.

5.4 Switched capacitor comparators

Switched capacitor configurations give comparator designers additional degrees of freedom that simplify offset cancellation, manipulation of multiple inputs and comparing differential signals using two-input comparators. Two types of switched capacitor comparators are commonly used:

1. Level Shifting Input Network
2. Charge Redistribution Input Network

5.4.1 Level shifting input network

In this structure, a single capacitance is used in series with the comparator inputs to store one of the input voltages (e.g. the reference voltage) [5]. It can also be used to store the offset the preamplifier for offset-cancellation. Examples are shown in Figure 5.19 for a single-ended implementation, and Figure 5.20 for a differential implementation. The timing diagram is depicted in Figure 5.21. During $\phi 1$, the reference is sampled on the capacitor, and the latch is in positive feedback generating the output corresponding to the previous sample. During $\phi 2$, the input is applied in series with the capacitance. The input to the preamplifier will be proportional to the input signal minus the reference voltage stored on the capacitor, and the output of the preamplifier is stored on the latch's terminals. The two clock phases are non-overlapping, and the latch clock's falling edge is advanced compared to the other clock edges.

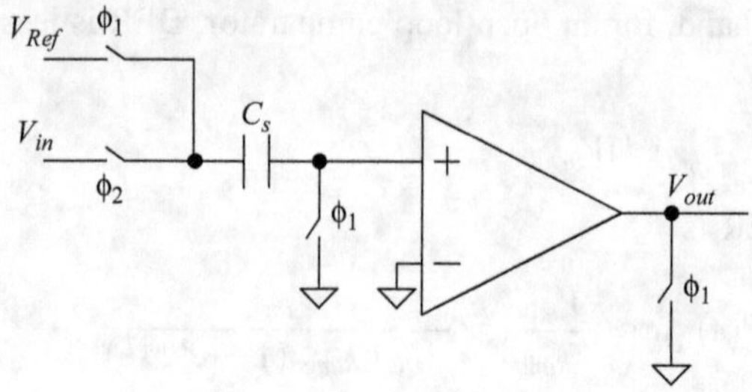

Figure 5.19 A switched-capacitor comparator using a level-shifting capacitor

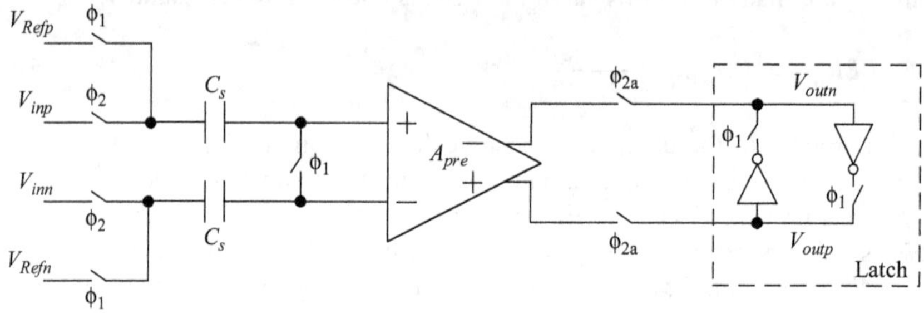

Figure 5.20 A switched-capacitor differential comparator using a level-shifting capacitor

Using a series capacitance results in some attenuation due to the voltage division between the series capacitance and the input capacitance of the comparator. In addition, the mismatch between the two capacitances contributes to the offset of the comparator. These considerations put a lower limit on the value of the series capacitance. However, if the capacitance is too large, it would require a longer time or lower impedance to charge and drive.

It is important to note that in this structure, the input propagates through the capacitance, the preamplifier and all the way to the latch regeneration nodes. The sampling bandwidth of the whole path must be large enough to achieve the desired settling. The sampling instant is determined by the falling edge of the clock $\phi 2a$. Alternatively, the input can be sampled on the capacitance during $\phi 1$, and the reference can be applied during $\phi 2$. This configuration may be preferable if the bandwidth required to sample the reference is less than that needed to sample the input.

The input-referred offset and noise of the comparator in Figure 5.20 contain components from the sampling network, the preamplifier and the latch. For the offset, this can be represented as

$$V_{OS}^2 = V_{OS\_precharge}^2 + V_{OS\_sample}^2 + \frac{V_{OS\_pre}^2}{A_s^2} + \frac{V_{OS\_latch}^2}{A_s^2 A_{pre}^2} \tag{5.45}$$

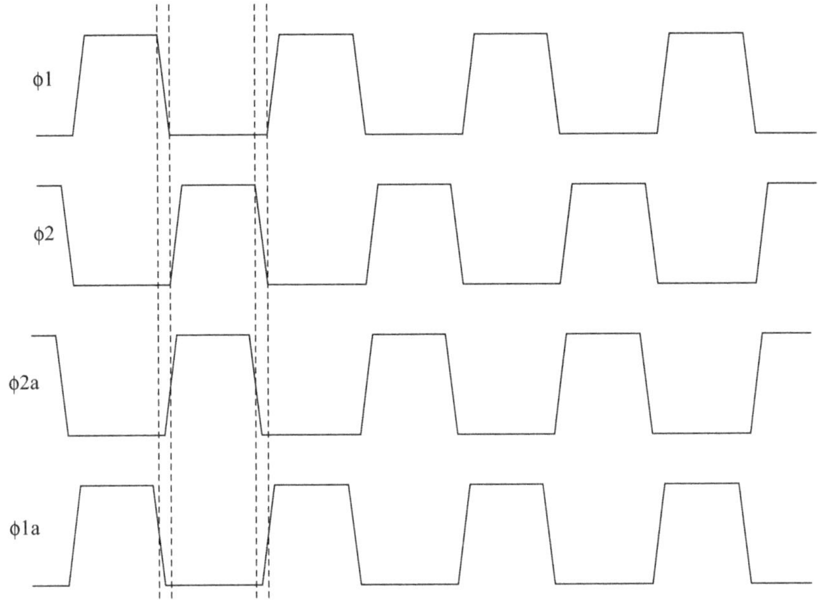

Figure 5.21 Timing diagram of a switched-capacitor differential comparator

where V_{OS} is the input-referred offset, $V_{OS\_precharge}$ is the input-referred offset due to pre-charging during $\phi 1$, V_{OS_sample} is the input-referred offset due to the sampling network during $\phi 2$, V_{OS_pre} is the offset of the preamplifier, A_s is the gain/attenuation of the sampling network, V_{OS_latch} is the offset of the latch, and A_{pre} is the gain/attenuation of the preamplifier. The sampling network gain during $\phi 2$ is given by

$$A_s = \frac{C_s}{C_s + C_{pre}} \tag{5.46}$$

where C_s is the series capacitance and C_{pre} is the input capacitance of the preamplifier.

Similarly, the input-referred noise power is given by

$$v_n^2 \approx 2\left(v_{n\_precharge}^2 + v_{n\_sample}^2\right) + \frac{v_{n\_pre}^2}{A_s^2} + \frac{v_{n\_latch}^2}{A_s^2 A_{pre}^2} \tag{5.47}$$

where $v_{n\_precharge}$ is the noise stored on the sampling capacitance during $\phi 1$, v_{n_sample} is the input-referred sampling noise during $\phi 2$, v_{n_pre} is the noise voltage of the differential preamplifier, and v_{n_latch} is the noise voltage of the differential latch. The factor of 2 in (5.47) is due to the differential operation. The input-referred sampling noise during $\phi 2$ is due to the sampling switches, and will be

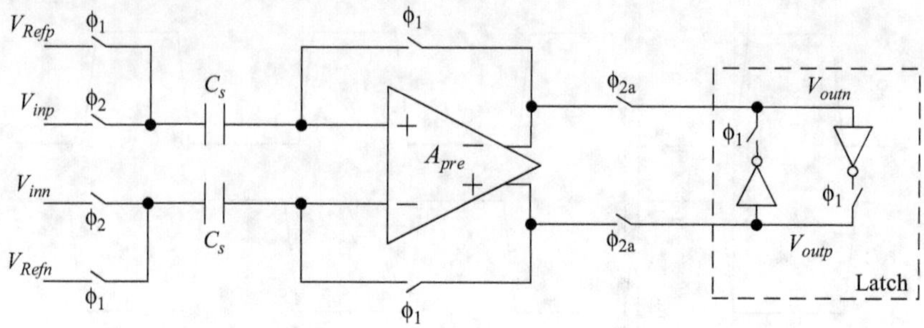

Figure 5.22 A switched-capacitor differential comparator using a level-shifting capacitor and offset cancellation of the preamplifier offset

band-limited by the preamplifier's bandwidth. The noise sampled on the sampling capacitance during $\phi 1$ is given by

$$v_{n\_precharge}^2 = \frac{kT}{C_s}\left(\frac{C_s + C_p}{C_s}\right) \tag{5.48}$$

where C_s is the series capacitance and C_p is the parasitics capacitance at the input of the preamplifier during $\phi 1$. From (5.47), we see that the noise is dominated by the kT/C noise and the noise of the preamplifier. If the gain of the preamplifier is low, the latch's noise contribution can also be significant.

A variation of this structure with offset cancellation is shown in Figure 5.22, where the offset of the preamplifier is stored on the capacitance during $\phi 1$, and then applied in series during $\phi 2$. In this case, the comparator's offset will be dominated by the latch, which can be attenuated by the gain of the preamplifier.

5.4.2 Charge redistribution input network

Another switched capacitor structure is shown in Figure 5.23, where the two inputs are sampled on two different sets of capacitors during the same phase. In the next phase, the sampling switches are opened, and the hold switches are closed [6, 7]. The charge redistribution takes place, and after some delay the latch is enabled. This structure has the advantage of a very fast sampling network, which consists of only sampling capacitances and switches. However, a delay is required between the sampling edge and enabling the latch, to accommodate the time needed for the charge redistribution and the propagation through the preamplifier and the latch. This increases the propagation delay compared to the level-shifting structure. Therefore, this structures trades propagation delay for sampling delay.

In addition to the high sampling bandwidth, the charge redistribution structure can also accommodate complex processing of more than two input signals, and enjoys smaller kick-back on the input signal. By decoupling the sampling process from the amplification and latching, the input is isolated from the latch's kick-back and the non-linearity of the preamplifier.

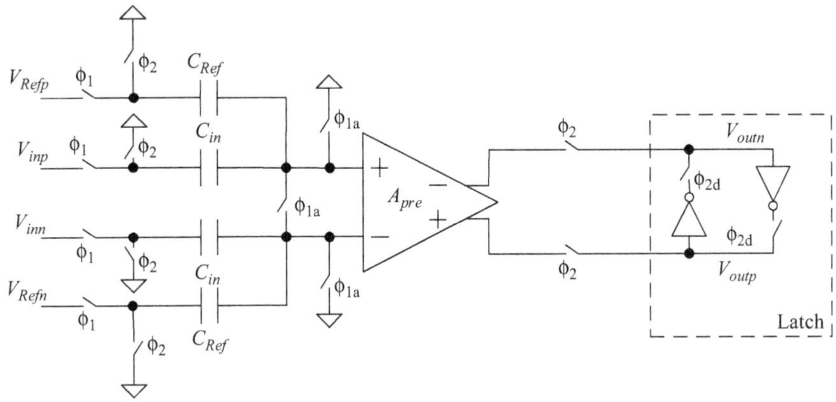

Figure 5.23 A switched-capacitor differential comparator using a charge redistribution sampling network. The clock $\phi 2d$ is delayed compared to $\phi 2$ [6, 7]

The input-referred offset and noise of the comparator in Figure 5.23 are composed of components from the sampling network, the preamplifier and the latch. For the offset, this can be represented as

$$V_{OS}^2 = V_{OS\_precharge}^2 + V_{OS\_sample}^2 + \frac{V_{OS\_pre}^2}{A_s^2} + \frac{V_{OS\_latch}^2}{A_s^2 A_{pre}^2} \tag{5.49}$$

where V_{OS} is the input-referred offset, $V_{OS\_precharge}$ is the input-referred offset due to pre-charging during $\phi 1$, V_{OS_sample} is the input-referred offset of the charge redistribution network during $\phi 2$, V_{OS_pre} is the offset of the preamplifier, A_s is the gain/attenuation of the charge redistribution network, V_{OS_latch} is the offset of the latch, and A_{pre} is the gain/attenuation of the preamplifier. The sampling network gain during $\phi 2$ is given by

$$A_s = \frac{C_{in}}{C_{in} + C_{Ref} + C_{pre}} \tag{5.50}$$

where C_{in} is the input sampling capacitance, C_{Ref} is the reference sampling capacitance, and C_{pre} is the input capacitance of the preamplifier. Since C_{in} and C_{Ref} are usually equal, the sampling gain of this comparator is significantly smaller than the gain of the level shifting comparator shown in Figure 5.20 and given in (5.46). This indicates that the input-referred offset will be worse for the charge redistribution comparator compared to the level shifting comparator.

The input-referred noise power is given by

$$v_n^2 = 2\left(v_{n\_precharge}^2 + v_{n\_sample}^2\right) + \frac{v_{n\_pre}^2}{A_s^2} + \frac{v_{n\_latch}^2}{A_s^2 A_{pre}^2} \tag{5.51}$$

where $v_{n\_precharge}$ is the noise stored on the sampling capacitance during $\phi 1$, v_{n_sample} is the input-referred sampling noise during $\phi 2$, v_{n_pre} is the noise voltage of the differential preamplifier and v_{n_latch} is the noise voltage of the differential latch. The factor of 2 is due to the differential operation.

The input-referred sampling noise during $\phi 2$ is due to the switches that are turned on during this phase, and will be band-limited by the bandwidth of the preamplifier. The noise sampled on the sampling capacitance during $\phi 1$ is given by

$$v_{n\_precharge}^2 = \frac{kT}{C_{in}} \left(\frac{C_{in} + C_{Ref} + C_p}{C_{in}} \right) \tag{5.52}$$

where C_p is the capacitance at the input of the preamplifier during $\phi 1$.

Equations (5.50)–(5.52) show that the input-referred noise will also be worse for the charge redistribution comparator compared to the level shifting comparator. This is due to the lower sampling gain and the higher sampling noise.

In summary, compared to the level shifting network, the charge redistribution structure enjoys a wider sampling bandwidth, better isolation and more flexibility during sampling, at the expense of longer propagation delay, worse offset, higher noise, and lower accuracy.

5.5 Offset cancellation

The accuracy of the comparator is limited by errors that can be static, dynamic, random, or systematic. They can be on the form of offset, gain, reference, aging, or settling errors. Careful layout and sizing techniques can reduce the systematic and random mismatches. However, the focus on speed and power consumption usually results in taking measures that compromise the accuracy. This is especially true in pipelined converters, where the redundancy and digital error correction allow for low accuracy in the flash design. However, in many cases, it is still desirable to reduce or cancel the offsets of the comparators.

In order to reduce the input-referred offset, cancellation is often employed in what is commonly called auto-zeroing. An example was shown in Figure 5.22. During the reset phase, the offset of the preamplifier is stored on the capacitor C_s. In the sampling phase, the capacitor is connected such that the stored offset is in series with the opposite polarity and hence gets cancelled.

Another technique to cancel the offset is using factory calibration/trimming. The offset is measured by applying a zero input. The offset can also be calibrated during normal operation by taking the comparator offline to be calibrated, replacing it with another spare comparator, then switching it back in after calibration. This, however, requires additional switches in the output path, which can degrade the propagation delay. In some cases, the offset can be measured in the background without disrupting the ADC normal operation using the residue of the stage in a pipeline converter as shown in Chapter 9 [8]. The offset is then cancelled by inserting an opposite offset at the input, or in the preamplifier or latch.

5.6 Loading and kick-back

The non-linear current drawn by the preamplifier and latch during the sampling phase of the level-shifting structure contributes to the sampling distortion of the input signal. In addition, the charge injection (kick-back) due to the latch resetting and regeneration can degrade the distortion if there is no adequate time for the "kick" to settle out. On the other hand, the loading and kick-back effects of the charge redistribution comparator are more benign because they are limited to the kick-back from the switches, while the preamplifier and latch are isolated from the input.

In some structures, the preamplifier is used not necessarily to provide gain, but to provide a capacitive input impedance, reduce the kick-back on the input and allow for the easy resetting of the latch. In other cases, the preamplifier can be used to create a non-linear current that is used for distortion enhancement by cancelling the current drawn by the main signal path [5] as mentioned in Chapter 4.

5.7 Comparator examples

In high speed comparators, the goal is often to maximize the speed while minimizing power consumption. As for accuracy, it is desirable to reduce the offset and other causes of errors, but the trend is to use factory calibration, offset cancellation, background calibration, and auto-zeroing techniques to handle the offset. In addition, the popularity of multi-step converters such as pipelined and SAR converters with redundancy, have made the accuracy of comparators a secondary concern, because large errors can be corrected using the digital error correction without needing calibration or trimming. Therefore, the focus of the comparator design shifts to speed and power consumption.

An example of a comparator that is composed of a preamplifier and a latch is shown in Figure 5.24. The preamplifier helps drive the latch with the required sampling bandwidth, while isolating the input from its kick-back. The latch is optimized for speed, power, and drive capability. The input propagates through the preamplifier to the regeneration nodes of the latch. The sampling instant is determined by the time the switches M_{NS}, M_{NS1}, and M_{NS2} are turned off.

The comparator offset in this example is composed of offsets due to the sampling network, the preamplifier and the latch. From (5.45) and (5.49), it can be expressed as

$$V_{OS}^2 \approx V_{OS\_precharge}^2 + V_{OS\_sample}^2 + \frac{V_{OS_{MN1/MN2}}^2}{A_s^2} + \frac{\left(V_{OS_{MP1/MP2}}^2 + V_{OS_{MN5/MN6}}^2\right)}{A_s^2 A_{pre}^2} \quad (5.53)$$

where $V_{OS\_precharge}$ is the input-referred offset due to pre-charging during $\phi 1$, V_{OS_sample} is the input-referred offset of the sampling network, A_s is the attenuation of the sampling network, A_{pre} is the gain of the preamplifier, $V_{OS_{MN1/MN2}}$ is the offset of the input pair of the preamplifier M_{N1}/M_{N2}, $V_{OS_{MN5/MN6}}$ is the offset of the

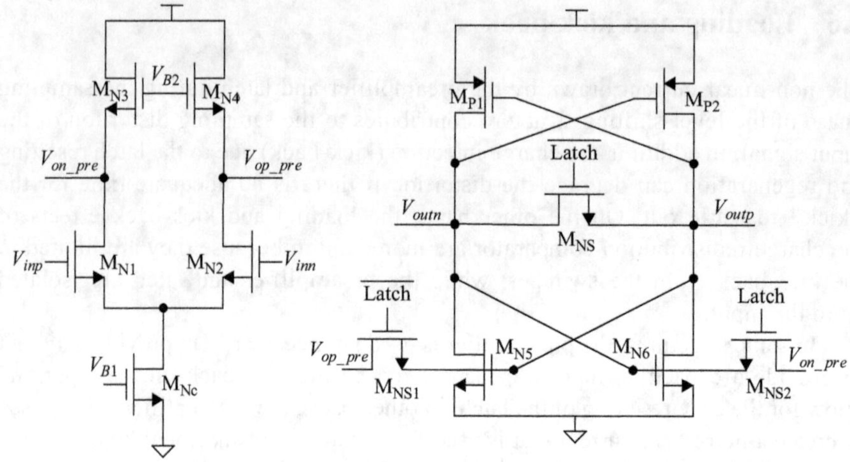

Figure 5.24 A schematic of a comparator comprises a preamplifier and a latch.

latch NMOS devices, and $V_{OS_{MP1/MP2}}$ is the offset of the latch PMOS devices. The offsets of NMOS and PMOS pairs are given by (5.3) and (5.4), respectively.

The input-referred noise power is given by

$$v_n^2 \approx 2\left(v^2_{n_{precharge}} + v^2_{n_{sample}}\right) + \frac{v^2_{n_{pre}}}{A_s^2} + \frac{v^2_{n_{latch}}}{A_s^2 A_{pre}^2}$$
$$\approx 2\left(v^2_{n_{precharge}} + v^2_{n_{sample}} + \frac{v^2_{n_{MN1}}}{A_s^2} + \frac{v^2_{n_{MN5}}}{A_s^2 A_{pre}^2} + \frac{v^2_{n_{MP1}}}{A_s^2 A_{pre}^2}\right) \tag{5.54}$$

where $v_{n\_precharge}$ is the noise stored on the sampling capacitance during the precharge phase, v_{n_sample} is the input-referred sampling noise during the amplification phase, A_{pre} is the preamplifier gain, $v_{n_{MN1}}$ is the gate-referred noise of the M_{N1} or M_{N2} device, $v_{n_{MN5}}$ is the gate-referred noise of the M_{N5} or M_{N6} device, and $v_{n_{MP1}}$ is the gate-referred noise of the M_{P1} or M_{P2} device. The gate-referred thermal noise power of a MOS transistor is given by

$$v_n^2 = \frac{2}{3} \times \frac{4kT}{g_m} \Delta f \tag{5.55}$$

where g_m is the transconductance, k is the Boltzmann constant, T is the absolute temperature, and Δf is the bandwidth.

The gain of the stand-alone preamplifier is given by

$$A_{pre} = \frac{g_{mN1}}{g_{mN3}} \tag{5.56}$$

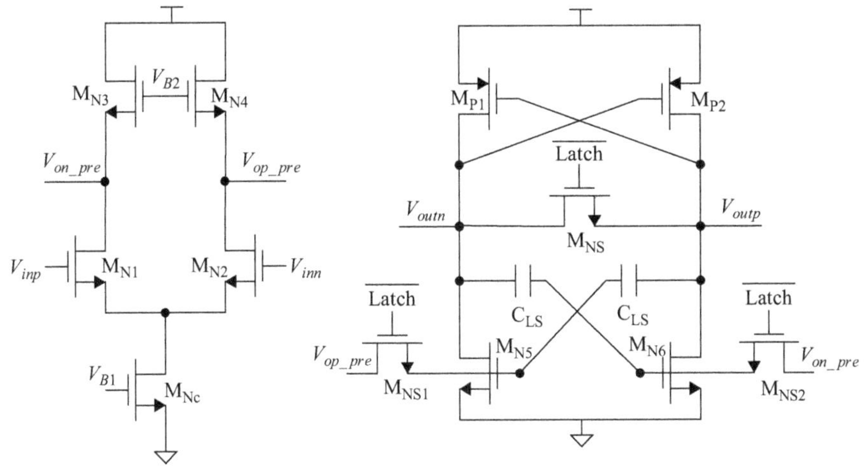

Figure 5.25 A schematic of a comparator comprises a preamplifier and a latch. © 2014 IEEE Reprinted, with permission, from Reference 5

However, when loaded with the latch, the output resistance will include the "crow-bar" switch of the latch (M_{NS}) and any other differential resistive load (R_L), to give

$$A_{pre} = \frac{g_{mN1}}{g_{mN3} + 2/R_L} \tag{5.57}$$

The bandwidth of the preamplifier determines the sampling bandwidth and is given as

$$BW_{3dB_{pre}} = \frac{g_{mN3} + 2/R_L}{2\pi C_L} \tag{5.58}$$

where C_L is the load capacitance. Finally, the unity gain bandwidth of the latch is

$$\omega_u = \frac{g_{mN5} + g_{mP1}}{C_L} \tag{5.59}$$

When the "Latch" signal is low, the latch is reset, and the input is sampled on the preamplifier and the latch regeneration nodes. When the "Latch" signal goes high, the latching starts, and the output moves toward a valid logic level. Usually, the input is disconnected in this phase, to avoid any disruption to the latching process.

Figure 5.25 shows a similar structure with the addition of level shifting capacitances (C_{LS}) [5]. These capacitances are used for DC level shifting, so that the V_{gs} of the NMOS and PMOS devices are maximized independently beyond the power supply value V_{DD}. In the absence of the capacitances C_{LS}, the V_{gs}'s of the NMOS and PMOS devices of the latch are limited by

$$V_{gsN6} + V_{sgP2} = V_{DD} \tag{5.60}$$

By adding the capacitances C_{LS}, we have

$$V_{gsN6} + V_{sgP2} = V_{DD} + V_{C_{LS}} \tag{5.61}$$

That is, using level shifting, the V_{gs} voltages can be maximized. For example, if $V_{C_{LS}}$ is equal to V_{DD}, the V_{gs} of each of the NMOS and PMOS devices can be made equal to V_{DD}, which makes their sum equal to $2V_{DD}$, instead of V_{DD}. This improves the g_m of the devices, thereby improving the noise and speed of the latch substantially.

It is important to note that in some of these implementations, the preamplifier is optimized to reduce the kick-back, help drive, and reset the latch, while maximizing the sampling bandwidth. It may not provide any amplification, as its gain may actually be less than 1.

Figure 5.26 shows an example of a CMOS comparator with a high-gain preamplifier that uses PMOS current sources as loads. The high output impedance formed by the PMOS load makes it important to employ common-mode feedback (CMFB) to control the output common-mode operating point [9]. The CMFB is accomplished by connecting the output of the preamplifier to the gate of the NMOS current source. If the output common-mode is too high, the NMOS current increases, thus lowering the output common-mode, and vice versa, therefore, keeping the amplifier working in its desired region of operation. The high resistance of the PMOS current source means the bandwidth will be much lower than

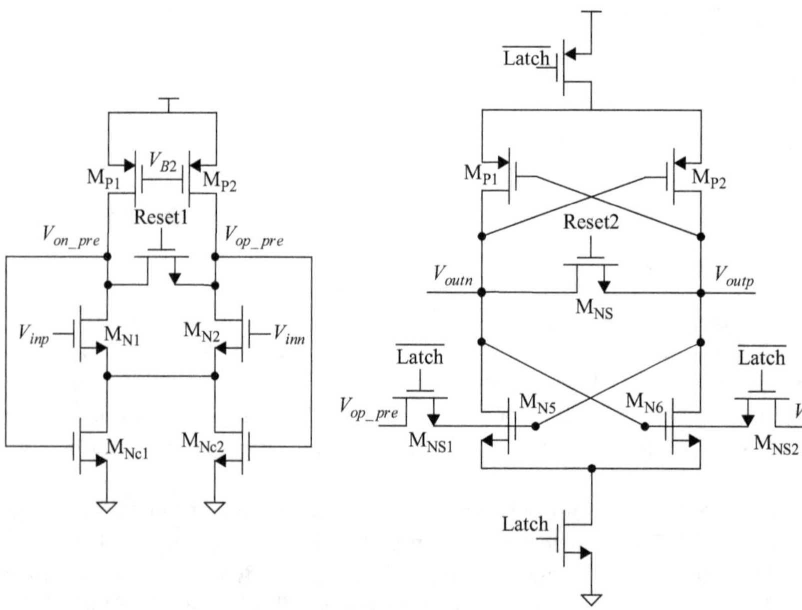

Figure 5.26 A schematic of a comparator comprises a preamplifier and a latch. © 2000 IEEE. Reprinted, with permission, from Reference 9

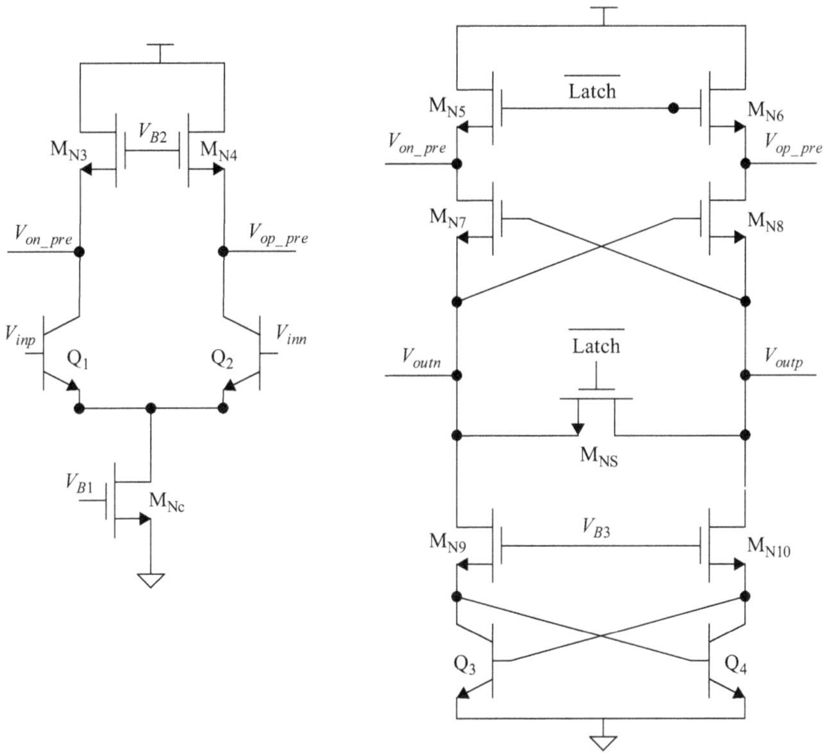

Figure 5.27 A schematic of a BiCMOS comparator comprises a preamplifier and a latch. © 2006 IEEE. Reprinted, with permission, from Reference 6

the preamplifier of Figure 5.24. However, in this implementation, the dynamic gain is used to amplify the input signal, without requiring complete settling. Unlike the structure of Figure 5.24, this preamplifier does not have enough bandwidth to handle a high-frequency input signal, and usually requires a held signal [9].

Figure 5.27 shows a BiCMOS comparator, where bipolar transistors are used as input devices in the preamplifier and in the latch. The high unity-gain frequency f_T, high gain, low parasitics, and low offset of the bipolar transistors improve the propagation delay, regeneration time, and the offset of the comparator [6]. These comparators enjoy higher speed, better accuracy, and better BER compared to their CMOS counterparts.

In the above-mentioned examples, a sampling network is employed to compare the differential input with a certain reference level. Alternatively, the comparator can have 4 inputs, in order to compare two differential inputs without needing sampling capacitors. An example of a 4-input comparator is shown in Figure 5.28. Although this type of comparators can be quite fast, it suffers from possible errors due to the common mode differences between the two differential inputs.

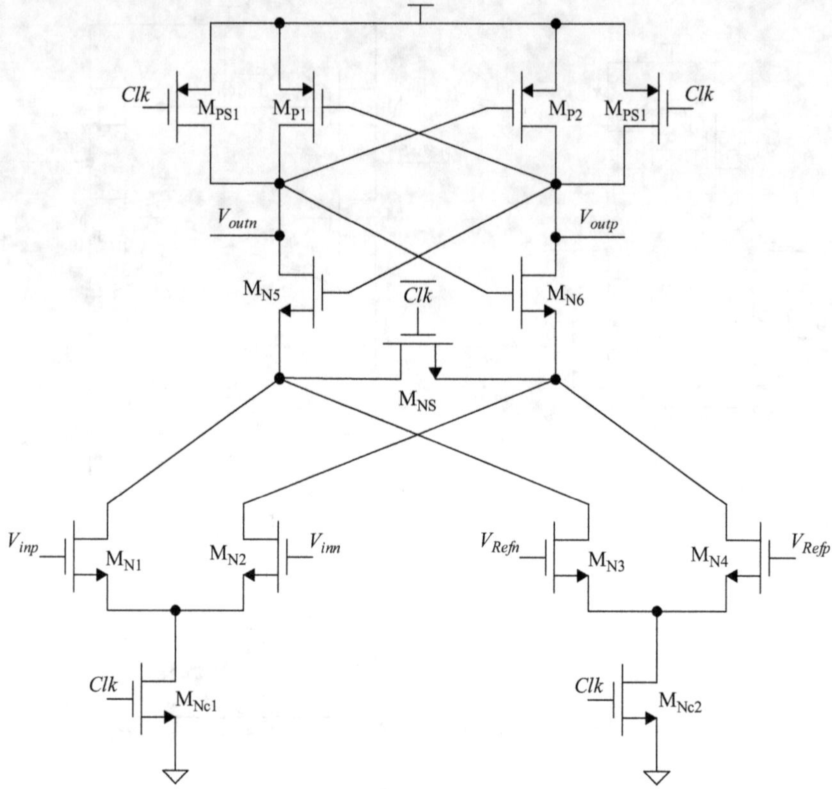

Figure 5.28 A schematic of a 4-input comparator

5.8 Conclusion

In this chapter, the design and performance metrics of comparators were discussed. We covered open loop, regenerative, and switched capacitor comparators. The preferred implementation of building the comparator as a cascade of one or more preamplifiers followed by a regenerative latch was presented. In some applications using multiple preamplifiers is considered optimum. In other applications, only one preamplifier is used for isolating the kick-back and driving the latch, while providing no amplification. Design trade-offs were explained with examples of techniques to cancel the offsets and improve the speed. Finally, we presented circuit examples of comparators in CMOS and BiCMOS processes.

Problems

1. A comparator is built using a single pole open loop amplifier with DC gain of 40 dB, and gain bandwidth product of 100 MHz.
   (a) What is the propagation delay of the comparator?

(b) If the minimum valid logic level is 1 V, what is the minimum amplitude that the comparator can resolve?
(c) If the input full-scale of the comparator is 1 V, what is the best BER achievable?

2. A comparator is built using a latch with a gain bandwidth product of 100 MHz. If the minimum valid logic level is 1 V, and using an input voltage equivalent to that of Problem 1,
   (a) What is the propagation delay of the comparator?
   (b) If the input full-scale of the comparator is 1 V, what is the BER achievable within 50 ns?
   (c) What is the BER if the time changes to 100 ns?
3. Using (5.39) and (5.40), derive the condition needed for the regenerative latch to be faster than the open loop amplifier, if they have the same unity gain frequency. Assume the same initial condition. Explain the result.
4. Using (5.39) and (5.40), derive the conditions needed for the regenerative latch to be slower than the open loop amplifier. Assume the same initial condition.
5. For a total gain G, derive the optimum number of cascaded amplifiers needed to achieve the total gain with the fastest response.
6. For a regenerative latch that is operating at 500 MS/s, what is the unity gain frequency needed to achieve a BER of 10^{-9}, 10^{-10}, and 10^{-15}? Discuss the results.
7. A comparator is composed of a preamplifier followed by a latch. The preamplifier has a gain of 20 dB and a 3-dB bandwidth of 1 GHz. The latch has a unity gain frequency of 1 GHz. If the input voltage is 1 mV, and the valid output logic level is 0.5 V. Find the comparator's propagation delay if the latch is enabled when its input is equal to:
   (a) 1 mV.
   (b) 2 mV.
   (c) 5 mV.
   (d) 10 mV.

   When should the latch be enabled to minimize the propagation delay?
8. A comparator is composed of a preamplifier followed by a latch. The preamplifier has a gain of 20 dB and a 3-dB bandwidth of 1 GHz. The latch has a unity gain frequency of 10 GHz. If the input voltage is 1 mV, and the valid output logic level is 0.5 V, Find the comparator's propagation delay if the latch is enabled when its input is equal to:
   (a) 1 mV.
   (b) 2 mV.
   (c) 5 mV.
   (d) 10 mV.

   When should the latch be enabled to minimize the propagation delay?
9. Repeat Problem 8 if the gain of the preamplifier is 1 (0 dB).
10. A comparator is composed of a latch with an offset of 10 mV. If the input-referred offset of the comparator should not exceed 1 mV, how many

preamplifiers (if any) should we use to drive the latch while minimizing the propagation delay? Ignore the offsets of the preamplifiers or assume they will be corrected during operation.

11. For the comparator structure shown in Figure 5.20, assume the noise voltage of the latch is 1 mV, and the noise of the preamplifier is 0.5 mV. The gain of the preamplifier is 20 dB, its 3-dB bandwidth is 100 MHz and its input capacitance is 50 fF. The sampling capacitance 0.1 pF and the time constant of the RC network is 10× smaller than the time constant of the preamplifier. The parasitic capacitance at the input of the preamplifier during $\phi 1$ is 50 fF. Estimate the total input-referred noise of the comparator.
12. For the comparator structure shown in Figure 5.23, assume the noise voltage of the latch is 1 mV, and the noise of the preamplifier is 0.5 mV. The gain of the preamplifier is 20 dB, its 3-dB bandwidth is 100 MHz and its input capacitance is 50 fF. The sampling capacitances are both 0.1 pF and the time constants of the RC networks are 10× smaller than the time constant of the preamplifier. The parasitic capacitance at the input of the preamplifier during $\phi 1$ is 50 fF. Estimate the total input-referred noise of the comparator.
13. Using a SPICE like simulator, model the comparator shown in Figure 5.20. You can use transistor-level circuits or macro-models. The capacitance C_s is 1 pF and the reference $V_{Refp} - V_{Refn}$ is 0.5 V. The preamplifier has a gain of 20 dB and a unity gain frequency of 100 MHz. The latch has a unity gain frequency of:
    (a) 50 MHz.
    (b) 100 MHz.
    (c) 200 MHz.

    Plot the output of the comparator as a function of time for differential inputs of 0.1 mV, 1 mV, 10 mV, and 100 mV.
14. For Problem 13, introduce offsets in C_s, the preamplifier, and the latch. Comment on the impact of each offset.
15. Repeat Problems 13 and 14 for the comparator shown in Figure 5.23 with equal sampling capacitances of 1 pF each. Discuss the results.
16. Derive the expressions for the gain and 3-dB bandwidth for the preamplifier shown in Figure 5.26. Compare the results with those for the preamplifier of Figure 5.24. If there is no time for the preamplifier to settle, what is its dynamic gain as a function of the settling time t_s? How do you think the two comparators compare in terms of sampling bandwidth, offset, and propagation delay if their latches were identical?
17. For the comparator in Figure 5.28, derive the expression for the differential output in terms of the inputs and the reference values. How would a difference in the common mode value affect the output?
18. A 5-bit 1 GS/s flash ADC with an input full-scale of 1 V. Each of its comparators is composed of a preamplifier with a gain of 10, and a latch with unity gain frequency of 20 GHz. If the valid logic level is 0.5 V, what is the BER? Repeat if the flash is only 4 bits. Discuss the results.

References

[1] A. Hastings, "The Art of Analog Layout," Prentice Hall, Upper Saddle River, NJ, 2001.

[2] P.E. Allen and D.R. Holberg, "CMOS Analog Circuit Design," Second Edition, Oxford University Press, New York, NY, 2002.

[3] J. Doernberg, P.R. Gray, and D.A. Hodges, "A 10-bit 5-Msample.s CMOS Two-Step Flash ADC," *IEEE Journal of Solid-State Circuits*, 24(2), pp. 241–249, Apr. 1989.

[4] M. Gustavsson, J.J. Wikner, and N.N. Tan, "CMOS Data Converters for Communications," Kluwer Academic Publishers, Dordrecht, The Netherlands, 2000.

[5] A.M.A. Ali, H. Dinc, P. Bhoraskar, *et al.*, "A 14b 1GS/s RF Sampling Pipelined ADC with Background Calibration," *IEEE Journal of Solid-State Circuits*, 49(12), pp. 2857–2867, Dec 2014.

[6] A.M.A. Ali, C. Dillon, R. Sneed, *et al.*, "A 14-bit 125 MS/s IF/RF Sampling Pipelined ADC with 100 dB SFDR and 50 fs Jitter," *IEEE Journal of Solid-State Circuits*, 41(8), pp. 1846–1855, Aug 2006.

[7] A.M.A. Ali, A. Morgan, C. Dillon, *et al.*, "A 16-bit 250-MS/s IF Sampling Pipelined ADC with Background Calibration," *IEEE Journal of Solid-State Circuits*, 45(12), pp. 2602–2612, Dec 2010.

[8] A.M.A. Ali, H. Dinc, P. Bhoraskar, *et al.*, "A 14-bit 2.5GS/s and 5GS/s RF Sampling ADC with Background Calibration and Dither," *IEEE VLSI Circuits Symposium*, pp. 206–207, 2016.

[9] K. Nagaraj, D.A. Martin, M. Wolfe, *et al.*, "A Dual-Mode 700-Msamples/s 6-bit 200-Msamples/s 7-bit A/D Converter in a 0.25-μm Digital CMOS Process," *IEEE Journal of Solid State Circuits*, 35(12), pp.1760–1768, Dec 2000.

Chapter 6
Amplifiers

In this chapter, the structure, analysis, and design of amplifiers are discussed. Amplifiers are used as sample-and-hold circuits (SHAs), as inter-stage (MDAC) amplifiers in pipelined ADCs, and in the integrators of sigma-delta converters. The performance of the amplifier is often critical for the whole ADC. For example, in a pipelined ADC, the MDAC's amplifier is the main building block that determines the performance and speed of the ADC's quantizer. If a feedback amplifier drives a capacitive load or a switched capacitor circuit, it can use an operational transconductance amplifier (OTA). If it drives a resistive load, it needs to use a low impedance operational amplifier (opamp). However, for simplicity, the "opamp" term is often used to describe both kinds of amplifiers.

6.1 Switched capacitor circuits

The reliance of modern analog design on CMOS and BiCMOS processes has led to a rise in the importance and popularity of switched capacitor circuits. This is due to the fact that MOS transistors are good switches and that capacitors enjoy better linearity, better matching, and lower temperature variation than resistors. Switched capacitor circuits rely on the capacitance ratios and the clock rate to determine the circuit's key characteristics. If designed properly, they can be relatively process-independent and achieve reasonably good accuracy without needing trimming or calibration [1]. The main requirements for these circuits are good switches, non-overlapping clocks, and enough time to settle to the desired accuracy. Switched capacitor circuits are also compatible with sampled-time (i.e. discrete-time) systems, which make them a natural and attractive choice for high-resolution ADCs.

6.1.1 Switched capacitor resistor

If we have a capacitor C that is switched between two nodes with voltages V_1 and V_2 at a clock rate (frequency) f_s, as shown in Figure 6.1, the charges q_1 and q_2 will be given by

$$q_1 = CV_1 \tag{6.1}$$

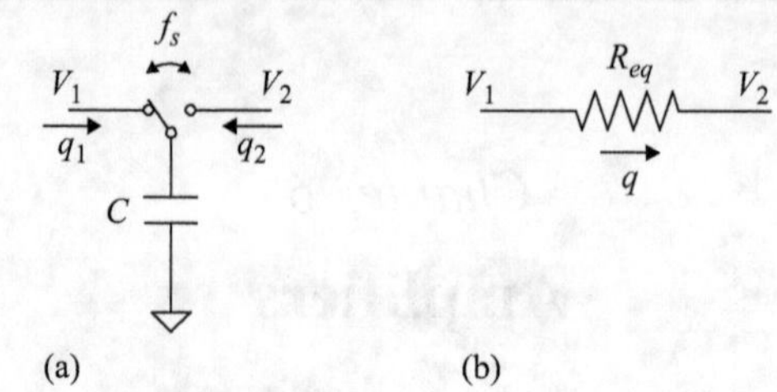

Figure 6.1 (a) A switched-capacitor and (b) its equivalent resistor

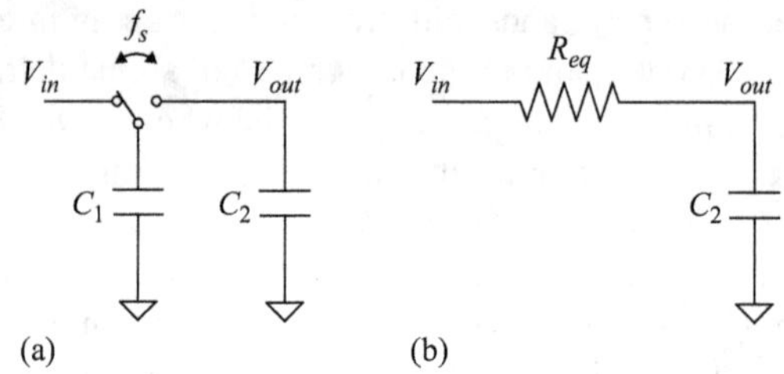

Figure 6.2 (a) A switched capacitor low-pass filter, and (b) its equivalent continuous-time filter

and

$$q_2 = CV_2 \tag{6.2}$$

The charge transfer between the two nodes due to switching is given by q, where

$$q = q_1 - q_2 = C(V_1 - V_2) \tag{6.3}$$

The effective current flowing between the two nodes is given by

$$I = \frac{q}{T_s} = C(V_1 - V_2)f_s \tag{6.4}$$

where T_s is the clock period. Therefore, the switched capacitor is equivalent to a resistor R_{eq} that is given by

$$R_{eq} \approx (V_1 - V_2)/I \approx 1/f_sC \tag{6.5}$$

This approximation is valid for frequencies that are much lower than the clock rate. Therefore, this switched capacitor can be used in the circuit shown in Figure 6.2 to form a low-pass filter, such that

$$\frac{V_{out}}{V_{in}} \approx \frac{1}{1 + sR_{eq}C_2} \approx \frac{1}{1 + sC_2/f_sC_1} \tag{6.6}$$

The equivalent resistance is given by

$$R_{eq} = \frac{1}{C_1 f_s} \tag{6.7}$$

The cut-off frequency f_c of this filter is

$$f_c = \frac{f_s C_1}{2\pi C_2} \tag{6.8}$$

If this filter is implemented by an explicit resistor R_{eq}, the cut-off frequency will be

$$f_c = \frac{1}{2\pi R_{eq} C_2} \tag{6.9}$$

By comparing (6.8) and (6.9), we can see that, in the switched capacitor (SC) case, the cut-off frequency depends on the clock rate and the capacitance ratio, which can be made very accurate. While in the traditional RC case, the cut-off frequency depends on the absolute values of the resistance and the capacitance, which can change substantially with process and temperature. Therefore, we can achieve higher accuracy with the SC filter, as long as we operate in the discrete-time domain. The continuous-time approximation is valid only for frequencies that are much lower than the clock rate.

The accurate operation of an SC circuit requires non-overlapping clocks, as shown in Figure 6.3. The non-overlap ensures a break-before-make switching, otherwise leakage, glitches, and charge injection will degrade the accuracy of the circuit. We can generate non-overlapping clocks using the circuit shown in Figure 6.4. In this case, the non-overlap time is given by t_d, which is equal to the

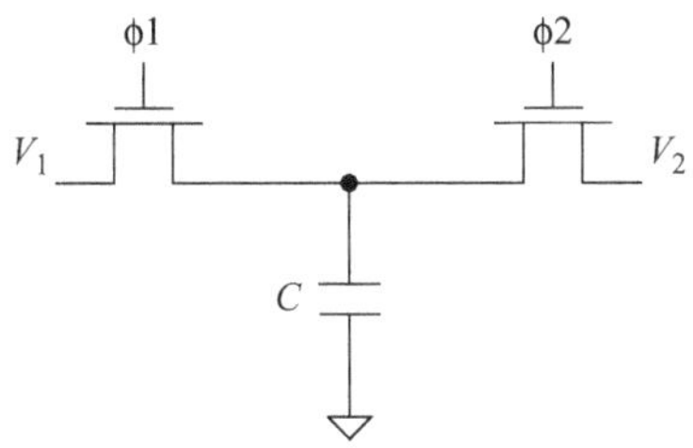

Figure 6.3 A switched-capacitor using non-overlapping clocks

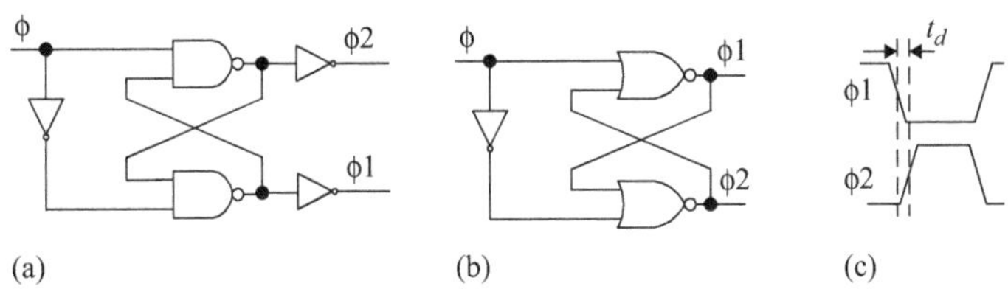

Figure 6.4 Non-overlap clock generators using (a) NAND gates, (b) NOR gates, (c) The resulting clocks with non-overlap time of t_d

typical gate delay. Another example is shown in Figure 6.5, where the non-overlap time is increased to $3t_d$ by inserting two inverters in the feedback path of the non-overlap circuit.

6.1.2 Switched capacitor active filters

Another example of a switched capacitor filter is the inverting integrator. An RC inverting integrator is shown in Figure 6.6(a), where if we assume the amplifier is ideal, the transfer function is given by

$$\frac{V_{out}}{V_{in}} = -\frac{1}{sR_1C_2} \tag{6.10}$$

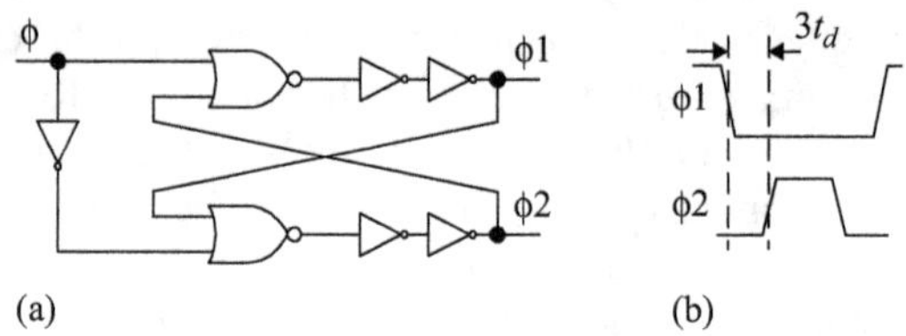

Figure 6.5 (a) A non-overlap clock generator using NOR gates. (b) The resulting non-overlapping clocks with longer non-overlap time ($3t_d$)

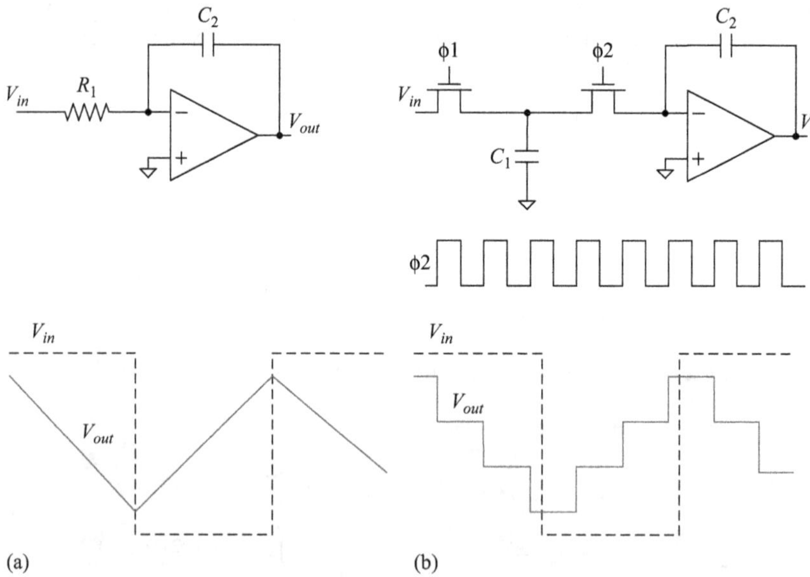

Figure 6.6 (a) An inverting continuous-time integrator (low-pass filter). (b) A switched capacitor equivalent. The inputs and outputs waveforms are shown for both cases

This circuit gives continuous-time operation with a time constant given by

$$\tau = R_1 C_2 \tag{6.11}$$

On the other hand, an SC equivalent is shown in Figure 6.6(b) and has an approximate transfer function of

$$\frac{V_{out}}{V_{in}} = -\frac{C_1 f_s}{sC_2} \tag{6.12}$$

This SC circuit gives a discrete-time (sampled-time) operation with a time constant given by

$$\tau = \frac{C_2}{C_1 f_s} \tag{6.13}$$

The discrete-time nature of switched capacitor circuits makes a z-domain representation more appropriate for filter analysis [2], where z is defined as

$$z = e^{j\omega T_s} \tag{6.14}$$

T_s is the clock period, ω is the angular frequency, and z^{-1} represents a unit delay. The transfer function in the z-domain can be calculated by analyzing the charges flowing in and out of the capacitance in the two clock phases. During $\phi 1$, the charge q_1 is given by

$$q_1[n] = C_2 V_{in}[n] \tag{6.15}$$

During $\phi 2$, the charge q_2 flowing out of the capacitance is

$$q_2[n] = -C_2 V_{out}[n] + C_2 V_{out}[n-1] \tag{6.16}$$

Since $\phi 2$ is high half a clock cycle after $\phi 1$, then:

$$q_2[n] = q_1[n - 1/2] \tag{6.17}$$

Substituting (6.15) and (6.16) into (6.17), we get

$$q_2[n] = -C_2 V_{out}[n] + C_2 V_{out}[n-1] = C_1 V_{in}[n - 1/2] \tag{6.18}$$

Therefore, in the z-domain, we obtain

$$-C_2 V_{out} + C_2 V_{out} z^{-1} = C_1 V_{in} z^{-1/2} \tag{6.19}$$

and the transfer function is given by

$$\frac{V_{out}}{V_{in}} = -\frac{C_1}{C_2} \frac{z^{-1/2}}{(1 - z^{-1})} \tag{6.20}$$

This represents an inverting integrator in the discrete-time domain. Replacing z by its continuous-time representation gives

$$\frac{V_{out}}{V_{in}} = -\frac{C_1}{C_2}\frac{e^{-j\omega T_s/2}}{(1 - e^{-j\omega T_s})} \tag{6.21}$$

For frequencies much smaller than f_s, ωT_s is very small, and hence

$$e^{-j\omega T_s} \approx 1 - j\omega T_s \tag{6.22}$$

Substituting from (6.22) into (6.21) gives

$$\frac{V_{out}}{V_{in}} \approx -\frac{C_1}{j\omega T_s C_2} \approx -\frac{f_s C_1}{j\omega C_2} \approx -\frac{1}{j\omega R_{eq} C_2} \tag{6.23}$$

which is the continuous-time approximation given in (6.10). That is,

$$\frac{V_{out}}{V_{in}} \approx -\frac{1}{sR_{eq}C_2} \approx -\frac{f_s C_1}{sC_2} \tag{6.24}$$

One drawback of the SC filter shown in Figures 6.6(b) and 6.7 is its sensitivity to the parasitics at the top plate of the capacitor. These parasitics degrade the accuracy, matching and speed of the circuit. In the presence of parasitics, instead of (6.24), the transfer function becomes

$$\frac{V_{out}}{V_{in}} \approx -\frac{f_s(C_1 + C_{p1} + C_{p2})}{sC_2} \tag{6.25}$$

This indicates that the parasitic capacitances have changed the time constant and the equivalent resistance of the circuit.

An alternative switched capacitance resistor that is less sensitive to parasitics is shown in Figure 6.8. In this case, the charge q_1 flowing during $\phi 1$ is

$$q_1 = C_1(V_1 - V_2) \tag{6.26}$$

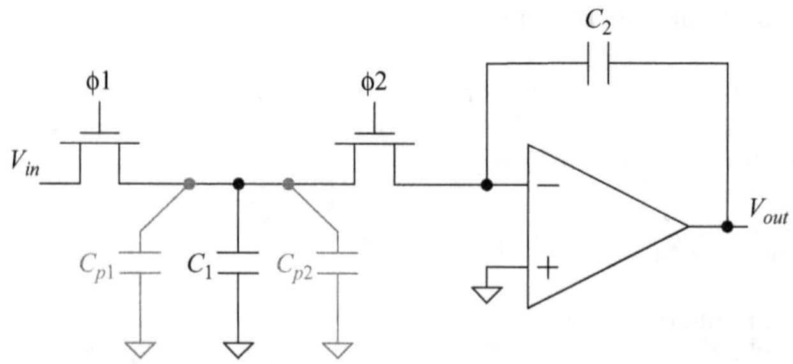

Figure 6.7 A switched capacitor filter with the parasitic capacitances shown. This structure is sensitive to the parasitic capacitances

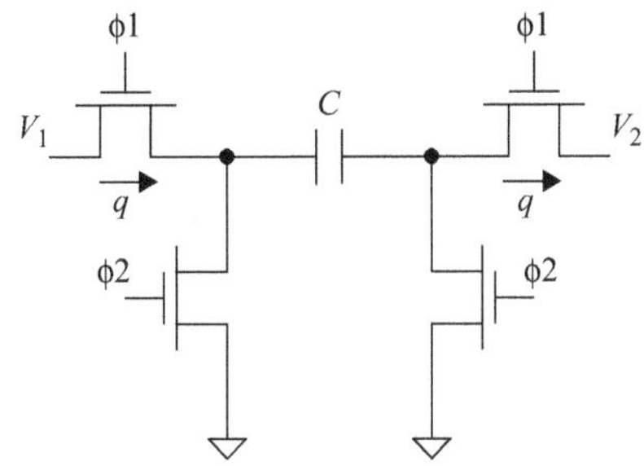

Figure 6.8 A parasitic-insensitive switched-capacitor resistor

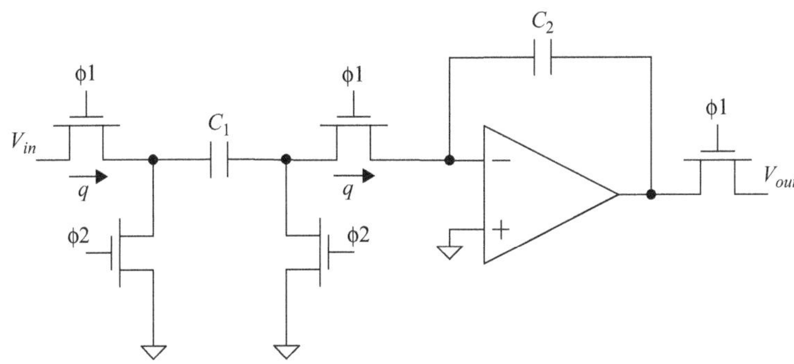

Figure 6.9 A parasitic-insensitive inverting switched-capacitor integrator (low-pass filter)

The capacitance gets discharged during ϕ2. Therefore, the equivalent resistance is still given by

$$R_{eq} \cong (V_1 - V_2)/I = 1/f_s C \tag{6.27}$$

Using this SC resistor in an inverting integrator configuration is shown in Figure 6.9, where the parasitic capacitance C_{p1} is over-ridden by the driving voltage V_{in} and C_{p2} is driven to zero by the feedback loop of the circuit. Therefore, this integrator is often called a parasitic-insensitive inverting integrator. The transfer function is

$$\frac{V_{out}}{V_{in}} = -\frac{C_1}{C_2}\frac{1}{(1 - z^{-1})} \tag{6.28}$$

which can be approximated in the continuous-time domain, for frequencies that are much smaller than the clock rate, as:

$$\frac{V_{out}}{V_{in}} \approx -\frac{C_1}{j\omega T_s C_2} \approx -\frac{f_s C_1}{j\omega C_2} \approx -\frac{1}{j\omega R_{eq} C_2} \tag{6.29}$$

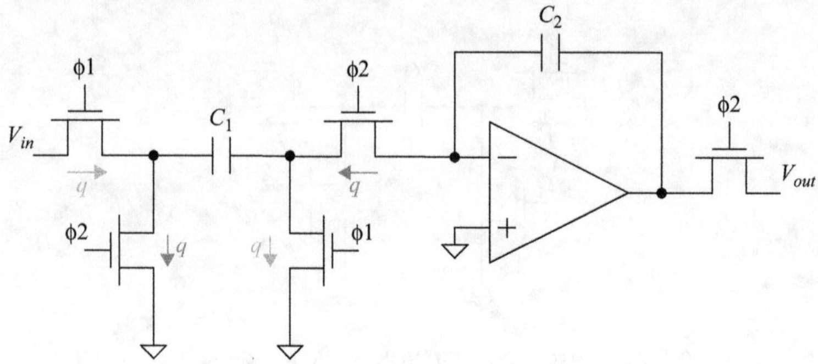

Figure 6.10 A non-inverting switched-capacitor integrator (low-pass filter)

Similarly, a parasitic-insensitive non-inverting integrator can be configured as shown in Figure 6.10. During ϕ1, the input is sampled on the capacitance and the charge q flows to the right from the input to charge the capacitance C_1. During ϕ2, the charge q flows leftward out of the capacitance to ground, and hence charges the capacitance C_2. Unlike the inverting parasitic-insensitive integrator, the output of this circuit is sampled during ϕ2, which is half a cycle after the input is sampled. This gives the transfer function

$$\frac{V_{out}}{V_{in}} = \frac{C_1}{C_2} \frac{z^{-1/2}}{(1 - z^{-1})} \tag{6.30}$$

and its continuous-time approximation for low frequencies

$$\frac{V_{out}}{V_{in}} \approx \frac{C_1}{j\omega T_s C_2} \approx \frac{f_s C_1}{j\omega C_2} \approx \frac{1}{j\omega R_{eq} C_2} \tag{6.31}$$

6.1.3 Switched capacitor amplifiers

We have focused so far on filters and integrators to explain the operation of switched capacitor circuits, but the same analysis can be extended to switched capacitor amplifiers. If the feedback capacitor is reset/discharged every cycle, the memory from the previous cycle disappears and the integration process is interrupted. This makes the circuit behave as an amplifier instead. This is shown in Figure 6.11 for a non-inverting amplifier, whose transfer function is

$$\frac{V_{out}}{V_{in}} = \frac{C_1 z^{-1/2}}{C_2} \tag{6.32}$$

If we ignore the half cycle delay, the transfer function will be

$$\frac{V_{out}}{V_{in}} \cong \frac{C_1}{C_2} \tag{6.33}$$

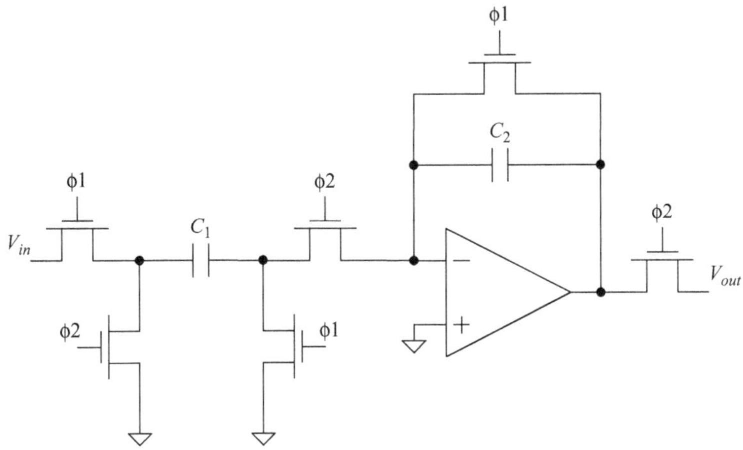

Figure 6.11 A switched-capacitor non-inverting amplifier

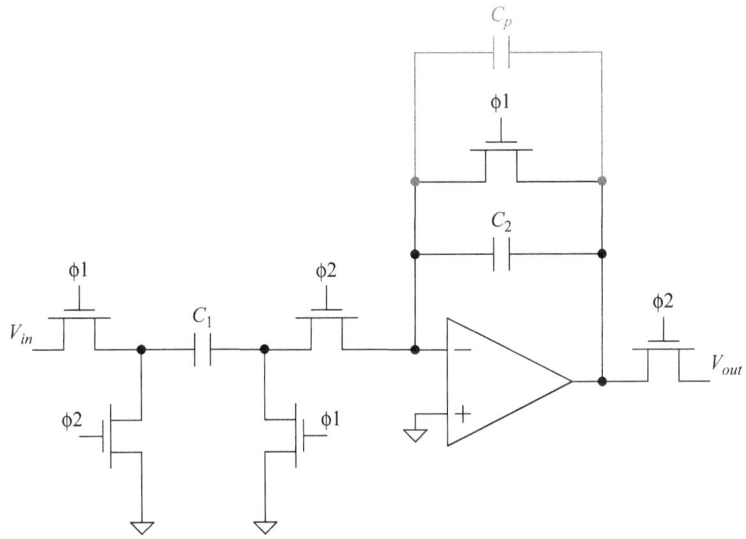

Figure 6.12 A switched-capacitor non-inverting amplifier showing the effect of the parasitic capacitance across the feedback capacitor

It is important to note that although these circuits are insensitive to the C_1 and C_2 parasitics to ground, they are sensitive to the parasitics across C_1 and C_2. This is the case with the circuits of Figures 6.9–6.11, because a parasitic across a capacitor will effectively change the value of that capacitor. This is illustrated for the amplifier circuit in Figure 6.12, where a parasitic across the capacitance C_2 that is not

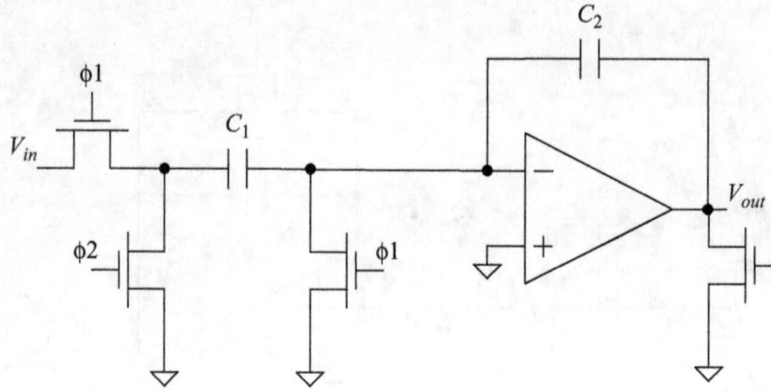

Figure 6.13 A simplified form of the non-inverting switched-capacitor amplifier

matched with an equivalent proportional parasitic across C_1 will cause a gain error in the transfer function that is given by

$$\frac{V_{out}}{V_{in}} = \frac{C_1 z^{-1/2}}{C_2 + C_p} \tag{6.34}$$

Since the gain depends on the capacitance ratio, it can be made to achieve good accuracy by using identical units, such that the parasitics match proportionally without changing the gain measurably.

A simplified form of this circuit is shown in Figure 6.13. This is an SC amplifier that can also be used as a sample-and-hold amplifier (SHA), in which case the sampling switched uses an advanced clock as shown in Figure 6.14, and discussed in Chapter 4. The reduction of the number of switches connected to the amplifier's summing node reduces the parasitic capacitances on the summing node and hence improves the feedback factor, the accuracy, and the speed. If the operational amplifier is assumed to be ideal, the gain of this SC amplifier is

$$G = \frac{C_1}{C_2} \tag{6.35}$$

A flip-around version of this amplifier is shown in Figure 6.15, where the feedback capacitor is used as a sampling capacitor during the sampling phase, and as a feedback capacitor during the hold phase. The gain of this flip-around amplifier is

$$G = \frac{C_1 + C_2}{C_2} \tag{6.36}$$

If a unity gain is required, the capacitance C_1 is removed, which gives a structure that can be used as a SHA and has an inherently accurate gain of 1.

The circuit of Figure 6.15 enjoys a higher feedback factor for the same gain compared to the circuit shown in Figure 6.14. It requires additional switches, however, which can impact its speed. It also suffers from possible common-mode glitches, if the input common-mode is different from the output common-mode. These common-mode glitches can complicate the design and slow down the differential settling.

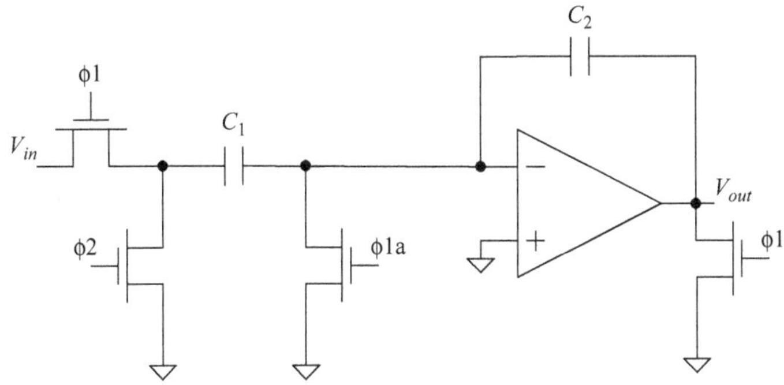

Figure 6.14 A simplified form of the non-inverting switched-capacitor amplifier, with an advanced sampling switch to minimize charge-injection from the input switch. It can be used as a S/H amplifier

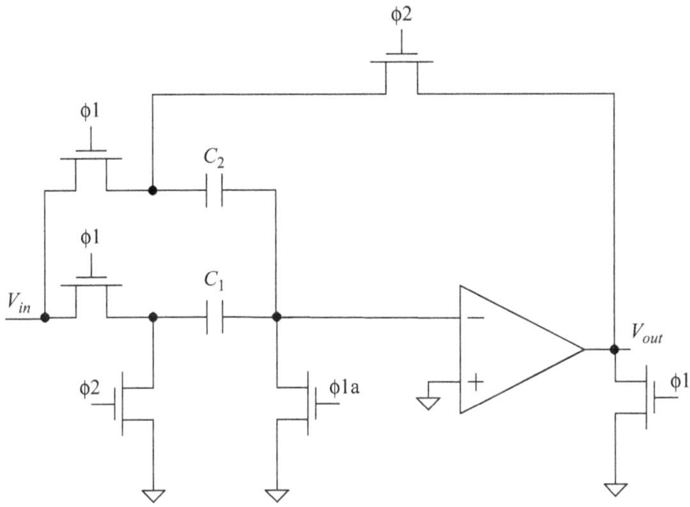

Figure 6.15 A non-inverting switched-capacitor amplifier with a flip-around capacitor

6.1.4 Non-idealities of a switch capacitor amplifier

If the operational amplifier has a finite DC open loop gain A_{DC}, the analysis of the switched capacitor amplifier's gain for the circuit in Figure 6.16 gives

$$\frac{V_{out}}{V_{in}} = \frac{C_1}{C_2}\frac{1}{\left(1+\frac{(C_1+C_2+C_{ps})/C_2}{A_{DC}}\right)} = \frac{C_1/C_2}{\left(1+\frac{1}{\beta A_{DC}}\right)} = \frac{G}{\left(1+\frac{1}{\beta A_{DC}}\right)} \tag{6.37}$$

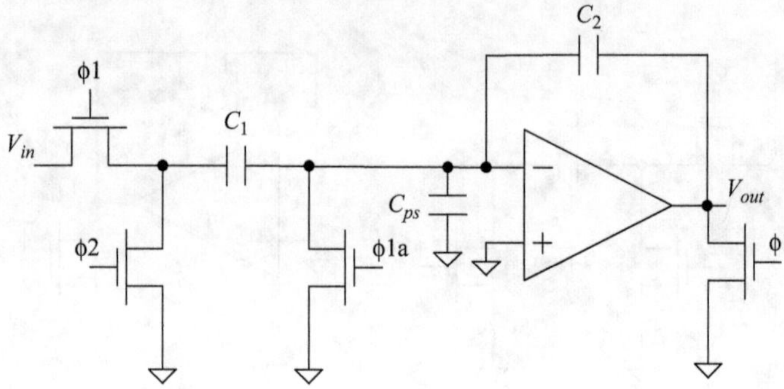

Figure 6.16 A simplified form of the non-inverting switched-capacitor amplifier

where C_{ps} is the parasitic capacitance on the summing node of the amplifier, and β is the feedback factor given by

$$\beta = \frac{C_2}{C_1 + C_2 + C_{ps}} \tag{6.38}$$

The expression (6.37) can be derived using charge conservation. During $\phi 1$, q_1 flows to charge C_1 and is given by

$$q_1 = V_{in} C_1 \tag{6.39}$$

During $\phi 2$, charge conservation at the summing node indicates that the charge will be redistributed, such that

$$q_1 + V^- C_1 + V^- C_{ps} = (V_{out} - V^-) C_2 \tag{6.40}$$

where V^- is the summing node voltage, which is equal to

$$V^- = -\frac{V_{out}}{A_{DC}} \tag{6.41}$$

Substituting from (6.39) and (6.41) into (6.40), we get

$$V_{in} C_1 - \frac{V_{out} C_1}{A_{DC}} - \frac{V_{out} C_{ps}}{A_{DC}} = \left(V_{out} + \frac{V_{out}}{A_{DC}} \right) C_2 \tag{6.42}$$

Rearranging the terms gives (6.37), which is

$$A_{cl_{DC}} = \frac{V_{out}}{V_{in}} = \frac{C_1}{C_2} \frac{1}{\left(1 + \frac{(C_1 + C_2 + C_{ps})/C_2}{A_{DC}} \right)} = \frac{C_1/C_2}{\left(1 + \frac{1}{\beta A_{DC}} \right)} = \frac{G}{\left(1 + \frac{1}{\beta A_{DC}} \right)} \tag{6.43}$$

Comparing (6.43) with (6.35), we see that the finite open loop DC gain A_{DC} creates an error in the overall closed loop gain. Therefore, it is desirable to maximize the open loop DC gain A_{DC} to minimize the error. This gain formula of (6.43) assumes complete settling. Therefore, it is often called the *closed loop DC gain* $A_{cl_{DC}}$ of the circuit.

In addition to the above static (DC) inaccuracy, the switched capacitor amplifier suffers from dynamic inaccuracy due to the settling error. This is partially due to the limited bandwidth of the operational amplifier (or OTA) and due to the parasitic resistances of the switches. If the operational amplifier has a single pole with unity gain angular frequency of ω_u, we will derive in Section 6.2.3 that the amplifier closed loop gain A_{cl} is given by

$$A_{cl} = \frac{V_{out}}{V_{in}} = \frac{C_1/C_2}{\left(1 + \dfrac{1}{\beta A_{DC}}\right)} \frac{1}{(1 + s/\beta\omega_u)} = \frac{A_{cl_{DC}}}{(1 + s/\beta\omega_u)} \tag{6.44}$$

Therefore, the closed loop bandwidth BW_{cl} of the switched amplifier circuit is

$$BW_{cl} \cong \beta f_u \tag{6.45}$$

where β is the feedback factor, and $f_u = \omega_u/2\pi$ is the unity gain frequency of the operational amplifier. For the circuit of Figure 6.16, β is given by

$$\beta = \frac{C_2}{C_1 + C_2 + C_{ps}} \tag{6.46}$$

where C_1 is the sampling capacitance, C_2 is the feedback capacitance, and C_{ps} is the parasitic capacitance at the amplifier's summing node. The unity gain frequency is typically given by

$$f_u = \frac{g_m}{2\pi C_L} \tag{6.47}$$

where g_m is the transconductance of the OTA and C_L is its total load capacitance. This will be covered in more detail later in this chapter for the different amplifier structures.

If we assume the amplifier is a single-pole system, the settling error ε over time t_s will be approximately given by

$$\varepsilon = e^{-t_s\beta\omega_u} = e^{-t_s 2\pi BW_{cl}} \tag{6.48}$$

From (6.45), we observe that the closed loop bandwidth is determined by the opamp unity gain frequency and the feedback factor. Equation (6.48) indicates that the settling error depends on the closed loop bandwidth and the settling time, which is dictated by the clock frequency.

It is clear from the above discussion that the accuracy of the SC amplifier is determined by the open loop gain of the operational amplifier, its unity gain bandwidth, and the feedback factor. These parameters affect the settling error and the DC accuracy of the amplifier. In a pipelined ADC's inter-stage amplifiers, the gain accuracy is very important, because it impacts the quantizer's linearity as discussed in the Chapter 7. On the other hand, in S/H amplifiers, the gain accuracy is not very critical, but the gain linearity is more important.

Example 1: Starting from the non-inverting amplifier of Figure 6.13, how can we build an inverting amplifier?

Answer 1: An example of an inverting amplifier is shown in Figure 6.17. The phases of some of the switches are reversed, such that during ϕ2 the capacitances are reset, and during ϕ1 the input is applied and the output is sampled.

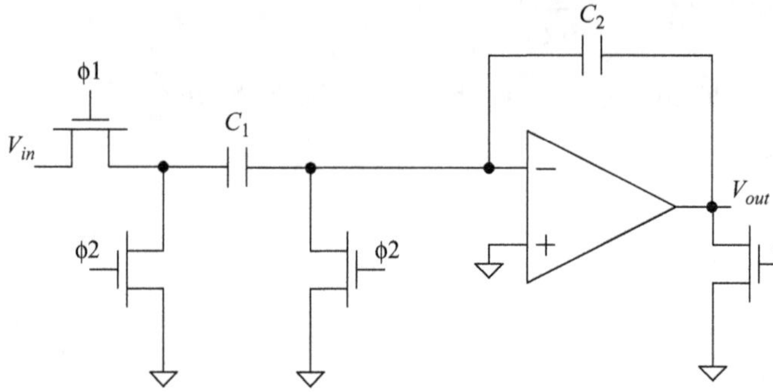

Figure 6.17 An example of a switched-capacitor inverting amplifier

The DC transfer function of the inverting amplifier is given by

$$\frac{V_{out}}{V_{in}} = \frac{-C_1/C_2}{\left(1 + \dfrac{1}{\beta A_{DC}}\right)} \tag{6.49}$$

Example 2: How can we build a subtracting SC amplifier?

Answer 2: We can combine the inverting and non-inverting amplifiers to obtain a subtracting amplifier that is shown in Figure 6.18. During ϕ1, the input V_{in1} is sampled, and the feedback capacitance and amplifier are reset. During ϕ2, V_{in2} is applied, and the output is sampled. This configuration subtracts V_{in2} from V_{in1}, such that

$$V_{out} = \frac{C_1/C_2}{\left(1 + \dfrac{1}{\beta A_{DC}}\right)}(V_{in1} - V_{in2}) \tag{6.50}$$

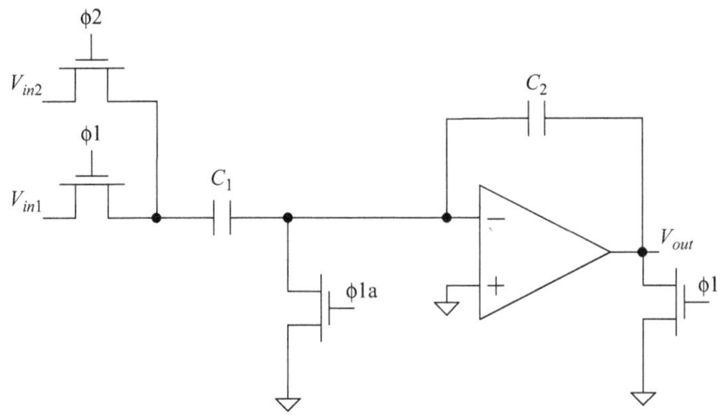

Figure 6.18 An example of a switched capacitor subtracting amplifier

6.2 Amplifier design

In the previous section, the OTA was presented as a black box. In this section, the structure and design parameters of the OTA are discussed. In the context of switched capacitor and capacitive load circuits, the operational amplifiers are usually implemented as OTAs. Unlike operational amplifiers, OTAs provide a current output with a relatively high output impedance instead of a voltage output with a low output impedance. After settling, the impedance of the load capacitor is very high, which removes the need for the low output impedance in the operational amplifier.

Some of the important design parameters of an amplifier are:

- DC gain
- Slew rate
- Unity gain frequency
- Closed loop bandwidth
- Phase margin
- Common-mode rejection ratio (CMRR)
- Power supply rejection ratio (PSRR)
- Noise
- Input common-mode range
- Output dynamic range

6.2.1 DC gain

The DC gain is defined as the ratio of the output voltage to the input voltage after complete settling. That is, at very low frequency that approaches DC. The DC gain determines the accuracy of the amplifier, and is given by

$$A_{DC} = \left.\frac{v_o}{v_{in}}\right|_{f\to 0} = \left.\frac{v_o}{v_{in}}\right|_{t\to\infty} \tag{6.51}$$

where v_o and v_{in} are the output and input voltages, respectively. In this section, small letters are used for these voltages to represent their signal components as opposed to their DC bias values. In switched capacitor circuits, it is important to distinguish between the "DC" terms, which refer to the signal components after adequate settling time, as opposed to the amplifier's DC bias voltages, which have no signal content.

6.2.2 Slew rate

The slew rate is defined as the maximum rate of change of the output voltage with time. This is given by

$$SR = \frac{\Delta V}{\Delta t} = \frac{I}{C} \tag{6.52}$$

The slew rate determines the upper limit on the large-signal settling speed, when all the current sourcing or sinking ability of the OTA is used to charge the capacitances inside and outside the amplifier. Typically, the settling behavior is composed of a large-signal settling period followed by a small-signal settling period. It is desirable to minimize the slewing period as it is highly non-linear in nature [10].

6.2.3 Small-signal settling

As the slewing and the large signal settling end, and the output approaches its final value, the amplifier's small-signal settling is governed by its bandwidth, feedback factor, and phase margin. The settling behavior will depend heavily on the number of poles of the system.

6.2.3.1 Single-pole system

For a single-pole system, the pole frequency, unity gain frequency, and closed loop bandwidth are shown in the Bode plot of Figure 6.19 for the generic feedback amplifier depicted in Figure 6.20. If we assume a single-pole system, the open loop transfer function is given by

$$A(s) = \frac{A_{DC}}{1 + s/\omega_p} \tag{6.53}$$

where A_{DC} is the open loop DC gain of the amplifier, and ω_p is the dominant pole. The closed loop gain from the amplifier's input to its output will be

$$A_{cl\_gen} = \frac{v_o}{v^+ - v^-} = \frac{A(s)}{1 + \beta A(s)} \tag{6.54}$$

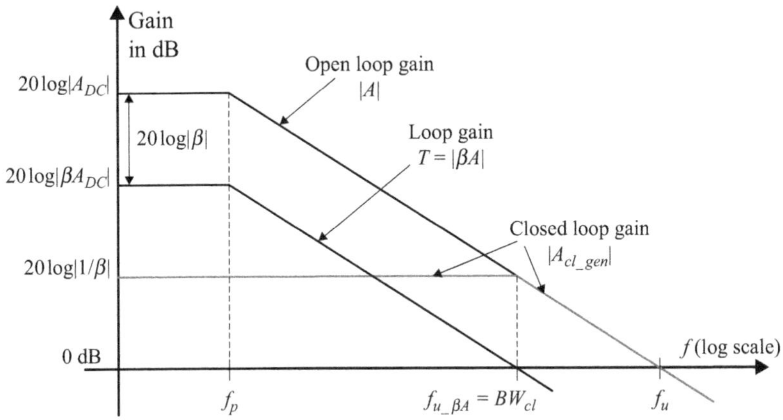

Figure 6.19 Bode plot showing the open loop gain (A), loop gain (βA), the dominant pole frequency (f_p), the closed loop bandwidth (BW_{cl}), the unity gain frequency (f_u) of the amplifier, and the unity gain frequency of the loop gain $f_{u\_\beta A}$

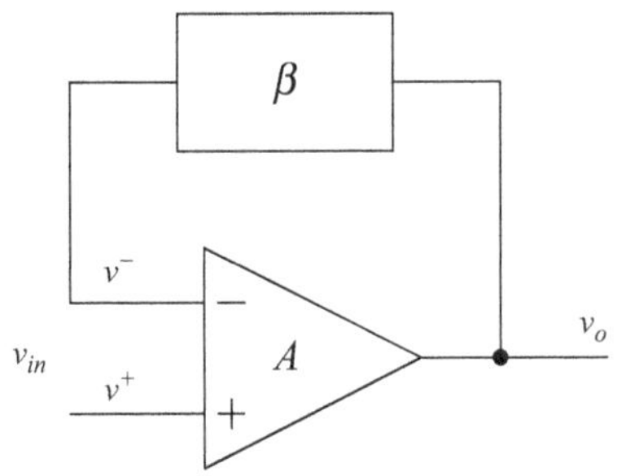

Figure 6.20 A representation of a feedback amplifier

where β is the feedback factor. The "*gen*" suffix denotes a *generic* feedback amplifier structure shown in Figure 6.20. Substituting (6.53) into (6.54) gives

$$A_{cl\_gen} = \frac{\dfrac{A_{DC}}{1+s/\omega_p}}{1+\beta\dfrac{A_{DC}}{1+s/\omega_p}} = \frac{A_{DC}}{1+s/\omega_p+\beta A_{DC}} = \frac{\dfrac{A_{DC}}{1+\beta A_{DC}}}{1+s/\omega_p(1+\beta A_{DC})} \tag{6.55}$$

which can be rearranged to give

$$A_{cl\_gen} = \frac{A_{cl\_gen_{DC}}}{1+\dfrac{s}{\beta A_{DC}\omega_p}} = \frac{A_{cl\_gen_{DC}}}{1+\dfrac{s}{\beta\omega_u}} = \frac{A_{cl\_gen_{DC}}}{1+\dfrac{s}{2\pi BW_{cl}}} \tag{6.56}$$

and

$$A_{cl\_gen} = \frac{A_{cl\_gen_{DC}}}{1 + \frac{s}{2\pi\beta f_u}} = \frac{A_{cl\_gen_{DC}}}{1 + \frac{s}{2\pi BW_{cl}}} \tag{6.57}$$

where the closed loop bandwidth is given by

$$BW_{cl} = f_u \times \beta \tag{6.58}$$

and the closed loop DC gain of the amplifier is given by

$$A_{cl\_gen_{DC}} = \frac{1/\beta}{1 + 1/\beta A_{DC}} = \frac{A_{DC}}{1 + \beta A_{DC}} \tag{6.59}$$

The loop gain (T) is defined as:

$$T = \beta A(s) \tag{6.60}$$

and the DC loop gain is given by

$$T_{\mathrm{DC}} = \beta A_{DC} \tag{6.61}$$

It is important to note that the unity gain frequency of the open loop amplifier f_u is different from the unity gain frequency of the loop gain $f_{u\_\beta A}$, such that

$$f_{u\_\beta A} \approx BW_{cl} \approx f_u \times \beta \tag{6.62}$$

In the circuit shown in Figure 6.16, which is redrawn in a simpler format in Figure 6.21, the location of the input is different from that in the generic amplifier of Figure 6.20. Therefore, we expect the numerator to change, while the denominator (i.e. the characteristic equation) should stay the same. The DC closed loop gain for the circuit in Figure 6.21 is

$$A_{cl_{DC}} = \frac{v_o}{v_{in}} = \frac{C_1/C_2}{1 + 1/\beta A_{DC}} \tag{6.63}$$

The closed loop gain is given by

$$A_{cl} = \frac{A_{cl_{DC}}}{1 + \frac{s}{2\pi\beta f_u}} \tag{6.64}$$

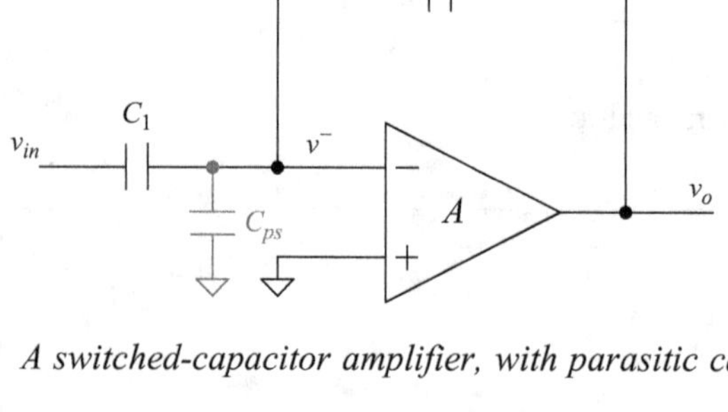

Figure 6.21 A switched-capacitor amplifier, with parasitic capacitor C_{ps}

where the feedback factor is given by

$$\beta = \frac{C_2}{C_1 + C_2 + C_{ps}} \tag{6.65}$$

Equations (6.63)–(6.65) agree with (6.37)–(6.45). The single-pole system is unconditionally stable, and the small-signal settling is determined by the unity gain frequency and the feedback factor. Increasing the unity gain frequency or the feedback factor improves the closed loop bandwidth and hence the settling behavior.

6.2.3.2 Two-pole system

In a second-order system, the Bode plot is shown in Figure 6.22. A second pole (f_{p2}) exists, which will be shown to impact the settling behavior. The open loop gain is given by [2]

$$A(s) = \frac{A_{DC}}{\left(1 + s/\omega_{p1}\right)\left(1 + s/\omega_{p2}\right)} \tag{6.66}$$

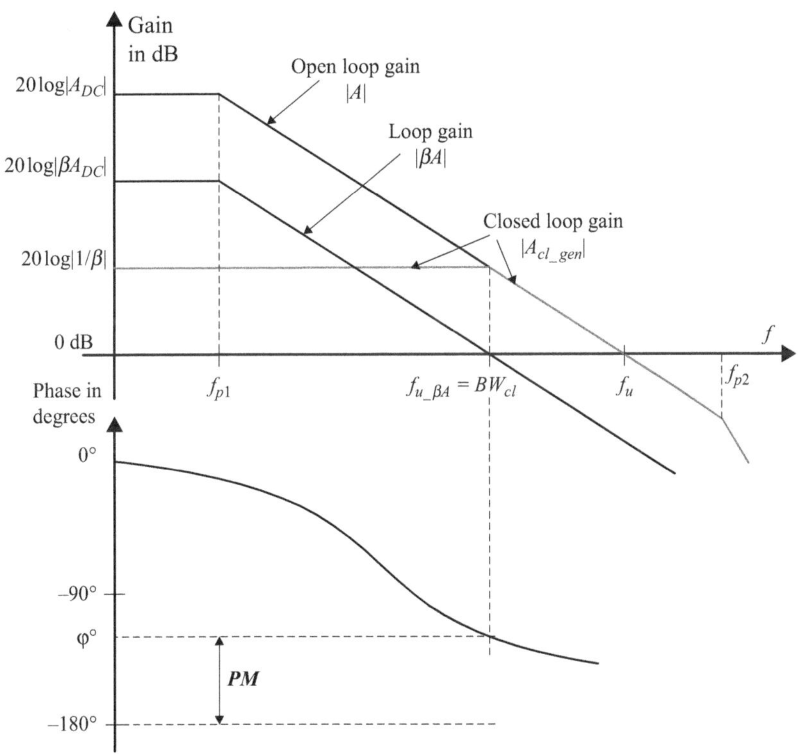

Figure 6.22 Bode plot showing the open loop gain (A), loop gain (βA), the dominant pole frequency (f_{p1}), the non-dominant pole (f_{p2}), the closed loop bandwidth (BW_{cl}), and the unity gain frequency (f_u) of a two-pole amplifier

where ω_{p1} is the first pole and ω_{p2} is the second pole. The first pole is usually designed to be much smaller than the second pole. Hence, ω_{p1} is sometimes called the dominant pole, and ω_{p2} is the non-dominant pole. The closed loop transfer function of the amplifier in Figure 6.20 is given by

$$A_{cl\_gen} = \frac{v_o}{v^+ - v^-} = \frac{A(s)}{1 + \beta A(s)} \tag{6.67}$$

Substituting (6.66) into (6.67) gives

$$A_{cl\_gen}(s) = \frac{\omega_{p1}\omega_{p2}A_{DC}}{s^2 + (\omega_{p1} + \omega_{p2})s + (1 + \beta A_{DC})\omega_{p1}\omega_{p2}} \tag{6.68}$$

Similar to (6.59), the closed loop DC gain is still given by

$$A_{cl\_gen_{DC}} = \frac{A_{DC}}{1 + \beta A_{DC}} \tag{6.69}$$

Equation (6.68) can be expressed in terms of the closed loop DC gain as

$$A_{cl\_gen}(s) = \frac{(1 + \beta A_{DC})\omega_{p1}\omega_{p2}A_{cl\_gen_{DC}}}{s^2 + (\omega_{p1} + \omega_{p2})s + (1 + \beta A_{DC})\omega_{p1}\omega_{p2}} \tag{6.70}$$

which can be represented as

$$A_{cl\_gen}(s) = \frac{\omega_0^2 A_{cl\_gen_{DC}}}{s^2 + 2\xi\omega_0 s + \omega_0^2} = \frac{\omega_0^2 A_{cl\_gen_{DC}}}{s^2 + \left(\frac{\omega_0}{Q_p}\right)s + \omega_0^2} \tag{6.71}$$

where ω_0 is the resonance frequency, Q_p is the quality factor, ξ is the damping ratio, and

$$\xi = \frac{1}{2Q_p} \tag{6.72}$$

The resonance frequency ω_0 is given by

$$\omega_0^2 = (1 + \beta A_{DC})\omega_{p1}\omega_{p2} \tag{6.73}$$

The damping ratio ξ is given by

$$\xi = \frac{\omega_{p1} + \omega_{p2}}{2\sqrt{(1 + \beta A_{DC})\omega_{p1}\omega_{p2}}} \tag{6.74}$$

From (6.71), the characteristic equation CE is given by

$$CE(s) = s^2 + 2\xi\omega_0 s + \omega_0^2 \tag{6.75}$$

This is a second-order equation with two roots. There are three scenarios:

If $\xi > 1$, $Q_p < 0.5$, the two roots are real, and the system is *over-damped.*
If $\xi < 1$, $Q_p > 0.5$, the two roots are complex and the system is *under-damped.*
If $\xi = 1$ and $Q_p = 0.5$, the two roots coincide, and the system is *critically damped.*

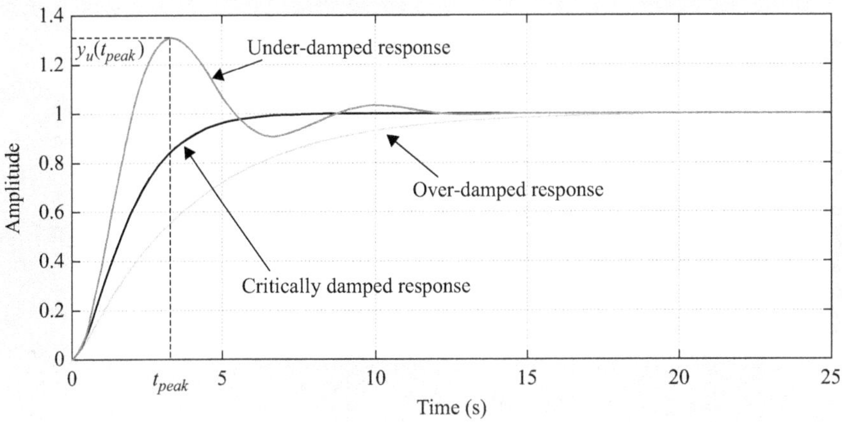

Figure 6.23 Step responses of a two-pole system

The three types of step responses are shown in Figure 6.23. The step response of an under-damped second-order system is given by [2]:

$$y_u(t) = \left[1 + \frac{1}{\sqrt{1-\xi^2}} e^{-\xi\omega_0 t} \sin\left(\omega_0\sqrt{1-\xi^2}t + \emptyset\right)\right] u(t) \tag{6.76}$$

where

$$\emptyset = \tan^{-1}\frac{\sqrt{1-\xi^2}}{\xi} \tag{6.77}$$

In under-damped operation, the location of the peak overshoot is

$$t_{peak} = \frac{\pi}{\omega_0\sqrt{1-\xi^2}} \tag{6.78}$$

and the magnitude of the peak overshoot is

$$y_u\left(t_{peak}\right) = 1 + e^{-\pi\xi/\sqrt{1-\xi^2}} \tag{6.79}$$

When over-damped, the step response can be better represented as:

$$y_u(t) = 1 - \frac{1}{2\sqrt{\xi^2-1}}\left(\frac{e^{-\left(\xi-\sqrt{\xi^2-1}\right)\omega_0 t}}{\xi-\sqrt{\xi^2-1}} - \frac{e^{-\left(\xi+\sqrt{\xi^2-1}\right)\omega_0 t}}{\xi+\sqrt{\xi^2-1}}\right) \tag{6.80}$$

When critically damped, the step response is given by

$$y_u(t) = [1 - (1+\omega_0 t)e^{-\omega_0 t}] \tag{6.81}$$

If $\beta A_{DC} >> 1$, $\omega_{p1} << \omega_{p2}$ and $s >> \omega_{p1}$, we define the closed loop angular bandwidth ω_{cl} as

$$\omega_{cl} \approx \beta A_{DC}\omega_{p1} = 2\pi \times BW_{cl} = \omega_{u\_\beta A} = 2\pi f_{u\_\beta A} \approx \beta\omega_u \tag{6.82}$$

From (6.71), the closed loop gain will be given by

$$A_{cl\_gen}(s) \approx \frac{\beta A_{DC}\omega_{p1}\omega_{p2}A_{cl\_gen_{DC}}}{s^2 + \omega_{p2}s + \beta A_{DC}\omega_{p1}\omega_{p2}} = \frac{\omega_{cl}\omega_{p2}A_{cl\_gen_{DC}}}{s^2 + \omega_{p2}s + \omega_{cl}\omega_{p2}} \tag{6.83}$$

From (6.73), the resonance frequency can be approximated by

$$\omega_0^2 \approx \beta A_{DC}\omega_{p1}\omega_{p2} \approx \omega_{cl}\omega_{p2} \tag{6.84}$$

From (6.74), the damping ratio will be

$$\xi \approx \frac{\omega_{p2}}{2\sqrt{\beta A_{DC}\omega_{p1}\omega_{p2}}} \approx \frac{1}{2}\sqrt{\frac{\omega_{p2}}{\omega_{cl}}} \tag{6.85}$$

Since critical damping requires $\xi = 1$. Therefore, from (6.85), the following condition needs to be satisfied for critical damping:

$$\omega_{p2} \approx 4\omega_{cl} \approx 4\beta A_{DC}\omega_{p1} \tag{6.86}$$

That is, for critical damping, the non-dominant pole needs to be located at about 4× the closed loop bandwidth.

The phase margin is defined for the loop gain function as the difference between the phase at the unity loop gain point and −180°. That is the *PM* is given by

$$PM = \phi|_{T=0\,\text{dB}} - (-180°) \tag{6.87}$$

where $\phi|_{T=0\,\text{dB}}$ is the phase where the loop gain $T = 0$ dB, and T is equal to βA. This is illustrated in the Bode plot shown in Figure 6.22. It is important to note that the phase margin is measured at the unity gain frequency of the loop gain $f_{u\_\beta A}$, not the unity gain frequency of the open loop amplifier f_u.

For a second-order system the *PM* is given by [2]:

$$PM = \cos^{-1}\left(\sqrt{1 + 4\xi^4} - 2\xi^2\right) \tag{6.88}$$

In terms of the poles, it can be approximated by [3]

$$PM \approx 180 - \tan^{-1}\left(\frac{\omega_{u\_\beta A}}{\omega_{p1}}\right) - \tan^{-1}\left(\frac{\omega_{u\_\beta A}}{\omega_{p2}}\right) \approx 180 - \tan^{-1}\left(\frac{\omega_{cl}}{\omega_{p1}}\right) - \tan^{-1}\left(\frac{\omega_{cl}}{\omega_{p2}}\right) \tag{6.89}$$

Critical damping corresponds to a damping ratio of 1, and hence $PM = 76°$. A smaller phase margin indicates under-damping, and a larger phase margin corresponds to over-damping.

It is sometimes desirable to have a slightly under-damped performance, where the damping factor is about $1/\sqrt{2}$, and hence a *PM* of about 60°. This represents the smallest t_{peak} time in (6.78) and a "maximally flat" or *Butterworth* response [3].

The fastest settling response, where the settling time is minimum, depends on the allowed settling error. For settling errors greater than 4.3%, the maximally flat response yields the fastest settling, which corresponds to a *PM* of about 60° and a damping ratio of $1/\sqrt{2}$. However, for smaller settling errors, the maximally flat peaking is not acceptable, and we need the peak magnitude to conform to the settling error. Using (6.79) above, gives the error to be

$$\varepsilon = e^{-\pi\xi/\sqrt{1-\xi^2}} \tag{6.90}$$

This leads to a non-dominant pole location of [3]

$$\frac{\omega_{p2}}{\omega_{p1}} \approx \frac{4(1+\beta A_{DC})}{1+(\pi/\ln\varepsilon)^2} \tag{6.91}$$

The phase margin for fastest settling is given by

$$PM_{fastest} \approx 90 - \tan^{-1}\left[\frac{1+(\pi/\ln\varepsilon)^2}{4}\right] \tag{6.92}$$

It is interesting to note that as the settling error decreases and approaches zero, the *PM* for fastest settling will approach 76°, which is the critically damped condition. That is, critical damping represents the fastest settling mode for arbitrarily small settling errors. However, practically, fastest settling tends to occur for slightly under-damped behavior.

We should also note that the closed loop gain of the circuit in Figure 6.21 will have the same denominator (Characteristic Equation) as the generic amplifier of Figure 6.20, given in (6.70) and (6.71). The numerator, however, will have the term $A_{cl\_gen_{DC}}$ in (6.70) and (6.71) replaced with $A_{cl_{DC}}$, where $A_{cl_{DC}}$ is given by (6.63).

6.2.3.3 Three-pole system

In the presence of more than two poles, the phase margin can reach and surpass 0°, leading to positive feedback and instability. If the frequencies of the higher-order poles are significantly larger than the first two poles, the system can be approximated by a two-pole system. Otherwise, an equivalent second pole can be estimated from simulation by the frequency at which the phase is equal to −135°. If the poles are real, the equivalent second pole can be approximated as follows [6]

$$\frac{1}{\omega_{p2_{eq}}} \approx \sum_{i=2}^{N_p}\frac{1}{\omega_{pi}} - \sum_{i=1}^{N_z}\frac{1}{\omega_{zi}} \tag{6.93}$$

where $\omega_{p2_{eq}}$ is the equivalent second pole, ω_{pi} is the *i*th pole, ω_{zi} is the *i*th zero, N_p is the number of poles, and N_z is the number of zeros.

In the presence of more than two poles, compensation is usually needed to prevent instability, improve the phase margin, and optimize the frequency response. This is frequently accomplished by reducing the bandwidth, lowering the frequency of the first pole and possibly increasing the frequencies of the non-dominant poles [2, 5, 7]. A commonly used compensation technique is the Miller compensation that is discussed later in this chapter.

6.2.4 CMRR and PSRR

The CMRR describes the amplifier's ability to reject common-mode changes. It is the ratio of its differential gain (A_{diff}) to its common-mode gain (A_{CM}). It is given by

$$CMRR = \frac{A_{diff}}{A_{CM}} \tag{6.94}$$

The power supply rejection ratio is defined as the ratio of the differential gain (A_{diff}) to the gain from the supply (or ground) to the output (A_{dd}). It is given by

$$PSRR = \frac{A_{diff}}{A_{dd}} \tag{6.95}$$

6.2.5 Noise

Finally, the amplifier noise is a very important design consideration. The equivalent input thermal noise voltage v_n for a MOS device is given by [2]

$$v_n^2 = \frac{2}{3} \times \frac{4kT}{g_m} \Delta f \tag{6.96}$$

where g_m is the transconductance, k is the Boltzmann constant, T is the absolute temperature, and Δf is the bandwidth. On the other hand, the flicker ($1/f$) noise voltage is

$$v_n^2 = \frac{KF}{WLC_{ox}^2} \frac{\Delta f}{f} \tag{6.97}$$

where C_{ox} is the oxide capacitance per unit area, W is the channel width of the transistor, L is the channel length, KF is a flicker noise constant, and f is the frequency.

For bipolar transistors, the equivalent input noise voltage v_n is

$$v_n^2 = 4kT\left(r_B + \frac{1}{2g_m}\right)\Delta f \tag{6.98}$$

where r_B is the base resistance. The first term is due to thermal noise, while the second term is due to shot noise of the collector current when referred to the input.

The equivalent input noise current i_n for a bipolar transistor is given by

$$i_n^2 = 2q\left(I_B + \frac{I_C}{\beta^2} + \frac{KF_B I_B}{A_{EB} f}\right)\Delta f \tag{6.99}$$

where β is the short circuit current gain of the bipolar transistor $= I_C/I_B$, A_{EB} is the emitter area, f is the frequency, KF_B is a flicker noise coefficient, I_B is the base current, I_C is the collector current, q is the electron charge, and r_B is the base resistance. The first two terms are due to shot noise, while the last term is flicker noise.

When a bipolar transistor is driven by source resistance R_s, the total noise voltage v_{nT} is given by the combination of the noise voltage and the noise current as follows

$$v_{nT}^2 = v_n^2 + R_s^2 i_n^2 \tag{6.100}$$

The noise figure *NF* of the bipolar transistor is given by

$$NF = 1 + \frac{v_{nT}^2}{4kTR_s\Delta f} \tag{6.101}$$

Example 3: The clock rate is 1 GS/s and the settling requirement of an amplifier needs to be 14-bit accurate. The amplifier is configured in a switched capacitor circuit similar to Figure 6.16 with a gain of 4. Assuming the parasitic capacitance at the summing node is equal to the feedback capacitance, what is the DC gain and unity gain frequency needed for the amplifier?

Answer 3: Assume the amplifier is a single-pole system for simplicity. Using (6.37) to (6.45) or (6.63) to (6.65), the transfer function is

$$\frac{V_{out}}{V_{in}} = \frac{C_1/C_2}{\left(1 + \frac{1}{\beta A_{DC}}\right)} \frac{1}{(1 + s/\beta\omega_u)} \tag{6.102}$$

The closed loop bandwidth BW_{cl} is

$$BW_{cl} \cong \beta f_u \tag{6.103}$$

The settling error ε over time t_s will be approximately

$$\varepsilon = e^{-t_s\beta\omega_u} \tag{6.104}$$

The closed loop DC gain is given by

$$A_{cl_{DC}} = \frac{C_1/C_2}{1 + 1/\beta A_{DC}} \tag{6.105}$$

and

$$\beta = \frac{C_2}{C_1 + C_2 + C_{ps}}$$

Since the gain is 4, then $C_1/C_2 = 4$, and since $C_{ps} = C_2$, then

$$\beta = \frac{1}{4 + 1 + 1} = \frac{1}{6}$$

Since the amplifier needs to be 14-bit accurate, then the total error is

$$\varepsilon_{Total} = 2^{-14}$$

Dividing this error equally between the DC error and the settling error gives

$$\varepsilon_{DC} = 2^{-15}$$

$$\varepsilon_{settling} = 2^{-15}$$

From (6.101), the DC error is given by

$$\varepsilon_{DC} \approx \frac{1}{\beta A_{DC}} = 2^{-15}$$

Therefore, the DC gain will be

$$A_{DC} \approx \frac{2^{15}}{\frac{1}{6}} \approx 6 \times 2^{15}$$

That is equal to

$$A_{DC} \approx 20\log\left(6 \times 2^{15}\right) \approx 106\ \text{dB}$$

Since, from (6.103), the settling error is given by

$$\varepsilon_{settling} \approx e^{-\beta\omega_u t_s} \approx 2^{-15}$$

Therefore,

$$\omega_u \approx \frac{\ln 2^{15}}{\beta t_s} \approx \frac{15 \ln 2}{t_s/6}$$

Since the settling time is given by

$$t_s = \frac{1}{2f_s} = \frac{1}{2\ 10^9} = 0.5\ \text{ns}$$

Therefore,

$$\omega_u \approx \frac{6 \times 15 \ln 2}{0.5 \times 10^{-9}} \approx 125\ \text{Grad/s}$$

and the unity gain frequency will need to be

$$f_u \approx 20\ \text{GHz}$$

The closed loop bandwidth is given by

$$BW_{cl} = f_u \times \beta \approx 3.2\ \text{GHz}$$

That is, to settle to the required accuracy, the amplifier's gain-bandwidth product needs to be approximately equal to 20 GHz, its closed loop bandwidth should be about 3.2 GHz and its DC gain needs to be about 106 dB. That is extremely difficult to achieve!

6.3 Operational amplifiers

6.3.1 The differential pair

One of the simplest structures that can be used as an OTA is the differential pair. A CMOS implementation of the differential pair is shown in Figure 6.24. The DC gain is given by

$$A_{DC} = \frac{v_{outp} - v_{outn}}{v_{inp} - v_{inn}} = g_{mN1} R_o \tag{6.106}$$

where g_{mN1} is the transconductance of the input devices M_{N1}/M_{N2} and R_o is the output resistance at the output node. It can be represented as

$$R_o = R_{dsN1} || R_{dsP1} \tag{6.107}$$

where R_{dsN1} and R_{dsP1} are the output resistances (source-drain resistances) of the devices M_{N1} and M_{P1}, respectively. The small signal equivalent circuit of a MOS device is shown in Figure 6.25.

The open loop 3-dB bandwidth of the differential pair is determined by the pole at the output node, which is given by

$$\omega_p = \frac{1}{R_o C_L} \tag{6.108}$$

where C_L is the *total* capacitance at the output node.

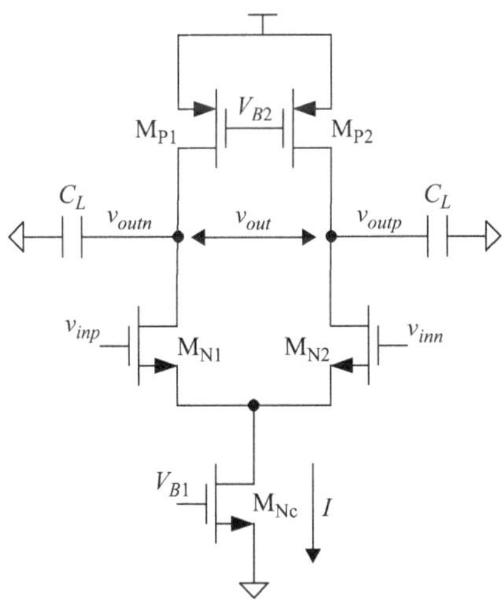

Figure 6.24 A simplified schematic of a differential pair. Transistors M_{N1} and M_{N2} are the input devices (or input pair). Transistors M_{P1} and M_{P2} are the load current sources. Common-mode feedback (CMFB) is not shown

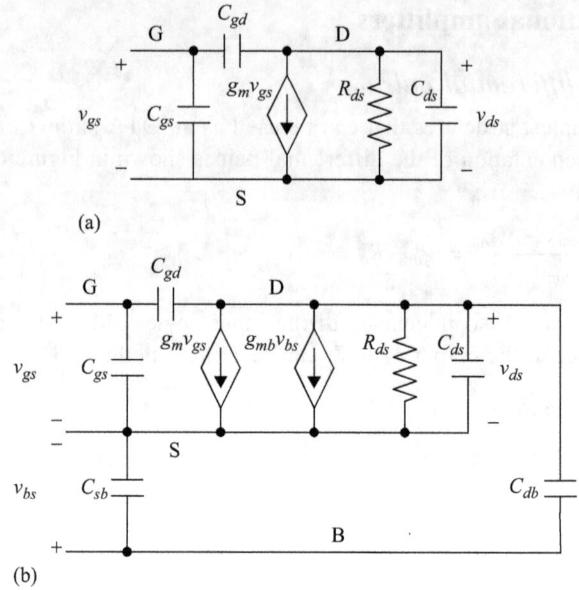

Figure 6.25 *(a) A simplified model of a 3-terminal MOS transistor when the bulk (back-gate) is connected to the source. (b) A simplified model of a 4-terminal MOS transistor*

When driven by a voltage source, the differential pair is a single-pole system. The unity gain frequency is approximately equal to the gain-bandwidth product and given by

$$\omega_u \approx \omega_p A_{DC} \approx \frac{g_{mN1}}{C_L} \tag{6.109}$$

where ω_u is the unity gain angular frequency in radians/second, and f_u is the unity gain frequency in Hz, which is given by

$$f_u \approx f_p A_{DC} \approx \frac{g_{mN1}}{2\pi C_L} \tag{6.110}$$

When used in a closed loop switched capacitor circuit as shown in Figure 6.16, the gain A_{DC} and the unity gain frequency f_u are used to estimate the gain and bandwidth of the closed loop amplifier as described earlier and repeated here for convenience. The closed loop gain A_{cl} is given by

$$A_{cl} = \frac{A_{cl_{DC}}}{1 + \dfrac{s}{2\pi\beta f_u}} = \frac{A_{cl_{DC}}}{1 + \dfrac{s}{2\pi BW_{cl}}} \tag{6.111}$$

where the closed loop bandwidth BW_{cl} is

$$BW_{cl} = f_u \times \beta \approx \frac{g_{mN1} \times \beta}{2\pi C_L} \tag{6.112}$$

The feedback factor β is obtained from Figure 6.16 to be

$$\beta = \frac{C_2}{C_1 + C_2 + C_{ps}} \tag{6.113}$$

The closed loop DC gain is given by

$$A_{cl_{DC}} = \frac{C_1/C_2}{1 + 1/\beta A_{DC}} \approx \frac{C_1/C_2}{1 + 1/\beta g_{mN1} R_o} \tag{6.114}$$

The gain of the differential pair is relatively low. In fine lithography processes (such as 65 nm or 28 nm), the intrinsic gain of an NMOS device is in the order of 5–10 (i.e. 14–20 dB). This is barely 2–3 bit accurate. A much higher gain is usually needed to achieve reasonable accuracy.

The slew rate of the differential pair is given by:

$$SR = \frac{I}{C_L} \tag{6.115}$$

where I is the tail current value.

The input common-mode range is determined by the acceptable input range that will keep the devices operating in saturation. As the input common-mode level decreases, the tail current source's V_{ds} is reduced until it eventually goes out of saturation, if its V_{ds} is less than V_{ds_sat}. That is

$$V_{in\_min} = V_{gsN1} + V_{dsNc\_sat} \tag{6.116}$$

where V_{gsN1} is the V_{gs} of the device M_{N1}, and V_{dsNc_sat} is the V_{ds_sat} of the tail device M_{Nc}. On the other hand, the maximum input voltage is determined by the device M_{N1} going out of saturation:

$$V_{in\_max} = V_{gsN1} + V_{out\_min} - V_{dsN1\_sat} \tag{6.117}$$

The output range is determined by the devices M_{N1} and M_{P1}. The minimum output voltage is determined by either

$$V_{out\_min} = V_{dsNc\_sat} + V_{dsN1\_sat} \tag{6.118}$$

or

$$V_{out\_min} = V_{in\_max} - V_{gsN1} + V_{dsN1\_sat} \tag{6.119}$$

whichever is greater.

The maximum output voltage is determined by

$$V_{out\_max} = V_{DD} - V_{sdP1\_sat} \tag{6.120}$$

The low frequency single-ended *CMRR*, defined as the ratio between the single-ended gain for a differential input A_{diff} and the single-ended gain for a common-mode input A_{CM}, is determined by the output impedance of the tail current source R_{dsNc} and is given by

$$CMRR = \frac{A_{diff}}{A_{CM}} \approx g_{mN1}R_{dsNc} \tag{6.121}$$

Increasing the output impedance of the tail current source improves the *CMRR*. When perfectly balanced, the common-mode signal cancels out at the differential output resulting in an infinite differential *CMRR*. However, in the presence of imbalance, a portion of this common-mode signal can "leak" to the differential output, which degrades the differential accuracy and the common-mode rejection of the differential output.

The single-ended input-referred noise power is

$$v_n^2 = v_{nN1}^2 + v_{nP1}^2\left(\frac{g_{mP1}^2}{g_{mN1}^2}\right) \tag{6.122}$$

where v_{nN1} and v_{nP1} are the gate-referred noise voltages of the devices M_{N1} and M_{P1}, respectively.

The differential input-referred noise is

$$v_{n\_diff}^2 = v_{nN1}^2 + v_{nN2}^2 + v_{nP1}^2\frac{g_{mP1}^2}{g_{mN1}^2} + v_{nP2}^2\frac{g_{mP2}^2}{g_{mN2}^2} \tag{6.123}$$

When the two sides are matched, we get

$$v_{n\_diff}^2 \approx 2v_n^2 \approx 2\left(v_{nN1}^2 + v_{nP1}^2\frac{g_{mP1}^2}{g_{mN1}^2}\right) \tag{6.124}$$

6.3.2 The Miller effect

If the differential pair is driven by a high-impedance source, as shown in Figure 6.26, another pole will appear on the input. This pole will be formed by the source resistance and the Miller capacitance, which appears at the input as

$$C_m = (1 - A)C_{gd} \tag{6.125}$$

Therefore,

$$C_m \approx g_{mN1}R_oC_{gdN1} \tag{6.126}$$

For a large source resistance R_s, the Miller pole will be the dominant one and is given by

$$\omega_{p1} \approx \frac{1}{g_{mN1}R_oC_{gdN1}R_s} \tag{6.127}$$

The Miller effect causes pole splitting that creates a low-frequency pole given by (6.127) above, while pushing out the non-dominant pole at the output node to be

$$\omega_{p2} \approx \frac{g_{mN1} C_{gdN1}}{C_{in} C_{gdN1} + C_{in} C_L + C_{gdN1} C_L} \tag{6.128}$$

where C_{in} is the capacitance between the input node and ground. As C_{gd} increases, the magnitude of the non-dominant pole's frequency ω_{p2} increases, and hence the pole splitting. Moreover, a right hand zero appears at

$$\omega_z \approx \frac{g_{mN1}}{C_{gdN1}} \tag{6.129}$$

Therefore, the Miller capacitance increases the input capacitance, creates a new dominant pole, and causes pole splitting in the frequency response. It also creates a right hand zero. These effects change the frequency response substantially and hence must be properly understood.

6.3.3 *The cascode amplifier*

Starting from the differential pair, a higher gain can be achieved by using a cascode configuration as shown in Figure 6.27. The cascode devices M_{N3} and M_{P3} increase the output impedance and the gain of the amplifier.

The DC gain is given by

$$A_{DC} = \frac{v_{outp} - v_{outn}}{v_{inp} - v_{inn}} = g_{mN1} R_o \tag{6.130}$$

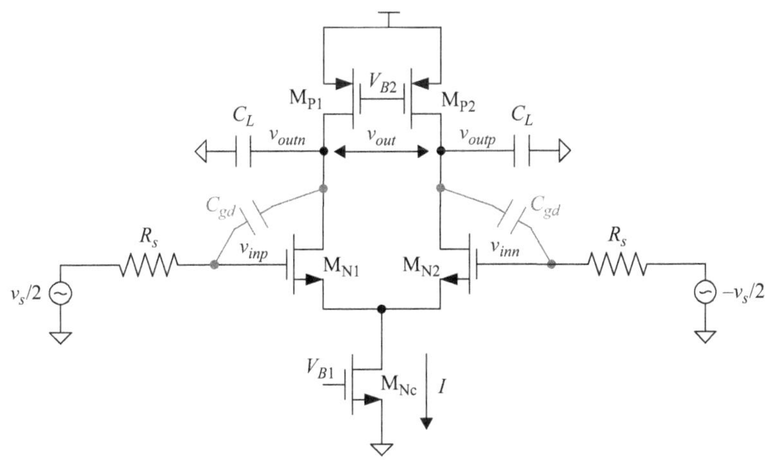

Figure 6.26 A simplified schematic of a differential pair driven by a source with impedance R_s. The parasitic C_{gd} capacitances of the input devices are shown

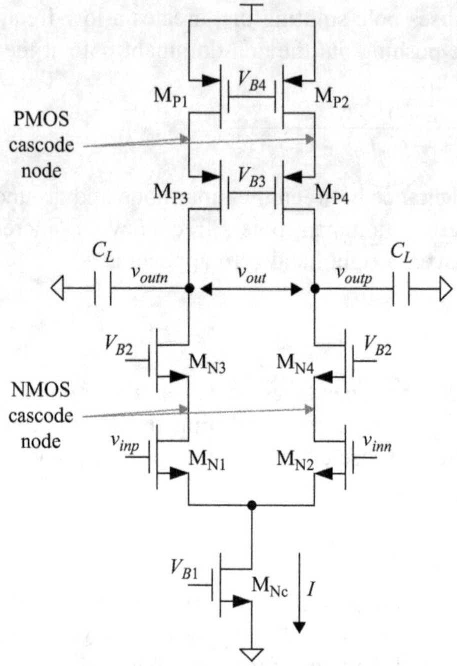

Figure 6.27 A simplified schematic of a cascode differential amplifier. Both the NMOS and PMOS sides are cascoded

The output resistance R_o is

$$R_o = R_{oN}||R_{oP} \tag{6.131}$$

where the output resistance on the NMOS side R_{oN} is

$$R_{oN} \approx R_{dsN1} g_{mN3} R_{dsN3} \tag{6.132}$$

and the output resistance on the PMOS side R_{oP} is

$$R_{oP} \approx R_{dsP1} g_{mP3} R_{dsP3} \tag{6.133}$$

Therefore, the gain and the output resistance have increased by approximately a factor of $g_m R_{ds}$. Therefore, if the intrinsic gain is in the order of 5–10, the cascode amplifier gain increases from 5–10 to 25–100, and from 14–20 dB to 28–40 dB. This improves the accuracy to the 4–6 bit level.

The slew rate is still given by I/C_L, and the dominant pole is

$$\omega_{p1} = \frac{1}{R_o C_L} \tag{6.134}$$

The unity gain frequency is still given by

$$\omega_u \approx \omega_p A_{DC} \approx \frac{g_{mN1}}{C_L} \tag{6.135}$$

Therefore, the cascode amplifier increases the gain substantially without reducing the unity gain frequency. However, a second pole is added to the circuit at the source node of the device M_{N3}. This makes the cascode amplifier a two-pole circuit, and the non-dominant pole is

$$\omega_{p2} = \frac{g_{mN3}}{C_{cascode}} \tag{6.136}$$

where

$$C_{cascode} \approx C_{dbN1} + C_{gdN1} + C_{sbN3} + C_{gsN3} \tag{6.137}$$

The non-dominant pole is determined by the cascode node on the NMOS side because, in this case, the NMOS side is in the signal path. On the other hand, the cascode node on the PMOS side has a much smaller effect on the frequency response because it is not in the signal path. Conversely, if the input is applied on the PMOS devices as shown in Figure 6.28, the cascode node on the PMOS side will be the node of the non-dominant pole, while that on the NMOS side will be less influential.

The improvement in the gain with cascoding comes at the expense of the output dynamic range, which needs to accommodate three devices on the NMOS

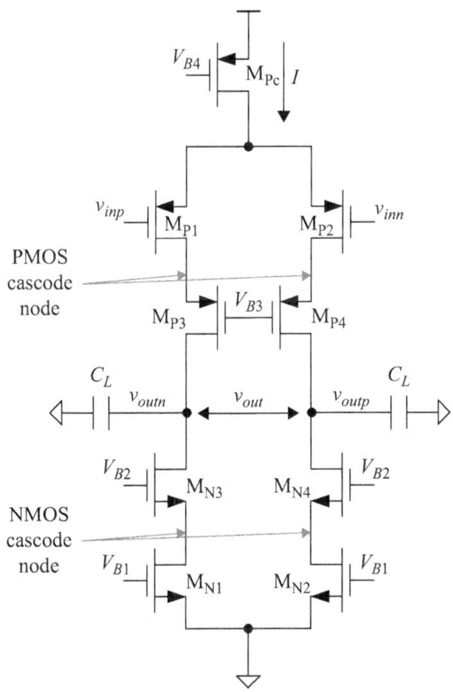

Figure 6.28 A simplified schematic of a cascode differential amplifier with a PMOS input pair. Both the NMOS and PMOS sides are cascoded

side, instead of two for the differential pair, and two devices on the PMOS side, instead of one. Therefore,

$$V_{out\_min} = V_{dsNc\_sat} + V_{dsN1\_sat} + V_{dsN3\_sat} \tag{6.138}$$

and

$$V_{out\_max} = V_{DD} - V_{sdP3\_sat} - V_{sdP4\_sat} \tag{6.139}$$

However, in order to achieve the dynamic range expressed in (6.138) and (6.139), it is important to tightly control the bias of the cascode devices. In Figure 6.27, the bias voltages V_{B2} and V_{B3} need to be set so as to minimize the V_{ds} consumed by the devices M_{N1} and M_{P1}. Those V_{ds}'s need to be the minimum values required to keep those devices in saturation. Larger values of V_{ds} can be used, but they will be at the expense of the dynamic range. Moreover, the bias voltages need to track changes in temperature and process, so that the devices M_{N1} and M_{P1} stay in saturation, with a well-controlled and small overhead. An example of this kind of tightly controlled biasing is discussed in Reference 4.

The single-ended input-referred noise of the cascode amplifier is given by

$$v_n^2 = v_{nN1}^2 + v_{nP1}^2\left(\frac{g_{mP1}^2}{g_{mN1}^2}\right) + v_{nN3}^2\left(\frac{1}{g_{mN1}^2 R_{dsN1}^2}\right) + v_{nP3}^2\left(\frac{1}{g_{mN1}^2 R_{dsP1}^2}\right) \tag{6.140}$$

The differential input-referred noise is

$$v_{n\_diff}^2 = 2\left[v_{nN1}^2 + v_{nP1}^2\left(\frac{g_{mP1}^2}{g_{mN1}^2}\right) + v_{nN3}^2\left(\frac{1}{g_{mN1}^2 R_{dsN1}^2}\right) + v_{nP3}^2\left(\frac{1}{g_{mN1}^2 R_{dsP1}^2}\right)\right] \tag{6.141}$$

Therefore, the noise contributions of the cascode devices are significantly smaller than those of the input and current source devices. This is due to the degeneration effect of the resistances R_{dsN1} and R_{dsP1}, which reduces the contribution of the cascode devices to the amplifier's noise. Therefore, the single-ended noise of the cascode amplifier is almost the same as that of a differential pair and is given by

$$v_n^2 \approx v_{nN1}^2 + v_{nP1}^2\left(\frac{g_{mP1}^2}{g_{mN1}^2}\right) \tag{6.142}$$

The differential input noise is

$$v_{n\_diff}^2 \approx 2\left[v_{nN1}^2 + v_{nP1}^2\left(\frac{g_{mP1}^2}{g_{mN1}^2}\right)\right] \tag{6.143}$$

It is interesting to note that the effect of the Miller capacitance at the input in the cascode amplifier, when driven by a high impedance driver, is significantly different and more benign than a differential pair. The relatively low gain from the input to the cascode node reduces the Miller effect and the input capacitance of the amplifier.

6.3.4 The active cascode amplifier

To increase the gain further, active cascoding is often employed, which is shown in Figure 6.29. An example showing a simple implementation of the active cascode devices is shown in Figure 6.30. The output impedance and gain are further enhanced by the active cascode devices as follows

$$A_{DC} = \frac{v_{outp} - v_{outn}}{v_{inp} - v_{inn}} = g_{mN1} R_o \tag{6.144}$$

The output resistance R_o is

$$R_o = R_{oN} || R_{oP} \tag{6.145}$$

where the output resistance on the NMOS side is

$$R_{oN} \approx R_{dsN1} g_{mN3} R_{dsN3} g_{mNa} R_{dsNa} \tag{6.146}$$

and the output resistance on the PMOS side is

$$R_{oP} \approx R_{dsP1} g_{mP3} R_{dsP3} g_{mPa} R_{dsPa} \tag{6.147}$$

Therefore, the gain and the output resistance have increased by approximately a factor of $(g_m R_{ds})^2$ compared to a differential pair, and by a factor of $g_m R_{ds}$

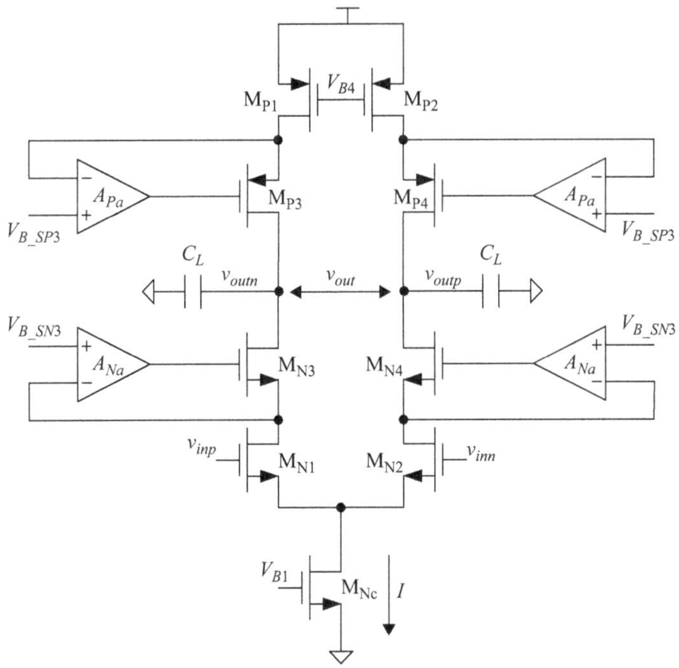

Figure 6.29 A simplified schematic of a differential amplifier with active cascodes. Both the NMOS and PMOS sides are cascoded

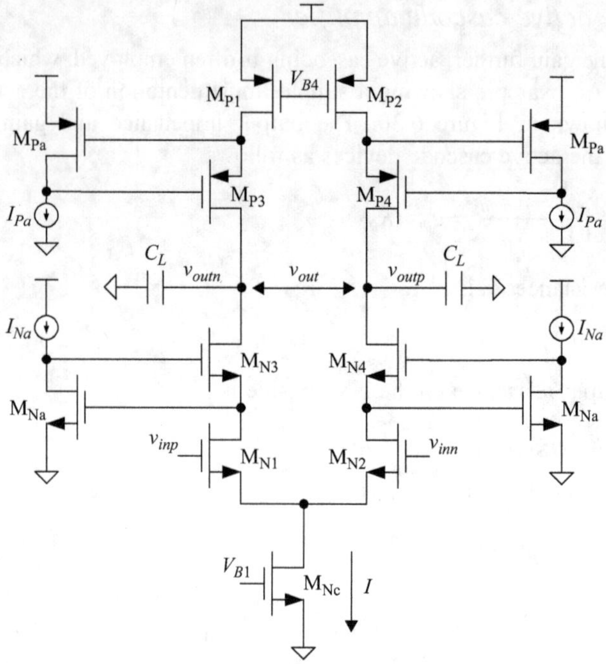

Figure 6.30 A simplified schematic of a differential amplifier with active cascodes. Both the NMOS and PMOS sides are cascoded. The active cascode auxiliary amplifiers are simple common-source amplifiers

compared to a cascode amplifier. Therefore, if the intrinsic gain is in the order of 5–10, the active cascode amplifier gain increases from 5–10 to 125–1000, and from 14–20 dB to 42–60 dB. This is almost 7–10 bits of accuracy.

It is interesting to note that for the overall frequency response of the active cascode amplifier to stay relatively unaffected by the active cascode pole, its frequency does not need to be larger than the unity gain frequency of the whole amplifier. This is shown in Figure 6.31.

As long as the unity gain frequency of the active cascode auxiliary amplifier f_{ua} is larger than the 3-dB bandwidth of the original amplifier f_{p1_orig}, then the active cascode should not affect the small signal frequency response roll off significantly [9]. That is

$$f_{ua} > f_{p1\_orig} \tag{6.148}$$

For stability, the unity gain frequency of the active cascode needs to be lower than its non-dominant pole, which is also the non-dominant pole of the whole amplifier f_{p2} at the cascode node. That is,

$$f_{ua} < f_{p2} \tag{6.149}$$

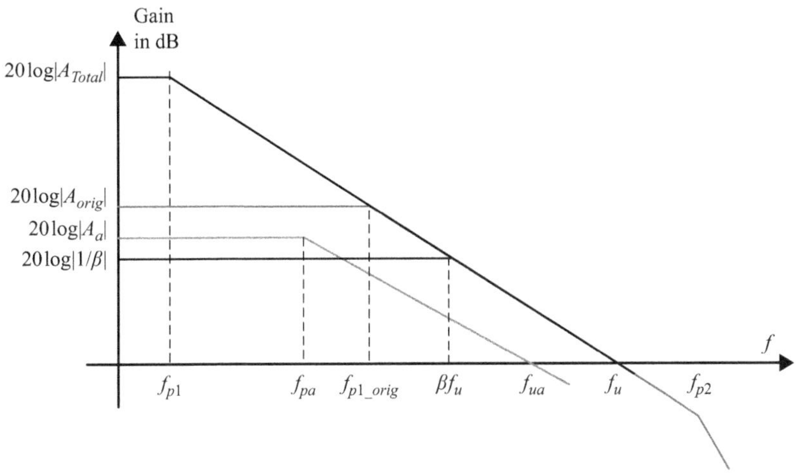

Figure 6.31 A Bode plot of an active cascode amplifier. © 1990 IEEE. Reprinted, with permission, from Reference 9

Therefore, combining (6.148) and (6.149), we get

$$f_{p1\_orig} < f_{ua} < f_{p2} \tag{6.150}$$

It is important to note that the non-dominant pole frequency has decreased because of the Miller effect of the active cascode device's gain and the additional devices at the cascode node.

The effect of the active cascode on the frequency response is complicated by the fact that it introduces a pole-zero doublet [8] at the unity gain frequency of the auxiliary amplifier that can degrade the frequency response if its frequency f_{pz} is not high enough. This is given as

$$f_{pz} \approx f_{ua} \approx \frac{g_{ma}}{2\pi C_{gN3}} \tag{6.151}$$

This doublet frequency needs to be larger than the circuit's closed loop bandwidth BW_{cl}. That is,

$$BW_{cl} < f_{ua} < f_{p2} \tag{6.152}$$

Therefore,

$$\beta f_u < f_{ua} < f_{p2} \tag{6.153}$$

where β is the feedback factor.

Therefore, the speed requirements on the active cascode auxiliary amplifier are not very difficult, but the designer needs to pay attention to the location of the poles and zeros to ensure stability and fast settling.

From the noise standpoint, the active cascode devices degrade the overall single-ended noise as follows

$$v_n^2 = v_{nN1}^2 + v_{nP1}^2\left(\frac{g_{mP1}^2}{g_{mN1}^2}\right) + v_{nN3}^2\left(\frac{1}{g_{mN1}^2 R_{dsN1}^2}\right) + v_{nP3}^2\left(\frac{1}{g_{mN1}^2 R_{dsP1}^2}\right) + v_{nNa}^2\left(\frac{g_{mNa}^2 R_{dsNa}^2}{g_{mN1}^2 R_{dsN1}^2}\right) + v_{nPa}^2\left(\frac{g_{mPa}^2 R_{dsPa}^2}{g_{mN1}^2 R_{dsP1}^2}\right) \quad (6.154)$$

And the differential noise is given by

$$v_{n\_diff}^2 = 2\left[v_{nN1}^2 + v_{nP1}^2\left(\frac{g_{mP1}^2}{g_{mN1}^2}\right) + v_{nN3}^2\left(\frac{1}{g_{mN1}^2 R_{dsN1}^2}\right) + v_{nP3}^2\left(\frac{1}{g_{mN1}^2 R_{dsP1}^2}\right) + v_{nNa}^2\left(\frac{g_{mNa}^2 R_{dsNa}^2}{g_{mN1}^2 R_{dsN1}^2}\right) + v_{nPa}^2\left(\frac{g_{mPa}^2 R_{dsPa}^2}{g_{mN1}^2 R_{dsP1}^2}\right)\right] \quad (6.155)$$

By ignoring the noise contributions of the cascode devices, (6.154) can be approximated as

$$v_n^2 \approx v_{nN1}^2 + v_{nP1}^2\left(\frac{g_{mP1}^2}{g_{mN1}^2}\right) + v_{nNa}^2\left(\frac{g_{mNa}^2 R_{dsNa}^2}{g_{mN1}^2 R_{dsN1}^2}\right) + v_{nPa}^2\left(\frac{g_{mPa}^2 R_{dsPa}^2}{g_{mN1}^2 R_{dsP1}^2}\right) \quad (6.156)$$

Equation (6.155) can also be approximated to be

$$v_{n\_diff}^2 \approx 2\left[v_{nN1}^2 + v_{nP1}^2\left(\frac{g_{mP1}^2}{g_{mN1}^2}\right) + v_{nNa}^2\left(\frac{g_{mNa}^2 R_{dsNa}^2}{g_{mN1}^2 R_{dsN1}^2}\right) + v_{nPa}^2\left(\frac{g_{mPa}^2 R_{dsPa}^2}{g_{mN1}^2 R_{dsP1}^2}\right)\right] \quad (6.157)$$

Therefore, the noise contributions from the active cascode devices can be substantial, and they degrade the amplifier's overall noise. From (6.157), we see that their contribution can be comparable to the contribution of the input devices.

6.3.5 *The two-stage amplifier*

Two-stage amplifiers are often used to improve the output swing, and decouple the input from the output. In addition, they are capable of providing higher gains than single-stage amplifiers. An example is shown in Figure 6.32, which is the Miller compensated amplifier. Another example is shown in Figure 6.33, where the first stage is a cascode amplifier. Another possible variation, which is not shown, is an active cascode first stage.

The gain of the two-stage amplifier is given by

$$A_{DC} = A_{DC1}A_{DC2} \quad (6.158)$$

which gives

$$A_{DC} = g_{mN1}R_{o1}g_{mN5}R_{o2} \quad (6.159)$$

If the first stage is an active cascode stage [13, 14], the overall gain of the amplifier is increased by another factor of g_mR_{ds} due to the second stage's gain,

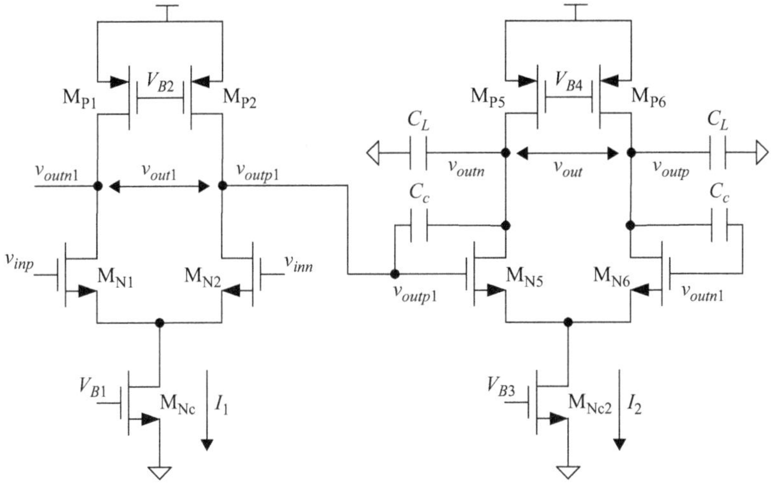

Figure 6.32 A two-stage amplifier with Miller compensation using the capacitors (C_c). Both stages are differential pairs

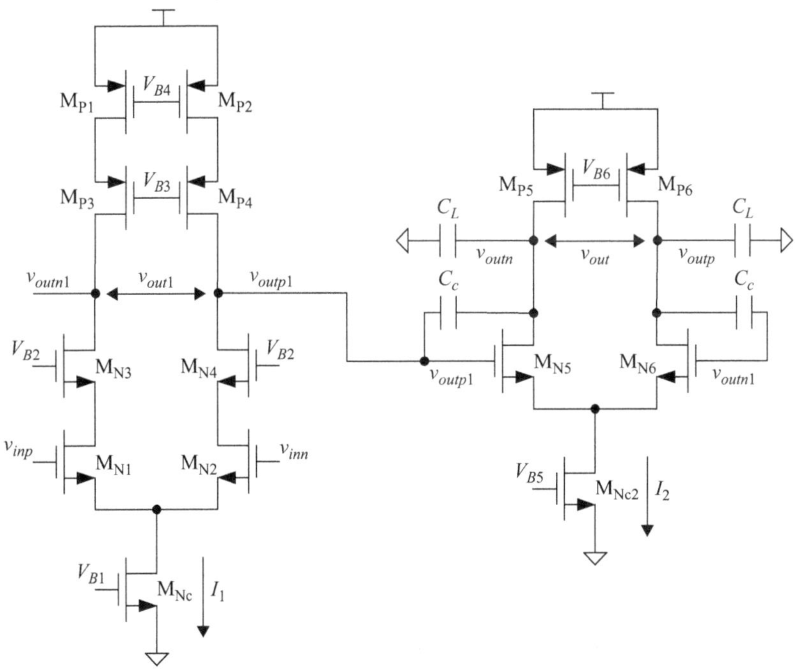

Figure 6.33 A two-stage amplifier with Miller compensation using the capacitors (C_c). The first stage is a cascode amplifier, and the second stage is a differential pair

compared to a single-stage active cascode amplifier. Therefore, if the device's intrinsic gain is in the order of 5–10, the gain of the two-stage amplifier with an active cascode first stage increases to about 625–10000, which is in the range of 56–80 dB. This improves the accuracy to the range of 9–13 bits of accuracy.

The dominant pole of the two-stage amplifier is formed by the Miller compensation capacitance and the output resistance of the first stage. This is given by

$$\omega_{p1} \approx \frac{1}{A_{DC2} C_c R_{o1}} \approx \frac{1}{g_{mN5} R_{o2} C_c R_{o1}} \tag{6.160}$$

Therefore, using (6.159) and (6.160), the unity gain frequency is

$$\omega_u = \frac{g_{mN1}}{C_c} \tag{6.161}$$

The non-dominant pole at the output node can be obtained using (6.128) for an amplifier with Miller capacitance, while noting that C_c is usually large enough such it is practically short circuit at the high frequency of the non-dominant pole. Therefore, the non-dominant pole is approximately given by

$$\omega_{p2} \approx \frac{g_{mN5} C_c}{C_{in2} C_c + C_{in2} C_L + C_c C_L} \approx \frac{g_{mN5}}{C_{in2} + C_L} \tag{6.162}$$

where C_{in2} is the capacitance between the input of the second stage and ground. A right hand zero also exists due to the compensation Miller capacitance and is given by

$$\omega_z = \frac{g_{mN5}}{C_c} \tag{6.163}$$

This zero degrades the settling behavior, which makes it necessary to push it to very high frequencies. Typically, the zero is set to be

$$\omega_z \geq 10\, \omega_u \tag{6.164}$$

which means

$$g_{mN5} \geq 10\, g_{mN1} \tag{6.165}$$

Alternatively we can move the zero to the left hand plane (LHP) by inserting a series resistance R_z with the compensation capacitance as shown in Figure 6.34. This moves the zero to be at

$$\omega_z = \frac{1}{\left(\frac{1}{g_{mN5}} - R_z\right) C_c} \tag{6.166}$$

If $R_z > 1/g_{mN5}$, the zero moves to the LHP and helps the settling. Alternatively, if the zero is placed on top of the non-dominant pole, they cancel as follows

$$\omega_z = \frac{1}{\left(\frac{1}{g_{mN5}} - R_z\right) C_c} = -\omega_{p2} \approx -\frac{g_{mN5}}{C_L} \tag{6.167}$$

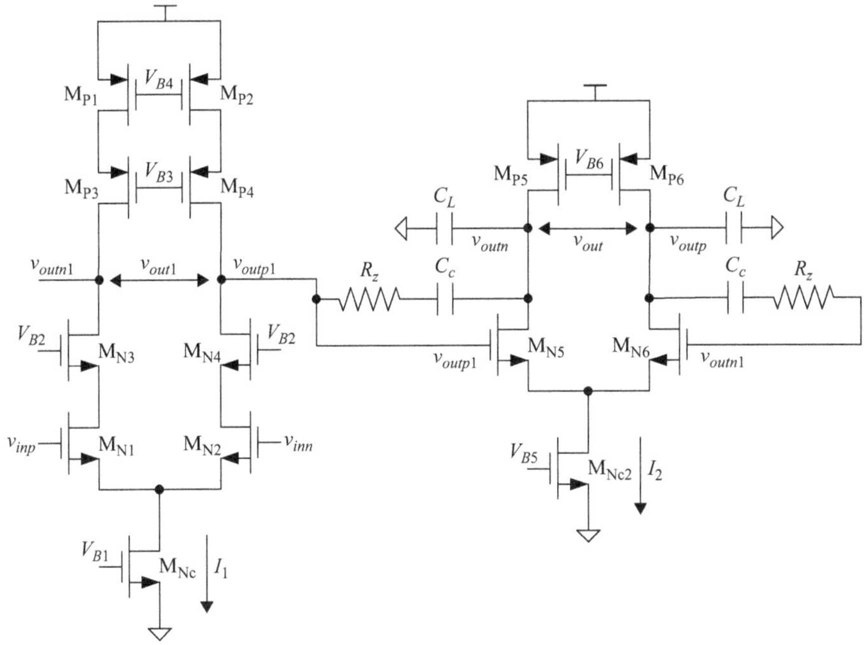

Figure 6.34 A two-stage amplifier with Miller compensation using the capacitors (C_c). Resistor R_z is added for zero compensation

The value of R_z needed for this cancellation is given by

$$R_z \approx \frac{C_c + C_L}{g_{mN5}C_c} \tag{6.168}$$

If the first stage is cascoded, another non-dominant pole exists at the cascode node, and is given by

$$\omega_{p3} = \frac{g_{mN3}}{C_{cascode}} \tag{6.169}$$

The slew rate of a two-stage amplifier is limited by the two tail current sources in the two stages. The slew rate of the first stage is

$$SR = \frac{I_1}{C_c} \tag{6.170}$$

The slew rate of the second stage is

$$SR = \frac{I_2 - I_1}{C_L} \tag{6.171}$$

In (6.171), I_1 is subtracted from I_2 because of the current taken by C_c. The amplifier's overall slew rate will be determined by the worse of the two slew rates in (6.170) and (6.171).

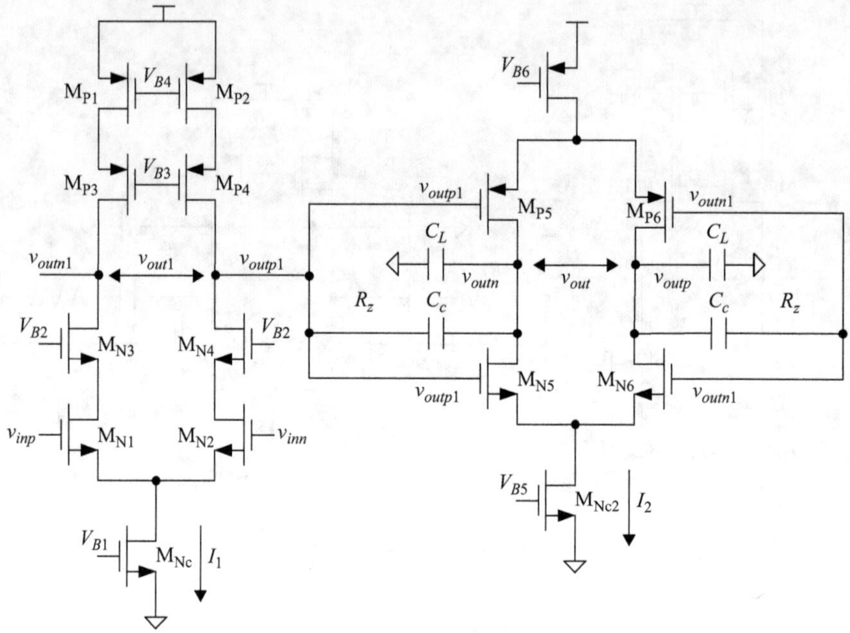

Figure 6.35 A two-stage amplifier. The first stage is a cascode amplifier, the second stage is a push-pull amplifier [15]

The noise of a two-stage amplifier is usually dominated by the noise of the first stage, as the noise of the second stage is scaled down by the gain of the first stage.

Another example of a two-stage amplifier is shown in Figure 6.35, where the second stage is a push-pull stage. The push-pull amplifier utilizes the g_m of both the PMOS and the NMOS amplifiers. The gain of the second stage is

$$A_{DC2} = (g_{mN5} + g_{mP5})R_{o2} \tag{6.172}$$

The overall gain is given by

$$A_{DC} = g_{mN1}R_{o1}(g_{mN5} + g_{mP5})R_{o2} \tag{6.173}$$

The dominant pole is

$$\omega_{p1} = \frac{1}{(g_{mN5} + g_{mP5})R_{o2}C_cR_{o1}} \tag{6.174}$$

The unity gain frequency is given by

$$\omega_u = \frac{g_{mN1}}{C_c} \tag{6.175}$$

The non-dominant pole at the output node is given by

$$\omega_{p2} \approx \frac{(g_{mN5} + g_{mP5})}{C_{in2} + C_L} \tag{6.176}$$

For the same current in the second stage, the push-pull structure lowers the first pole's frequency and increases the gain. If the g_m of the PMOS transistor is comparable to the NMOS's, the push-pull structure can improve the overall settling performance and the speed, by increasing the second pole's frequency. However, if the g_m of the PMOS transistor is much worse than the NMOS's, the increase in the load capacitance of the first stage can sometimes reduce the second pole's frequency response, and hence degrade the speed instead.

A BiCMOS two-stage amplifier is shown in Figure 6.36. In this design the designer took advantage of the bipolar transistor as a cascode device, because of its higher g_m compared to an NMOS transistor. This increases the frequency of the non-dominant pole of the cascode node, according to (6.169), which improves the settling behavior. Moreover, the higher intrinsic gain of the bipolar transistor improves the output resistance of the NMOS side. To have a comparable output

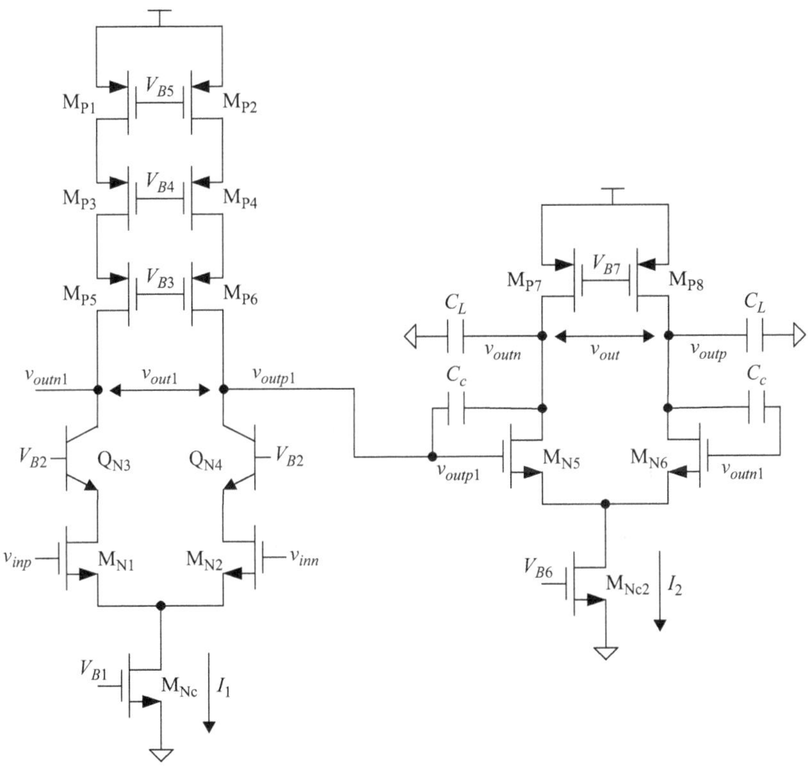

Figure 6.36 A two-stage BiCMOS amplifier with Miller compensation using the capacitors (C_c). The first stage is a cascode amplifier. A BJT is used as a cascode device on the NMOS side, and double cascoding is done on the PMOS side. © 2010 IEEE. Reprinted, with permission, from Reference 12

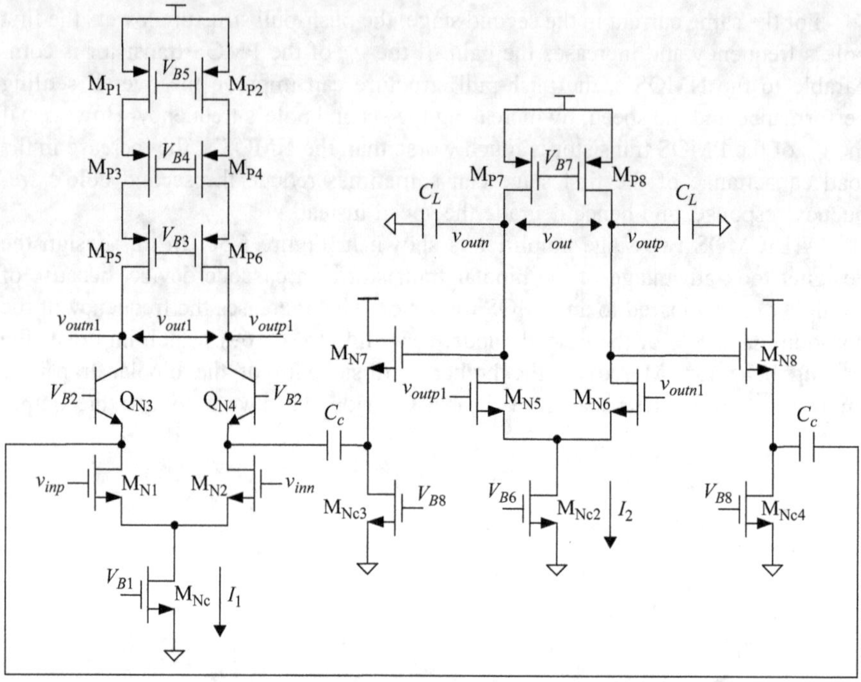

Figure 6.37 A two-stage BiCMOS amplifier with compensation using the capacitors (C_c) and the followers M_{N7} and M_{N8}. The compensation is done at the cascode node of the first stage, instead of its output. The first stage is a cascode amplifier. A BJT is used as a cascode device on the NMOS side, and double cascoding is done on the PMOS side [12]. © 2010 IEEE. Reprinted, with permission, from Reference 12

resistance on the PMOS side, double cascoding is employed, which increases the resistance by another $g_m R_{ds}$ factor.

Figure 6.37 shows another example of a BiCMOS amplifier, where the compensation capacitances are connected to the NPN cascode node. This helps improve the RHP zero of the second stage. The high g_m of the NPN device helps reduce the delay of the compensation path, and pushes the non-dominant pole frequency up. Moreover, the compensation capacitances are buffered using source followers, which increases the frequency of the RHP zero substantially [11, 12].

Figure 6.38 shows an example of a BiCMOS two-stage amplifier, where the NPN transistors are used as the input devices of the second stage. The high g_m of the NPNs in the second stage increases the frequency of the non-dominant pole of the second stage and lowers its power consumption. However, the relatively high value of the base current of the NPN transistor and its low input resistance makes it necessary to buffer the second stage using emitter followers between the two stages.

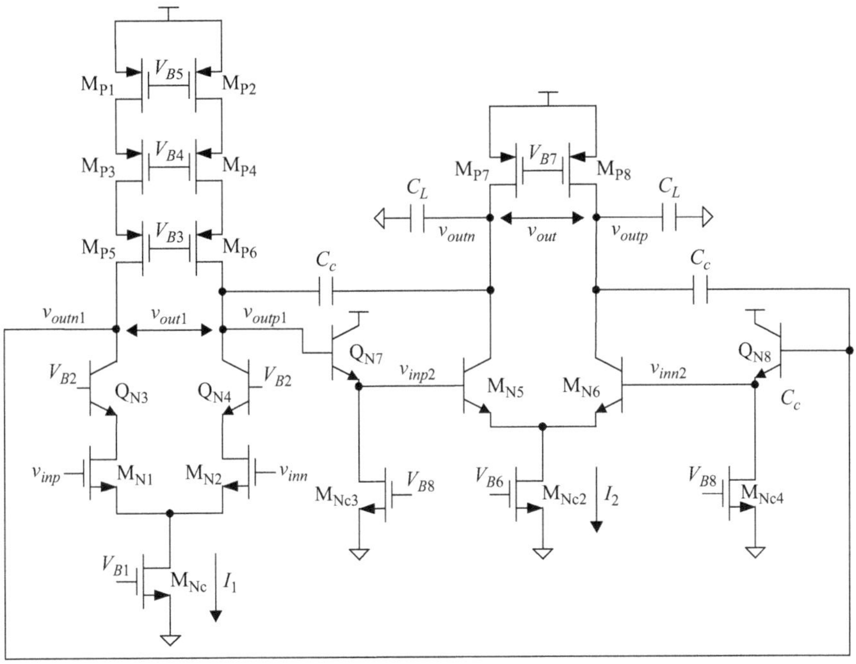

Figure 6.38 A two-stage BiCMOS amplifier with compensation using the capacitors (C_c). BJTs are used as input pairs for the second stage. Emitter followers are used as buffers between the first and second stages. The first stage is a cascode amplifier. A BJT is used as a cascode device on the NMOS side, and double cascoding is done on the PMOS side [1, 16]. © 2006 IEEE. Reprinted, with permission, from Reference 1

6.3.6 *The common-mode feedback*

The common-mode levels of the high impedance nodes in all the previously mentioned amplifiers need to be well controlled in order to optimize the amplifier's performance and maximize its dynamic range and linearity. This is usually done using common-mode control circuits, which are called common-mode feedback (CMFB).

An example of the CMFB is shown in Figure 6.39. The output common-mode is measured by buffering and averaging the output nodes. The measured common-mode value is compared to the desired value. The output is used to control the current source on either the PMOS or the NMOS side. For example, if the measured common-mode value is larger than the desired value, the PMOS current is reduced. The reduction of the PMOS current relative to the NMOS current causes the common-mode voltage to gradually drop. This loop is usually slow compared to the differential speed, and may cause common-mode stability and settling issues.

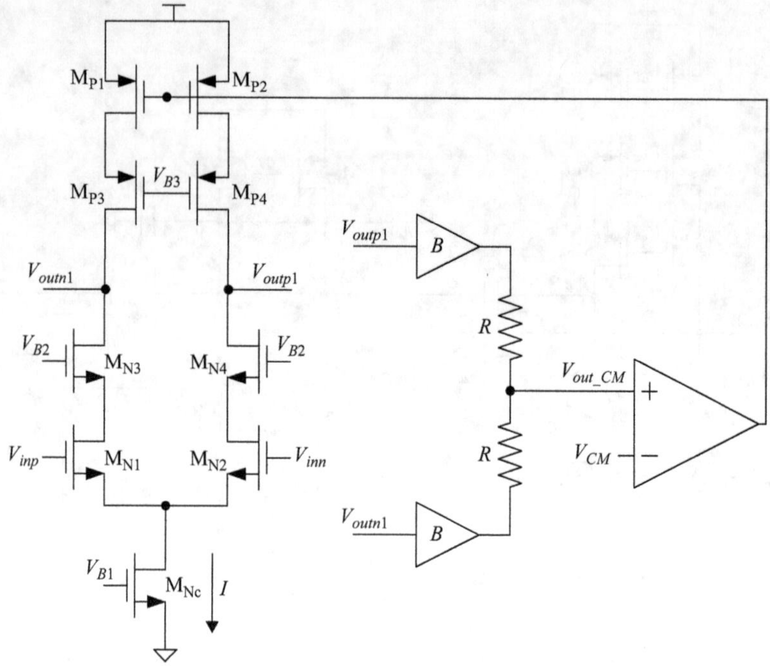

Figure 6.39 An example of a common-mode feedback circuit used to measure the output common-mode voltage and control the PMOS current source accordingly

Another common-mode feedback example is shown in Figure 6.40, using switched capacitor circuits. The common-mode capacitors C_{CM} are charged to the desired value, and refreshed by the switched capacitor C_B. The bias voltages applied across the SC are set to the desired common-mode voltage (V_{CM}) and the gate bias voltage of the PMOS current source (V_{B4}). If the common-mode voltage increases, the gate voltage of the PMOS current source will increase, which in turn lowers the PMOS current, and hence lowers the common-mode.

Compared to the circuit of Figure 6.39, this CMFB loop is faster and easier to control. It, however, controls the *difference* between the output common-mode voltage and the PMOS gate voltage, and not the absolute value of the common-mode voltage. It tends to be prone to inaccuracies because of charge injection and device mismatches.

To reduce the impact of the common-mode capacitance C_{CM} on the differential settling, buffering can be used. An example is shown in Figure 6.41, where source-follower buffers are employed to reduce the loading effect of the common-mode capacitance C_{CM}. This helps improve the differential settling behavior, while marginally affecting the settling of the CMFB loop. This is accomplished at the

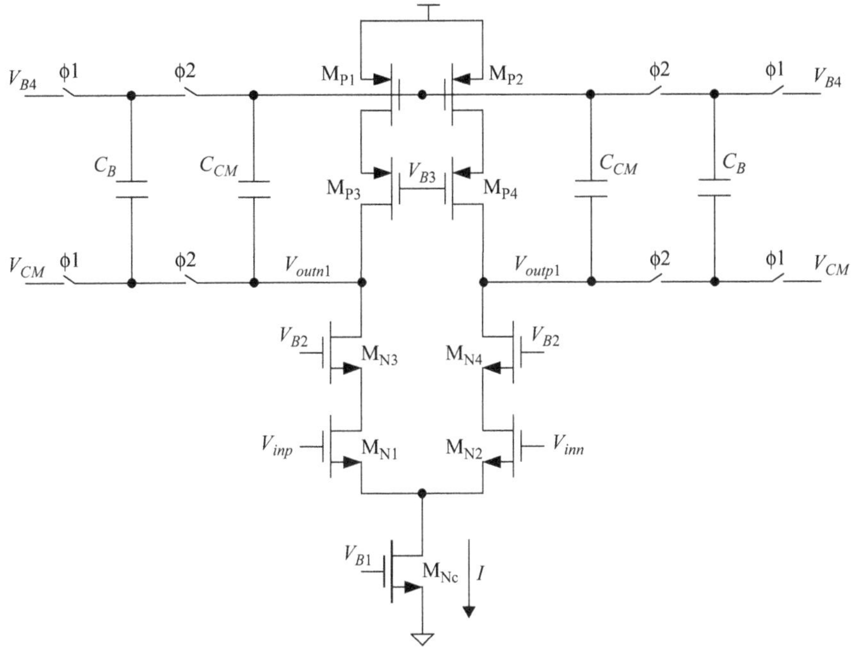

Figure 6.40 An example of a common-mode feedback circuit using switched-capacitor circuits

expense of additional power consumption in the source followers. Also, the bias voltage used needs to accommodate the V_{gs} drop in the source follower, by setting it to $V_{CM} - V_{gsN5}$, instead of V_{CM} [17].

A tightly controlled common-mode level helps maximize the dynamic range of the amplifier, prevent linearity issues, and enhance the differential settling. Designing a stable and fast CMFB loop, without degrading the differential behavior, can be a challenging design problem for very high speed and high performance amplifiers. However, it is a key aspect of the amplifier design that should not be ignored.

6.4 Conclusion

In this chapter, the amplifier structures that are most commonly used in high speed ADCs are discussed. Switched capacitor filters and amplifiers were briefly covered. The gain and settling behavior of amplifiers in open loop and closed loop circuits were analyzed. Various design parameters were explained and some state-of-the-art examples were presented.

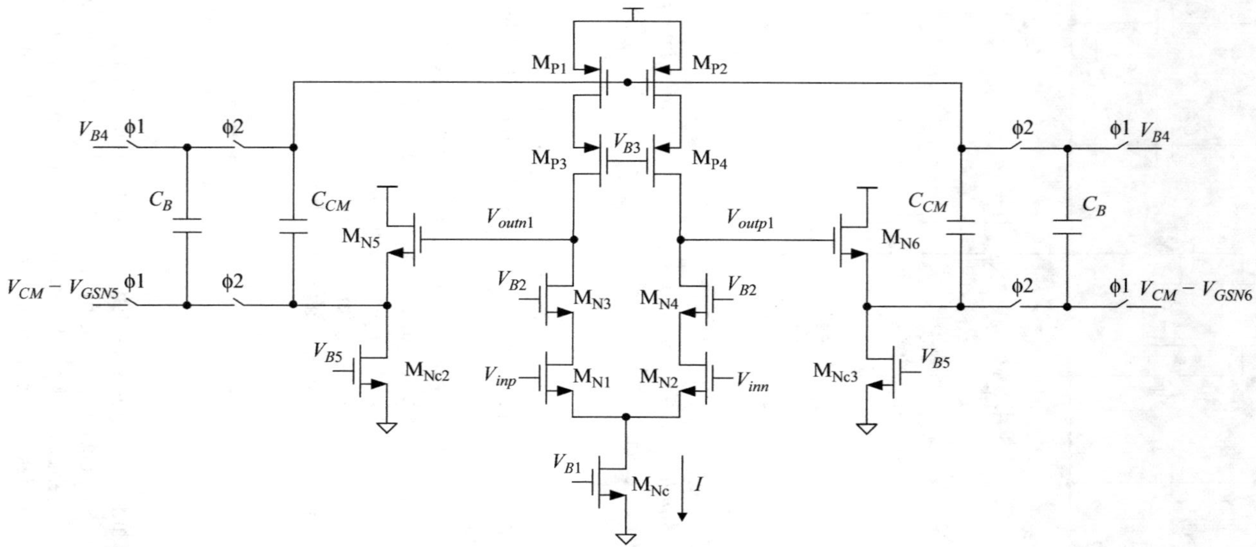

Figure 6.41 An example of a common-mode feedback circuit using switched-capacitor circuits that are buffered using source followers [17]

Problems

1. Analyze the integrators of Figure 6.9 and 6.10 if the operational amplifier has a finite open loop gain A.
2. Repeat Problem 1 if the operational amplifier has a finite open loop gain A and unity gain frequency f_u.
3. Using a SPICE-like simulator, simulate the switched capacitor integrators shown in Figures 6.7, 6.9, and 6.10. Use a clock frequency of 100 MHz, $C_1 = C_2 = 1$ pF. You can use transistor-level circuits or macro-models. Assume an ideal operational amplifier.
4. Repeat Problem 3 if the amplifier has a finite DC gain of 60 dB and unity gain frequency of 10 GHz. Discuss the errors.
5. Repeat Problem 3 if the amplifier has a finite DC gain of 60 dB and unity gain frequency of 100 MHz. Discuss the errors.
6. Draw a switched capacitor inverting amplifier with a gain of 4. If the clock frequency is 50 MHz, and the settling accuracy needs to be 0.25 LSB at the 10-bit level, what should the unity gain frequency of the operational amplifier be? What is the DC gain?
7. Repeat Problem 6 if the operational amplifier has 2 poles. What are the locations of the two poles?
8. An operational amplifier with 2 poles and 60 dB DC gain is used in a switched capacitor amplifier circuit similar to Figure 6.16, with a gain of 8. Ignore the parasitic capacitances. The amplifier is under-damped with 1% overshoot. If the amplifier settles to 14-bit accuracy at a clock rate of 200 MHz, what are the locations of the 2 poles? What is the phase margin?
9. For Problem 8, what can we change to improve the settling time? What is the new clock rate that can be used for the same accuracy?
10. Using a SPICE-like simulator, simulate the switched capacitor amplifiers shown in Figures 6.14, 6.15, and 6.17. Use a clock frequency of 100 MHz, $C_1 = 2C_2 = 1$ pF. You can use transistor-level circuits or macro-models. Assume an ideal operational amplifier.
11. Repeat Problem 10 using an operational amplifier with a DC gain of 60 dB. Discuss the errors.
12. Repeat Problem 10 using an operational amplifier with a DC gain of 60 dB and unity gain frequency of 1 GHz. Discuss the errors.
13. Repeat Problem 10 using an operational amplifier with a DC gain of 60 dB and unity gain frequency of 500 MHz. Discuss the errors.
14. For the switched capacitor amplifier shown in Figure 6.14 with a gain of 4. If the input-referred noise of the operational amplifier is 500 μV, what is the noise referred to the input of the closed loop amplifier during the gain phase? What is the output referred noise during the gain phase?
15. For the switched capacitor amplifier shown in Figure 6.14 with a gain of 4. The input-referred noise of the operational amplifier is 500 μV. What is the input-referred noise during the input sampling phase? What is the input-referred noise during the gain phase?

16. For the amplifier shown in Figure 6.40, what is the low-frequency equivalent of the CMFB circuit formed by capacitances C_{CM}, C_B, the switches, and the bias voltages? Can you propose an alternative structure that does not use switched capacitors? What are the pros and cons of each structure?
17. Discuss the advantages and disadvantages of the CMFB structure shown in Figure 6.39.
18. For the cascode amplifier of Figure 6.27, derive the expression for the input capacitance of the amplifier. If the cascode amplifier is driven by a large source impedance, how does it compare with the differential pair of Figure 6.26? Discuss.
19. Using a SPICE-like simulator with an arbitrary CMOS process, simulate the cascode amplifier of Figure 6.27. Size the devices properly to achieve a gain of 30 dB and to be critically damped. What is the current and unity gain frequency of the amplifier? What is the input noise voltage?
20. For the amplifier of Problem 19, employ active cascoding as shown in Figure 6.30. What is the impact on the total current, phase margin, unity gain frequency, and input noise?
21. Discuss the pros and cons of the active cascode implementation of Figure 6.30. How would you improve that amplifier?
22. If we assume that the intrinsic gain is the same for all devices, how does the active cascode amplifier of Figure 6.30 compare with the two-stage amplifier of Figure 6.33? Compare the two amplifiers in terms of gain, power, dynamic range, noise, and speed.
23. Compare the two BiCMOS amplifiers of Figure 6.36 and 6.37. Use analysis or simulation.
24. Compare the two BiCMOS amplifiers of Figure 6.37 and 6.38. Use analysis or simulation.

References

[1] A.M.A. Ali, C. Dillon, R. Sneed, *et al.*, "A 14-bit 125 MS/s IF/RF Sampling Pipelined ADC With 100 dB SFDR and 50 fs Jitter," *IEEE Journal of Solid-State Circuits*, 41(8), pp. 1846–1855, Aug 2006.

[2] K.R. Laker and W.M.C. Sansen, "Design of analog integrated circuits and systems," McGraw Hill, New York, NY, 1994.

[3] H.C. Yang and D.J. Allstot, "Considerations for Fast Settling Operational Amplifiers," *IEEE Transactions on Circuits and Systems*, 37(3), pp. 326–334, Mar 1990.

[4] A.M.A. Ali, "Stably-biased cascode networks," US Patent 7,023,281, Apr 2006.

[5] P.R. Gray, P.J. Hurst, S.H. Lewis, and R.G. Meyer, "Analysis and Design of Analog Integrated Circuits," Fourth Edition, John Wiley & Sons, New York, NY, 2001.

[6] D.A. Johns and K. Martin, "Analog Integrated Circuit Design," John Wiley & Sons, New York, NY, 1997.

[7] P.R. Gray and R.G. Meyer, "MOS Operational Amplifier Design – A Tutorial Overview," *IEEE Journal of Solid-State Circuits*, SC-17(6), pp. 969–982, Dec 1982.

[8] B.Y. Kamath, R.G. Meyer, and P.R. Gray, "Relationship between Frequency Response and Settling Time of Operational Amplifiers," *IEEE Journal of Solid-State Circuits*, SC-9(6), pp. 347–352, Dec 1974.

[9] K. Bult and G.J.G.M. Geelen, "A Fast-Settling CMOS Op Amp for SC Circuits with 90-dB DC Gain," *IEEE Journal of Solid-State Circuits*, 25(6), pp. 1379–1384, Dec 1990.

[10] K. Nagaraj, "CMOS Amplifiers Incorporating a Novel Slew Rate Enhancement Technique," *IEEE Custom Integrated Circuits Conference (CICC)*, pp. 11.6.1–11.6.5, 1990.

[11] P.J. Hurst, S.H. Lewis, J.P. Keane, *et al.*, "Miller Compensation Using Current Buffers in Fully Differential CMOS Two-Stage Operational Amplifiers," *IEEE Transactions on Circuits and Systems-I: Regular Papers*, 51(2), pp. 275–285, Feb 2004.

[12] A.M.A. Ali, A. Morgan, C. Dillon, *et al.*, "A 16-bit 250-MS/s IF Sampling Pipelined ADC with Background Calibration," *IEEE Journal of Solid-State Circuits*, 45(12), pp. 2602–2612, Dec 2010.

[13] A.M.A. Ali, H. Dinc, P. Bhoraskar, *et al.*, "A 14b 1GS/s RF Sampling Pipelined ADC with Background Calibration," *IEEE ISSCC Digest of Technical Papers*, pp. 482–483, Feb 2014.

[14] A.M.A. Ali, H. Dinc, P. Bhoraskar, *et al.*, "A 14b 1GS/s RF Sampling Pipelined ADC with Background Calibration," *IEEE Journal of Solid-State Circuits*, 49(12), pp. 2857–2867, Dec 2014.

[15] A.M.A. Ali, H. Dinc, P. Bhoraskar, *et al.*, "A 14-bit 2.5GS/s and 5GS/s RF Sampling ADC with Background Calibration and Dither," *IEEE VLSI Circuits Symposium*, pp. 206–207, 2016.

[16] A.M.A. Ali, "Differential amplifiers with enhanced gain and dynamic range," US Patent 7,253,686, Aug 2007.

[17] A.M.A. Ali, "Amplifier networks with controlled common-mode level and converter systems for use therewith," US Patent 7,746,171, Jun 2008.

[6] D. A. Johns and K. Martin, "Analog Integrated Circuit Design," John Wiley & Sons, New York, 1997.
[7] P. R. Gray and R. G. Meyer, "MOS Operational Amplifier Design – A Tutorial Overview," IEEE Journal of Solid-State Circuits, SC-17, pp. 969–[illegible], Dec. 1982.
[8] B. Y. Kamath, R. G. Meyer, and P. R. Gray, "Relationship between Frequency Response and Settling Time of Operational Amplifiers," IEEE Journal of Solid-State Circuits, SC-9, pp. 347–352, Dec. 1974.
[9] K. Bult and G. J. G. M. Geelen, "A Fast-Settling CMOS Op Amp for SC Circuits with 90-dB DC Gain," IEEE Journal of Solid-State Circuits, vol. 25, pp. 1379–1384, Dec. 1990.
[10] [illegible]

Chapter 7

Pipelined A/D converters

In a pipelined ADC, the conversion process is broken down into multiple steps taking place in *pipelined* stages. The stages operate on different samples of the input signal simultaneously, which enables high throughput at the expense of latency. Each stage quantizes its input into a number of bits, and passes the residual quantization error, after amplification, to the following stages for finer quantization. Referred to the input, the quantization step size and noise get progressively smaller as the input sample propagates down the pipeline. As long as the quantization error is accurately generated and amplified in every stage, before passing it to the following stage, pipelined ADCs can achieve high resolutions and linearity.

The assembly-line-like approach used in pipelined ADCs enables high throughput and relatively high resolution, making the pipelined ADC arguably one of the most important architectures in the high speed and high performance space. Sampling rates up to 2.5 GS/s and resolutions up to 16 bits with better than 100 dB linearity (SFDR) have been reported [1–7].

Moreover, the algorithmic and feed-forward nature of the pipeline architecture makes it amenable to linearity correction using digital signal processing techniques that enhance performance and speed. This helps take advantage of the fine lithography CMOS processes by employing digital assistance (calibration) techniques that utilize the relatively efficient digital circuits to improve analog performance and lower power consumption.

7.1 Architecture

The typical pipelined ADC architecture is shown in Figure 7.1. The ADC consists of a sample-and-hold circuit followed by a cascade of pipeline stages. Each stage is composed of a sub-ADC, a DAC, a subtractor and a residue amplifier (RA). The sub-ADC, which has traditionally been a flash ADC, digitizes the input signal with a number of bits equal to the resolution of the stage. The DAC generates an analog representation of these bits, which is then subtracted from the input to generate the residue. The residue is amplified by the RA in order to relax the accuracy requirements of the following stages.

In Figure 7.1, the resolution of the first stage is k_1, the second stage is k_2, and so on. If the gain of the first stage's RA is $G_1 = 2^{k_1}$, then the residue will ideally span the full-scale range of the following stage. This, however, is not practical, as

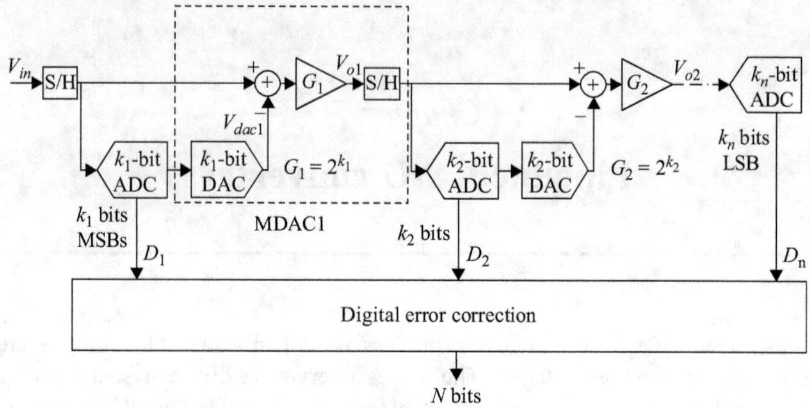

Figure 7.1 Basic pipelined ADC architecture without redundancy

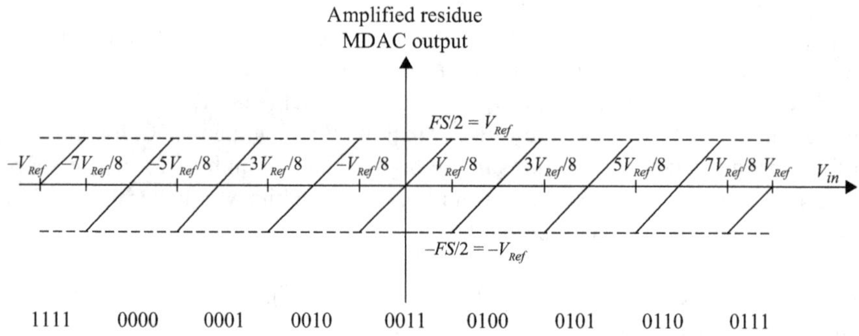

Figure 7.2 Ideal residue for the pipelined ADC shown in Figure 7.1 with $k_1 = 3$ and inter-stage gain of 8. An extra code is used for symmetry and to fold the end sub-ranges

any error in the sub-ADC will cause the residue to over-range beyond the following stages' full-scale, and hence lead to missing codes and gross errors. This is shown in Figures 7.2 and 7.3 for the case of $k_1 = 3$ bits. The ideal residue with a gain of 2^{k_1} is shown in Figure 7.2, and the residue in the presence of a comparator error is shown in Figure 7.3.

In order to reduce the sensitivity to the sub-ADC (comparator) offsets, redundancy is employed by using an inter-stage gain that is less than 2^{k_1}. For example, as shown in Figure 7.4, if the gain is $2^{(k_1-1)}$, the residue will span only half the full-scale range, and the other half can be used as a *correction range* to accommodate the sub-ADC errors. This is called *digital error correction*, and is one of the attractive aspects of pipelined ADCs. That is, the ADC performance is relatively insensitive to errors in the sub-ADCs, as long as they are within the bounds of the correction range. Figure 7.5 depicts the ideal residue when 1-bit

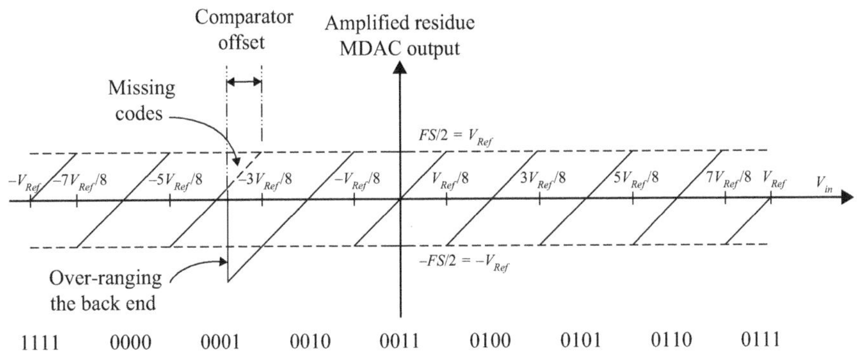

Figure 7.3 Residue for the pipelined ADC shown in Figure 7.1 with $k_1 = 3$, and inter-stage gain of 8, in the presence of comparator offset. The over-ranging causes missing codes

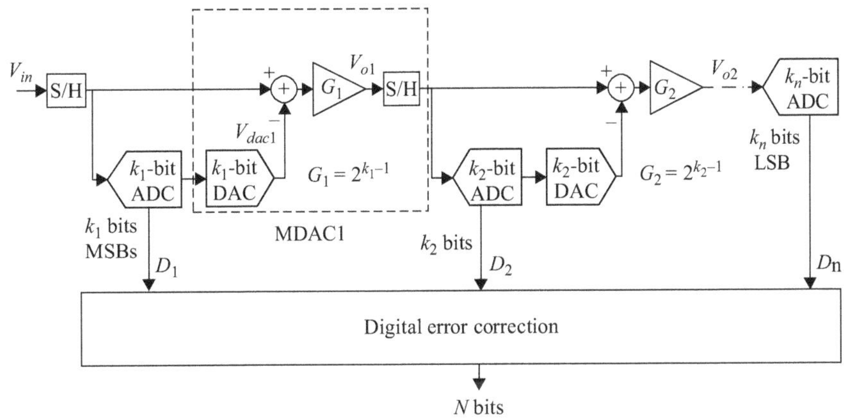

Figure 7.4 Basic pipelined ADC architecture with redundancy

redundancy is employed, and Figure 7.6 shows this residue in the presence of a comparator offset error. The flash error correction range is half the full-scale, and any comparator error would be correctable by the following stages as long as the residue is still bounded within the full-scale range of the back-end stages.

From Figure 7.5, we see that the sub-ADC is a mid-tread ADC with an odd number of quantization steps (sub-ranges). As discussed in Chapter 1, this prevents chatter around the zero input, which is a major transition. In addition, the last two sub-ranges are limited to half the size of the other sub-ranges. This mid-tread implementation with odd number of codes and "folding" of the last sub-ranges achieves symmetry around zero, simplifies the digital processing and limits the output of the stage to be always within half a sub-range. The extra code requires an extra comparator in the stage and an extra bit in the output. An example of the coding used is illustrated in Figures 7.5 and 7.6.

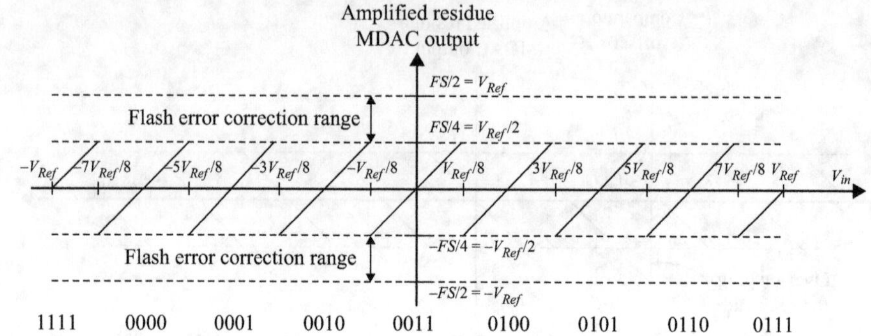

Figure 7.5 Ideal residue for the pipelined ADC shown in Figure 7.4 with $k_1 = 3$ after amplification with a gain of 4. An extra code is used for symmetry and to fold the end sub-ranges

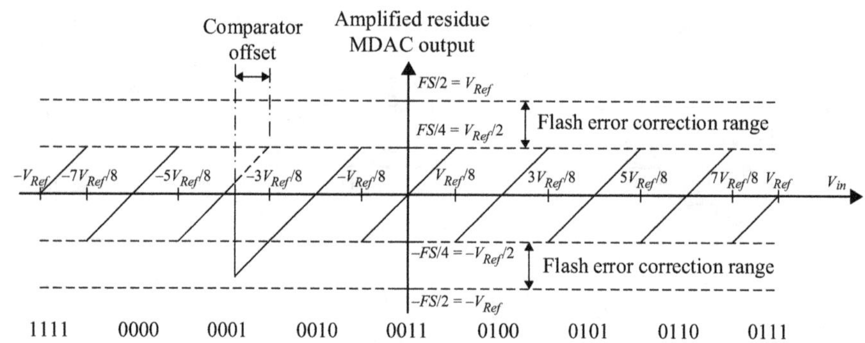

Figure 7.6 Residue for the pipelined ADC shown in Figure 7.3 with $k_1 = 3$, after amplification with a gain of 4, in the presence of comparator offset. Since the error is within the correction range bounds, no over-ranging should occur in the back end and it gets corrected by the following stages

The digital error correction block combines the bits generated by all stages, align them appropriately in time, and add them up with the proper weighting. A block diagram of the digital error correction is shown in Figures 7.7(a), and an example of the processing of the "raw" (or sub-ADC) bits for a 14-bit ADC with 1-bit redundancy is shown in Figure 7.7(b). The overlap between the bits of the various stages is due to the 1-bit redundancy and all the inter-stage gains being $2^{(k_i-1)}$ instead of 2^{k_i} for all values of i.

Without loss of generality, if we focus on the first stage, we see that its residue is given by

$$V_{o1} = G_1(V_{in} - V_{dac1}) = G_1\left(V_{in} - \frac{D_1 \times V_{Ref}}{2^{k_1-1}}\right) \tag{7.1}$$

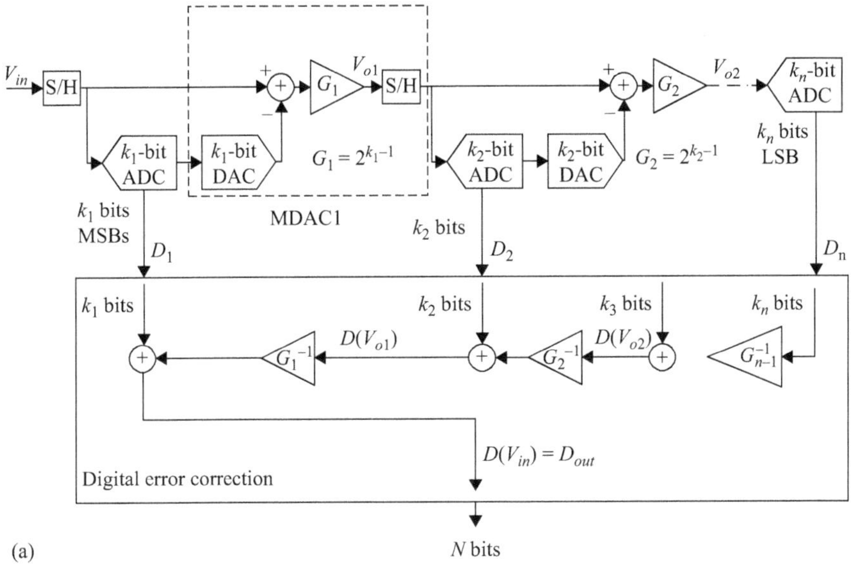

(a)

Stage-1: $k_1 = 3$:	$d_{13} d_{12} d_{11a}$
Stage-2: $k_2 = 3$:	$d_{11b} d_{10} d_{9a}$
Stage-3: $k_3 = 3$:	$d_{9b} d_8 d_{7a}$
Stage-4: $k_4 = 3$:	$d_{7b} d_6 d_{5a}$
Stage-5: $k_5 = 3$:	$d_{5b} d_4 d_{3a}$
Stage-6: $k_6 = 4$:	$d_{3b} d_2 d_1 d_0$
Output: 14 bits:	$d_{13}\ d_{12}\ d_{11}\ d_{10}\ d_9\ d_8\ d_7\ d_6\ d_5\ d_4\ d_3\ d_2\ d_1\ d_0$

(b)

Figure 7.7 (a) Combining the bits of the pipeline stages with redundancy and digital error correction. The gain on the digital side must match the gains on the analog side. (b). Combining the bits of the pipeline stages with redundancy and digital error correction for a 14-bit converter. When the gain is a power of 2, multiplication and division in the digital domain are achieved by bit shifting

where G_1 is the gain of the first stage, which is typically equal to 2^{k_1-1}, D_1 is the digital code of the first stage, which is given by

$$D_1 = 0, \pm 1, \pm 2, \cdots, \pm 2^{k_1-1},$$

V_{Ref} is the reference voltage, and k_1 is the number of bits in the first stage.

In the digital domain, the reconstruction of the digital signal starts with the output of the back-end flash D_n, which is the digital representation of the $(n-1)$th stage's residue $V_{o(n-1)}$. Therefore, we set $D\left(V_{o(n-1)}\right) = D_n$ and progress to the *i*th stage digital residue $D(V_{oi})$ which is given by

$$D\left(V_{o(i-1)}\right) = D(V_{oi}) \times G_i^{-1} + D_i \tag{7.2}$$

Applying this expression progressively from the back end to the front culminates in the ADC digital output D_{out} which is given as

$$D_{out} = D(V_{in}) = D(V_{o1}) \times G_1^{-1} + D_1 \tag{7.3}$$

This reconstruction of the ADC digital output from the "raw" bits is represented in Figure 7.7(a) and (b).

The DAC, subtractor, and amplifier are usually combined together into one block, called the multiplying DAC (MDAC). Moreover, instead of having a dedicated sample-and-hold amplifier (SHA), the sample-and-hold operation is sometimes performed separately by the MDAC and flash sampling networks. This is called a SHA-less architecture, which saves a substantial amount of power and noise, and was first proposed in [2]. It is shown in Figures 7.8, where the S/H operation is incorporated in the MDAC and the flash (sub-ADC). In this case, the absence of a held signal at the input of the first stage leads to a possible mismatch between the flash and the MDAC sampled inputs. The resulting mismatch error is aggravated as the input frequency increases. If the input signal is sinusoidal, and given by

$$x(t) = A\cos 2\pi f_{in} t \tag{7.4}$$

where f_{in} is the input frequency, and A is the input amplitude. The error Δx due to timing mismatch Δt between the flash and MDAC sampling instants is given by

$$\Delta x \propto 2\pi f_{in} A \Delta t \tag{7.5}$$

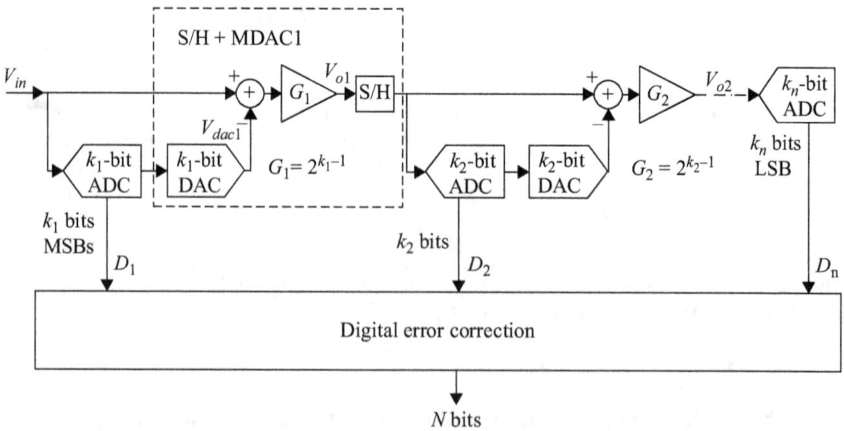

Figure 7.8 SHA-less pipelined ADC architecture with the S/H integrated into the MDAC and the stage sub-ADC

Therefore, when employing a SHA-less architecture, care must be taken to match the bandwidth and timing between the flash and MDAC paths. Mismatches between the two paths result in flash errors that consume a portion of the correction range, and get worse as the input frequency increases. However, the power and performance benefits of the SHA-less architecture are substantial. It was demonstrated to be capable of IF and RF sampling by [3], with subsequent improvements in [5, 6]. Matching that support input frequencies up to 3–5 GHz has been reported in [7].

Unlike the sub-ADC's accuracy, which is relaxed by the redundancy, the MDAC's accuracy is critical to the overall ADC performance. An error in the inter-stage gain or DAC results in non-linearity of the whole ADC. An example of the residue with inter-stage gain error (IGE) is shown in Figure 7.9. If the amplifier gain is less than the ideal value that is used in the digital back end (G_1 in (7.1)), the residue will have a slope that is smaller than the ideal value, which results in breaks in the transfer characteristic that can lead to missing codes, as shown in Figure 7.10. If the inter-stage gain is too large, it can lead to non-monotonicity. The INL signature of IGEs is the saw-tooth pattern shown in Figure 7.11. A positive INL slope indicates an inter-stage gain that is too small. A negative slope indicates a gain that is too large.

DAC errors cause vertical displacement of the residue segments (sub-ranges) relative to each other as shown in Figure 7.12. This represents code-dependent error in the V_{dac} term of (7.1) and can lead to breaks in the transfer characteristic as shown in Figures 7.13 and to INL errors like those shown in Figure 7.14. It is important to note that while the inter-stage amplifier needs to be as accurate as the following stage's resolution, the first stage DAC needs to be as accurate as the whole ADC. As for the following stages, their accuracy requirements are relaxed by the gains of the stages preceding them.

In addition to its impact on the INL, the IGE degrades both the SINAD and SFDR of the ADC. This can be particularly evident when they are plotted versus the input amplitude. The SINAD and SFDR can change substantially with the input amplitude, especially as the input crosses the sub-range boundaries. This has been discussed in Chapter 2. The worst performance was shown to be at an amplitude

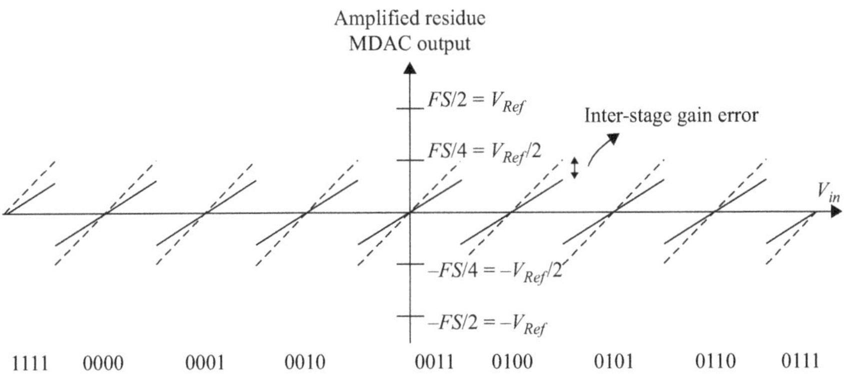

Figure 7.9 The stage-1 residue with an inter-stage gain error

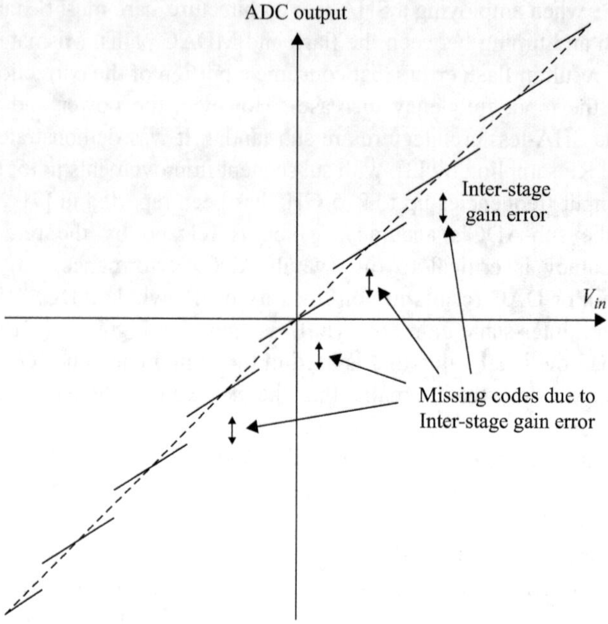

Figure 7.10 The transfer function of the ADC with inter-stage gain errors that are large enough to cause missing codes

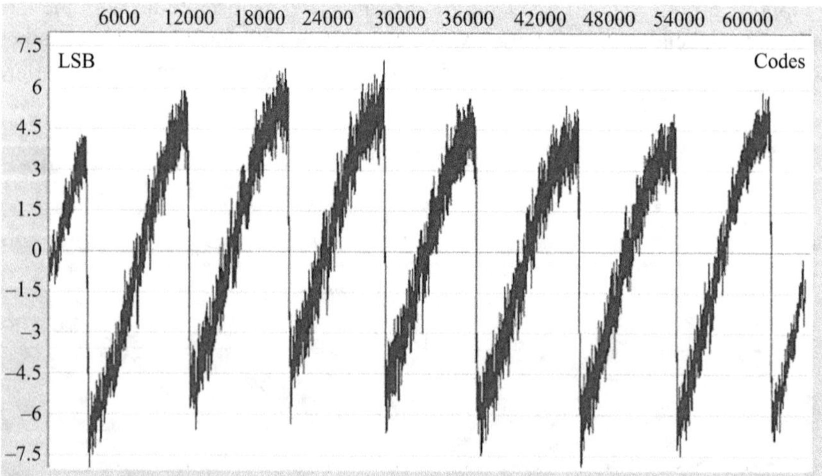

Figure 7.11 The INL of a pipelined ADC with inter-stage gain errors in stage-1. The inter-stage gain is smaller than the ideal value

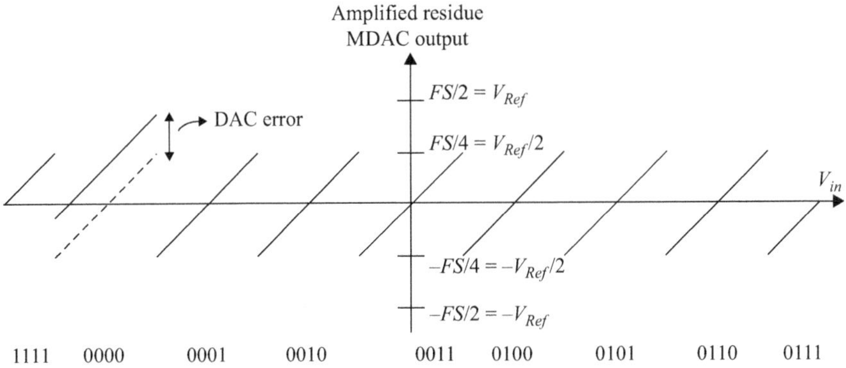

Figure 7.12 The stage-1 residue showing a DAC error

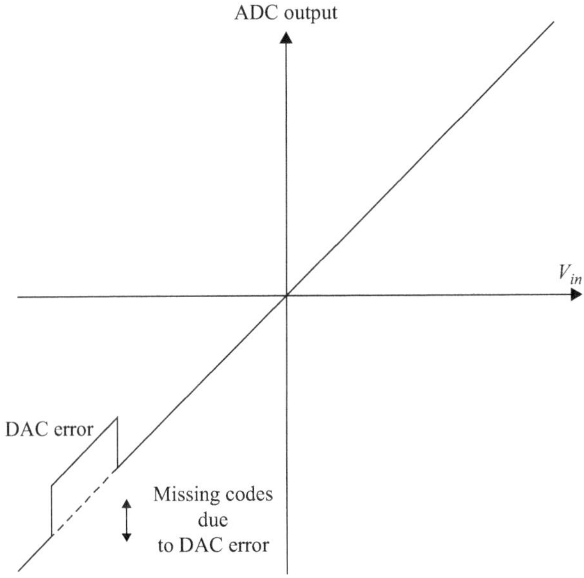

Figure 7.13 The transfer function of the ADC with missing codes due to a DAC error

that is slightly larger than a single sub-range, because it represents the largest error-to-signal ratio.

Figure 7.15 shows a plot of the SFDR versus input amplitude for a pipelined ADC with inter-stage gain errors in the first two stages. For this particular ADC, the first stage was 3 bits, and the second stage was 4 bits. The amplitude of a single sub-range of stage-1 is equal to -18 dBFS (i.e. $-6k_1 = -18$ dBFS). We see on the plot that the worst SFDR point is around an input amplitude of about -18 dBFS.

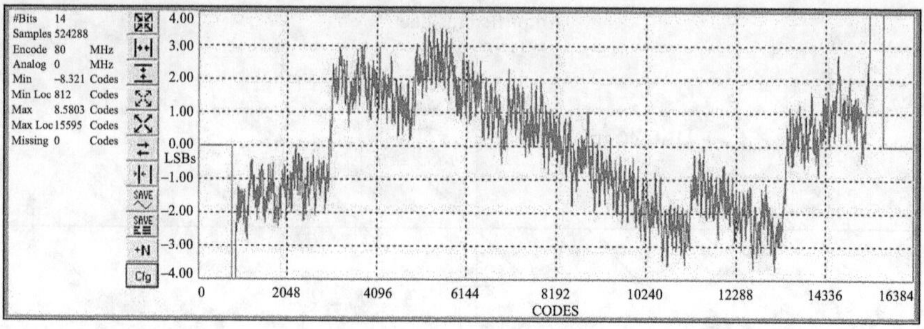

Figure 7.14 The INL of a pipelined ADC with DAC errors

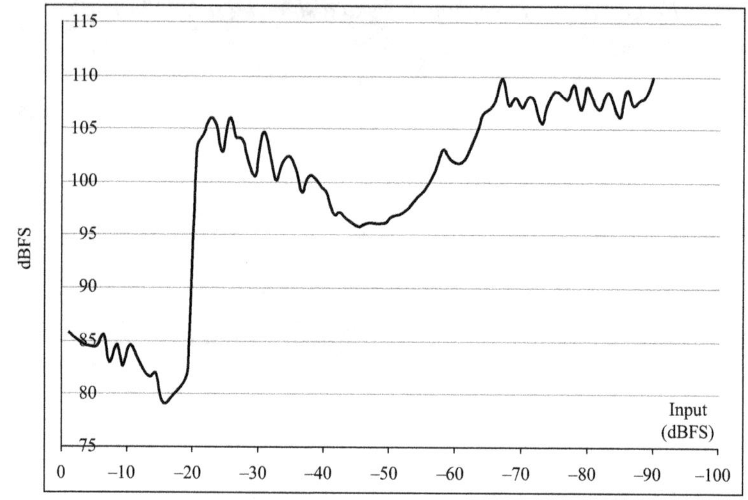

Figure 7.15 A sweep of SFDR versus input amplitude showing the impact of inter-stage gain errors in stage-1 at about −18 dBFS and stage-2 at about −42 dBFS

Since the second stage is 4 bits, its single sub-range amplitude is about −24 dB below the first stage's amplitude, which gives −18 − 24 = −42 dBFS. We can clearly see another dip in SFDR at around −42 dBFS. The second dip has a much better SFDR value because of the lower impact of stage-2 errors on the overall performance. This is discussed in Chapter 2 when analyzing the relationship between INL patterns and SFDR, and is discussed again in Chapter 9 as part of the dithering discussion.

An interesting property of pipelined ADCs is that the residues of the pipeline stages approach a uniform probability distribution as the signal propagates down the pipeline [8, 9]. Therefore, the quantization noise tends to be independent of the input signal for the back-end stages. This helps whiten the noise floor of the digital output spectrum, which is a nice property that usually requires dither to achieve. However, it is important to note that the noise due to the stage's non-linearity, which causes the INL jumps described earlier, is not white, especially for the first few stages.

This tendency for the residues to be uniformly distributed helps any injected dither signal to be uniformly distributed despite the non-idealities that might change their exact levels. Unless the dither has exactly binary levels, it tends to approach a uniform distribution as it propagates down the pipeline. Dithering is discussed in Chapter 9.

Example 1: A 14-bit ADC with 3 bits/stage and inter-stage gains of 4. The ADC needs to have an accuracy of 1 LSB. What are the accuracy/noise requirements in the flash, DAC and amplifier of the first two stages? How accurate are these estimates?

Answer 1: For quick estimation of stage-1: The flash needs to be 4-bit accurate, because the correction range is half the full-scale. The DAC needs to be accurate to 1 LSB at the 14-bit level, and the amplifier's accuracy is relaxed by its gain, which is 4. So it needs to be accurate to 4 LSB, which is 12-bit accurate.

For quick estimation of stage-2: The flash still needs to be 4-bit accurate, because the correction range is half the full-scale. The accuracy of the MDAC is relaxed by the preceding gain, which is 4. So, the DAC needs to be accurate to 4 LSB at the 14-bit level, which is 12-bit accurate. The amplifier's accuracy is relaxed by its gain and the gain of the preceding stage, which is 16. So it needs to be accurate to 16 LSB, which is 10-bit accurate. This is shown in Figure 7.16.

The point of this example is to illustrate the accuracy and noise relaxation in the sub-ADCs and the following stages. However, it is important to note that there is over-simplification in the answers given above. In practice, the *linearity* relaxation for stage-2 and beyond is only valid if we assume the errors from the multiple stages do not add up at the same location. If they do, we have to account for that, such that the total error is within the target specification. Moreover, the DAC's required accuracy can be different between small-signal errors (neighboring codes) versus large-signal linearity (cumulative errors). In addition, the amplifier's accuracy can be relaxed by the number of bits resolved instead of just the preceding gains (i.e. by 3 bits instead of 2 for stage-1 amplifier). This is discussed in more detail later in Section 7.3.2 of this chapter.

Similar to linearity, the noise contributions from the pipeline stages are reduced as we go down the pipeline. An inter-stage gain of G preceding a stage relaxes its input-referred noise power contribution by G^2. In general, the inter-stage amplifier gain substantially relaxes the accuracy and noise requirements of the following stages. This allows for scaling down the following stages and hence significant power saving.

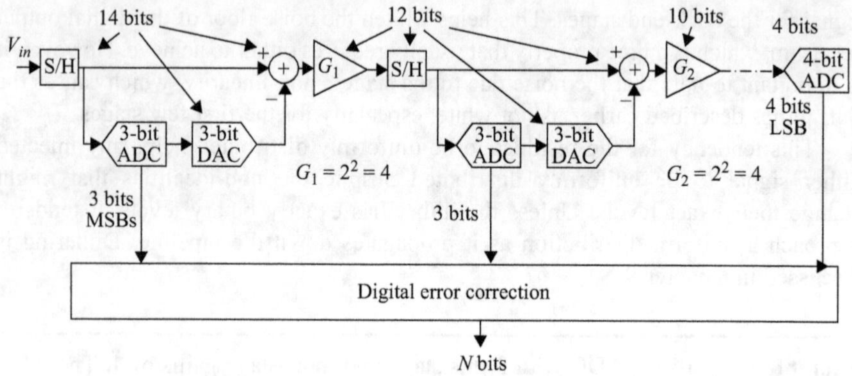

Figure 7.16 A pipelined ADC with the accuracy/noise relaxation of the various blocks shown. These are rough estimates that do not accurately represent the relaxation in linearity

7.2 Switched capacitor MDACs

In Chapter 6 we discussed switched capacitor circuits including integrators and amplifiers. We showed that we can build a subtracting amplifier using the structure shown in Figure 7.17. The output is given by

$$V_{out} = \frac{C_1/C_2}{1 + (C_1 + C_2 + C_p)/C_2 A}(V_{in1} - V_{in2}) \tag{7.6}$$

where A is the open loop DC gain and C_p is the parasitic capacitance at the summing node. This can be expressed as

$$V_{out} = \frac{C_1/C_2}{1 + K/A}(V_{in1} - V_{in2}) \tag{7.7}$$

where $K = 1/\beta$, and β is the feedback factor given by $\beta = C_2/(C_1 + C_2 + C_p)$, C_1 is the sampling capacitance, and C_2 is the feedback capacitance.

In Chapter 3, we discussed switched capacitor DACs. An example of a switched capacitor 3-bit DAC, controlled by the thermometer code D, is shown in Figure 7.18. The DAC output is given by

$$V_{out} = \sum_{i=1}^{8} \frac{D_i V_{Ref} C_i/C_f}{1 + (C_t + C_f + C_p)/(C_f A)} = \sum_{i=1}^{8} \frac{D_i V_{Ref} C_i/C_f}{1 + 1/\beta A} \tag{7.8}$$

where D_i is the ith bit of the 9-level DAC thermometer code (D_i: ± 1), V_{Ref} is the reference voltage, C_i is the individual DAC capacitance, C_f is the feedback capacitance, C_p is the parasitic capacitance at the summing node, C_t is the total DAC capacitance, and β is the feedback factor.

Combining the subtraction and DAC operations of Figures 7.17 and 7.18, we can build an MDAC as shown in Figure 7.19. Please note the reversal of the V_{Ref}

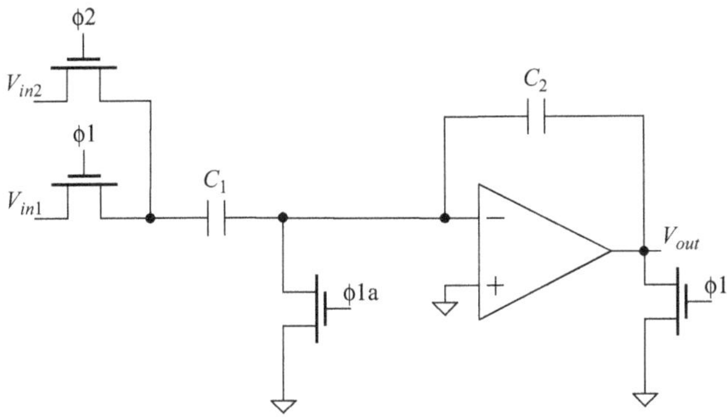

Figure 7.17 A switched capacitor subtracting amplifier

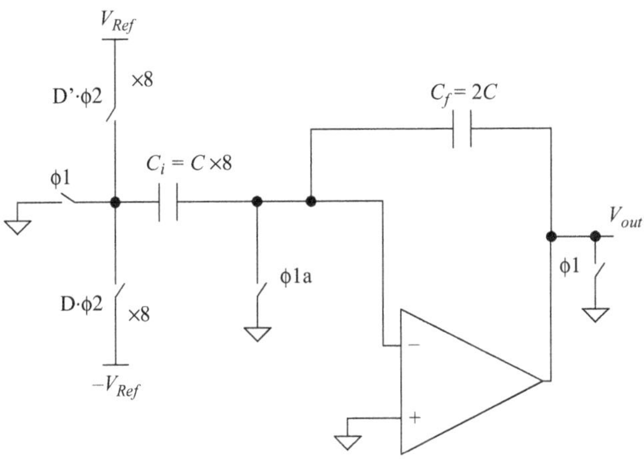

Figure 7.18 A switched capacitor DAC. This is a 3-bit unary DAC with D the digital word represented in thermometer code. The amplifier gain is equal to 4

polarity between the DAC and MDAC circuits in order to adjust for the opposite signs between the two cases. The output of the MDAC is given by

$$V_{out} = \frac{V_{in}C_t/C_f - \sum_{i=1}^{8} \mathrm{D}_i V_{Refi} C_i/C_f}{1 + (C_t + C_f + C_p)/(C_f A)} \tag{7.9}$$

where D_i is the ith bit of the 9-level DAC thermometer code (D_i: ±1), V_{Refi} is the reference voltage of the ith code, C_i is the individual DAC capacitance, C_p is the parasitic capacitance at the summing node, and C_t is the total sampling/DAC capacitance. The code-dependence of the reference voltage is introduced to capture possible reference errors in the DAC.

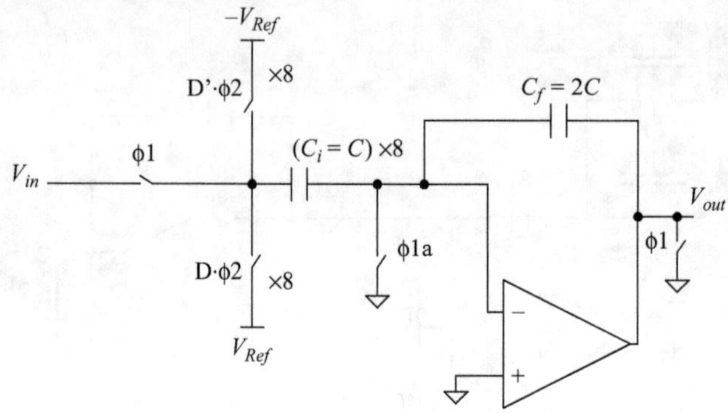

Figure 7.19 A switched capacitor 3-bit MDAC. The amplifier gain is equal to 4

The DAC function is captured by the summation in (7.9). If we assume the DAC is ideal and the capacitors are perfectly matched, and if we introduce small signal settling error in the amplifier assuming a single-pole system as discussed in Chapter 6, (7.9) becomes

$$V_{out} = \frac{V_{out}|_{\text{ideal}}}{1 + (C_t + C_f + C_p)/(C_f A)} \left(1 - e^{-\beta\omega_u t_s}\right) \tag{7.10}$$

where t_s is the small signal settling time, β is the feedback factor, which is given by

$$\beta = \frac{C_f}{C_t + C_f + C_p} = \frac{1}{K}$$

and ω_u is the unity gain frequency of the open loop operational amplifier. Therefore,

$$V_{out} = \frac{V_{out}|_{\text{ideal}}}{1 + K/A} \left(1 - e^{-\beta\omega_u t_s}\right) = V_{out}|_{\text{ideal}} - V_{out}K/A - V_{out}|_{\text{ideal}} e^{-\beta\omega_u t_s} \tag{7.11}$$

which illustrates the static gain error term $(V_{out}K/A)$ due to the amplifier's finite open loop gain A, and the dynamic gain error term $\left(V_{out}|_{\text{ideal}} e^{-\beta\omega_u t_s}\right)$ due to the small signal settling error. It is important to note that if the settling error is due to more than one pole, it may be under-damped, over-damped or critically damped, but it will still be linear and cause a gain error, as long as it is within the range of the small signal settling.

The fully differential form of the 3-bit MDAC is shown in Figure 7.20. When used as a first stage sampler, the input switch is bootstrapped to improve its linearity. In addition, an input buffer is often integrated with the ADC to reduce the kick-back on the input and improve the overall linearity, as discussed in Chapter 4. This is shown in Figure 7.21(a). Figure 7.21(b) depicts the timing diagram of the MDAC in Figure 7.21 (a). In phase 1 (ϕ1), the input is sampled on the sampling capacitances and the flash. The sampling instant is determined by the falling edge of the advanced clock ϕ1a on the sampling switches. Shortly afterwards, the flash

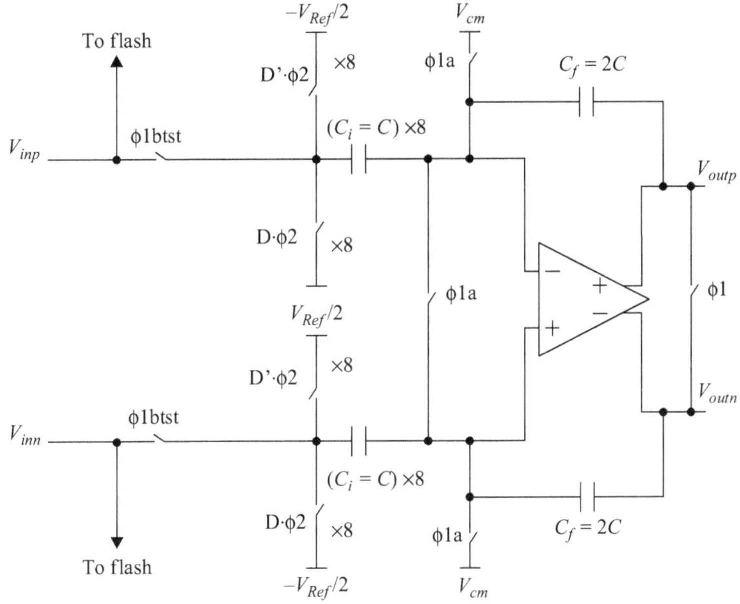

Figure 7.20 A fully differential switched capacitor 3-bit MDAC. The amplifier gain is equal to 4. The input switch is bootstrapped

comparators start latching and deliver their output thermometer code to the MDAC. The DAC capacitances are switched to $V_{Ref}/2$ or $-V_{Ref}/2$ based on the flash code. The hold phase (ϕ2) is used by both the flash and the MDAC, the longer the propagation delay of the flash comparators, the less time is available for the MDAC and reference settling. It is important to allocate these time periods in a manner that optimizes the overall performance and power consumption.

The reset switch clocked by ϕ_rst is used to briefly reset the capacitances after the hold phase and before they are connected to the input. This is meant to discharge the capacitances of the charge acquired during the gain/hold phase. This total charge represents the DAC code and is highly non-linear because of the DAC's quantization noise. If left without resetting, it would cause a non-linear charge injection that would significantly degrade the sampling linearity, if not given adequate time for settling during the sampling phase. At high sampling rates, the sampling time is usually not long enough for this non-linear charge injection to settle, and hence this brief resetting before sampling is required. However, this reset pulse consumes a portion of the sampling or hold time, which degrades the sampling or hold linearity. An alternative is to use kick-back calibration as discussed in Chapter 9.

In the implementations of Figures 7.19–7.21, the same set of capacitances is used for both sampling the input and the DAC, which is called the *shared capacitance* MDAC. Alternatively, a different set of capacitances can be used for each function, which is called the *split-capacitance* MDAC, and is shown in Figure 7.22.

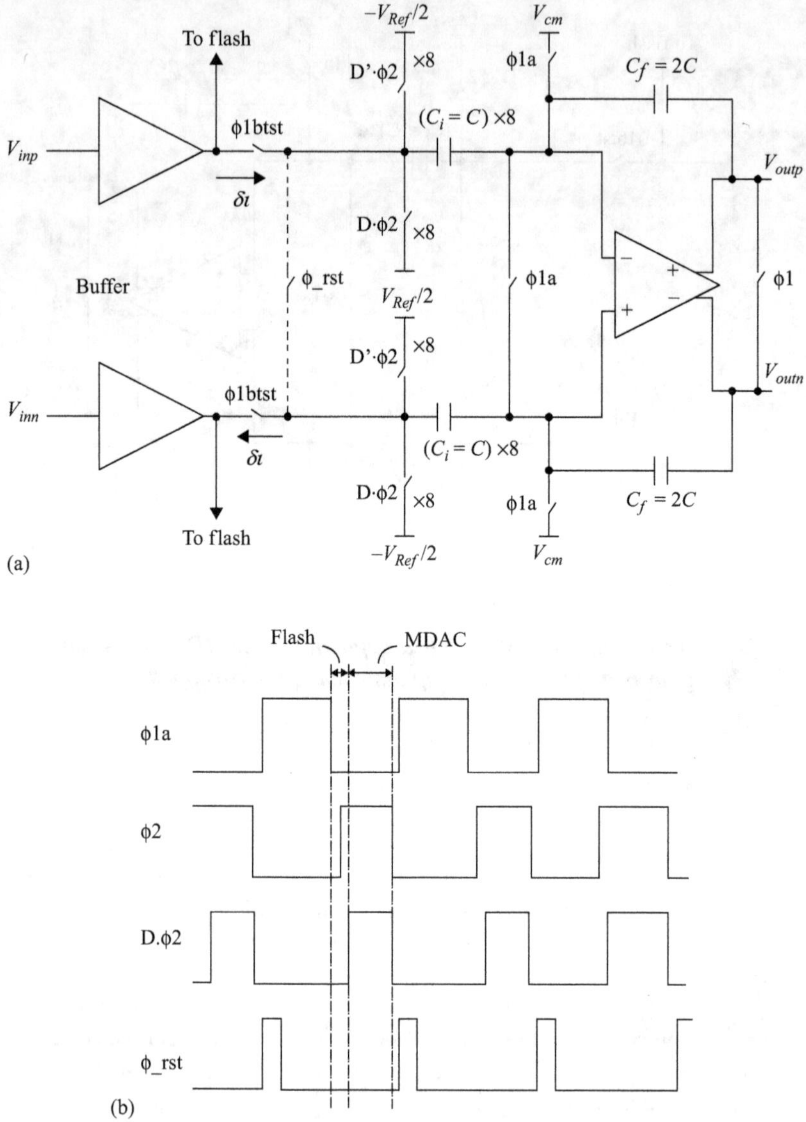

Figure 7.21 (a) A fully differential switched capacitor MDAC driven by an input buffer. The non-linear current drawn by the MDAC during sampling is denoted by δi. (b) A simplified timing diagram of the MDAC. © 2014 IEEE. Reprinted, with permission, from Reference 6

Splitting the capacitances eliminates the signal-dependent kick-back on the input and the reference. However, it degrades the feedback factor substantially from $\beta = C_f/(C_t + C_f + C_p)$ to $\beta = C_f/(2C_t + C_f + C_p)$. This leads to substantial degradation in speed and power consumption. It also degrades the noise

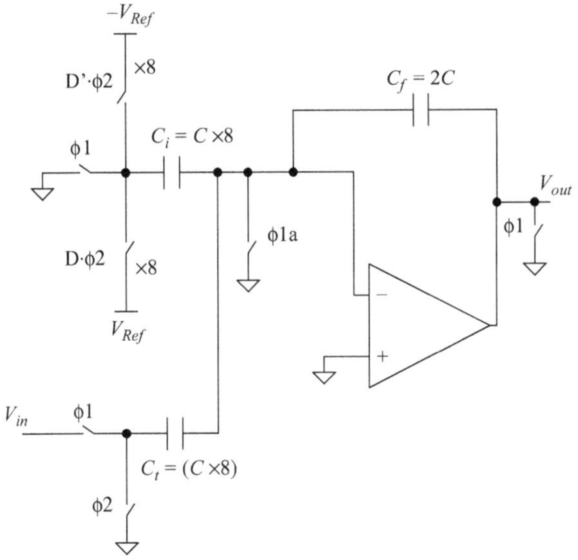

Figure 7.22 A 3-bit split-capacitance MDAC. The amplifier gain is equal to 4

performance as will be shown later in Section 7.3.3. Choosing between the two implementations is an important design decision that influences the performance and power efficiency of the whole ADC.

7.2.1 Reference buffer

One key component of the MDAC is the reference buffer, whose accuracy and settling behavior are critical to achieving the desired DAC and amplifier accuracies. Typically, source followers or emitter followers, which were analyzed in detail in Chapter 4, are used as reference buffers. In a differential implementation, similar to the one shown in Figure 7.20, two followers can be used: one for $V_{Ref}/2$ and the other for $-V_{Ref}/2$ [19]. Generally, both references are set relative to a common mode voltage V_{cm} that is appropriate for the MDAC. That is, the two reference values are $V_{cm} + V_{Ref}/2$ and $V_{cm} - V_{Ref}/2$.

A more efficient implementation stacks the two followers as shown in Figure 7.23 in order to reduce power consumption [20, 21]. The NMOS follower device (M1) drives the positive reference $V_{Ref}/2$, and the PMOS follower device (M2) drives the negative reference $-V_{Ref}/2$. The diode-connected device (M3) improves the differential settling substantially by providing another low impedance path in parallel with the output impedance of the follower device. The gates of the follower devices are driven by level-shifted and filtered DC references (V_{Refp_DC} and V_{Refn_DC}).

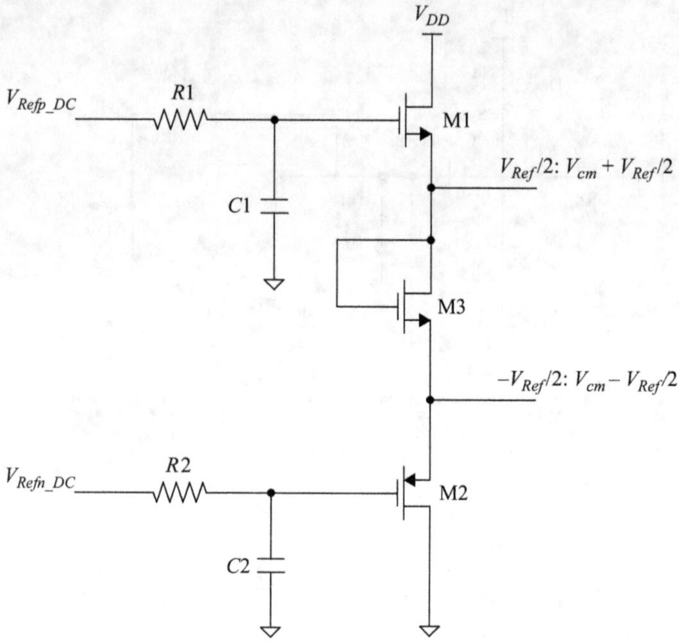

Figure 7.23 A reference buffer using stacked source followers [20, 21]

7.3 Performance limitations

1. Sampling non-linearity
2. Quantizer non-linearity
3. Noise
4. Jitter

7.3.1 Sampling non-linearity

The sampling non-linearity refers to the distortion introduced to the signal during the sampling process. This includes distortion in the external ADC input driver, in the internal input buffer and in the input sampling network. This is discussed in detail in Chapter 4.

Factors that influence the sampling linearity include the driver impedance, the linearity of the sampling network and the charge injection (kick-back) during sampling. The linearity is improved by reducing the impedance of the driver, improving the linearity of the buffer, improving the linearity of the sampling network, reducing the sampling capacitance, reducing the non-linear charge injection, and increasing the acquisition time.

7.3.2 *Quantization non-linearity*

In a pipelined ADC, there are multiple possible contributors to the quantizer's non-linearity. The accuracy of the quantizer depends heavily on the accuracy of the MDACs. Using the analyses of Chapter 6 and (7.9) and (7.10), and assuming single-pole small-signal settling, the MDAC's output is given by

$$V_{out} \approx \left(\frac{V_{in}C_t/C_f - \sum_{i=1}^{8} \mathrm{D}_i V_{Refi} C_i/C_f}{1 + K/A} \right) \left(1 - e^{-t_s \beta \omega_u}\right) \tag{7.12}$$

where $\beta = 1/K$ is the feedback factor, D_i is the ith bit of the 9-level DAC code (D_i: ± 1), based on the flash decision, ω_u is the amplifier's unity gain angular frequency, C_i is the individual sampling capacitances, C_t is the total sampling/DAC capacitance, C_f is the feedback capacitance, V_{Refi} is the reference voltage for the ith capacitance/code, and t_s is the available small-signal settling time.

From (7.12), we see that DAC errors occur if there is mismatch between the individual DAC capacitances or if the sampled reference value changes with the DAC code, for example, because of code-dependent reference settling. DAC errors can also be caused by code-dependent charge injection.

Regarding the DAC settling, the reference value sampled on the individual capacitances has to propagate through the resistance of the DAC switch. Settling errors due to the reference buffer's output resistance or the resistance of the DAC switch can lead to DAC errors that depend on the sampling rate. This is illustrated in Figure 7.24, where the reference buffer's settling dominates the large signal settling of the MDAC's output, and causes it to go temporarily in the wrong direction before reverting back to the right value. This is undesirable and strains the amplifier's slewing and settling. These large reference and DAC errors are highly non-linear and reduce the time available for the amplifier's small-signal settling. It is preferred to delay the amplifier's operation until the reference is settled reasonably close to its final value to avoid signal swings at the output that are too large. We can also see that the additional delay due to the DAC switches is relatively small as shown by the second bottom curve in Figure 7.24.

Inter-stage gain errors occur if the ratio between the total DAC capacitance and the feedback capacitance is not accurate. A small open loop gain (A) of the operational amplifier can also impact the accuracy of the closed loop gain, as shown in (7.11) and (7.12). In addition, the amplifier's settling error can result in gain error if it is within the small signal settling region described by (7.12) above. Reference errors can also lead to inter-stage gain errors if the reference of one stage is different from that of the following or preceding stage. This can be due to settling errors in the reference that are different between the two clock phases. Finally, inter-stage gain errors can also be caused by charge injection from the various switches and capacitors in the circuit, which are usually aggravated if there is inadvertent overlap or error in the clock timing.

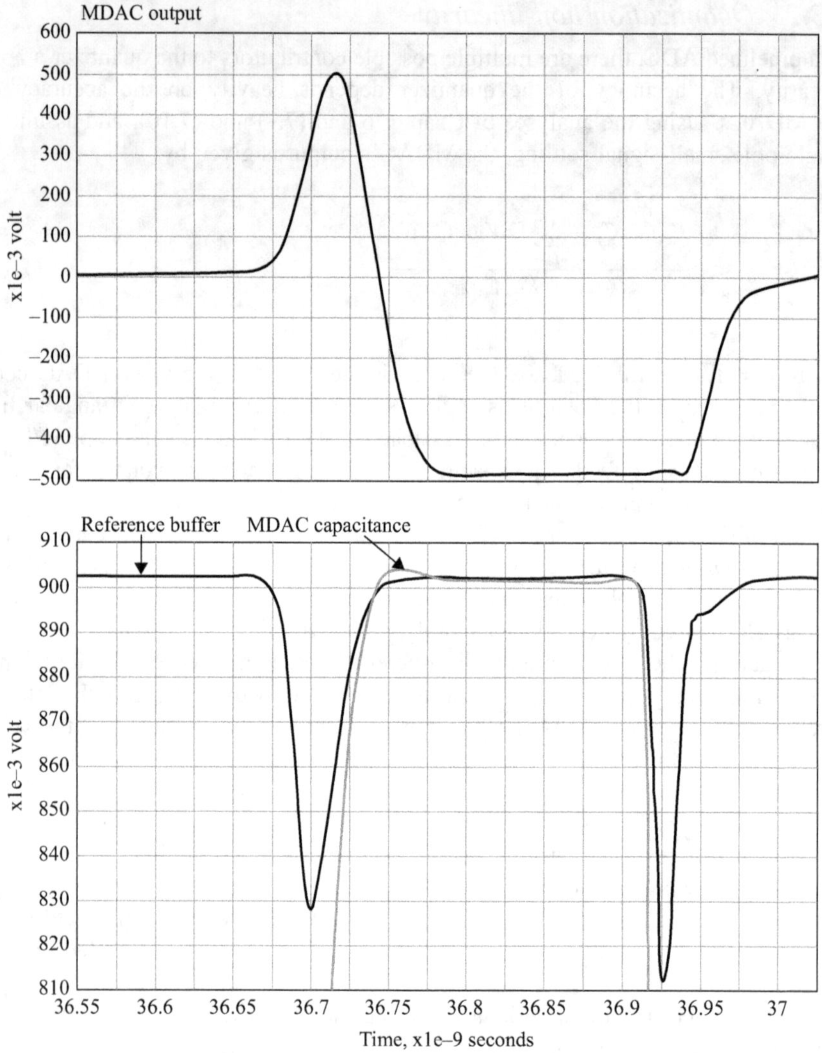

Figure 7.24 An example of the MDAC settling showing the MDAC output (top curve), the output of the reference buffer (bottom curve) and the signal at the MDAC capacitances (second bottom curve)

So to summarize, DAC errors can be caused by:

- Capacitor mismatch between the DAC capacitances: If $C_i \neq C_j$.
- Code-dependent reference error, which can be a settling error: If $V_{Refi} \neq V_{Refj}$, for different DAC codes i and j.
- Code-dependent charge injection.

Inter-stage gain errors can be caused by:

- Capacitor mismatch: If $\Sigma C_i/C_f \neq G$.
- Reference error, which can be a settling error: If V_{Ref} of stage $k \neq V_{Ref}$ of stage $k-1$ or $k+1$.
- Small open loop gain (A): If the term K/A is not small enough to achieve the desired accuracy.
- Amplifier settling error: This can be large signal settling error, which is highly non-linear, or it can be small signal settling error, which can be linear and manifested as a gain error. The small signal settling error is expressed in (7.12) by the error term $e^{-t_s\beta\omega_u}$ for a single pole system.
- Charge injection.

7.3.2.1 DAC's accuracy

Typically the DAC function of the MDAC is performed by a switched capacitor network similar to the one shown in Figure 7.20, and as expressed in (7.9). The DAC switches are controlled by the output bits of the sub-ADC. The accuracy of the switched capacitor DAC is determined by several factors, which include

- Capacitor mismatch
- Settling errors

The DAC capacitors are usually implemented using polysilicon, MIM or MOM capacitors, as described in Chapter 3. The constant nature of these capacitances makes it possible to calibrate the mismatch errors with fixed coefficients in the digital back end using factory calibration. To achieve the desired DAC accuracy, the capacitor mismatch errors need to satisfy the following requirement

$$\Delta_{DAC_i} < 2^{k_i} \times 2^{-(N+j_{SS})} \times \prod_{n=0}^{i-1} G_n \tag{7.13}$$

where Δ_{DAC_i} is the DAC capacitor mismatch ($\Delta C/C$) of the ith stage, k_i is the number of bits of the ith stage, N is the resolution of the ADC, G_n is the gain of the nth stage, with G_0 being the gain preceding the first stage, which is typically equal to 1, and j_{SS} is the additional bits of required accuracy *between neighboring codes,* or the required *small-signal* accuracy. For example, for a 14-bit ADC with an accuracy of 0.25 LSB between neighboring codes, N is 14, and j is equal to 2. If the required accuracy is 2 LSBs, then j_{ss} is equal to -1.

Settling errors due to the DAC switch resistors or the reference buffer's settling can cause DAC errors if they are code-dependent. Unlike errors due to capacitor mismatch, these errors depend on the sampling rate and possibly temperature and supply. This makes it necessary to minimize these settling errors in the analog domain or employ background calibration to track the changing conditions. The DAC settling error on each capacitance needs to satisfy the following requirement

$$\varepsilon_{DAC_i} < 2^{k_i} \times 2^{-(N+j_{SS})} \times \prod_{n=0}^{i-1} G_n \tag{7.14}$$

where ε_{DAC_i} is the DAC settling error of the ith stage, k_i is the number of bits of the ith stage, N is the resolution of the ADC, G_n is the gain of the nth stage, with G_0 being the gain preceding the first stage, which is typically equal to 1, and j_{SS} is the additional bits of small-signal accuracy required *between neighboring codes.*

It is important to note that the DAC accuracy requirements given in (7.13) and (7.14) are relaxed by the number of bits in the stage. This is true for the errors between neighboring codes, which impact the small-signal linearity. However, if the errors accumulate, the total error can be significantly larger. These cumulative errors can show up in the INL as large-signal non-linearity. Therefore, it is desirable to design the DAC with tighter accuracy than that given by (7.13) and (7.14), without the "2^{k_i}" relaxation. In this case, the allowed DAC capacitance mismatch of the ith stage Δ_{DAC_i} will be given by

$$\Delta_{DAC_i} < 2^{-(N+j_{LS})} \times \prod_{n=0}^{i-1} G_n \tag{7.15}$$

and the allowed DAC settling error of the ith stage ε_{DAC_i} is given by

$$\varepsilon_{DAC_i} < 2^{-(N+j_{LS})} \times \prod_{n=0}^{i-1} G_n \tag{7.16}$$

where j_{LS} is the desired large signal accuracy of the overall peak-to-peak INL, and G_n is the gain of the nth stage, with G_0 being the gain preceding the first stage, which is typically equal to 1. Equations (7.15) and (7.16) represent the requirements on the DAC's capacitance mismatch and settling errors to achieve a specific large signal linearity j_{LS}.

7.3.2.2 Inter-stage amplifier's accuracy

The amplifier's accuracy is determined by multiple factors, which include

- Open loop DC gain
- Capacitor matching
- Settling errors

The open loop DC gain of the amplifier contributes to the error as described in (7.11) and (7.12). The higher the DC gain, the lower the error. The required open loop gain A_i for the ith stage is given by

$$A_i > \frac{2^{-k_i} \times 2^{N+j}}{\beta_i \times \prod_{n=0}^{i-1} G_n} \tag{7.17}$$

where β_i is the feedback factor of the ith stage, k_i is the number of bits in the ith stage, N is the resolution of the ADC, G_n is the gain of the nth stage, with G_0 being the gain preceding the first stage, and j is the additional *peak-to-peak* DNL/INL accuracy required. In this case, the errors are in the form of a saw-tooth pattern, and do not accumulate like DAC errors. For example, for a 14-bit ADC with a required accuracy of 0.25 LSB peak-to-peak INL, N is equal to 14, and j is equal to 2. If the required accuracy is 4 LSBs, then j is equal to -2. Similarly, the amplifier's allowed settling error ε_{AMP_i} for the ith stage is

$$\varepsilon_{AMP_i} < 2^{k_i} \times 2^{-(N+j)} \times \prod_{n=0}^{i-1} G_n \tag{7.18}$$

where k_i is the number of bits in the first stage, N is the resolution of the ADC, G_n is the gain of the nth stage, with G_0 being the gain preceding the first stage, which is typically equal to 1, and j is the additional *peak-to-peak* DNL/INL accuracy required. As discussed in Chapter 6, and shown in (7.12), once the settling error is determined, the unity gain bandwidth of the amplifier can be found using the formula:

$$\varepsilon_{AMP_i} = e^{-\beta_i \omega_{u_i} t_{si}} \tag{7.19}$$

where t_{si} is the available small-signal settling time for the ith stage, ω_{u_i} is the unity gain angular frequency of the ith stage, $K = 1/\beta$ and β_i is the feedback factor of the ith stage. Alternatively, the settling error can be represented as

$$\varepsilon_{AMP_i} = e^{-2\pi BW_{cl_i} t_{s_i}} \tag{7.20}$$

where BW_{cl_i} is the closed-loop bandwidth of the ith stage. Therefore, for a specified settling error and a given small-signal settling time, we can obtain the unity gain frequency ω_{u_i} and the closed loop bandwidth BW_{cl_i}, which are given by

$$\omega_{u_i} = \frac{1}{\beta_i t_{s_i}} \ln\left(\frac{1}{\varepsilon_{AMP_i}}\right) \tag{7.21}$$

and

$$BW_{cl_i} = \frac{1}{2\pi t_{s_i}} \ln\left(\frac{1}{\varepsilon_{AMP_i}}\right) \tag{7.22}$$

Armed with the open loop gain from (7.17) and the unity gain frequency and closed loop bandwidth from (7.21) and (7.22), we have the specifications needed to design the MDAC's amplifier. An example of this process was discussed in Chapter 6 in Section 6.2.

Continuing with the amplifier's requirements, the allowed capacitor mismatch between the feedback capacitance and the DAC capacitances of the ith stage Δ_{AMP_i} is given by

$$\Delta_{AMP_i} < 2^{k_i} \times 2^{-(N+j)} \times \prod_{n=0}^{i-1} G_n \tag{7.23}$$

where k_i is the number of bits in the first stage, N is the resolution of the ADC, G_n is the gain of the nth stage, with G_0 being the gain preceding the first stage, which is typically equal to 1, and j is the additional *peak-to-peak* DNL/INL accuracy required. Similarly, the allowed reference settling error of the ith stage $\varepsilon_{Ref_{GAIN_i}}$ is given by

$$\varepsilon_{Ref_{GAIN_i}} < 2^{k_i} \times 2^{-(N+j)} \times \prod_{n=0}^{i-1} G_n \tag{7.24}$$

It is clear from (7.13)–(7.24) that for a certain DNL error or small-signal linearity, the required inter-stage gain and DAC accuracies are comparable. As we use more bits in the first stage of the pipeline, the number of segments (sub-ranges) and the inter-stage gain increase, which relaxes the accuracy requirements of the MDAC. This is an important factor in determining the optimum resolution of the pipeline stages as discussed later in this chapter. It is also important to remember that the relaxed DAC accuracies are valid for neighboring codes only, but not if the errors accumulate over a large number of codes, as shown in (7.15) and (7.16).

7.3.2.3 Reference's accuracy

The reference accuracy is determined by:

- DC error
- Settling error

The DC error of the reference results in an overall gain error of the whole ADC that can be corrected in the digital back end. Dynamic errors in the reference can result in DAC errors, if they are code dependent, or in inter-stage gain errors, if they are stage-dependent. For *code-dependent* reference settling, to achieve an accuracy j_{SS} between neighboring codes, the requirement on the reference settling of the ith stage $\varepsilon_{Ref_{DAC_i}}$ is

$$\varepsilon_{ref_{DAC_i}} < 2^{k_i} \times 2^{-(N+j_{SS})} \times \prod_{n=0}^{i-1} G_n \tag{7.25}$$

where k_i is the number of bits in the stage, N is the resolution of the ADC, G_n is the gain of the nth stage, with G_0 being the gain preceding the first stage, which is typically equal to 1, and j_{SS} is the number of bits of small-signal (DNL) accuracy. However, if the reference settling errors accumulate over a large number of codes, the reference settling requirement of the ith stage $\varepsilon_{ref_{DAC_i}}$ may need to be tightened to

$$\varepsilon_{ref_{DAC_i}} < 2^{-(N+j_{LS})} \times \prod_{n=0}^{i-1} G_n \tag{7.26}$$

where N is the resolution of the ADC, G_n is the gain of the nth stage, with G_0 being the gain preceding the first stage, which is typically equal to 1, and j_{LS} is the number of bits of large-signal (INL) accuracy. For the inter-stage gain error, the reference settling requirement $\varepsilon_{ref_{GAIN_i}}$ is given by

$$\varepsilon_{ref_{GAIN_i}} < 2^{k_i} \times 2^{-(N+j)} \times \prod_{n=0}^{i-1} G_n \tag{7.27}$$

where k_i is the number of bits in the first stage, N is the resolution of the ADC, G_n is the gain of the nth stage, with G_0 being the gain preceding the first stage, which is typically equal to 1, and j is the additional *peak-to-peak* DNL/INL accuracy required.

Equations (7.25)–(7.27) indicate that the DAC settling requirement can be tougher than the inter-stage gain settling requirement. However, this tight requirement on the DAC errors is needed only with code-dependent settling errors that are cumulative. The inter-stage gain errors are independent of the DAC code. Therefore it is quite possible, and common, to see a reference settling error that is code-independent, and hence causes an IGE, but not a DAC error. If the settling error is not code-dependent, the DAC settling criteria in (7.26) can be satisfied, while the seemingly looser gain settling requirement of (7.27) may not be satisfied.

7.3.3 Noise and jitter

Noise is one of the most important and fundamental design parameters. As described in Chapter 2, reducing the noise usually requires an increase in the power consumption along a 3 dB/octave curve. That is, an improvement of 3 dB in noise requires doubling the power consumption. The main noise contributors are:

- Device noise in the signal path (thermal, shot, and flicker noise)
  - This includes noise during both phases (sample and hold).
  - The input-referred noise contribution is lower for stages down the pipeline because of the gains of the previous stages.
  - Scaling of the back-end stages is usually employed to lower their power consumption.
- Noise in the clock path (jitter). The impact of this noise increases as the input frequency increases.
- Noise due to quantization.
- Noise due to non-linearity (DNL/INL).

The thermal noise contribution of a sampling network with a sampling capacitor C_s is given by kT/C_s. However, when multiple capacitor are connected to the summing node of an MDAC amplifier, as shown in the MDAC schematics of Figure 7.20, the charge conservation at the summing node leads to an increase in the total input-referred noise power, due to the contributions of the additional RC networks, to be given by

$$v_{ns}^2 \approx 2\frac{kT}{C_s}\frac{\sum C_{sum-node}}{C_s} \tag{7.28}$$

where v_{ns} is the input-referred noise voltage during the sampling phase, k is the Boltzmann constant, T is the absolute temperature, C_s is the total sampling capacitor, which is equal to $8C$ in Figure 7.20, and the factor of 2 is to account for the differential nature of the circuit. For example, if the sampling capacitance of stage-1 is C_{s1}, the feedback capacitance is C_{f1}, the parasitic capacitance at the summing node is C_{p1}, then the input-referred noise power at the sampling phase will be given by

$$v_{ns}^2 = 2\frac{kT}{C_{s1}}\left(\frac{C_{s1} + C_{p1} + C_{f1}}{C_{s1}}\right) \tag{7.29}$$

If an input buffer is used to drive the ADC sampling network, as shown in Figure 7.21, the noise power of the buffer needs to be added to the above component to give

$$v_{ns}^2 = v_{n\_buffer}^2 + 2\frac{kT}{C_{s1}}\frac{\sum_{sum-node} C_1}{C_{s1}} \tag{7.30}$$

For the gain/hold phase, the input-referred noise power is given by

$$v_{nh}^2 \approx v_{nsh}^2 + v_{na}^2 + 2\frac{kT}{C_{s2}}\frac{\sum_{sum-node} C_2}{C_{s2}}\left(\frac{C_{f1}}{C_{s1}}\right)^2$$

and

$$v_{nh}^2 \approx v_{nsh}^2 + S_{na} \times \alpha \times BW_{cl1}\left(\frac{\sum_{sum-node} C_1}{C_{f1}}\right)^2\left(\frac{C_{f1}}{C_{s1}}\right)^2 + 2\frac{kT}{C_{s2}}\frac{\sum_{sum-node} C_2}{C_{s2}}\left(\frac{C_{f1}}{C_{s1}}\right)^2 \tag{7.31}$$

where v_{nh} is the input-referred noise voltage in the hold phase, v_{nsh} is the input-referred noise voltage of the first stage's switched capacitor networks in the hold phase, v_{na} is the input-referred noise of the amplifier, S_{na} is the noise spectral density of the amplifier, BW_{cl1} is the closed loop bandwidth of the amplifier, α is a parameter to account for the difference between the noise bandwidth and the 3-dB bandwidth, for a single pole system α is equal to $\pi/2$, C_{s2} is the total sampling capacitance of the second stage, and C_{f2} is the feedback capacitance of the second stage.

The first term on the right hand side of (7.31) represents the noise power of all the switched capacitor networks of the first stage. The RC time constants of the various capacitance branches in the MDAC are usually much smaller than the settling time constant of the amplifier. Since the closed-loop bandwidth BW_{cl1} is limited by the amplifier, and not by the RC time constants of the capacitance branches, this noise power v_{nsh}^2 is usually significantly less than the kT/C of the sampling phase, and the noise power of the amplifier.

The third term on the right hand side represents the kT/C noise of the second stage's sampling network, which is scaled down by the square of the gain of the first stage. This allows for the reduction of the capacitance in the second stage compared to the first stage. This also indicates that the first stage hold noise is dominated by the noise of the first stage's amplifier.

The second term on the right hand side of (7.31) is the input-referred noise power of the amplifier. The noise analysis for different amplifier structures was discussed in Chapter 6 in detail. For example, for a two-stage amplifier whose first stage is a differential pair or a cascode amplifier, and with NMOS input devices, the noise spectral density can be represented using (6.124) and (6.143) and is given by

$$S_{na} \cong 2S_{nN} + 2S_{nP}\left(\frac{g_{mP}}{g_{mN}}\right)^2 \tag{7.32}$$

where S_{nN} and S_{nP} are the noise power spectral densities of the NMOS and PMOS devices respectively, and g_{mN} and g_{mP} are their transconductances. Therefore, if we ignore the flicker noise, we get

$$S_{na} \cong 2\frac{2}{3}\frac{4kT}{g_{mN}} + 2\frac{2}{3}\frac{4kT}{g_{mP}}\left(\frac{g_{mP}}{g_{mN}}\right)^2 \tag{7.33}$$

where g_m is the transconductance, k is the Boltzmann constant, T is the absolute temperature. The closed-loop BW of the amplifier is typically given by the unity gain frequency multiplied by the feedback factor. Therefore, using (6.161) for the unity gain frequency of the Miller-compensated two-stage amplifier, the closed-loop bandwidth is given by

$$BW_{cl} \approx f_u\beta \approx \frac{g_{mN}}{2\pi C_c}\frac{C_f}{\sum\limits_{sum-node} C} \tag{7.34}$$

Substituting (7.33) and (7.34) into (7.31), and focusing on the amplifier's noise, we get

$$v_{na}^2 \approx \frac{4}{3}\frac{\alpha 4kT}{g_{mN}}\left(1+\frac{g_{mP}}{g_{mN}}\right)\frac{g_{mN}}{2\pi C_c}\frac{C_{f1}}{\sum\limits_{sum-node} C_1}\left(\frac{\sum\limits_{sum-node} C_1}{C_{s1}}\right)^2 \tag{7.35}$$

Assuming a single pole amplifier, this can be simplified to give

$$v_{na}^2 \approx \frac{4}{3}\frac{kT}{C_c}\left(1+\frac{g_{mP}}{g_{mN}}\right)\frac{C_{f1}}{C_{s1}}\left(\frac{\sum\limits_{sum-node} C_1}{C_{s1}}\right) \tag{7.36}$$

which can also be represented in terms of the feedback factor β as

$$v_{na}^2 \approx \frac{4}{3}\frac{kT}{C_c}\left(1+\frac{g_{mP}}{g_{mN}}\right)\frac{1}{\beta}\left(\frac{C_{f1}}{C_{s1}}\right)^2 \tag{7.37}$$

This shows that the amplifier's noise power does not depend directly on the amplifier's g_{mN}. As the g_{mN} changes, the noise changes but so does the bandwidth, and hence the integrated noise stays approximately the same. The noise does, however, depend on the ratio g_{mP}/g_{mN}. Therefore, like the sampling noise, the amplifier's noise follows a kT/C formula where C is equal to the Miller compensation capacitance C_c in the case of a two-stage amplifier.

From (7.37), we can also see that reducing the feedback factor β increases the noise. Moreover, for the same closed loop bandwidth, reducing the feedback factor requires the reduction of the Miller compensation capacitance, to keep the power consumption fixed. This increases the amplifier's noise even more substantially according to (7.37). Alternatively, we can increase the amplifier's currents instead, which increases the power consumption. Either way is not desirable, and hence it is important to increase the feedback factor as much as possible. Equation (7.37) also shows that the gain of the first stage has some impact on the noise. A higher gain improves the noise of the first stage.

Example 2: Compare the impact on the noise for three cases:
Case 1: A gain of 4 using the MDAC of Figure 7.20.
Case 2: A gain of 8 using the MDAC of Figure 7.20, with 16 capacitances in the MDAC instead of 8.
Case 3: A gain of 4, but using a separate set of capacitances for sampling the input and the MDAC, as shown in Figure 7.22.

Assume everything else is the same, and ignore the parasitic capacitance at the summing node.

Answer 2:
Case 1: A gain of 4 means C_s is equal to $4C_f$. Therefore, the sampling noise is

$$v^2_{ns\_case1} = 2\frac{kT}{C_{s1}}\frac{5}{4}$$

and the amplifier's noise in the hold phase is

$$v^2_{na\_case1} = \frac{4}{3}\frac{kT}{C_c}\left(1 + \frac{g_{mP}}{g_{mN}}\right)\frac{1}{4}\left(\frac{5}{4}\right) = \frac{4}{3}\frac{kT}{C_c}\left(1 + \frac{g_{mP}}{g_{mN}}\right)\frac{5}{16}$$

Case 2: A gain of 8 means C_s is equal to $8C_f$. Therefore, the sampling noise is

$$v^2_{ns\_case2} = 2\frac{kT}{C_{s1}}\frac{9}{8}$$

and the amplifier's noise in the hold phase would be

$$v^2_{na\_case2} = \frac{4}{3}\frac{kT}{C_c}\left(1 + \frac{g_{mP}}{g_{mN}}\right)\frac{1}{8}\left(\frac{9}{8}\right) = \frac{4}{3}\frac{kT}{C_c}\left(1 + \frac{g_{mP}}{g_{mN}}\right)\frac{9}{64}$$

From these results, we see that a gain of 8 improves both the sampling and the hold noise, compared to a gain of 4. The sampling noise improves by 10%, and the amplifier's noise seems to improve by 55%. However, that is not entirely accurate. Since the feedback factors are different between cases 1 and 2, we should note that C_c for a gain of 8 will need to be different from that used for a gain of 4, if the bandwidth and power consumption are to stay constant. If we assume C_c is reduced in the second case compared to the first case by a factor equal to the ratio of the feedback factors, we get

$$v^2_{na\_case2} = \frac{4}{3}\frac{kT}{C_{c2}}\left(1+\frac{g_{mP}}{g_{mN}}\right)\frac{1}{8}\left(\frac{9}{8}\right) = \frac{4}{3}\frac{kT}{C_c}\left(1+\frac{g_{mP}}{g_{mN}}\right)\frac{9}{64}\frac{9}{5}$$

Therefore, the amplifier's noise in case 2 is still better than case 1, but to a much lesser extent. The sampling noise is better by 10%, and the amplifier's noise is better by 19%.

Case 3: A gain of 4, with another set of DAC capacitances C_D equal to C_s means C_s is equal to $4C_f$ and C_D is also equal to $4C_f$. Therefore, the sampling noise is

$$v^2_{ns\_case3} = 2\frac{kT}{C_{s1}}\frac{9}{4}$$

and the amplifier's noise in the hold phase will be

$$v^2_{na\_case3} = \frac{4}{3}\frac{kT}{C_c}\left(1+\frac{g_{mP}}{g_{mN}}\right)\frac{1}{4}\left(\frac{9}{4}\right) = \frac{4}{3}\frac{kT}{C_c}\left(1+\frac{g_{mP}}{g_{mN}}\right)\frac{9}{16}$$

If we take into account the change needed in C_c to account for the different feedback factors, while keeping the power constant, the amplifier's noise will be

$$v^2_{na\_case3} = \frac{4}{3}\frac{kT}{C_{c2}}\left(1+\frac{g_{mP}}{g_{mN}}\right)\frac{1}{4}\left(\frac{9}{4}\right) = \frac{4}{3}\frac{kT}{C_c}\left(1+\frac{g_{mP}}{g_{mN}}\right)\frac{9}{16}\frac{9}{5}$$

Therefore, the sampling noise is worse than the first case by 80%, and the amplifier's noise is worse than the first case by 220%. That is, the noise in the split capacitance case is 3.2 times worse than the first case. That is 5 dB degradation in the amplifier's noise!

If we keep C_c fixed and allow for an increase in power, the noise for case 3 will be worse by 80% instead of 220%. That is still about 2.6 dB degradation in noise, plus an increase in power that is approximately equal to the ratio of the feedback factors, that is 9/5, which is an 80% increase in power.

These results indicate that there is a substantial cost to using two separate sets of capacitances for input sampling and the DAC operation (split-capacitance MDAC). The impact of the additional capacitances on the feedback factor results in a significant degradation in noise and power.

The cases in Example 2 illustrate the importance of the feedback factor, which has a direct and strong impact on the noise, settling and power consumption. The MDAC designer needs to maximize the feedback factor by reducing the summing node parasitics and avoiding using additional capacitances connected to the summing node for the DAC operation, or any other purpose, if possible.

Equations (7.30), (7.36) and (7.37) demonstrate that the first stage's noise is determined to a large extent by the choice of the sampling capacitance C_{s1}. This applies to both the sampling and hold phases. Therefore, choosing the capacitance value represents one of the most important decisions in designing an ADC. Since the sampling noise, and to some extent the hold noise, are proportional to kT/C_s, and if the gain of each stage is given by G, then the noise power of the kth stage is scaled down by $G^{2(k-1)}$. If the sampling capacitance of each stage is scaled down by a factor S compared to the preceding stage, then the noise power of the kth stage will be scaled up by a factor equal to $S^{(k-1)}$. Therefore, the input-referred noise power of the kth stage will be given by

$$v_{nk_{in}}^2 = v_{nk}^2 \frac{S^{k-1}}{G^{2(k-1)}} \tag{7.38}$$

where $v_{nk_{in}}^2$ is the input-referred noise power of the kth stage, and v_{nk}^2 is the noise power of the kth stage, S is the scaling factor between stages, and G is the gain of each stage.

Knowing the sampling and hold noise, the total noise of the first stage, which includes the sampling noise of the second stage is given by

$$v_{n1}^2 = v_{ns}^2 + v_{nh}^2 \tag{7.39}$$

The total noise of the pipeline can be expressed as

$$v_{nt}^2 = v_{ns}^2 + v_{nh}^2 + v_{nh2}^2 + v_{nh3}^2 + \cdots \tag{7.40}$$

where v_{ns} is the input-referred noise voltage of the sampling phase, v_{nh} is the input-referred noise voltage of the stage-1 hold phase, v_{nh2} is the input-referred noise voltage of the stage-2 hold phase, v_{nh3} is the input-referred noise of the hold phase of stage-3, and so on.

To simulate the noise of a pipeline ADC, each phase can be simulated separately for every stage using AC noise analysis, and the total noise power is added as shown in (7.40). Another way to simulate the noise in switched capacitor circuits is using strobed periodic noise analysis, as mentioned in Chapter 4 for simulating the jitter. This analysis takes into account the modulation of the noise with the clock, and hence can give a reasonable representation of the total noise of a switched capacitor circuit after the sampling and hold operations.

In addition to the device noise in the input signal path, the clock jitter is another non-ideality that affects the sampling process. This contribution to the noise happens almost exclusively in the first stage where the input sampling takes place. This noise component is discussed in detail in Chapter 4 and needs to be accounted for in the total noise budget of the ADC.

Another source of noise is the quantization non-linearity, which degrades the DNL and INL, and hence the SNDR and the SFDR. In the case of a pipelined ADC, this is dominated by the MDAC errors, and behaves in a way that is similar to the quantization noise.

Therefore, the total noise in the ADC can be represented as

$$v_n^2 = v_n^2\big|_{\text{input}} + v_n^2\big|_{\text{jitter}} + v_n^2\big|_{\text{INL/DNL}} + v_n^2\big|_{\text{quant}} \tag{7.41}$$

where $v_n|_{\text{input}}$ is the noise voltage due to devices in the input signal path, $v_n|_{\text{jitter}}$ is the noise voltage due to the clock jitter, $v_n|_{\text{INL/INL}}$ is the noise voltage due to the quantizer's non-linearity, and $v_n|_{\text{quant}}$ is the noise voltage due to ideal quantization. The main takeaway is that the impact of all the noise contributions is reduced significantly as we go down the pipeline. If the gain in every stage is G, the noise of the kth stage will be reduced by a factor of $G^{2(k-1)}$. This leads to a significant relaxation in the accuracy and noise requirements as we go down the pipeline.

7.4 Pipelined ADC design considerations

As we approach designing a pipelined ADC, we face multiple crucial decisions that can have significant impact on the performance, power and even viability of the design. Some of the important design considerations are:

- Sampling capacitor value.
- ADC full-scale.
- Number of bits/stage.
- Stage scaling factor.
- Input buffer or not.

7.4.1 Sampling capacitance value and input full-scale

The value of the sampling capacitance is one of the most important design decisions. It directly affects the thermal noise of the ADC and is a strong determinant of the power consumption. The noise power is inversely proportional to the capacitance value along a 3 dB/octave (10 dB/decade) curve. The power consumption increases linearly with the capacitance value along a 6 dB/octave (20 dB/decade) curve. Since the SNDR is calculated as the ratio of the signal power to the noise power, another parameter that is closely related is the ADC full-scale range. Increasing the full-scale range improves the SNDR by 20 dB/decade. The SNDR can be expressed as

$$\text{SNDR} = \frac{0.5V_{FS}^2/4}{v_n^2\big|_{\text{input}} + v_n^2\big|_{\text{jitter}} + v_n^2\big|_{\text{INL/DNL}} + v_n^2\big|_{\text{quant}}} \tag{7.42}$$

Thus, it is desirable to maximize the full scale of the ADC. However, a large full-scale may degrade linearity and be difficult to drive, because it requires a high gain amplifier prior to the ADC, which may also degrade the noise of the whole signal chain. Moreover, a large full scale may be difficult to support in fine lithography processes from the reliability standpoint.

Analyzing the impact of the capacitance value on the power consumption is complicated and must take into account the power consumption of the MDAC, reference buffer, input buffer (if there is one) and clocks. If the sampling capacitance is too small, the designer may not be able to achieve the SNDR specification, or he/she may have to increase the power in the amplifier or the following stages to make up for the high kT/C noise caused by the small capacitance. This may increase the overall power consumption. On the other hand, a capacitance that is too large will result in lower noise, but higher power consumption in the active circuits. An example of the trade-off is shown in Figure 7.25 for a 14-bit 125 MS/s ADC with 75 dB SNDR. An optimum value for the capacitance is found to be around 6 pF [3].

7.4.2 Number of bits per stage

The number of bits per stage is another important design parameter for power optimization. As the number of bits per stage increases, the noise and linearity requirements of the following stages are relaxed. In addition, the gain accuracy requirement of the first stage is also relaxed. The feedback factor degrades slightly, but the noise of the stage is improved as shown in (7.37) and Example 2. Generally speaking, as the number of bits increases, the amplifier tends to be more power efficient up to a point, beyond which the power consumption increases substantially. Once that point is reached, reducing the number of bits, and hence

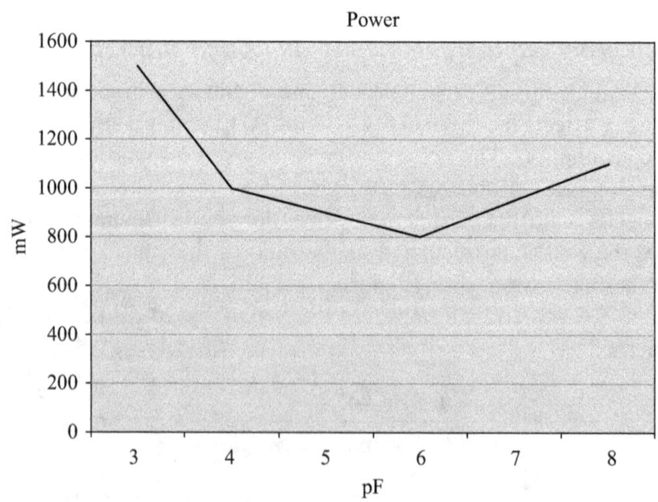

Figure 7.25 A simulation of the total power consumption as a function of the first stage sampling capacitance for a 14-bit 125MS/s ADC. An optimum point exists at 6 pF [3]

increasing the feedback factor, would enable faster settling and lower power consumption.

However, as the number of bits increases, the power consumption in the sub-ADC (flash) increases, the parasitics increase, the clocking power increases and the feedback factor degrades. In any particular case, there is usually a sweet spot for the power consumption versus the number of bits. It is critical to take all factors into account using realistic power estimates of all the contributors. An example is shown in Figure 7.26 for a 14-bit 125 MS/s ADC [3].

In general, a 4-bit/stage is often shown to be an optimum resolution per stage for noise-limited pipelined ADCs [3, 4]. However, other considerations may force the designer to deviate from the optimum resolution per stage. For example, in some cases, a large error correction range is required. This is needed in SHA-less architectures where high IF, or even RF, sampling needs to be supported. The mismatch between the MDAC and flash timing and bandwidth causes errors that consume a portion of the correction range and gets worse with increasing the input frequency. To accommodate high input frequency sampling, it is desirable to reduce the resolution in the first stage from the optimum value of 4 bits to 3 bits [3, 5, 6]. This is an example of situations where the practical considerations of the ADC design may interfere with the power efficiency in a manner that is not necessarily captured by the figure of merit (*FOM*).

Another consideration is the inherent simplicity and efficiency of the 1.5 bit/stage. The residue of a 1.5-bit stage is shown in Figure 7.27. The stage has two comparators, and the DAC has 3 codes, with the middle code being zero. Since this requires only a single DAC capacitance, the DAC is inherently linear. In addition,

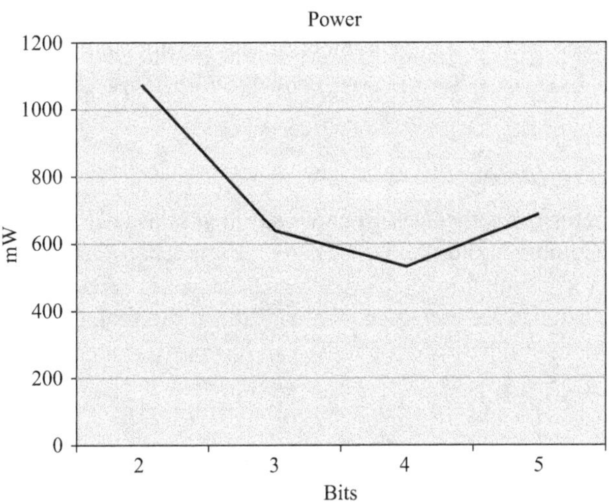

Figure 7.26 A simulation of the total power consumption as a function of the first stage sampling capacitance for a 14-bit 125MS/s ADC. An optimum point exists at 4 bits/stage [3, 4]

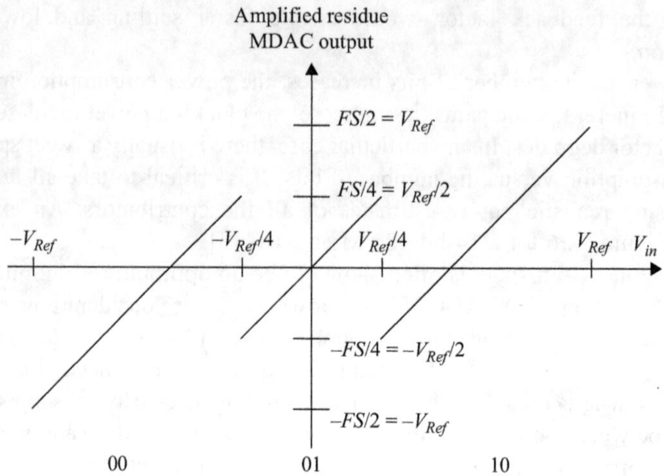

Figure 7.27 Output residue of a 1.5-bit stage

the lower gain value gives a better feedback factor and inherently allows for higher speeds for a fixed gain-bandwidth product that is limited by the process. However, the 1.5-bit stage suffers from some drawbacks. Figure 7.27 shows the residue of the stage reaching the full-scale even in the absence of comparator offsets. This strains the RA and requires it to have a wide dynamic range. Folding the last stages by adding two additional comparators, gives a 2-bit stage whose residue is shown in Figure 7.28. This limits the amplifier's ideal range to half the full-scale, but the DAC will no longer be inherently linear. Moreover, the small number of bits resolved per stage leads to a large number of stages, which limits the attractiveness of the 1.5 bit/stage to relatively low resolutions that are less than or equal to 8–10 bits.

7.4.3 Scaling factor

The scaling factor is another design parameter that is usually employed to reduce the power consumption and the die size. As the capacitance scaling factor increases, the power consumption of the following stages is reduced. However, their noise and mismatches are increased. From (7.38), if the scaling factor of stage i is given by S_i, the total input-referred noise can be approximated by

$$v_n^2(in) \approx v_{ns}^2 + v_{na}^2\left(1 + S_1/G_1^2 + S_1S_2/G_1^2G_2^2 + \cdots\right) \tag{7.43}$$

where $v_n(in)$ is the input-referred noise voltage. For a 3-bit/stage pipelined ADC, and a gain of 4 in every stage, if the scaling factor is also 4, the total input-referred noise is shown in Figure 7.29 to be given by

$$v_n^2(in) \approx v_{ns}^2 + v_{na}^2(1 + 1/4 + 1/16 + \cdots) \tag{7.44}$$

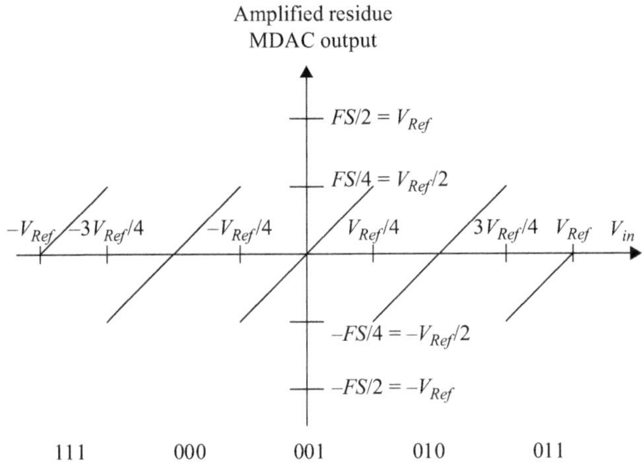

Figure 7.28 Output residue of a 2-bit stage with folded sub-ranges

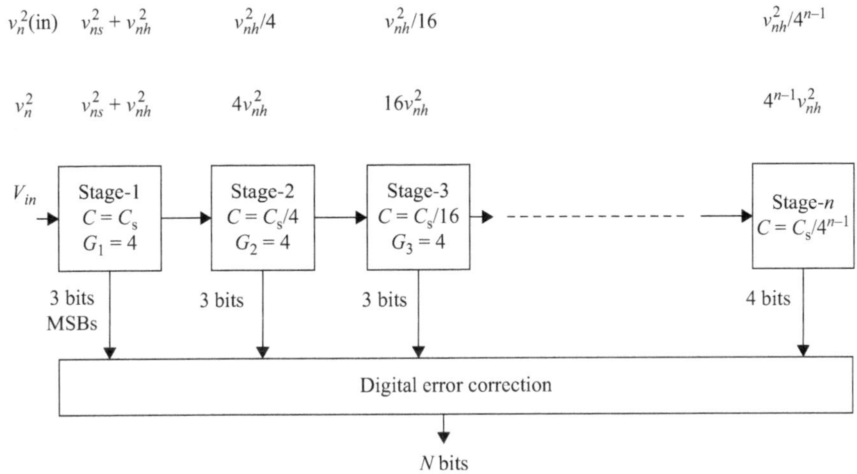

Figure 7.29 A block diagram of a pipelined ADC where each stage is scaled down by a factor of 4 compared to the preceding stage, and has a gain of 4. Also shown is the noise power of each stage v_n^2 and the input-referred noise power $v_n^2(in)$

It is important to note that even though the scaling factor S is equal to the gain G, a significant noise reduction occurs when the noise is referred to the input. This is because, when input-referred, the noise is scaled down by G^2, while the noise of the stage increases by a factor of S only (not S^2). Since the noise is scaled down by the square of the gain, and the power consumption is reduced by the scaling factor,

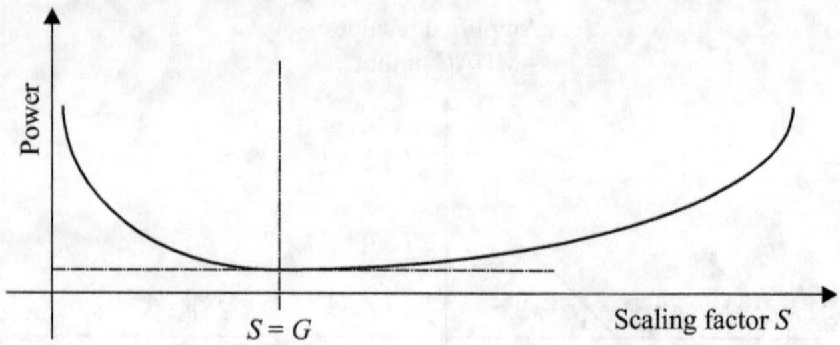

Figure 7.30 A simulation of the total power consumption as a function of the scaling factor for a 14-bit 125MS/s ADC. An optimum point exist at S = G = 4 [3, 4]

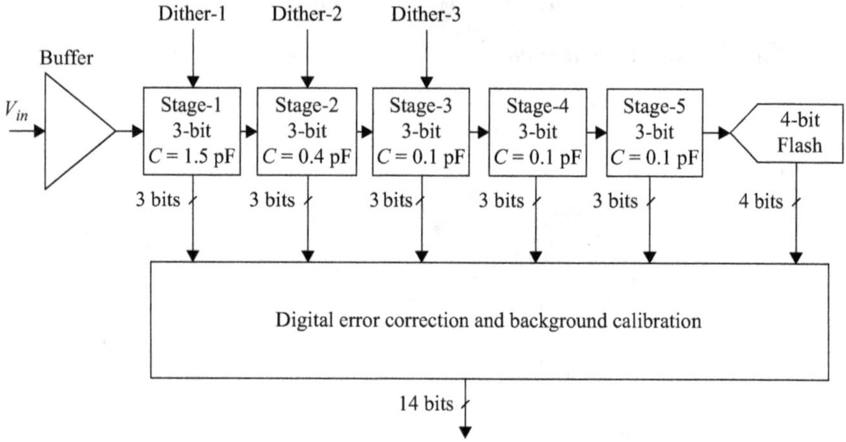

Figure 7.31 A block diagram of a 14-bit, 1GS/s pipelined ADC where each stage is scaled down by a factor of 4 compared to the preceding stage, and has a gain of 4. Dither is injected in the first 3 stages for calibration as discussed in Chapter 9. Scaling stops after stage-3 because the capacitances are too small to scale down any further [6]. © 2014 IEEE. Reprinted, with permission, from Reference 6

it can be shown that a near optimum point tends to be a scaling factor that is approximately equal to the gain value. That is,

$$S \approx G \tag{7.45}$$

This is shown in Figure 7.30 for a 14-bit 125 MS/s ADC [3] where the scaling factor is equal to the gain of each stage, which was 4 in that case. However, if the capacitance value is so small that further scaling is not possible, or would result in unacceptable mismatch errors, then scaling could be stopped or reduced. An example of a scaled pipeline is shown in Figure 7.31 where the first three stages are scaled, but it was not practical to scale the capacitance any further [6].

7.4.4 Input buffer

Employing an integrated input buffer reduces the kick-back due to charge injection and makes it easier to drive the ADC. It provides a large input impedance and a low output impedance to drive the sampling capacitance. However, the buffer adds additional power, increases the noise and limits the linearity of the sampling process. In many cases, the advantages of the buffer far outweigh the drawbacks, and it becomes almost a necessity to achieve the desired performance, especially for high sampling rates [6, 7].

7.5 Accuracy and speed challenge

The pipelined ADC operation relies on the MDAC with its inter-stage amplifier to accurately pass the quantization error of the stage with the appropriate amplification to the following stages. It represents a major bottleneck for the accuracy and speed. In fact, when doing feasibility of a pipelined ADC with a certain resolution and sampling rate, one of the main questions that need to be answered is the feasibility and power efficiency of the MDAC amplifier that is required to achieve the target performance. As the resolution and sampling rate increase, the challenge of designing the amplifier increases substantially. In addition, as we move to fine lithography CMOS processes, the speed of the transistors increase, but their intrinsic gain ($g_m R_{ds}$) and dynamic range are reduced substantially, which makes it more difficult to achieve the required amplifier performance.

To overcome this problem, pipelined ADC designers have taken several directions, which include:

- Using digital assistance (calibration), where the amplifier design is relaxed significantly, while correcting the resulting errors in the digital domain. This is the most popular and most active area of research in pipelined ADCs.
- Using analog methods for gain enhancement. Examples are correlated double sampling, summing node sensing, and correlated level shifting [10–14]. In spite of their elegance and effectiveness, these methods usually suffer from analog overhead that impacts the speed, noise, and/or power consumption.
- Using different kinds of amplifiers. Examples include MDACs that rely on zero crossing detection and integrator-based MDACs [15–18]. These methods are promising, but they transfer the problem from a gain-speed problem to a timing accuracy problem. In spite of their promise and efficiency, they have not managed to replace the traditional amplifier as the dominant structure for pipelined ADCs.

7.6 Conclusion

In spite of its limitations, the pipelined ADC architecture still runs supreme as the dominant architecture for high speed and high resolution ADCs. Through different incarnations and modifications, such as incorporating SAR ADCs in a pipelined ADC structure, using different methods of amplification and employing digital assistance, it has managed to survive and flourish. Moreover, by taking advantage of interleaving, it is well suited to continue to enjoy a prominent place in the high speed and high performance converter space.

Problems

1. What is the inter-stage gain that is causing the INL in Figure 7.11? What are the most likely causes?
2. What are the possible causes of the DAC errors in the INL of Figure 7.14? What are the magnitudes of the errors on the analog side? Note the symmetry between the two sides and the increase in the magnitude of the errors as we move away from the center.
3. Using a behavioral modeling language, build a model of a 10-bit pipelined ADC with 4 bits/stage and 1-bit redundancy. How many pipeline stages are needed and how many bits are in the back-end flash?
   (a) Plot the output FFT for an input sine wave of unity amplitude and frequency of 100 MHz. Pick a suitable sampling rate for the ADC.
   (b) Plot the INL and DNL of the ADC.
4. Using a behavioral modeling language, build a model of a 10-bit pipelined ADC with 1.5 bit/stage. How many stages are needed assuming a 2-bit back-end flash? What happens if we kept using 1.5 bit/stage all the way down the pipeline to achieve the 10-bit resolution without using the back-end flash? Compare the output codes in both cases.
   (a) Plot the output FFT for an input sine wave of unity amplitude and frequency of 100 MHz. Pick a suitable sampling rate for the ADC.
   (b) Plot the INL and DNL of the ADC.
5. In the pipeline models of Problems 3 and 4, introduce IGE and DAC errors and observe the effects on the output FFT and INL.
6. Simulate a 3-bit MDAC in SPICE. Introduce capacitance mismatches and observe the resulting error in the residue. Compare the results with the discussion in Section 3.2.
7. In the pipelined ADC of Problem 3, plot the residue of the first stage as a function of the input. Introduce errors in the flash, inter-stage gain, and DAC. Plot the residue with each error. Repeat for the pipelined ADC of Problem 4.
8. For a two-stage 8-bit pipelined ADC with 4 bits in the first stage and 1-bit redundancy, draw the residue of the first stage after amplification as a function of the input voltage for two cases: a mid-tread sub-ADC and mid-rise sub-ADC. What are the differences between the two cases? Reconstruct the final output word using the bits from the two stages in both cases.
9. For a 12-bit pipelined ADC with 4 bits/stage and 1-bit redundancy, if the required accuracy is 0.25 LSB, what is the approximate required accuracy/noise of each of the following:
   (a) Stage-1 flash ADC?
   (b) Stage-1 inter-stage amplifier?
   (c) Stage-1 DAC?

   (d) Stage-2 flash ADC?
   (e) Stage-2 inter-stage amplifier?
   (f) Stage-2 DAC?
10. Repeat Problem 9 assuming the combined effects of both stage-1 and stage-2.
11. For the first two stages in Problem 9, what is the allowed:
   (a) DAC capacitor mismatch?
   (b) Amplifier capacitor mismatch?
   (c) Open loop gain of the operational amplifier?
   (d) DAC reference error?
   (e) Inter-stage gain error?
   (f) Amplifier reference error?
   (g) DAC settling error?
   (h) Reference settling error?

   Assume only one error at a time.
12. Repeat Problem 11 assuming all the errors exist together and budget for all of them combined to achieve a net accuracy of 0.25 LSB.
13. If the sampling phase noise is 500 μV for a 4-bit MDAC with a gain of 8, what is the expected noise if the gain is changed to 4 and 16? Ignore the parasitic capacitances.
14. If the sampling phase noise is 500 μV for a 4-bit MDAC with a gain of 8, what is the expected noise if an additional capacitance is added to the summing node such that:
   (a) The additional capacitance is equal to the sampling capacitance?
   (b) The additional capacitance is equal to half the sampling capacitance?

   Ignore the parasitic capacitances.
15. If the input-referred amplifier noise in the gain phase of a 4-bit MDAC with a gain of 8 is 250 μV, what is the expected noise if the gain is changed to 4 and 16? Assume the operational amplifier is not changed and ignore the parasitic capacitances.
16. If the input-referred amplifier noise in the gain phase of a 4-bit MDAC with a gain of 8 is 250 μV, what is the expected noise if an additional capacitance is added to the summing node such that:
   (a) The additional capacitance is equal to the sampling capacitance?
   (b) The additional capacitance is equal to half the sampling capacitance?

   Ignore the parasitic capacitances and assume the amplifier is not changed.
17. Repeat Problem 15 if the amplifier's Miller capacitance is changed such that its bandwidth stays the same.
18. Repeat Problem 16 if the amplifier's Miller capacitance is changed such that its bandwidth stays the same.
19. Discuss the factors causing the pattern shown in Figure 7.25. What happens if the capacitor is too large or too small? Give examples.
20. Discuss the factors causing the pattern shown in Figure 7.26. What happens if the number of bits/stage is too large or too small? Give examples.
21. Discuss Figure 7.30. What happens if the scaling factor is too large or too small? Give examples.

References

[1] S.H. Lewis and P.R. Gray, "A Pipelined 5-Msample/s 9-bit Analog-to-Digital Converter," *IEEE Journal of Solid State Circuits*, SC-22(6), pp. 954–961, Dec 1987.

[2] I. Mehr and L. Singer, "A 55-mW, 10-bit, 40-Msample/s Nyquist-Rate CMOS ADC," *IEEE Journal of Solid State Circuits*, 35(3), pp. 318–325, Mar 2000.

[3] A.M.A. Ali, C. Dillon, R. Sneed, *et al.*, "A 14-bit 125 MS/s IF/RF Sampling Pipelined ADC with 100 dB SFDR and 50 fs Jitter," *IEEE Journal of Solid-State Circuits*, 41(8), pp. 1846–1855, Aug 2006.

[4] S. Devarajan, L. Singer, D. Kelly, *et al.*, "A 16-bit, 125 MS/s, 385 mW, 78.7 dB SNR CMOS Pipeline ADC," *IEEE Journal of Solid-State Circuits*, 44(12), pp. 3305–3313, Dec 2009.

[5] A.M.A. Ali, A. Morgan, C. Dillon, *et al.*, "A 16-bit 250-MS/s IF Sampling Pipelined ADC with Background Calibration," *IEEE J. Solid-State Circuits*, 45(12), pp. 2602–2612, Dec 2010.

[6] A.M.A. Ali, H. Dinc, P. Bhoraskar, *et al.*, "A 14-bit 1GS/s RF Sampling Pipelined ADC with Background Calibration," *IEEE Journal of Solid State Circuits*, 49(12), pp. 2857–2867, Dec 2014.

[7] A.M.A. Ali, H. Dinc, P. Bhoraskar, *et al.*, "A 14-bit 2.5GS/s and 5GS/s RF Sampling ADC with Background Calibration and Dither," *IEEE VLSI Circuits Symposium*, pp. 206–207, 2016.

[8] B. Levy, "A Propagation Analysis of Residual Distributions in Pipeline ADCs, *IEEE Transactions on Circuits and Systems-I: Regular Papers*, pp. 2366–2376, 2011.

[9] J. Guerber, M. Gande, and U.-K. Moon, "The analysis and application of redundant multistage ADC resolution improvements through PDF residue shaping," *IEEE Transactions on Circuits and Systems-I: Regular Papers*, 58(10), pp. 733–742, 2012.

[10] K. Nagaraj, "Switched-Capacitor Circuits with Reduced Sensitivity to Amplifier Gain," *IEEE Transactions on Circuits and Systems*, CAS-34, pp. 571–574, May 1987.

[11] A.M.A. Ali and K. Nagaraj, "Background Calibration of Operational Amplifier Gain Error in Pipelined A/D Converters," *IEEE Transactions on Circuits and Systems II*, 50(9), pp. 631–634, 2003.

[12] C. Enz and G. Temes, "Circuit Techniques for Reducing the Effects of Opamp Imperfections: Autozeroing, Correlated Double Sampling and Chopper Stabilization," *Proceedings of IEEE*, 84(11), pp. 1584–1614, Nov 1996.

[13] J. Li and Un-Ku Moon, "A 1.8-V 67-mW 10-bit 100-MS/s Pipelined ADC Using Time-Shifted CDS Technique," *IEEE Journal of Solid-State Circuits*, 39(9), pp. 1468–1476, Sept 2004.

[14] B.R. Gregoire and U.-K. Moon, "An Over-60 dB True Rail-to-Rail Performance Using Correlated Level Shifting and an Opamp with Only 30 dB

Loop Gain," *IEEE Journal of Solid-State Circuits*, 43(12), pp. 2620–2630, Dec 2008.

[15] T. Oh, H. Venkatram and U.-K. Moon, "A Time-Based Pipelined ADC Using Both Voltage and Time Domain Information," *IEEE Journal of Solid-State Circuits*, 49(4), pp. 961–971, Apr 2014.

[16] D. Gubbins, B. Lee, and U.-K. Moon, "Continuous-Time Input Pipeline ADCs," *IEEE Journal of Solid-State Circuits*, 45(8), Aug 2010.

[17] J.K. Fiorenza, T. Sepke, P. Holloway, *et al.*, "Comparator-Based Switched-Capacitor Circuits for Scaled CMOS Technologies," *IEEE Journal of Solid-State Circuits*, 41(12), pp. 2658–2668, Dec 2006.

[18] L. Brooks and H.-S. Lee, "A Zero-Crossing-Based 8-bit 200MS/s Pipelined ADC," *IEEE Journal of Solid-State Circuits*, 42(12), pp. 2677–2687, Dec 2007.

[19] A.M.A. Ali, "High-performance, low-noise reference generators," US Patent No 7,215,182, May 2007.

[20] G.W. Patterson and A.M.A. Ali, "Fast, efficient reference networks for providing low-impedance reference signals to signal converter systems," US Patent No 7,636,057, Dec 2009.

[21] G.W. Patterson and A.M.A. Ali, "Fast, efficient reference networks for providing low-impedance reference signals to signal processing systems," US Patent No 7,652,601, Jan 2010.

Chapter 8
Time-interleaved converters

The time-interleaved (TI) architecture is an attractive choice when achieving the desired sampling rate is not possible or efficient with a single ADC. Time-interleaved ADCs were first introduced in 1980 by Black and Hodges [1]. In principle, they enable high sampling rates by using multiple ADC channels in parallel. However, the inter-channel mismatches have traditionally limited the performance of this class of converters to low resolutions. Recently, there has been a resurgence of interest in time-interleaved ADCs to push the state of the art in performance, speed, and power consumption. This has been driven by the need for higher sampling rates while facing practical limits and prohibitive costs in process technology.

In this chapter, we cover the analysis, performance limitations, and some implementation approaches of time-interleaved ADCs. We augmented the theoretical analysis with an intuitive and practical perspective. However, the analysis is needed to understand the operation of this class of converters and to enable the understanding and development of methods and algorithms to improve their performance.

8.1 Time-interleaving

In time-interleaved ADCs, multiple ADCs (M), each with a sampling rate of f_s/M, are used in parallel to create an equivalent ADC with a sampling rate of f_s. The structure is shown conceptually in Figure 8.1, where the input is sampled by one of M converters in a round-robin fashion. The output bits are combined and time-aligned to obtain an output data rate of f_s. The special case of two-way interleaving ($M = 2$), which is sometimes called "ping-ponging," is shown in Figure 8.2, and the timing diagram is in Figure 8.3. Another example is four-way interleaving ($M = 4$), which is shown in Figure 8.4, with the timing diagram shown in Figure 8.5.

The timing diagrams of Figures 8.3 and 8.5 depict time-interleaving of independent ADC cores, where the sampling and hold phases are equal in duration. In some cases, the ADC designer may take advantage of interleaving to optimize the ADC core's timing more elaborately. This includes changing the relative durations of the sampling and holding phases, or dividing the timing into more phases. For example, in a four-way interleaved ADC, the timing of each core may be divided into 4 phases instead of 2. The 4 phases fit nicely with four-way interleaving and can accommodate the different durations needed for the

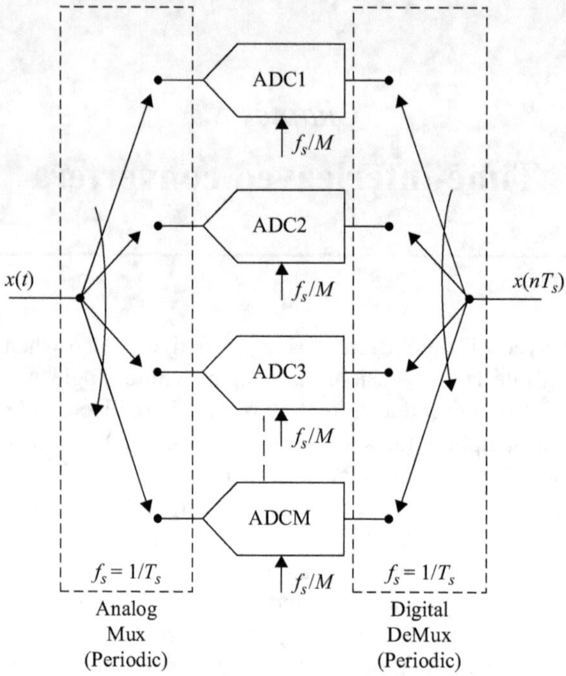

Figure 8.1 A conceptual representation of a time-interleaved ADC with M channels

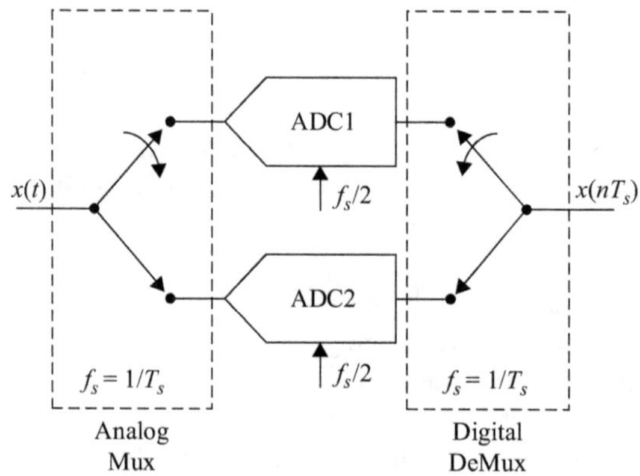

Figure 8.2 A conceptual representation of a two-way interleaved ADC (ping-ponged ADC)

intermediate tasks of a pipelined ADC core, such as: one phase for sampling, one phase for the flash operation, and two phases for the MDAC operation. This optimizes the allocation of time while ensuring that the overall throughput of the interleaved ADC is f_s.

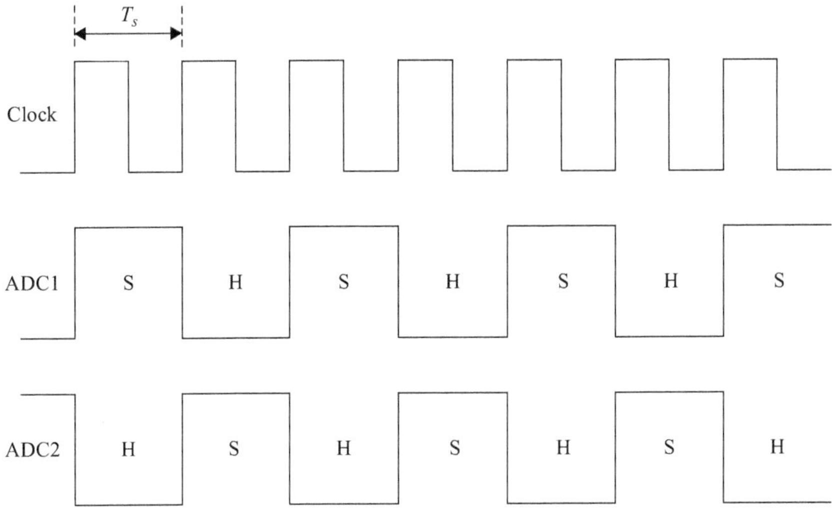

Figure 8.3 A simplified timing diagram for a two-way interleaved ADC. The glitch-free sampling time is not reduced and is equal to T_s

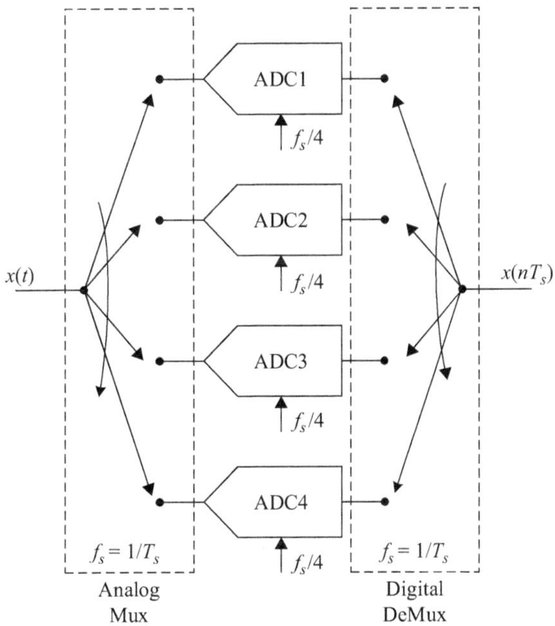

Figure 8.4 A conceptual representation of a four-way interleaved ADC

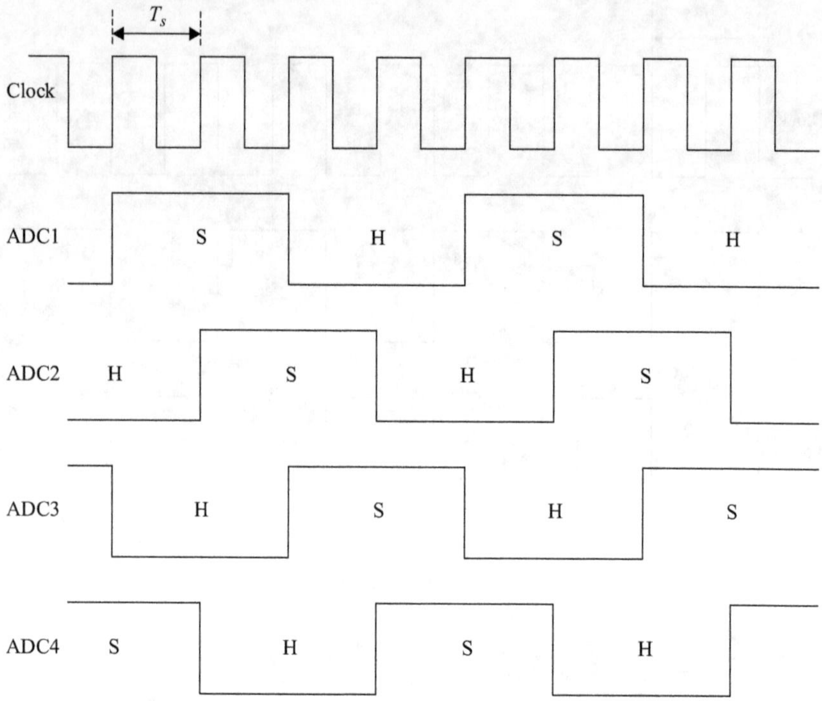

Figure 8.5 A simplified timing diagram for a four-way interleaved ADC. The glitch-free sampling time is not reduced and is equal to T_s

An interesting observation in the two-way interleaving case is that the input sampling time is equal to T_s (not $T_s/2$) and hence is not compromised by the interleaving process. In fact, the sampling time is equal to that of a non-interleaved converter at half the sampling rate ($f_s/2$), and is *double* the sampling time of a non-interleaved converter operating at the full sampling rate (f_s). This is an advantage of ping-ponging over using a straight non-interleaved ADC at the same sampling rate.

On the other hand, for four-way interleaving, although the sampling time for each core is $2T_s$, the input is glitched at twice that rate. For example, in Figure 8.5, ADC2 starts sampling in the middle of the sampling phase of ADC1. Similarly, ADC1 gets connected to the input in the middle of the sampling phase of ADC4. This means that, with four-way interleaving, the glitch-free sampling time is only T_s, and is effectively half the sampling time of the individual cores operating at $f_s/4$. Therefore, if needed, we may have to take measures to minimize the impact of the kick-back from one channel onto the other.

If the input signal is $x(t)$, the ith channel ideally samples the input at time instants that are given by [2–4]

$$t + t_i + kMT_s = t + iT_s + kMT_s \tag{8.1}$$

where t is an arbitrary starting time that can be set to zero for convenience, T_s is the sampling period of the interleaved ADC, i is the channel identifier (order), M is the number of channels, and k is an integer representing the sample number. So,

$$i = 0, 1, \ldots, M - 1 \qquad \text{and} \qquad k = 0, 1, 2, \ldots$$

Therefore, the signal processed by the ith channel $x_i(t)$ is

$$x_i(t) = x(t + t_i) \tag{8.2}$$

If the Fourier transform of $x(t)$ is given by $X(f)$, then the Fourier transform of the input signal "seen" by the ith channel is given by

$$X_i(f) = X(f)e^{j\omega t_i} \tag{8.3}$$

After sampling, and using (8.2), this ith signal is represented as

$$x_{si}(t) = \sum_{k=-\infty}^{\infty} x(-t_i + kMT_s)]\delta(t + t_i - kMT_s) \tag{8.4}$$

From the sampling discussion in Chapter 1, the Fourier transform of the ith channel's sampled signal is

$$X_{si}(f) = \frac{1}{MT_s}\sum_{k=-\infty}^{\infty} X_i(f - kf_s/M) \tag{8.5}$$

Substituting (8.3) in (8.5), we get

$$X_{si}(f) = \frac{1}{MT_s}\sum_{k=-\infty}^{\infty} X(f - kf_s/M)e^{j\left(\omega - \frac{2\pi k f_s}{M}\right)t_i} \tag{8.6}$$

The interleaved digital output is reconstructed by adding up the M sequences from all channels with the appropriate delays at the full rate T_s. That is,

$$x_s(t) = \sum_{i=0}^{M-1} x_{si}(t - iT_s) \tag{8.7}$$

which in the frequency domain gives

$$X_s(f) = \sum_{i=0}^{M-1} X_{si}(f)e^{-j\omega iT_s} \tag{8.8}$$

Substituting (8.6) into (8.8), we get:

$$X_s(f) = \frac{1}{MT_s}\sum_{i=0}^{M-1}\left(\sum_{k=-\infty}^{\infty} G_i X\left(f - \frac{kf_s}{M}\right)e^{j\left(\omega - \frac{2\pi k}{MT_s}\right)t_i}\right)e^{-j\omega iT_s} \tag{8.9}$$

which can be rearranged to give

$$X_s(f) = \frac{1}{T_s} \sum_{k=-\infty}^{\infty} \left(\frac{1}{M} \sum_{i=0}^{M-1} e^{j\left(\omega - \frac{2\pi k}{MT_s}\right)t_i} e^{-j\omega i T_s} \right) X\left(f - \frac{kf_s}{M}\right) \tag{8.10}$$

If the individual channels are matched perfectly and are equally spaced in time, then:

$$t_i = iT_s \tag{8.11}$$

and the ideally interleaved output is given by

$$X_s(f) = \frac{1}{T_s} \sum_{k=-\infty}^{\infty} \left(\frac{1}{M} \sum_{i=0}^{M-1} e^{-j\left(\frac{2\pi ki}{M}\right)} \right) X\left(f - \frac{kf_s}{M}\right) \tag{8.12}$$

Since,

$$\sum_{i=0}^{M-1} e^{-j\left(\frac{2\pi ki}{M}\right)} = \begin{cases} M & \text{for } k = 0, M, 2M, \ldots \\ 0 & \text{Otherwise} \end{cases} \tag{8.13}$$

then the ideally interleaved output is given by

$$X_s(f) = \frac{1}{T_s} \sum_{k=-\infty}^{\infty} X(f - kf_s) \tag{8.14}$$

which is the spectrum of a signal that is sampled at a sampling rate of $f_s = 1/T_s$ and does not suffer from any interleaving errors, as discussed in Chapter 1.

By interleaving the sampling time of each converter and aggregating the output samples with the proper timing alignment, the M ADC cores, each sampling at a rate of f_s/M, should behave as a single ADC with a sampling rate equal to f_s. From the performance standpoint, the resulting ADC will ideally have the same SNR, SINAD, and SFDR as the individual ADCs. However, as discussed in Chapter 2, the noise spectral density (NSD) will improve by a factor equal to $10 \log M$. This is due to the increase in the sampling rate combined with the same total SINAD, which results in a reduction of the NSD.

That is, for time-interleaved ADCs:

$$SINAD = SINAD_i \tag{8.15}$$

$$f_s = Mf_{si} \tag{8.16}$$

and

$$NSD = NSD_i - 10 \log M \tag{8.17}$$

where $SINAD_i$, f_{si}, and NSD_i are the SINAD, sampling rate, and NSD of each channel, respectively. The NSD is a negative quantity, so the negative sign in (8.17) indicates an improvement in the NSD.

Table 8.1 A comparison between the time-interleaved ADC and the parallel ADC relative to a single channel ADC

	Single channel ADC	Time-interleaved ADC	Parallel ADC
SINAD	$SINAD_i$	$SINAD_i$	$SINAD_i + 10\log M$
Sampling rate	f_{si}	Mf_{si}	f_{si}
Noise spectral density	NSD_i	$NSD_i - 10\log M$	$NSD_i - 10\log M$

The time-interleaved architecture should be distinguished from the *parallel* architecture where M ADCs sample the input simultaneously in parallel, and their outputs get added. Unlike time-interleaved ADCs, the parallel structure would result in an ADC with a sampling rate that is equal to the individual ADCs, but with a SINAD that is improved by $10\log M$. This is summarized in Table 8.1.

That is, in the parallel ADCs case:

$$SINAD = SINAD_i + 10\log M \tag{8.18}$$

$$f_s = f_{si} \tag{8.19}$$

and

$$NSD = NSD_i - 10\log M \tag{8.20}$$

Therefore, the time-interleaved ADC utilizes parallelism to increase the sampling rate, while the parallel ADC uses parallelism to increase the performance (SINAD) and lower the noise of the ADC, while keeping the sampling rate fixed. It is interesting to note the equivalence between an M-times increase in the sampling rate and the $10\log M$ improvement in SINAD, as discussed in Chapter 2 in the context of the Figures–of-Merit (FOM).

The time-interleaved architecture enables faster sampling rates with the power consumption linearly increasing with sampling rate. That is, the power is linearly proportional to the number of cores M and the sampling rate f_s. It is interesting to note that, for a certain process lithography (x nm), the power consumption tends to increase linearly with sampling rate, as presented in the FOM discussion in Chapter 2. This applies to the range of sampling rates that is relatively straightforward to achieve in that process node. However, as the speed increases further and starts to approach the limit of the process capability, the power usually increases in a non-linear fashion as shown in Figure 8.6. This non-linear region of the curve is inefficient, and could result in excessively high power consumption. Moving to a finer lithography process (x/2 nm) enables power reduction and the extension of the linear power curve to higher sampling rates. On the other hand, the interleaved architecture extends the linear relationship between power consumption and sampling rate farther on the graph, at least in theory. Therefore, higher sampling rates can be achieved with reasonable power consumption.

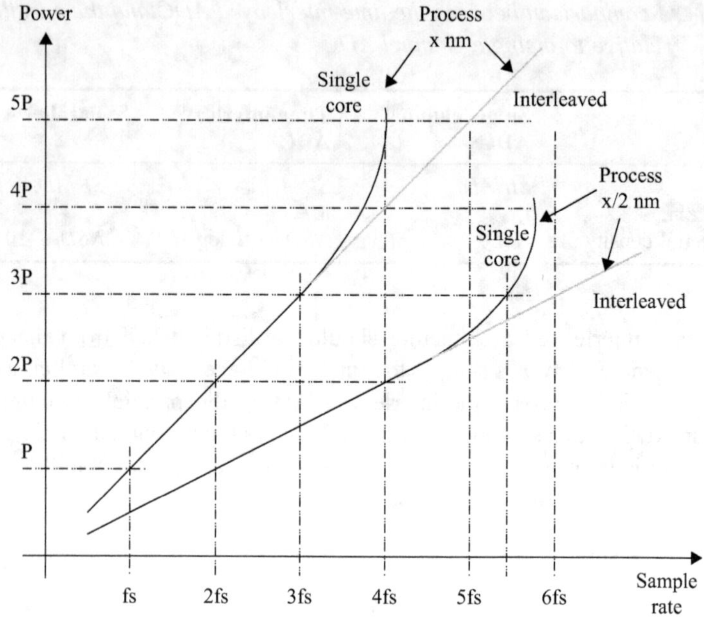

Figure 8.6 An illustration of the power consumption versus sampling rate as a function of process and architecture

However, in practice, the overhead of interleaving makes this linear power-speed relation hard to achieve, especially for large number of channels. Moreover, time-interleaved ADCs suffer from performance challenges and practical difficulties. If moderate or high performance is to be achieved, significant design effort and power overhead may be required, which can impact the ADC's complexity and power efficiency.

The main performance limitation of time-interleaved ADCs is due to mismatches between the interleaved channels, which result in interleaving spurs that can get worse with increasing the input frequency. Several types of mismatch exist, which include:

1. Offset mismatch.
2. Gain mismatch.
3. Timing mismatch.
4. Bandwidth mismatch.
5. Non-linearity mismatch.

In the presence of offset, gain, and timing mismatch, the *i*th channel signal is given by [3, 4]

$$x_i(t) = G_i x(t + t_i) + O_i \tag{8.21}$$

where G_i is the gain of the *i*th channel, O_i is its offset, and t_i is its sampling time, given by

$$t_i = iT_s + \delta t_i = iT_s + r_i T_s \tag{8.22}$$

The Fourier transform of $x_i(t)$ is

$$X_i(f) = G_i X(f) e^{j\omega t_i} + O_i \delta(f) \tag{8.23}$$

After sampling, the ith channel signal can be represented as

$$x_{si}(t) = \sum_{k=-\infty}^{\infty} [G_i x(-t_i + kMT_s) + O_i]\delta(t + t_i - kMT_s) \tag{8.24}$$

Similar to (8.5), its Fourier transform is given by

$$X_{si}(f) = \frac{1}{MT_s} \sum_{k=-\infty}^{\infty} X_i(f - kf_s/M) \tag{8.25}$$

Substituting (8.23) in (8.25), we get

$$X_{si}(f) = \frac{1}{MT_s} \sum_{k=-\infty}^{\infty} G_i X(f - kf_s/M) e^{j\left(\omega - \frac{2\pi k f_s}{M}\right)t_i} + \frac{1}{MT_s} \sum_{k=-\infty}^{\infty} O_i \delta(f - kf_s/M) \tag{8.26}$$

The interleaved digital output is reconstructed by adding up the M sequences from all channels with the appropriate delays at the full rate T_s. That is,

$$x_s(t) = \sum_{i=0}^{M-1} x_{si}(t - iT_s) \tag{8.27}$$

which in the frequency domain gives

$$X_s(f) = \sum_{i=0}^{M-1} X_{si}(f) e^{-j\omega i T_s} \tag{8.28}$$

Substituting (8.26) in (8.28), we get a representation of the interleaved signal

$$X_s(f) = \frac{1}{MT_s} \sum_{i=0}^{M-1} \left(\sum_{k=-\infty}^{\infty} G_i X\left(f - \frac{kf_s}{M}\right) e^{j\left(\omega - \frac{2\pi k}{MT_s}\right)t_i} \right) e^{-j\omega i T_s} + \frac{1}{MT_s} \sum_{i=0}^{M-1} \sum_{k=-\infty}^{\infty} O_i \delta(f - kf_s/M) e^{-j\omega i T_s} \tag{8.29}$$

Using the sifting property of the delta function, we get

$$X_s(f) = \frac{1}{MT_s} \sum_{i=0}^{M-1} \left(\sum_{k=-\infty}^{\infty} G_i X\left(f - \frac{kf_s}{M}\right) e^{j\left(\omega - \frac{2\pi k}{MT_s}\right)t_i} \right) e^{-j\omega i T_s} + \frac{1}{MT_s} \sum_{i=0}^{M-1} \sum_{k=-\infty}^{\infty} O_i \delta(f - kf_s/M) e^{-\frac{j2\pi k i}{M}} \tag{8.30}$$

Substituting (8.22) into (8.30), we obtain

$$X_s(f) = \frac{1}{MT_s}\sum_{i=0}^{M-1}\left(\sum_{k=-\infty}^{\infty} G_i X\left(f - \frac{kf_s}{M}\right)e^{j\left(\omega - \frac{2\pi k}{MT_s}\right)r_i T_s}\right)e^{-\frac{j2\pi ki}{M}}$$
$$+\frac{1}{MT_s}\sum_{i=0}^{M-1}\sum_{k=-\infty}^{\infty} O_i\delta(f - kf_s/M)e^{-\frac{j2\pi ki}{M}} \qquad (8.31)$$

Equations (8.30) and (8.31) give a representation of the time-interleaved sampled signal in the frequency domain, in the presence of inter-channel mismatches. The terms O_i, G_i, and r_i represent the offset, gain, and timing of the ith channel, respectively. They will be used in the next sections to analyze the impact of these mismatches on the interleaved spectrum.

8.2 Offset mismatch

One of the most benign non-idealities in ADCs is the offset, as discussed in Chapter 2. It can be easily fixed by adding or subtracting a fixed digital value from the ADC output. However, in interleaved ADCs, mismatches in the offset values of the different cores/channels result in spurs at frequencies equal to f_s/M and its multiples.

This can be deduced from (8.31) by removing the timing and gain mismatches. By setting $r_i = 0$ and $G_i = G = 1$ in (8.31), we obtain the expression for the time-interleaved sampled signal $X_s(f)$ in the presence of offset mismatch, which is given by

$$X_s(f) = \frac{1}{MT_s}\sum_{i=0}^{M-1}\left(\sum_{k=-\infty}^{\infty} X\left(f - \frac{kf_s}{M}\right)\right)e^{-\frac{j2\pi ki}{M}}$$
$$+\frac{1}{MT_s}\sum_{i=0}^{M-1}\sum_{k=-\infty}^{\infty} O_i\delta(f - kf_s/M)e^{-\frac{j2\pi ki}{M}} \qquad (8.32)$$

In (8.32), the first term on the right hand side is the ideal interleaved signal. Therefore,

$$X_s(f) = X_{s|ideal}(f) + \frac{1}{MT_s}\sum_{i=0}^{M-1}\sum_{k=-\infty}^{\infty} O_i\delta(f - kf_s/M)e^{-\frac{j2\pi ki}{M}} \qquad (8.33)$$

which can be rearranged to give

$$X_s(f) = X_{s|ideal}(f) + \frac{1}{T_s}\sum_{k=-\infty}^{\infty}\frac{1}{M}\sum_{i=0}^{M-1} O_i e^{-\frac{j2\pi ki}{M}}\delta\left(f - \frac{kf_s}{M}\right) \qquad (8.34)$$

The second term on the right hand side of (8.34) represents the spectral content $O(f)$ resulting from the offset mismatches. That is,

$$X_s(f) = X_{s|ideal}(f) + O(f) \tag{8.35}$$

The Delta-Dirac terms in (8.34) represent the interleaving spurs due to offset mismatches. These are given by

$$O(f) = \frac{1}{T_s}\sum_{k=-\infty}^{\infty} \beta(k)\delta\left(f - \frac{kf_s}{M}\right) \tag{8.36}$$

where $\beta(k)$ is given by

$$\beta(k) = \frac{1}{M}\sum_{i=0}^{M-1} O_i e^{-\frac{j2\pi ki}{M}} \tag{8.37}$$

Equation (8.35) indicates that the effect of the offset mismatches is an *additive* signal $O(f)$. From (8.36), we see that this additive signal is composed of spurs, whose magnitudes and frequency locations are independent of the input signal and depend only on the sampling rate, number of interleaved channels and the offset mismatches of the interleaved cores. The locations of these interleaving spurs are given by

$$f_{IL} = kf_s/M \quad \text{where } k = 0, 1, 2, 3\ldots, M-1 \tag{8.38}$$

The power in the spurs is given by:

$$P_N = \sum_{k=0}^{M-1} |\beta(k)|^2 = \frac{1}{M}\sum_{i=0}^{M-1} |O_i|^2 \tag{8.39}$$

which is in line with Parseval's Theorem.

It is important to note that the $k = 0$ "spur" represents the net offset of the whole time-interleaved ADC. It is the average value of the individual offsets, and hence is technically an offset (or DC) term, as opposed to an offset mismatch spur. This DC term can be removed by subtracting its value from the digital output. Examples of interleaving spurs due to offset mismatches are shown in Figure 8.7 for $M = 2$, and in Figure 8.8 for $M = 4$.

For a sinusoidal input signal with amplitude A, and in the presence of mismatched offsets denoted by O_i for the ith channel, we can find the SINAD and SFDR from (8.36) to (8.39) above. The SINAD is given by

$$SINAD = 10\log(P_s/P_N) = 10\log\left(\frac{A^2/2}{\frac{1}{M}\sum_{i=0}^{M-1}|O_i|^2}\right) \tag{8.40}$$

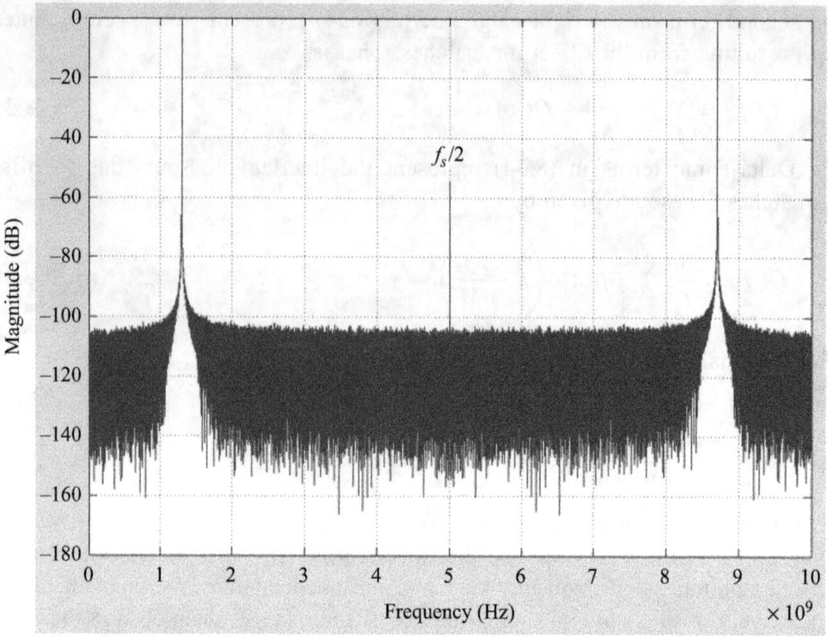

Figure 8.7 An example of a spectrum of a two-way interleaved ADC with offset mismatch. The overall offset of the ADC is zero

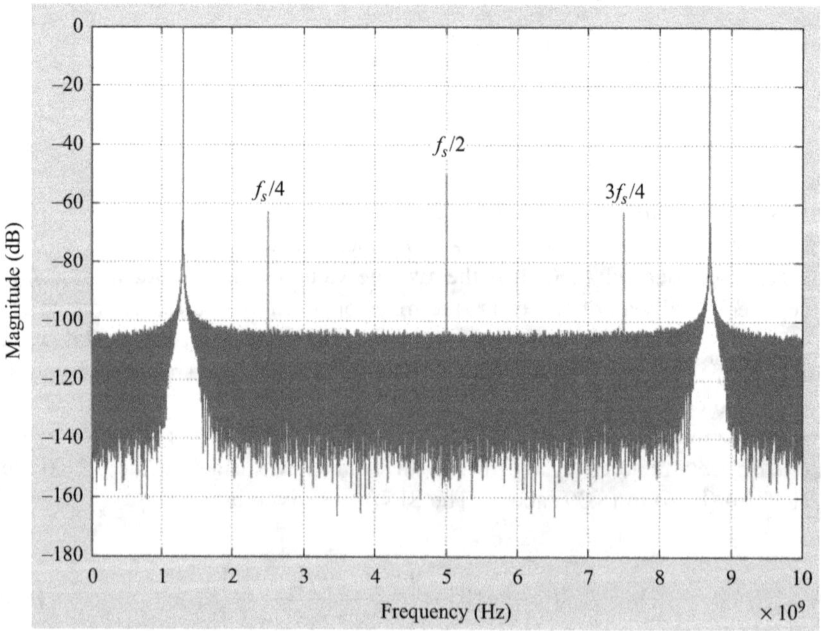

Figure 8.8 An example of a spectrum of a four-way interleaved ADC with offset mismatch. The overall offset of the ADC is zero

and the SFDR is

$$SFDR = 10\log\left(\frac{A^2/2}{\max\limits_{k=0:M-1}|\beta(k)|^2}\right) \tag{8.41}$$

8.2.1 Special cases

If the offset of the ith channel is denoted by O_i with a maximum O_{max} and a minimum O_{min}, the worst case occurs if all the offsets are divided equally between O_{max} and O_{min} with ΔO defined to be

$$\Delta O = (O_{max} - O_{min})/2 \tag{8.42}$$

and the average offset given as

$$O_{ave} = (O_{max} + O_{min})/2 \tag{8.43}$$

In this case, for a sinusoidal input signal with amplitude A, the SINAD is given by [4]

$$SINAD = 10\log(P_s/P_N) = 10\log\left(\frac{A^2/2}{\Delta O^2}\right) \tag{8.44}$$

If the offset mismatch information is represented as a random variable with zero mean and a variance σ_o^2, then the power of the additive signal generated due to the offset mismatches is given by [2, 4]

$$P_N = \frac{1}{M}\sum_{i=0}^{M-1}|O_i|^2 = \sigma_o^2 \tag{8.45}$$

The SINAD will be

$$SINAD = 10\log(P_s/P_N) = 10\log\left(\frac{A^2/2}{\sigma_o^2}\right) \tag{8.46}$$

Assuming equal amplitude among the spurs, the SFDR can be approximated as

$$SFDR = 10\log\left(\frac{A^2/2}{\sigma_o^2}\right) + 10\log(M-1) \tag{8.47}$$

8.2.2 Intuitive perspective

If we think of the offset as an additive DC term, the effect of interleaving will be to have multiple "DC" terms, one for each channel. The additive terms will, therefore, depend on which channel is sampling the input but are independent of the input signal. This is shown in Figure 8.9.

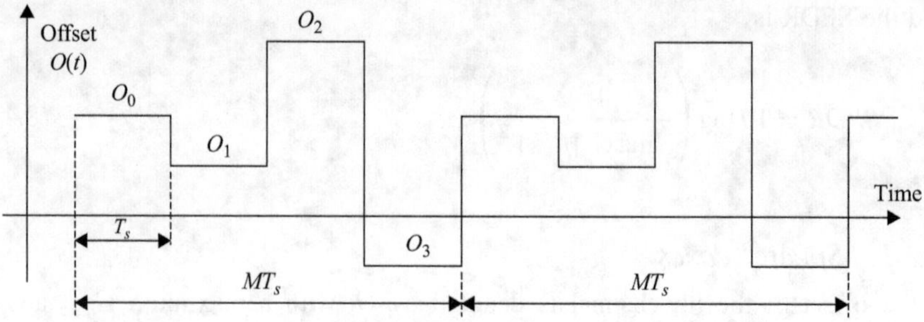

Figure 8.9 A representation of the offset signal as a function of time

Thus, the interleaved signal is given by

$$x_s(t) = x(t) + O(t) \tag{8.48}$$

The additive offset signal $O(t)$ is a periodic signal with a period MT_s. Therefore, from Fourier series analysis we expect it to have harmonics at multiples of $1/MT_s$. That is, it causes interleaving spurs at the locations:

$$f_{IL} = kf_s/M \tag{8.49}$$

which is the same result as (8.38). The power of the offset signal $O(t)$ shown in Figure 8.9 can be easily shown to be equal to:

$$P_N = \frac{1}{M}\sum_{i=0}^{M-1}|O_i|^2 \tag{8.50}$$

which is in line with (8.39). Therefore, we reach the same results for the effects of the offset mismatches on the interleaved signal using a different perspective.

8.3 Gain mismatch

In the case of the presence of only gain mismatches between the different channels, we remove the offset and timing mismatch terms from (8.31) to get [2, 5]

$$X_s(f) = \frac{1}{T_s}\sum_{k=-\infty}^{\infty}\frac{1}{M}\left(\sum_{i=0}^{M-1}G_iX\left(f-\frac{kf_s}{M}\right)\right)e^{-j2\pi ki/M} \tag{8.51}$$

where G_i is the gain of the ith channel. This can be rearranged to give

$$X_s(f) = \frac{1}{T_s}\sum_{k=-\infty}^{\infty}X\left(f-\frac{kf_s}{M}\right)\frac{1}{M}\left(\sum_{i=0}^{M-1}G_ie^{-j2\pi ki/M}\right) \tag{8.52}$$

If we represent the ith channel gain as

$$G_i = G + \delta G_i \tag{8.53}$$

then, the interleaved sampled signal $X_s(f)$ can be represented as

$$X_s(f) = X_{s|ideal}(f) + \frac{1}{MT_s}\sum_{i=0}^{M-1}\sum_{k=-\infty}^{\infty} \delta G_i \times X\left(f - \frac{kf_s}{M}\right)e^{-j2\pi ki/M} \tag{8.54}$$

The first term in the right hand side of (8.54) represents the ideal interleaved signal, while the second term represents the spurs due to the gain mismatches. For a sinusoidal signal with unity amplitude and frequency f_{in}, the interleaved signal is given by

$$X_s(f) = \frac{1}{T_s}\sum_{k=-\infty}^{\infty} \alpha(k)j\left[\delta\left(f + f_{in} - \frac{kf_s}{M}\right) - \delta\left(f - f_{in} - \frac{kf_s}{M}\right)\right]/2 \tag{8.55}$$

where the coefficients $\alpha(k)$ of the interleaving spurs are given by

$$\alpha(k) = \frac{1}{M}\sum_{i=0}^{M-1} G_i e^{-j2\pi ki/M} \tag{8.56}$$

If there are no gain mismatches,

$$\alpha(k) = \begin{cases} G & \text{for } k = 0,\ M,\ 2M, \ldots \\ 0 & \text{otherwise} \end{cases} \tag{8.57}$$

However, in the presence of gain mismatches:

$$\alpha(k) \neq 0, \ \text{ for } k \neq 0,\ M,\ 2M, \ldots \tag{8.58}$$

From (8.55), the interleaving spurs appear with magnitudes equal to $\alpha(k)$ and at frequencies equal to

$$f_{IL} = \pm f_{in} + \frac{k}{M} f_s \tag{8.59}$$

for $k = 1, 2, \ldots, M-1$.

The magnitudes of the interleaving spurs are given by

$$\alpha(k) = \frac{1}{M}\sum_{i=0}^{M-1} G_i e^{-j2\pi ki/M} \tag{8.60}$$

for $k = 1, 2, \ldots, M-1$.

The total power of the interleaved sampled signal is given by

$$P_{Total} = \sum_{k=0}^{M-1} \frac{|\alpha(k)|^2}{2} = \frac{1}{2M}\sum_{i=0}^{M-1} |G_i|^2 \tag{8.61}$$

The fundamental power is given by

$$P_s = \frac{|\alpha(0)|^2}{2} = \frac{1}{2}\left|\frac{\sum_{i=0}^{M-1} G_i}{M}\right|^2 = \frac{G^2}{2} \tag{8.62}$$

Therefore, the SINAD is given by

$$SINAD = \frac{P_s}{P_{Total} - P_s} \cong 10\log\left(\frac{G^2/2}{[(\sum_{i=0}^{M-1} G_i^2)/M - G^2]/2}\right) \tag{8.63}$$

That is,

$$SINAD \cong 10\log\left(\frac{G^2}{\frac{\sum_{i=0}^{M-1} G_i^2}{M} - G^2}\right) \cong 10\log\left(\frac{G^2}{\sigma_G^2}\right) \tag{8.64}$$

where σ_G^2 is the variance of the gain. The SFDR is given by

$$SFDR \cong 10\log\left(\frac{G^2}{\max\limits_{k=1:M-1} |\alpha(k)|^2}\right) \tag{8.65}$$

8.3.1 Special cases

The worst case occurs when all the gain values G_i are divided equally between G_{min} and G_{max} and the average gain G is given by [5]

$$G = \frac{G_{min} + G_{max}}{2} \tag{8.66}$$

and ΔG is

$$\Delta G = \frac{G_{max} - G_{min}}{2} \tag{8.67}$$

Then, the worst case SINAD is given by:

$$SINAD = \frac{P_s}{P_{Total} - P_s} = 20\log\left(\frac{G}{\Delta G}\right) \tag{8.68}$$

For the special case of a sine wave input with $M = 2$:

$$SINAD = SFDR = 20\log\left(\frac{G}{\Delta G}\right) \tag{8.69}$$

If the gain errors are a random variable with standard deviation σ_G and mean G, then the SINAD is given by [2–4, 10]:

$$SINAD \cong 10\log\left(\frac{\sigma_G^2 + MG^2}{(M-1)\sigma_G^2}\right) \tag{8.70}$$

which gives

$$SINAD \cong 10\log\left(\frac{G^2}{\sigma_G^2}\right) - 10\log\left(1 - \frac{1}{M}\right) \tag{8.71}$$

and the SFDR is

$$SFDR \cong 10\log\left(\frac{G^2}{\sigma_G^2}\right) + 10\log(M-1) \tag{8.72}$$

Examples of the interleaving spurs due to the gain mismatch are shown in Figure 8.10 for $M = 2$ and in Figures 8.11 and 8.12 for $M = 4$. It is interesting to see how the frequencies of the spurs depend on the fundamental frequency. A small change in the fundamental frequency can lead to a completely different pattern as we can see from Figures 8.11 to 8.12. The reader is advised to work through the formula of (8.59) to deduce the expected locations of the interleaving spurs for the cases of Figures 8.10–8.12. The frequency of the fundamental can be approximately deduced from the Figures.

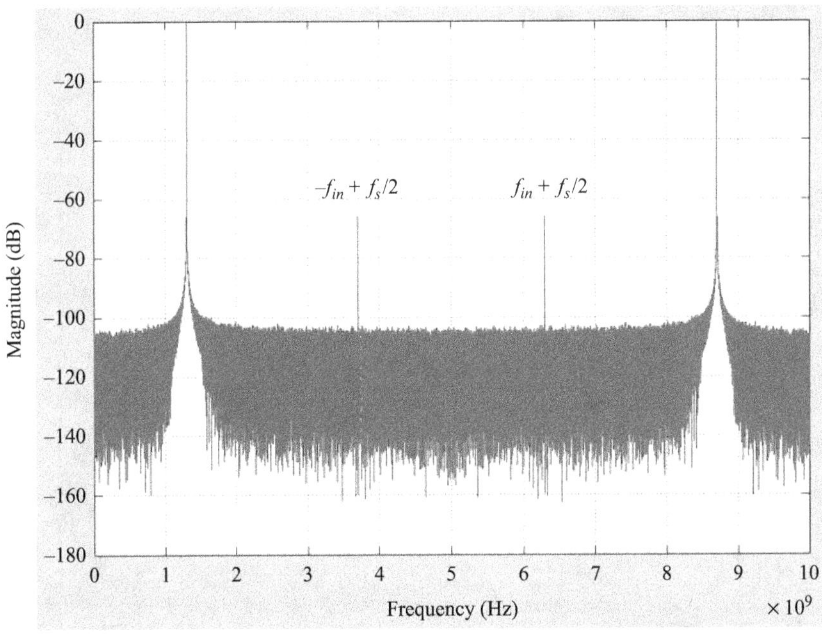

Figure 8.10 An example of a spectrum of a two-way interleaved ADC with gain mismatch only

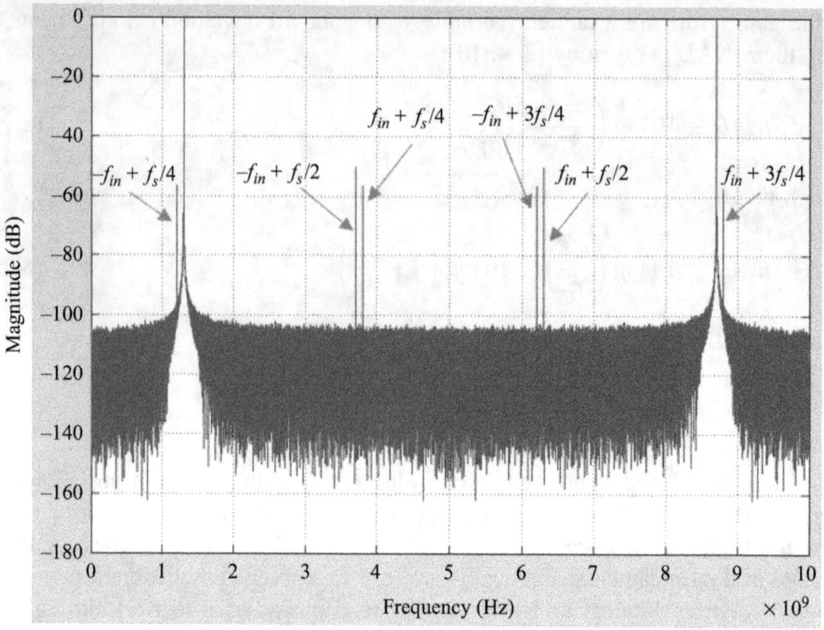

Figure 8.11 An example of a spectrum of a four-way interleaved ADC with gain mismatch

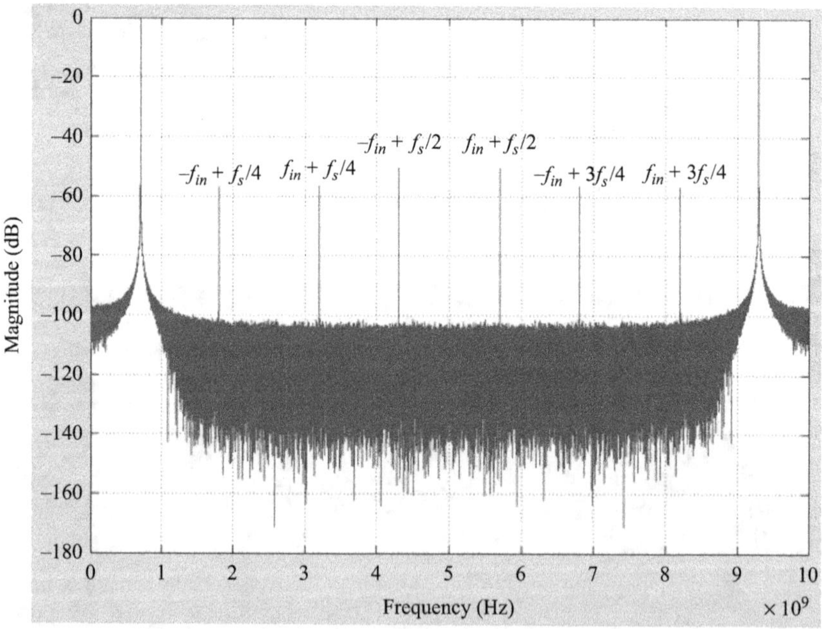

Figure 8.12 An example of a spectrum of a four-way interleaved ADC with gain mismatch

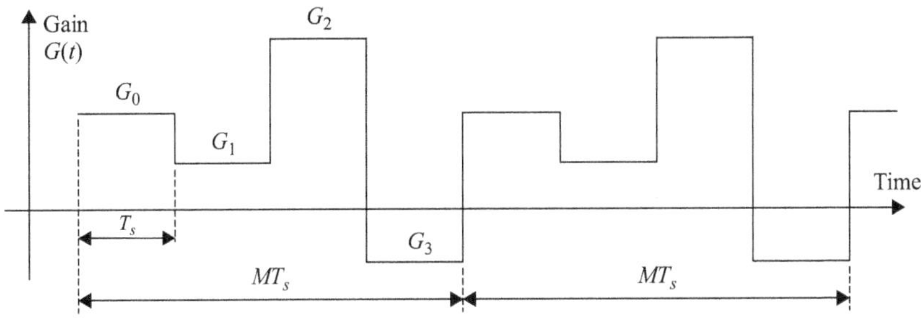

Figure 8.13 A representation of the signal representing the gain as a function of time

8.3.2 Intuitive perspective

We can understand the effect of the gain mismatch on the performance by representing the gain of the interleaved ADC as a function of time by the periodic signal $G(t)$, whose period is MT_s, and is shown in Figure 8.13. The impact of the gain mismatch is to modulate the input signal with the waveform shown in Figure 8.13. That is

$$x_s(t) = x(t) \times G(t) \tag{8.73}$$

This multiplication results in modulation of the input signal with the frequency content of the gain waveform. From Fourier series analysis, the frequency components of the periodic gain signal $G(t)$ is given by

$$f_{Gain} = kf_s/M \tag{8.74}$$

Therefore, the effect of the modulation described in (8.73) is to generate the sum and difference frequencies as follows:

$$f_{IL} = f_{Gain} \pm f_{in} = \pm f_{in} + \frac{k}{M} f_s \tag{8.75}$$

which is similar to (8.59). The power of the modulated signal is similar to that of an amplitude modulated signal and is given by

$$P_{Total} = \frac{1}{2M} \sum_{i=0}^{M-1} |G_i|^2 \tag{8.76}$$

which is similar to (8.61). Therefore, the SINAD is given by

$$SINAD \cong 10 \log \left(\frac{G^2}{\frac{\sum_{i=0}^{M-1} G_i^2}{M} - G^2} \right) \approx 10 \log \left(\frac{G^2}{\sigma_G^2} \right) \tag{8.77}$$

which is similar to (8.64). Therefore, we reach the same results for the effects of the gain mismatches on the interleaved signal using the modulation perspective.

8.4 Timing mismatch

Analyzing (8.31) for the timing mismatch, we set the offset and gain terms to zero to get an expression for the sampled signal with timing mismatch:

$$X_s(f) = \frac{1}{MT_s}\sum_{i=0}^{M-1}\left(\sum_{k=-\infty}^{\infty} X\left(f - \frac{kf_s}{M}\right)e^{j\left(\omega - \frac{2\pi k}{MT_s}\right)t_i}\right)e^{-j\omega iT_s} \tag{8.78}$$

where t_i is the sampling time of the ith channel, which can be expressed as

$$t_i = iT_s + r_iT_s \tag{8.79}$$

The mismatch of the ith channel is

$$\delta t_i = t_i - iT_s = r_iT_s \tag{8.80}$$

Substituting in (8.78) gives

$$X_s(f) = \frac{1}{MT_s}\sum_{i=0}^{M-1}\left(\sum_{k=-\infty}^{\infty} X\left(f - \frac{kf_s}{M}\right)e^{j\left(\omega - \frac{2\pi k}{MT_s}\right)r_iT_s}\right)e^{-\frac{j2\pi ki}{M}} \tag{8.81}$$

For a sinusoidal signal with unity amplitude and frequency f_{in}, the sampled signal becomes

$$X_s(f) = \frac{1}{2MT_s}\sum_{i=0}^{M-1}\sum_{k=-\infty}^{\infty} je^{-\frac{j2\pi ki}{M}}\left[e^{-j2\pi f_{in}r_iT_s}\delta\left(f + f_{in} - \frac{kf_s}{M}\right)\right.$$
$$\left. - e^{j2\pi f_{in}r_iT_s}\delta\left(f - f_{in} - \frac{kf_s}{M}\right)\right] \tag{8.82}$$

Therefore, the interleaved signal can be represented as

$$X_s(f) = \frac{1}{2T_s}\sum_{k=-\infty}^{\infty} j\left[\alpha^*(M-k)\delta\left(f + f_{in} - \frac{kf_s}{M}\right) - \alpha(k)\delta\left(f - f_{in} - \frac{kf_s}{M}\right)\right] \tag{8.83}$$

where the interleaving spur coefficients are given by

$$\alpha(k) = \frac{1}{M}\sum_{i=0}^{M-1}\left[e^{j2\pi f_{in}r_iT_s}\right]e^{-j2\pi ki/M} \tag{8.84}$$

And α^* is the complex conjugate of α. Therefore, from (8.83), the interleaving spurs will be located at

$$f_{IL} = \pm f_{in} + \frac{k}{M} f_s \tag{8.85}$$

for $k = 1, 2, \ldots, M-1$.

It is interesting to note the impact of the term $e^{j2\pi f_{in} r_i T_s}$ in (8.84), which did not exist in the case of the gain mismatch. This term indicates a phase shift between the interleaving spurs and the main signal. For small mismatches, this shift is approximately 90°, which is an important difference between the gain and timing mismatch that affects how they can be corrected, as discussed in Chapter 9.

From (8.83), the SINAD due to timing mismatch is given by

$$SINAD = \frac{P_s}{P_{Total} - P_s} \cong 10 \log \left(\frac{|\alpha(0)|^2}{\sum_{k=1}^{M-1} |\alpha(k)|^2} \right) \tag{8.86}$$

The SFDR is given by

$$SFDR \cong 10 \log \left(\frac{|\alpha(0)|^2}{\max\limits_{k=1:M-1} |\alpha(k)|^2} \right) \tag{8.87}$$

Since the interleaving spurs due to timing mismatches are located at the same frequencies as those of gain mismatches, they look very similar in the output magnitude spectrum to the spurs shown in the examples of Figures 8.10–8.12.

8.4.1 Special cases

If the maximum timing mismatch is t_{max}, the minimum timing mismatch is t_{min}, the number of channels with t_{max} equals the number of channels with t_{min}, and the average mismatch is

$$t_{ave} = \frac{t_{min} + t_{max}}{2} \tag{8.88}$$

and

$$\Delta t = \frac{t_{max} - t_{min}}{2} \tag{8.89}$$

Then, the worst case SINAD is approximately given by

$$SINAD \cong 20 \log \left(\frac{1}{2\pi f_{in} \Delta t} \right) \tag{8.90}$$

For example, for a ping-pong case ($M = 2$), if the time mismatch between the two channels is $\Delta\tau$, then $t_{max} = \Delta\tau$, $t_{min} = 0$, and the Δt is

$$\Delta t = \frac{\Delta\tau}{2} \tag{8.91}$$

Substituting in (8.90) gives the SINAD to be

$$SINAD \cong 20\log\left(\frac{1}{\pi f_{in}\Delta\tau}\right) \cong 20\log\left(\frac{1}{2\pi f_{in}\Delta t}\right) \tag{8.92}$$

and the SFDR is approximately given by

$$SFDR \cong 20\log\left(\frac{1}{\pi f_{in}\Delta\tau}\right) \cong 20\log\left(\frac{1}{2\pi f_{in}\Delta t}\right) \tag{8.93}$$

However, for a sine wave input and $M = 2$, we will show later in (8.144) that a more accurate expression is given by

$$SINAD = SFDR = 20\log\left[\cot(\pi f_{in}\Delta\tau)\right] = 20\log\left[\cot(2\pi f_{in}\Delta t)\right] \tag{8.94}$$

If $\pi f_{in}\Delta\tau \ll 1$, then

$$\cos(\pi f_{in}\Delta\tau) \approx 1, \sin(\pi f_{in}\Delta\tau) \approx \pi f_{in}\Delta\tau, \tag{8.95}$$

and

$$\cot(\pi f_{in}\Delta\tau) \approx \frac{1}{\pi f_{in}\Delta\tau} \tag{8.96}$$

Therefore,

$$SINAD = SFDR = 20\log\left[\cot(\pi f_{in}\Delta\tau)\right] \approx 20\log\left(\frac{1}{\pi f_{in}\Delta\tau}\right) \tag{8.97}$$

which matches the approximate expression given in (8.92)

If the timing errors are random variables with standard deviation σ_t and zero mean, then the SINAD is given by [2–4, 10]

$$SINAD \cong 20\log\left(\frac{1}{2\pi f_{in}\sigma_t}\right) - 10\log\left(1 - \frac{1}{M}\right) \tag{8.98}$$

and the SFDR is approximated by

$$SFDR \cong 20\log\left(\frac{1}{2\pi f_{in}\sigma_t}\right) + 10\log(M - 1) \tag{8.99}$$

which indicates that the SINAD degrades with the input frequency as expected.

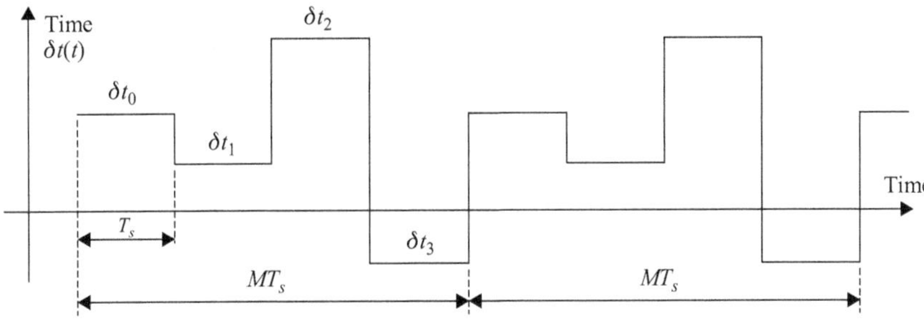

Figure 8.14 *A representation of the signal representing the timing as a function of time*

8.4.2 Intuitive perspective

One way to think of the timing mismatch is as a phase modulation of the input signal with the periodic timing signal $\delta t(t)$ whose period is MT_s and is shown in Figure 8.14. This can be represented as follows

$$x_s(t) = x(t - \delta t) \tag{8.100}$$

If the input signal is a sine wave given by

$$x(t) = \sin(2\pi f_{in} t) \tag{8.101}$$

Then the interleaved sampled signal is

$$x_s(t) = \sin(2\pi f_{in} t - 2\pi f_{in} \delta t) \tag{8.102}$$

which can be represented as

$$x_s(t) = \sin(2\pi f_{in} t)\cos(2\pi f_{in} \delta t) - \cos(2\pi f_{in} t)\sin(2\pi f_{in} \delta t) \tag{8.103}$$

For small mismatches, (8.103) can be approximated by

$$x_s(t) \cong \sin(2\pi f_{in} t) - (2\pi f_{in} \delta t)\cos(2\pi f_{in} t) \tag{8.104}$$

Therefore, the interleaving process modulates the input signal with the $\delta t(t)$ signal in a manner similar to gain modulation. However, unlike gain mismatch, the interleaving spurs due to timing mismatch are orthogonal to the main signal as shown in (8.104). Therefore, the frequency locations of the interleaving spurs due to timing mismatches will be similar to those due to gain mismatches and are given by

$$f_{IL} = \pm f_{in} + f_{\delta t} = \pm f_{in} + \frac{k}{M} f_s \tag{8.105}$$

which is similar to (8.85). From (8.104), the interleaving spur power is given by

$$P_N = \frac{(2\pi f_{in})^2}{2M} \sum_{i=0}^{M-1} |\delta t_i|^2 \tag{8.106}$$

For the special case, defined by (8.88) and (8.89), the interleaving noise power is given by

$$P_N \approx 2(\pi f_{in} \Delta t)^2 \tag{8.107}$$

where Δt is defined by (8.89). Therefore, the SINAD is

$$SINAD \approx 20 \log \left(\frac{1}{2\pi f_{in} \Delta t} \right) \tag{8.108}$$

For $M = 2$, the SINAD will be

$$SINAD \approx 20 \log \left(\frac{1}{\pi f_{in} \Delta \tau} \right) = 20 \log \left(\frac{1}{2\pi f_{in} \Delta t} \right) \tag{8.109}$$

Both (8.108) and (8.109) match (8.90) and (8.92). Therefore, we reach the same results for the effects of the timing mismatches on the interleaved signal using the phase modulation perspective.

8.5 Bandwidth mismatch

Bandwidth mismatch results in both gain and phase mismatches. The gain mismatch component depends on the input frequency and hence is more difficult to handle than the regular gain mismatch. The phase mismatch, however, tends to be similar to the timing mismatch.

The interleaving spurs will be located at

$$f_{IL} = \pm f_{in} + \frac{k}{M} f_s \tag{8.110}$$

for $k = 1, 2, \ldots, M - 1$.

In general, the effect of the bandwidth on the input signal is highly dependent on the input signal frequency, waveform, and the nature of the sampling circuit in terms of number of poles, etc. For simplification, we can study the case of a sampling network that can be represented by a simple RC circuit as discussed in Chapter 4 and shown in Figure 8.15.

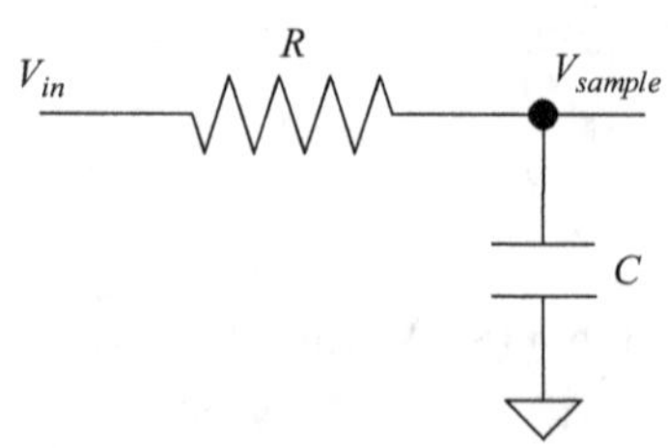

Figure 8.15 A simplified representation of a sampling network

In spite of its simplicity, this model covers a wide range of CMOS samplers in state-of-the-art ADCs. The transfer function of the sampler during the tracking phase is given by [5]

$$H(f) = \frac{1}{1 + j\omega RC} = \frac{1}{1 + jf/B_{in}} \tag{8.111}$$

where the 3-dB bandwidth of the circuit B_{in} (or its cut-off frequency) is

$$B_{in} = \frac{1}{2\pi RC} \tag{8.112}$$

If the input is a sinusoidal signal given as

$$x(t) = A\sin(2\pi f_{in} t)$$

then, the sampled or tracked voltage at the output is

$$x_{out}(t) = G \times A\sin(2\pi f_{in} t + \theta) \tag{8.113}$$

where the gain term is given by

$$G = \frac{1}{\sqrt{1 + \left(\frac{f_{in}}{B_{in}}\right)^2}} \tag{8.114}$$

and the phase is

$$\theta = -\arctan\left(\frac{f_{in}}{B_{in}}\right) \tag{8.115}$$

If there are M interleaved channels sampling the input, then the kth channel output will be

$$x_k(t) = G_k \times A\sin(2\pi f_{in} t + \theta_k) \tag{8.116}$$

where the gain of the kth channel is given by

$$G_k = \frac{1}{\sqrt{1 + \left(\frac{f_{in}}{B_{in_k}}\right)^2}} \tag{8.117}$$

and the phase of the kth channel is given by

$$\theta_k = -\arctan\left(\frac{f_{in}}{B_{in_k}}\right) \tag{8.118}$$

From (8.117) and (8.118), we see that even though the channels' DC gains are perfectly matched, the bandwidth mismatch results in a frequency-dependent gain mismatch. In addition, bandwidth mismatch also gives a phase skew in a manner similar to the timing mismatch. The effect of the mismatch can be substantially reduced if the bandwidth is much larger than the input frequency.

For the special case of two-channel interleaving (ping-pong), with $M = 2$, we get

$$x_1(t) = G_1 \cdot A \sin(2\pi f_{in} n T_s + \theta_1) \text{ for } n\text{: odd} \tag{8.119}$$

and

$$x_2(t) = G_2 \cdot A \sin(2\pi f_{in} n T_s + \theta_2) \text{ for } n\text{: even} \tag{8.120}$$

Therefore the sampled signal is given by

$$x_{out}(nT_s) = A_f \sin(2\pi f_{in} n T_s + \theta_f) + A_{IL} \sin(2\pi(-f_{in} + f_s/2) n T_s + \theta_{IL}) \tag{8.121}$$

where A_f is the amplitude of the fundamental and the A_{IL} is the amplitude of the interleaving spur, such that

$$A_f = \sqrt{G_c^2 \cos^2(\theta_d) + G_d^2 \sin^2(\theta_d)} \tag{8.122}$$

and

$$A_{IL} = \sqrt{G_c^2 \sin^2(\theta_d) + G_d^2 \cos^2(\theta_d)} \tag{8.123}$$

where the gain terms in (8.122) and (8.123) are given by

$$G_d = (G_1 - G_2)/2 \qquad \text{and} \qquad G_c = (G_1 + G_2)/2 \tag{8.124}$$

The phase terms in (8.122) and (8.123) are given by

$$\theta_d = (\theta_1 - \theta_2)/2 \qquad \text{and} \qquad \theta_c = (\theta_1 + \theta_2)/2 \tag{8.125}$$

The phases of the fundamental θ_f and the interleaving spur θ_{IL} in (8.121) are given by

$$\theta_f = \arctan\left[\frac{G_c \sin(\theta_c)\cos(\theta_d) + G_d \cos(\theta_c)\sin(\theta_d)}{G_c \cos(\theta_c)\cos(\theta_d) - G_d \sin(\theta_c)\sin(\theta_d)}\right] \tag{8.126}$$

and

$$\theta_{IL} = -\arctan\left[\frac{G_c \cos(\theta_c)\sin(\theta_d) + G_d \sin(\theta_c)\cos(\theta_d)}{G_c \sin(\theta_c)\sin(\theta_d) - G_d \cos(\theta_c)\cos(\theta_d)}\right] \tag{8.127}$$

The SINAD due to the bandwidth mismatch is

$$SINAD = 10\log\left(\frac{A_f^2}{A_{IL}^2}\right) = 10\log\left(\frac{G_c^2 \cos^2(\theta_d) + G_d^2 \sin^2(\theta_d)}{G_c^2 \sin^2(\theta_d) + G_d^2 \cos^2(\theta_d)}\right) \tag{8.128}$$

The SFDR is given by

$$SFDR = 10\log\left(\frac{A_f^2}{A_{IL}^2}\right) = 10\log\left(\frac{G_c^2\cos^2(\theta_d) + G_d^2\sin^2(\theta_d)}{G_c^2\sin^2(\theta_d) + G_d^2\cos^2(\theta_d)}\right) \tag{8.129}$$

Therefore, bandwidth mismatch causes both amplitude and phase modulation. The amplitude modulation is frequency-dependent and hence cannot be fixed with a simple gain correction. The complexity of the bandwidth mismatch effects make them the most challenging to fix. However, the impact of the bandwidth mismatch can be reduced by ensuring that the bandwidth is much larger than the input frequency.

8.6 Mismatch summary

In Table 8.2, we summarize the impact of the different kinds of inter-channel mismatch that have been discussed so far. It is important to note that the interleaving spurs, if left uncorrected, would limit the spur-free bandwidth of the interleaved ADC to the Nyquist bandwidth of a *single* channel (i.e. $f_s/2M$). If the interleaving spurs are too large to be acceptable, correcting them is necessary to be able to utilize the Nyquist bandwidth of the whole ADC ($f_s/2$). An example is shown in Figure 8.16, for a two-way interleaving case. The spur-free bandwidth in that case is equal to $f_s/4$, which is the Nyquist bandwidth of a single ADC core.

In spite of this limitation, interleaved ADCs can be very useful in over-sampling situations even if the interleaving spurs are not corrected. If the signal band is properly located in the frequency spectrum and its bandwidth is much smaller than the Nyquist bandwidth, the interleaving spurs can fall out of band and be filtered digitally. Interleaving would still achieve higher sampling rate that simplifies the anti-aliasing filtering and helps improve the performance as discussed in Chapters 1 and 2.

Table 8.2 The impact of different kinds of inter-channel mismatch on the performance of an M-way interleaved ADC, with sampling rate f_s. The input frequency is f_{in}, f_{IL} is the frequency of the interleaving spurs, and the factor k is given by: 1, 2, ..., M − 1

Type of mismatch	Effect on input	Spur location
Offset mismatch	Additive effect	$f_{IL} = \frac{k}{M}f_s$
Gain mismatch	Amplitude modulation	$f_{IL} = \pm f_{in} + \frac{k}{M}f_s$
Timing mismatch	Phase modulation	$f_{IL} = \pm f_{in} + \frac{k}{M}f_s$
Bandwidth mismatch	Frequency-dependent amplitude and phase modulation	$f_{IL} = \pm f_{in} + \frac{k}{M}f_s$

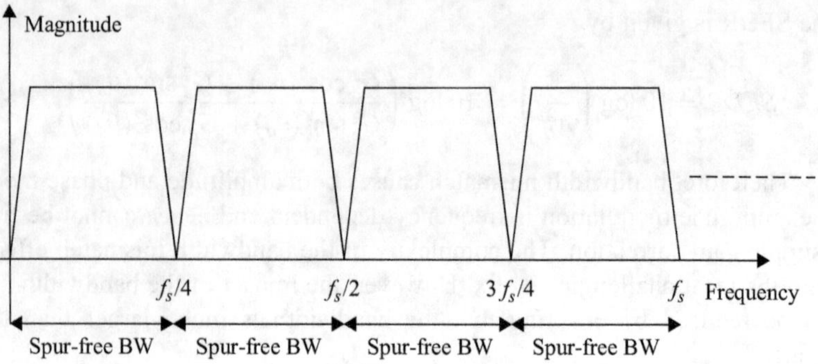

Figure 8.16 An illustration of the spur-free zones in a ping-pong interleaved ADC ($M = 2$). The interleaving spurs can be digitally filtered as long as they are outside the signal band

8.7 Ping-pong special cases

8.7.1 $M = 2$, gain mismatch only

If there is only gain mismatch, the phase error in (8.125) will be

$$\theta_d = 0 \tag{8.130}$$

From (8.122),

$$A_f = \frac{1}{2}\sqrt{G_c^2} = \frac{G_c}{2}$$

From (8.123),

$$A_{IL} = \frac{1}{2}\sqrt{G_d^2} = \frac{G_d}{2}$$

From (8.126),

$$\theta_f = \arctan\left[\frac{\sin(\theta_c)}{\cos(\theta_c)}\right] = \theta_c \tag{8.131}$$

From (8.127),

$$\theta_{IL} = -\arctan\left[\frac{G_d \sin(\theta_c)}{-G_d \cos(\theta_c)}\right] = \theta_c \tag{8.132}$$

Therefore the fundamental and the interleaving spur are in phase. The SINAD and SFDR due to the gain mismatch are given by

$$SINAD = SFDR = 10\log\left(\frac{G_c^2}{G_d^2}\right) \tag{8.133}$$

which matches (8.69) given before as

$$SINAD = SFDR = 20\log\left(\frac{G}{\Delta G}\right) \tag{8.134}$$

8.7.2 $M = 2$, phase mismatch only

If there is only timing mismatch. Equation (8.124) gives

$$G_d = 0 \tag{8.135}$$

Equation (8.122) gives

$$A_f = \frac{1}{2}\sqrt{G_c^2\cos^2(\theta_d)} = \frac{G_c\cos(\theta_d)}{2} \tag{8.136}$$

Equation (8.123) gives

$$A_{IL} = \frac{1}{2}\sqrt{G_c^2\sin^2(\theta_d)} = \frac{G_c\sin(\theta_d)}{2} \tag{8.137}$$

The phase of the fundamental in (8.126) is given by

$$\theta_f = \arctan\left[\frac{\sin(\theta_c)}{\cos(\theta_c)}\right] = \theta_c \tag{8.138}$$

The phase of the interleaving spur from (8.127) is

$$\theta_{IL} = -\arctan\left[\frac{\cos(\theta_c)}{\sin(\theta_c)}\right] = \theta_c - \frac{\pi}{2} \tag{8.139}$$

That is, the interleaving spur due to timing mismatch is orthogonal to the fundamental. It is also orthogonal to the interleaving spur due to the gain mismatch. The SINAD and SFDR due to the phase mismatch are given by

$$SINAD = SFDR = 10\log\left(\frac{A_f^2}{A_{IL}^2}\right) = 20\log(\cot\theta_d) \tag{8.140}$$

where,

$$\theta_d = (\theta_1 - \theta_2)/2 \tag{8.141}$$

If the phase mismatch is due to timing mismatch, the phase difference is given by

$$\theta_d = 2\pi f_{in}(t_1 - t_2)/2 = 2\pi f_{in}(t_1 - t_2)/2 = \pi f_{in}\Delta\tau \tag{8.142}$$

where

$$\Delta\tau = (t_1 - t_2), \qquad \text{and} \qquad \Delta t = (t_1 - t_2)/2 \tag{8.143}$$

Therefore,

$$SINAD = SFDR = 20\log(\cot \pi f_{in}\Delta\tau) = 20\log(\cot 2\pi f_{in}\Delta t) \tag{8.144}$$

This represents the derivation of the exact formula that was presented before in (8.94).

8.8 Non-linearity mismatch

In the previous sections, we discussed offset, gain, timing, and bandwidth mismatches in time-interleaved ADCs. In this section, we discuss an often overlooked kind of mismatch, which is the non-linearity mismatch. That is due to mismatches in the differential non-linearity (DNL) and integral non-linearity (INL) of the individual channels. It can also be thought of as mismatches in the non-linear distortion (harmonics) of the individual channels. The effect of these mismatches tends to be negligible if the non-linearity and the resulting harmonics are very small. However, as the linearity of the individual channels degrade, the mismatches in those non-linearities can be significant, and would need to be corrected as well.

If we represent the signal of the *i*th channel as a function of its non-linear terms, we obtain [12]

$$x_i^{\sim}(t) = \sum_q a_{i,q} x^q(t) \tag{8.145}$$

where q is the order of the non-linear distortion and $a_{i,q}$ is a coefficient that depends on the channel and the order of the non-linear term. In the frequency domain, the sampled signal is represented as [12]

$$X_{si}^{\sim}(f) = \frac{1}{MT_s}\sum_{k=-\infty}^{\infty} X_i^{\sim}\left(f - \frac{k}{MT_s}\right)e^{-j2\pi ik/M} \tag{8.146}$$

where

$$X_i^{\sim}(f) = \sum_q a_{i,q} X^{(*q)}(f) \tag{8.147}$$

and $X^{(*q)}(f)$ is the $(q-1)$-fold convolution of $X(f)$ with itself, given by

$$X^{(*q)}(f) = \begin{cases} \delta(f) & \text{for } q = 0 \\ X(f) & \text{for } q = 1 \\ \left(\dfrac{1}{2\pi}\right)^{q-1}[X(f) * X(f) * \ldots * X(f)] & \text{for } q \geq 2 \end{cases} \tag{8.148}$$

where $X(f) * X(f) * \ldots * X(f)$ is the $(q-1)$-fold convolution of $X(f)$ with itself. The interleaved output is given by

$$X^{\sim}(f) = \frac{1}{MT_s} \sum_{k=-\infty}^{\infty} \sum_{i=0}^{M-1} X_i^{\sim}\left(f - \frac{k}{MT_s}\right) e^{-j2\pi ik/M} \tag{8.149}$$

This shows that the interleaving spurs appear at integer multiples of $1/MT_s$. If we consider the qth order non-linearity (or harmonic) of the individual ADC, interleaving M channels with those mismatched non-linearities result in additional harmonics in the following locations:

$$f_{IL} = \pm q f_{in} + \frac{k}{M} f_s \tag{8.150}$$

for $k = 1, 2, \ldots, M-1$.

It is important to note that the higher-order non-linearities can contribute to the interleaving spurs at the locations of the lower-order non-linearities. This is due to the folding of higher-order non-linear terms onto lower-order harmonics. So the qth order non-linearity contributes to the spur at the following locations:

$$f_{IL} = \pm l f_{in} + \frac{k}{M} f_s \tag{8.151}$$

where:

$$l = \begin{cases} 0, 2, 4, \ldots, q-2 & \text{for even } q \\ 1, 3, 5, \ldots, q-2 & \text{for odd } q \end{cases}$$

Therefore, the mismatches in the non-linearity can indeed degrade the interleaving spurs, especially if the levels of the harmonics are high.

8.9 Improving performance

Time-interleaved ADCs have been a topic of research for a long time. Multiple approaches have been proposed and used to improve their performance. Those include analog, digital, and mixed-signal techniques. Offset and gain mismatches tend to be relatively easy to handle compared to the timing/phase mismatches, which are challenging to fix. Bandwidth mismatch is very difficult to correct due to its multiple manifestations. It is preferable to avoid its impact by maximizing the bandwidth. In this section we discuss some of the methods employed to improve the performance of interleaved ADCs.

8.9.1 Improving matching

The most obvious approach to reduce the interleaving spurs is to improve the matching between the interleaved channels. This is usually done using best

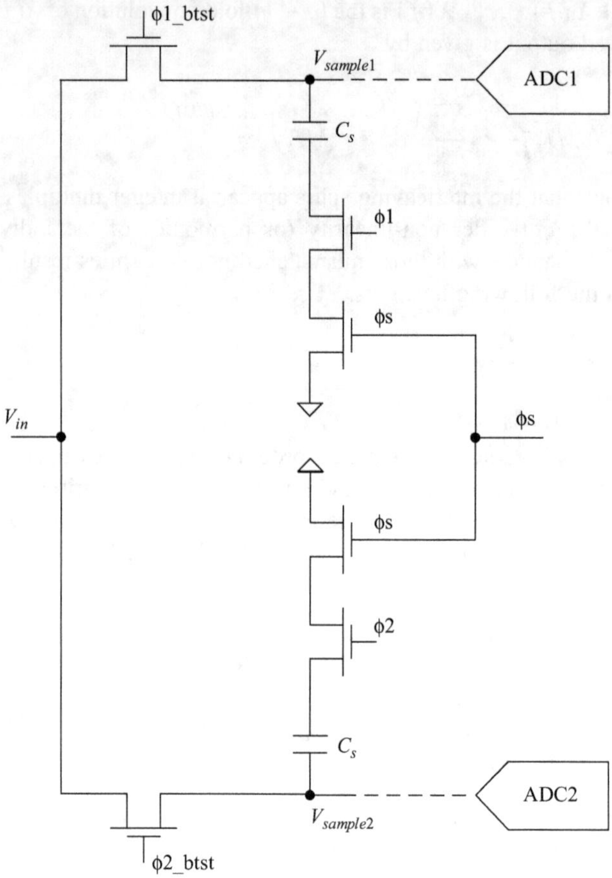

Figure 8.17 An illustration of using a common clock for sampling to reduce the timing mismatch between two interleaved ADCs

practices in circuit design and layout techniques. Physical proximity, sharing circuits when possible, and matching the routing are examples of methods that have been employed. However, it is unrealistic to expect that good practices alone will be enough to achieve moderate or high performance. As the number of channels increases, analog matching becomes very difficult beyond 4–6 bits of accuracy.

A commonly used technique to reduce the timing mismatch is to use a common clock for sampling in all the channels. This is shown in Figures 8.17 and 8.18, where the sampling instant of each channel is determined by the common clock ϕs operating at the full sampling rate. However, although this approach helps reduce timing mismatch errors, the routing of the ϕs signal to the various channels may still be slightly mismatched. Layout techniques, such as star-routing, help reduce those mismatches. However, the routing parasitic mismatches usually limit the matching to about 6–8 bits of accuracy.

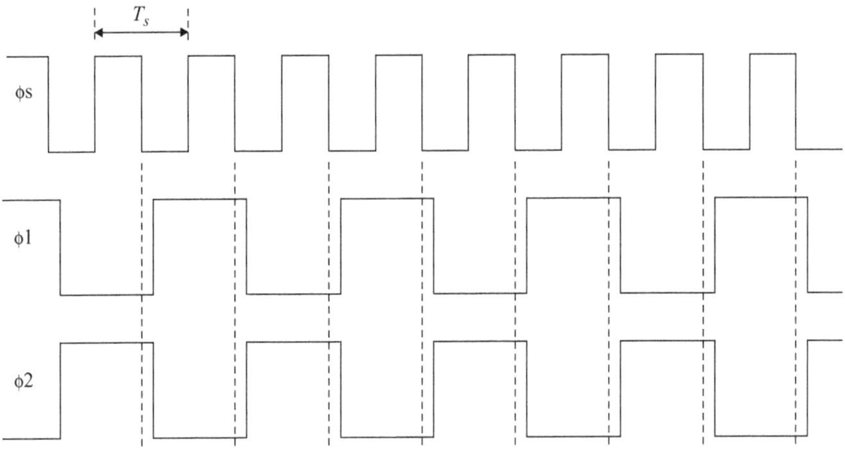

Figure 8.18 A simplified timing diagram showing a common clock for sampling to reduce the timing mismatch between two interleaved ADCs

Another often used approach is to share the inter-stage amplifiers between neighboring channels, when interleaving pipelined ADCs. This is usually done to reduce power consumption, although it has the added benefit of reducing the mismatches between the two channels that share the amplifiers [11].

8.9.2 Using a full-speed sample and hold

Since timing and bandwidth mismatches are the most difficult to fix, ADC designers often resort to using a front-end S/H circuit that operates at the full interleaved sampling rate (f_s). This common S/H circuit provides each channel with a held signal, which practically eliminates the phase mismatches. This is shown in Figure 8.19. Unfortunately, this comes at the expense of significant power consumption and noise degradation, because such a S/H circuit is usually a substantial contributor to the power and noise of the ADC.

8.9.3 Calibration

One of the commonly used methods to improve the performance of time-interleaved ADCs is to calibrate the mismatches. This can be done in the digital domain, analog domain, or both [9]. It can be done in the factory, in the foreground, or in the background.

8.9.3.1 Analog

In analog calibration, circuit parameters in the analog domain are adjusted to minimize mismatches. These can be offset or gain parameters that are adjusted by changing resistors, capacitors, or other circuit elements. The adjustment can be done by switching different components in and out, or by trimming their values. In addition, timing can be calibrated by adjusting clock delays using switchable

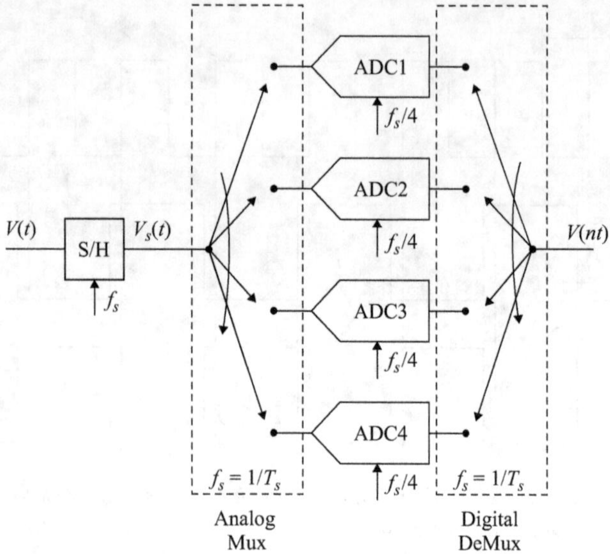

Figure 8.19 A conceptual block diagram showing the use of a common full-speed Sample and Hold circuit (S/H) to eliminate timing mismatch between the two channels

capacitors or delay elements in the clock path. They can also be adjusted by adjusting voltage levels that control the clock or threshold levels.

8.9.3.2 Digital

In digital calibration, the mismatches are corrected in the digital domain. This can be easily done for offset and gain mismatches by addition and multiplication of offset and gain parameters, respectively. It can also be employed to fix the timing mismatch, although this is usually much more challenging.

Since the timing adjustment is much smaller than the sampling period, digital interpolation and filtering are needed to estimate the signal variation over small time steps, and hence correct for small timing mismatches. This can be quite complex and can limit the bandwidth of correction compared to the Nyquist bandwidth, due to the limitation in the interpolation accuracy. Therefore, it is often more desirable to correct the timing and bandwidth mismatches in the analog domain, if high accuracy and wide correction bandwidth are needed [6–8].

8.9.3.3 Mixed signal

In mixed signal calibration, some of the mismatches (such as offset and gain) are corrected in the digital domain, while others (such as timing) are corrected in the analog domain. In addition, the timing calibration can be estimated digitally and corrected in an analog fashion. The digital and mixed signal calibration approaches are discussed in more detail in Chapter 9 as examples of the digitally assisted analog techniques employed in high speed ADCs.

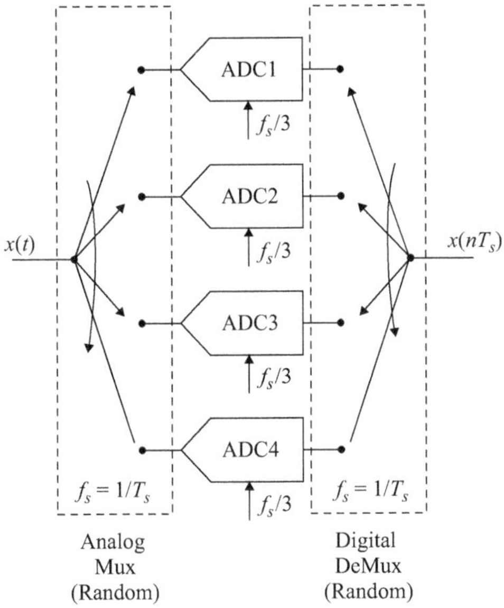

Figure 8.20 A conceptual block diagram showing the use of an extra channel to be used for randomization in a three-way interleaved ADC

8.9.4 Randomization

In spite of the best efforts to calibrate and fix channel mismatches, there may be limitations and residual errors that can't be effectively fixed. Randomization is proposed as a technique that improves the spur levels resulting from channel mismatches by smearing them in the noise floor. An example is shown in Figure 8.20 for a three-way interleaved ADC, where an additional channel is added to enable the random selection of the channels that are being used at any time. Conceptually, this should eliminate the interleaving spurs by spreading their energy in the noise floor. It is important, however, to note that randomization improves the SFDR but does not improve the SINAD. Since the residual mismatches are not fixed, their errors still exist. Randomization simply alleviates the periodicity of the channel selection and hence spreads the interleaving spurs in the noise floor.

At least one additional channel is needed in order to enable the random pattern of selection. For example, if we interleave two channels, a random pattern cannot be achieved without an additional third channel, because no channel is fast enough to process two successive samples. This represents a significant overhead in power, area, and complexity. An example of a 4-way interleaving case is shown in Figure 8.21 with randomization applied in Figure 8.22. The randomization tends to raise the noise floor and, if it is not fully random, can lead to noise humps (coloring) as shown in Figure 8.23. This is discussed further in Chapter 9.

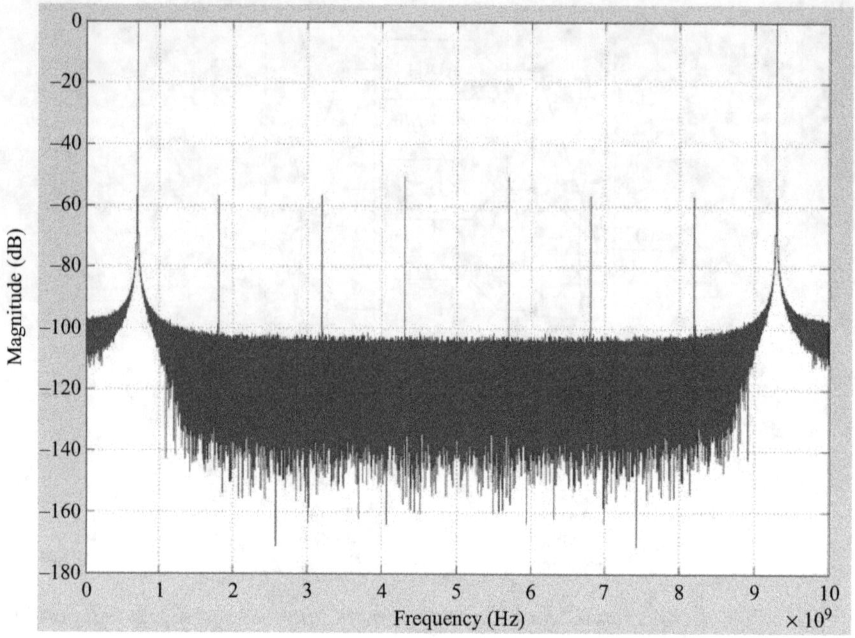

Figure 8.21 Interleaving spurs in a 4-way interleaved ADC

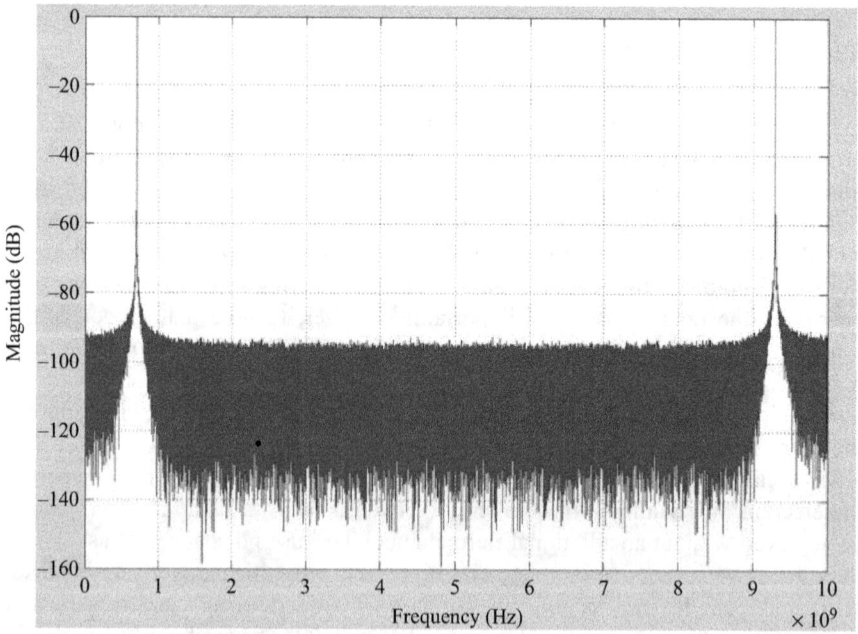

Figure 8.22 Randomization spreads the interleaving spurs in the noise floor. The noise floor is higher than that in Figure 8.21

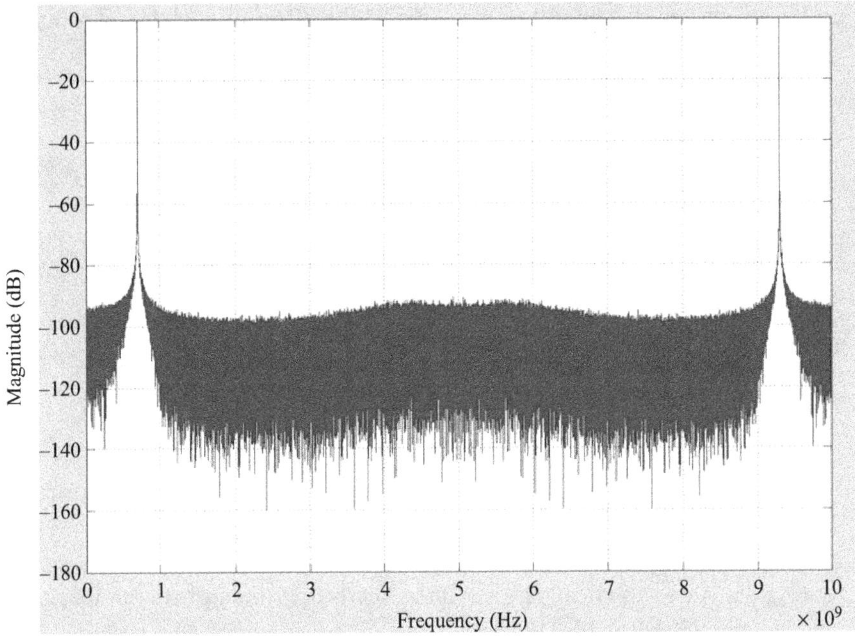

Figure 8.23 *The effect of "practical" randomization, which is not completely random. It spreads the interleaving spurs in the noise floor around the location of the original spurs, resulting in possible noise "humps"*

8.10 Conclusion

In this chapter, we discussed the time-interleaved ADC architecture. Detailed analysis and intuitive perspectives were presented. The different kinds of mismatches and their impact on performance were explained. Some of the implementation approaches and trade-offs were also discussed. In the next chapter, some calibration techniques used to fix the interleaving spurs are presented.

Problems

1. Using any programming language, model the impact of offset mismatch in two-way interleaving. Plot the FFT at the output for 1 mV offset mismatch.
2. Repeat Problem 1 for four-way and eight-way interleaving, assuming the offset alternates between 0.5 mV and −0.5 mV between the interleaved channels. Note the locations of the interleaving spurs. Does the level or frequency location of the interleaving spurs depend on the input amplitude or frequency? Discuss.

3. Using any programming language, model the effect of 0.1% gain mismatch on a 100 MHz input sine wave with unity amplitude, for:
   (a) Two-way interleaving.
   (b) Four-way interleaving (assume alternating gain between 0.5% and −0.5% between channels).
   (c) Eight-way interleaving (assume alternating gain between 0.5% and −0.5% between channels).
   (d) Plot the FFTs in every case.
   (e) Plot the phases of the main signal and the images in every case.
   (f) What are the SINAD and SFDR in every case?
4. Repeat Problem 3 for the following cases:
   (a) Input amplitude = 0.5 V, and input frequency = 100 MHz.
   (b) Input amplitude = 0.5 V, and input frequency = 200 MHz.
   (c) Input amplitude = 2 V, and input frequency = 200 MHz.
   (d) What are the SINAD and SFDR in every case?
5. Using any programming language, model the effect of 1 ps timing mismatch on a 100 MHz input sine wave with unity amplitude, for:
   (a) Two-way interleaving.
   (b) Four-way interleaving (assuming alternating timing between 0.5ps and −0.5ps between channels).
   (c) Eight-way interleaving (assuming alternating timing between 0.5ps and −0.5ps between channels).
   (d) Plot the FFTs in every case.
   (e) Plot the phases of the main signal and the images in every case.
   (f) What are the SINAD and SFDR in every case?
6. Add an extra channel, and apply randomization for every case in Problem 5.
7. Draw the block and timing diagrams of two-way interleaved 10-bit pipelined ADCs. Each core has 4 bits/stage and a four-bit backend flash.
8. Repeat Problem 7, if the inter-stage amplifiers are shared between the two channels.
9. Repeat Problem 7, for four-way interleaving.
10. Draw the block and timing diagrams of two-way interleaved 8-bit SAR ADCs.
11. Repeat Problem 10, for four-way and eight-way interleaving.
12. Discuss your observations from Problems 9 and 11.

References

[1] W.C. Black and D.A. Hodges, "Time Interleaved Converter Arrays," *IEEE Journal of Solid-State Circuits*, SC-15(6), pp. 1022–1029, Dec 1980.

[2] M. Gustavsson, J.J. Wikner, and N.N. Tan, "CMOS Data Converters for Communications," Kluwer Academic Publishers, Norwell, MA, 2000.

[3] C. Vogel and G. Kubin, "Analysis and Compensation of Nonlinearity Mismatches in Time-Interleaved ADC Arrays," *IEEE ISCAS*, pp. I-593–I-596, 2004.

[4] C. Vogel, "The Impact of Combined Channel Mismatch Effects in Time-Interleaved ADCs," *IEEE* Transactions on *Instrumentation and Measurement*, 54(1), pp. 415–427, Feb 2005.
[5] N. Kurosawa, H. Kobayashi, K. Maruyama, *et al.*, "Explicit Analysis of Channel Mismatch Effects in Time-Interleaved ADC Systems," *IEEE Transactions on Circuits and Systems-I: Fundamental Theory and Applications*, 48(3), pp. 261–271, Mar 2001.
[6] S.M. Jamal, D. Fu, N.C.-J. Chang, *et al.*, "A 10-b 120-Msample/s Time-Interleaved Analog-to-Digital Converter with Digital Background Calibration," *IEEE Journal of Solid-State Circuits*, 37(12), pp. 1618–1627, Dec 2002.
[7] T. Laakso, V. Valimaki, M. Karjalainen, *et al.*, "Splitting the Unit Delay – Tools for Fractional Delay Filter Design," *IEEE Signal Processing Magazine*, 13(1), pp. 30–60, Jan 1996.
[8] M. El-Chammas and B. Murmann, "A 12-GS/s 81-mW 5-bit Time-Interleaved Flash ADC with Background Timing Skew Calibration," *IEEE Journal of Solid-State Circuits*, 46(4), pp. 838–847, Apr 2011.
[9] M. El-Chammas, X. Li, S. Kimura, *et al.*, "A 12 Bit 1.6 GS/s BiCMOS 2×2 Hierarchical Time-Interleaved Pipeline ADC," *IEEE Journal of Solid-State Circuits*, 49(9), pp. 1876–1885, Sep 2014.
[10] Y.-C. JenQ, "Digital Spectra of Nonuniformly Sampled Signals: Fundamentals and High-speed Waveform Digitizers," *IEEE Transactions on Instrumentation and Measurement*, 37(2), pp. 245–251, Jun 1988.
[11] W. Bright, "8b 75MSample/s Parallel Pipelined ADC Incorporating Double Sampling," *IEEE ISSCC Digest of Technical Papers*, pp. 146–147, 1998.
[12] C. Vogel and G. Kubin, "Analysis and compensation of nonlinearity mismatches in time-interleaved ADC arrays," *IEEE ISCAS*, pp. I-593–I-596, 2004.

Chapter 9
Digitally assisted converters

The never-ending quest for converters with higher sampling rates, wider bandwidths, lower cost, and higher integration has progressively pushed the process technology of high speed ADCs to finer lithographies. Shorter length MOS devices have higher transconductance, smaller resistance, and lower parasitics. However, they suffer from smaller dynamic range, lower intrinsic gain, and lower output impedance. These limitations make it difficult to achieve good linearity and signal-to-noise ratio with reasonable power consumption, while pushing the sampling rates up. Digital assistance (i.e. calibration) has emerged as one of the primary tools to improve performance and reduce power consumption, especially in fine lithography processes where digital processing is more efficient. The goal has been to relax the requirements on the analog side, and correct for the resulting errors digitally.

Innovation in the field of digitally assisted converters has a long and rich history [1]. In the industry, it has evolved from factory calibration, which is used during production testing, into complex signal processing algorithms that operate continuously in the background to fix analog imperfections while adapting to a changing environment. Nevertheless, analog designers should not rejoice (or panic, depending on their perspective) yet. Employing digital assistance requires knowledge of signal processing and mixed-signal design to fully take advantage of the process technology's analog strength while circumventing its weaknesses. Successful implementation of a digitally assisted ADC involves understanding the analog imperfections and optimizing the analog circuits to be compatible with digital assistance. It has proven to be at least as challenging as traditional ADC design. So, in spite of the digital assistance, there will still be ample problems that need the creativity and resourcefulness of good analog designers.

Two of the architectures that have seen significant research in digital assistance are the pipelined and the time-interleaved ADCs. In pipelined ADCs, the MDAC is the keystone of the ADC's performance. It also determines the power consumption of the ADC because the MDAC is usually its major power consumer. Relaxing the design requirements of the amplifier has been an important area of research in digitally assisted ADCs to enable higher sampling rates, higher performance, and lower power consumption.

In time-interleaved converters, digital assistance is employed to correct the interleaving spurs caused by inter-channel mismatches. These include offset, gain, and timing mismatches. Estimating and correcting these mismatch errors in the background is a challenging design problem.

In general, the difficulty in the background calibration of ADCs is usually not in correcting the error, but in estimating it accurately and seamlessly without disrupting normal operation. The correction can be done in the analog or digital domains, with the digital correction being easier and more efficient. One exception is the correction of timing mismatches in interleaved ADCs, where the digital correction can suffer from timing resolution and bandwidth limitations that do not exist in the analog correction.

In this chapter, we cover some of the calibration techniques used in pipelined and time-interleaved ADCs. It is important to note that many of the calibration techniques that are discussed in the context of pipelined ADC are applicable to other multi-step ADC architectures as well. Moreover, they can also be applied to MASH sigma-delta ADCs, because of their similarity to the pipeline architecture as noted in Chapter 3.

9.1 Calibration of pipelined ADC non-linearity

The pipelined ADC architecture is discussed in Chapter 7, and shown in Figure 9.1. A typical switched capacitor MDAC implementation is depicted in Figure 9.2 in a single-ended form for simplicity. The performance of the pipelined ADC's quantization is determined to a large extent by the accuracy of its MDACs, with their residue amplifier (RA) being the main design bottleneck.

As discussed in Chapter 7, the ideal residue of a stage (say the ith stage) can be represented as

$$V_{oi} = G_i(V_{in_i} - V_{dac_i}) = G_i\left(V_{in_i} - \frac{D_i \times V_{Ref}}{2^{k_i-1}}\right) \tag{9.1}$$

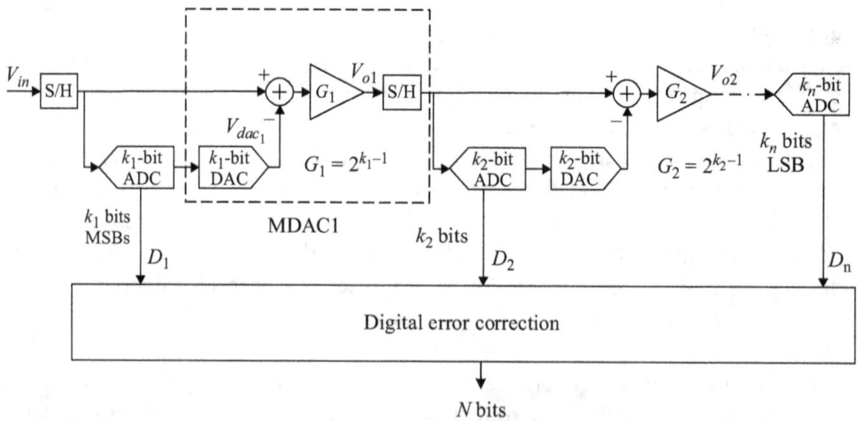

Figure 9.1 A basic pipelined ADC with redundancy

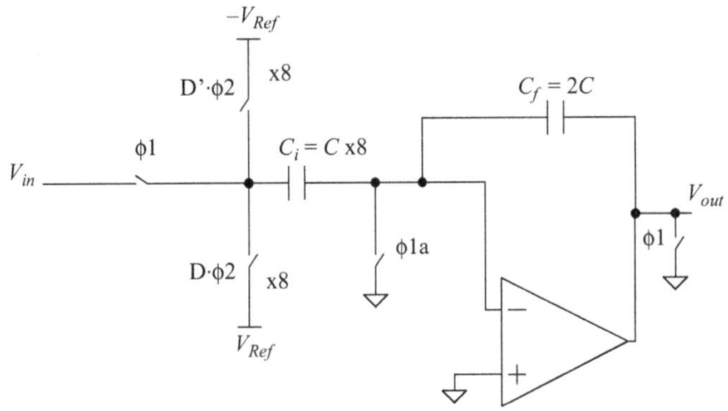

Figure 9.2 A typical switched capacitor MDAC shown in single-ended form for simplicity

where V_{in_i} is the input of the ith stage, G_i is the gain of the ith stage, and D_i is the digital code of the ith stage which is given by

$$D_i = 0, \pm 1, \pm 2, \ldots, \pm 2^{k_i - 1}, \tag{9.2}$$

V_{Ref} is the reference voltage, and k_i is the number of bits in the ith stage. In the presence of inter-stage gain error (IGE), the output can be represented as

$$V_{oi} = G_i^{'}(V_{in_i} - V_{dac_i}) = \frac{G_i^{'}}{G_i} V_{oi}\big|_{\text{ideal}} \tag{9.3}$$

where $G_i^{'}$ is the actual gain of the ith stage. Therefore, we can retrieve the ideal output value by applying the inverse of the gain to the output value, such that

$$V_{oi}\big|_{\text{ideal}} = \frac{G_i}{G_i^{'}} V_{oi} \tag{9.4}$$

Therefore, in the digital domain this is represented in terms of the digital residue $D(V_{oi})$ as follows

$$D(V_{oi})\big|_{\text{Cal}} = \frac{G_i}{G_i^{'}} D(V_{oi}) \tag{9.5}$$

where $D(V_{oi})|_{\text{Cal}}$ is the calibrated residue of the ith stage in the digital domain.

In the presence of DAC errors, (9.1) becomes

$$V_{oi} = G_i(V_{in_i} - V_{dac_i}^{'}) = G_i\left(V_{in_i} - \alpha_{dac_i}(D_i)\frac{D_i \times V_{Ref}}{2^{k_i-1}}\right) \tag{9.6}$$

where $\alpha_{dac_i}(D_i)$ is the code-dependent DAC error term. The correction in the digital domain can be done using code-dependent addition of correction terms as follows

$$D(V_{oi})\big|_{\text{Cal}} = D(V_{oi}) + G_i\left(\alpha_{dac_i}(D_i)\frac{D_i \times V_{Ref}}{2^{k_i-1}} - \frac{D_i \times V_{Ref}}{2^{k_i-1}}\right) \tag{9.7}$$

Therefore, using the concepts described in (9.5) and (9.7), the quantization errors can be corrected in every stage of the pipeline. There are numerous techniques that apply these concepts to detect and correct those quantization errors. In the rest of this section, several calibration techniques of the MDAC and its RA are discussed.

9.1.1 Factory and foreground calibration

The earliest and most commonly used form of calibration is the factory calibration. In ADCs with an algorithmic nature, such as pipelined, SAR, and cyclic ADCs, some quantizer's non-linearities can be easily corrected in the digital back end. For example, in pipelined ADCs, as discussed in Chapter 7, the bits from the various stages are combined with the proper weighting as shown in Figure 9.3. The inter-stage gain applied in the digital domain must match the inverse of the inter-stage gain on the analog side with the required accuracy. If the analog gain is a power of 2, the corresponding digital gain can be implemented by simple bit shifts as shown in Figure 9.4 for a 14-bit converter with 3-bit stages.

An IGE in the analog domain results in a saw-tooth pattern in the integral non-linearity (INL) as shown in Figures 9.5 and 9.6. The INL pattern in Figure 9.5 shows an inter-stage gain that is larger than the ideal value, while Figure 9.6 depicts an INL for an inter-stage gain that is smaller than the ideal value. Either way, the IGE can be corrected by applying the gain value that matches the inverse of the analog inter-stage gain to the stage's residue in the digital back end. This was shown in (9.5) and is depicted conceptually in Figures 9.7 and 9.8. In Figure 9.7,

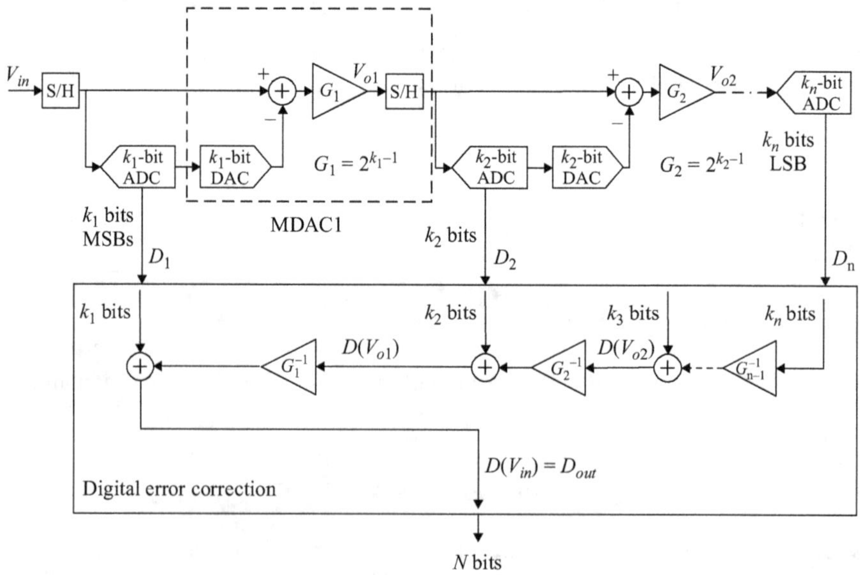

Figure 9.3 Combining the bits of the pipeline stages with redundancy and digital error correction. The gains on the digital side must match the gains on the analog side

Stage-1: $k_1 = 3$:	$d_{13}d_{12}d_{11a}$
Stage-2: $k_2 = 3$:	$d_{11b}d_{10}d_{9a}$
Stage-3: $k_3 = 3$:	$d_{9b}d_8d_{7a}$
Stage-4: $k_4 = 3$:	$d_{7b}d_6d_{5a}$
Stage-5: $k_5 = 3$:	$d_{5b}d_4d_{3a}$
Stage-6: $k_6 = 4$:	$d_{3b}d_2d_1d_0$
Output: 14-bits:	$d_{13}d_{12}d_{11}\ d_{10}d_9\ d_8d_7\ d_6d_5\ d_4d_3\ d_2d_1d_0$

Figure 9.4 Combining the bits of the pipeline stages with redundancy and digital error correction for a 14-bit converter. When the gain is a power of 2, multiplication and division in the digital domain is achieved by bit shifting

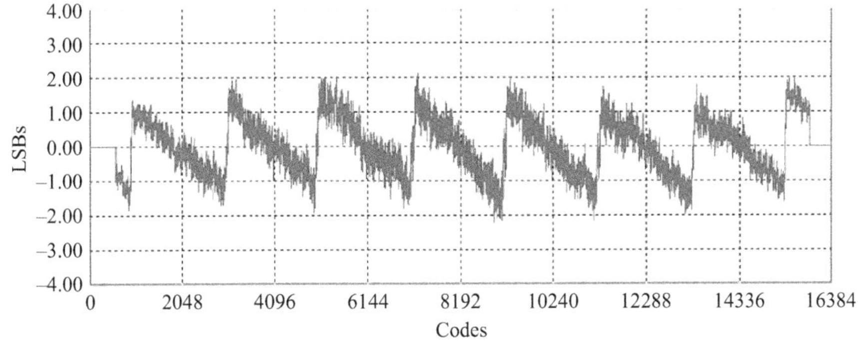

Figure 9.5 An INL showing an inter-stage gain error in stage-1, where the gain is larger than the ideal value

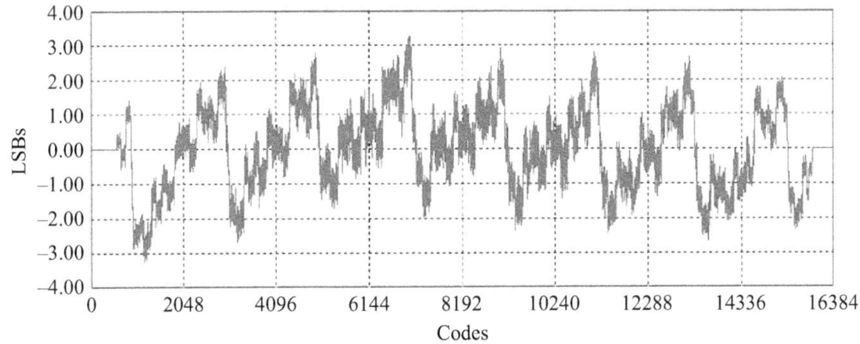

Figure 9.6 An INL showing an inter-stage gain error in stage-1, where the gain is smaller than the ideal value. We can also see inter-stage gain error in stage-2, which appears as saw-tooth pattern within each stage-1 sub-range

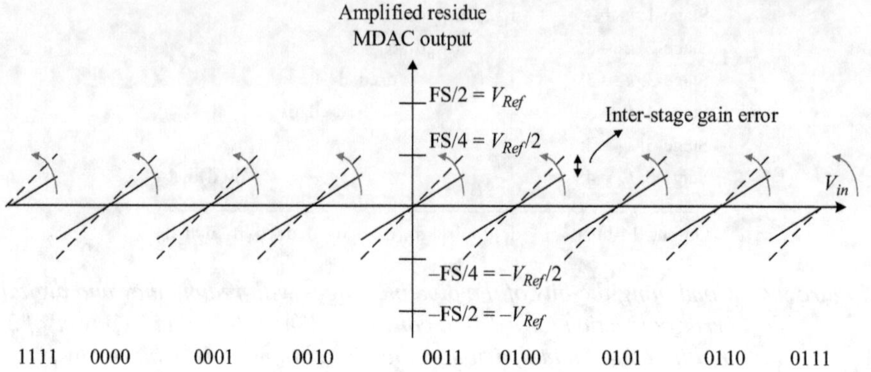

Figure 9.7 A stage-1 residue with a gain error. The arrows show the impact of a gain error fix in the digital domain

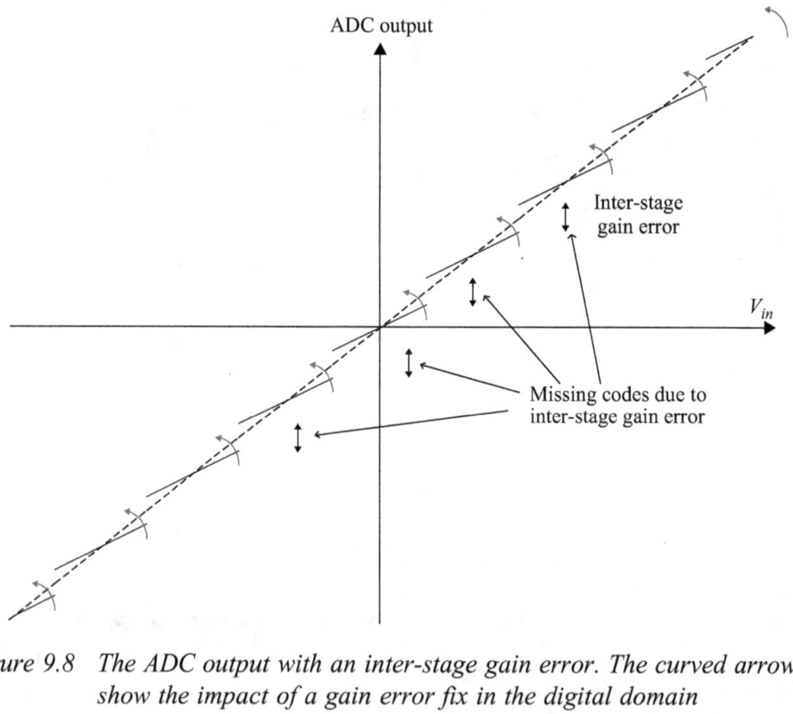

Figure 9.8 The ADC output with an inter-stage gain error. The curved arrows show the impact of a gain error fix in the digital domain

a residue with IGE is shown and the fix is done by multiplication with a gain correction factor that rotates the sub-ranges to match the correct slope. The resulting output characteristic is plotted in Figure 9.8.

Alternatively, since digital multipliers are expensive, the correction can be implemented using sub-range-dependent addition, where the sub-ranges are shifted to achieve alignment. This approach is shown in Figure 9.9 and fixes the IGEs, but

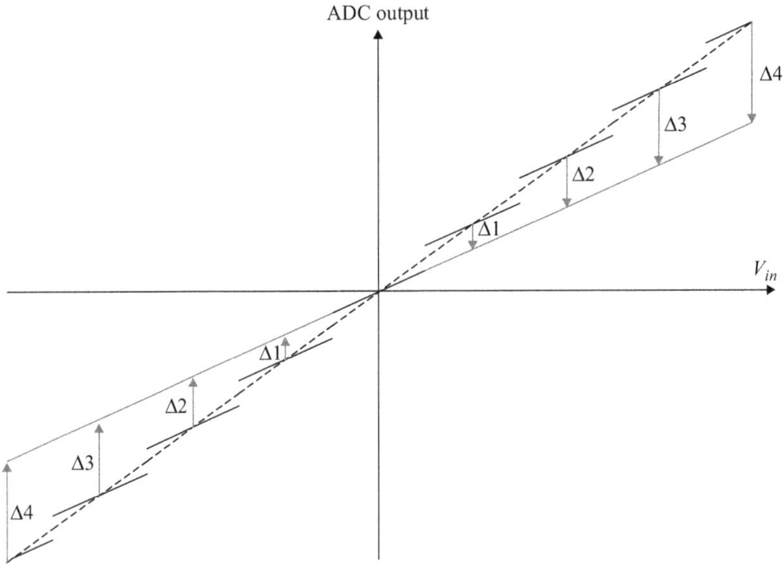

Figure 9.9 The ADC output with an inter-stage gain error. The arrows show the impact of a gain error fix in the digital domain using code-dependent additions instead of multiplications. The result is an overall gain error of the whole ADC shown as a difference in slope between the solid line and the dotted line

leads to a net overall gain error for the ADC that can be corrected using a single multiplier on the final digital word. Alternatively, the resulting ADC gain error can be corrected by adjusting the ADC reference on the analog side.

In Figures 9.5–9.7, we should note that the number of sub-ranges in the INL and residue plots is 9 sub-ranges (7 sub-ranges plus two half sub-ranges), not 8 as implied by using 3-bits in the first stage. This is due to the mid-tread implementation of the sub-ADC, and the extra comparator added to "fold" the end sub-ranges and limit of the output of the MDAC to stay within half the correction range, as mentioned in Chapter 7. The remaining unused half sub-ranges can be utilized to inject dither, as discussed later in this chapter.

In addition to IGEs, the MDAC can suffer from DAC errors due to capacitor mismatches, code-dependent settling errors, as discussed in Chapter 7. These errors cause shifts in the individual sub-ranges of the residue, as shown in Figure 9.10, and manifest themselves as shifts in some segments of the INL as shown in the example of Figure 9.11. They can be corrected by shifting the corresponding segments in the digital domain by the appropriate value. This was shown in (9.7).

Foreground calibration is often implemented in the factory during production test by observing the INL, or by forcing each comparator to switch between two different decisions to measure the corresponding "jump" in the output residue.

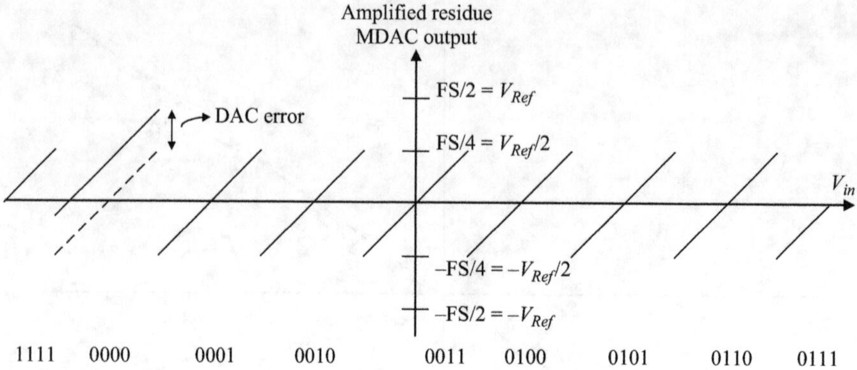

Figure 9.10 A residue showing a DAC error in stage-1. It can be fixed using code-dependent addition in the digital domain

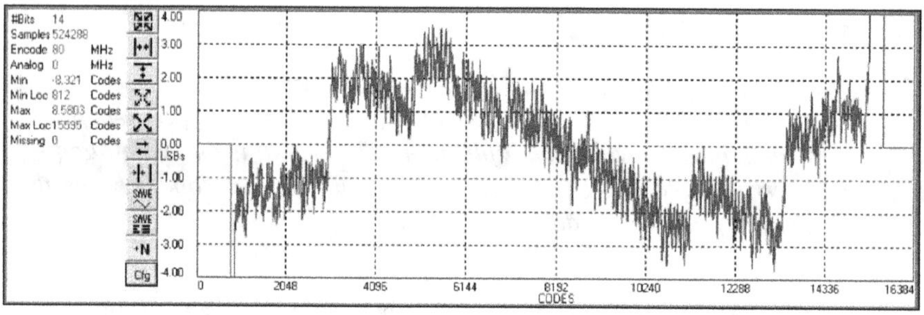

Figure 9.11 An INL showing DAC errors in stage-1

Factory calibration holds as long as the errors are constant and do not depend on temperature, supply, aging, etc. Capacitor mismatches tend to fall under this category and therefore are good candidates to fix using factory calibration, with digital coefficients that are fused permanently using non-volatile memory (NVM).

Foreground calibration can also be initiated by the user of the ADC in the form of "self-calibration." An example is described in Reference 1, where an internal calibration signal is applied, the DAC and IGE errors are measured, and the correction coefficients are applied without external intervention. These approaches require interrupting the normal operation of the ADC, which may be acceptable in some applications. They are effective in fixing constant errors, but not errors that vary with temperature, supply, aging, or sampling rate. These limitations created the need for adaptive calibration techniques that run continuously in the background to correct the errors while adapting to changes without the user's intervention or disrupting the ADC's operation. Examples of these background calibration algorithms are described in the following sections.

9.1.2 Correlation-based calibration

Correlation-based calibration is an effective technique for the background correction of IGEs in pipelined ADCs that are due to insufficient open loop gain in the amplifier, linear settling errors, and capacitance mismatches [2–5]. It can be also extended to correct for memory errors, kick-back errors, and the amplifier's non-linearity. Moreover, it can be applied to other multi-step architectures, such as SAR ADCs, to correct for their DAC errors.

Instead of disrupting the operation of the ADC, this approach injects a pseudo-random (PN or pseudo-noise) calibration signal that is uncorrelated with the input signal. This PN signal (sometimes called dither signal) is typically added to the input signal in either the MDAC *or* the flash of the stage to be calibrated, but not both. Since it passes through the same path as the DAC signal, it encounters the same non-idealities and hence can detect the IGE. Using a statistical correlator, or the Least Mean Square (LMS) algorithm, the PN signal is "correlated out" in the digital back end, and the IGE is estimated in the process. The LMS algorithm for estimating the gain error is given as follows

$$G_e[n+1] = G_e[n] + \mu \times V_d[n] \times (V_R[n] - V_d[n] \times G_e[n]) \tag{9.8}$$

where $G_e[n]$ is the nth sample estimate of the inter-stage gain, μ is the algorithm's step size, V_d is the ideal PN (dither) digital signal, and V_R is the digital residue (output) of the stage being calibrated. The step size (μ) controls the accuracy and convergence time of the algorithm. A small μ leads to higher accuracy and longer convergence time.

Figure 9.12 depicts a pipelined ADC with correlation-based calibration of the IGE in the first two stages, where two PN signals (dither signals) are injected

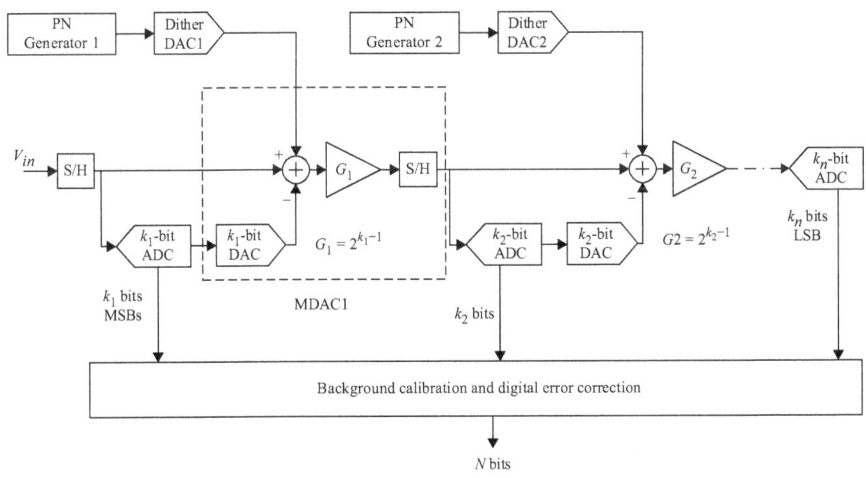

Figure 9.12 A block diagram showing the structure of a pipelined ADC with correlation-based calibration of the inter-stage gain error in the first two stages

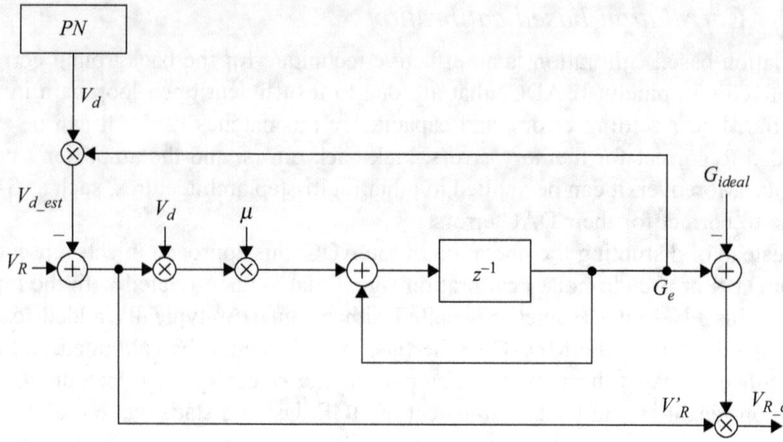

Figure 9.13 A block diagram showing the calibration and dither subtraction performed in the digital block using the LMS algorithm

the MDACs. These two dither signals must be uncorrelated with the input signal and with each other. If more stages need to be calibrated, additional uncorrelated PN signals will need to be injected in every stage that needs calibration. Although the convergence can proceed simultaneously for all the stages, the correction of a back-end stage must be applied before its bits propagate to a front-end stage.

Figure 9.13 shows the gain estimation and dither subtraction using the LMS algorithm described in (9.8). The digital residue signal V_R is obtained from the back-end pipeline with all the needed correction applied to make it as accurate a representation of the residue as possible. The dither is multiplied by the estimate of the gain and is subtracted from the residue signal. The result is multiplied by the ideal dither and by μ, then passes through an accumulator to give an estimate for the inter-stage gain G_e. The dither estimate is subtracted from the residue, which is corrected by the estimate of the gain error to give the calibrated residue V_{R_cal}. This LMS operation constitutes a feedback loop whose bandwidth is controlled by the step size μ. A large μ gives a large bandwidth, fast convergence, and low accuracy, and vice versa. Unlike feed-forward correlation approaches [2, 3], the LMS algorithm gives a smooth convergence of the gain estimate toward the final value, without the update "jumps" associated with the window-based feed-forward approaches [4, 5].

The injection of the calibration dither signal in the MDAC can be accomplished using the dither capacitances (C_{di}), as shown in Figure 9.14. The number of the dither capacitances (N_d) depends on the number of dither levels. During $\phi 1$, the dither capacitances are discharged to remove the previous charges and eliminate any memory. During $\phi 2$, they are connected to V_{Ref} or $-V_{Ref}$ depending on the

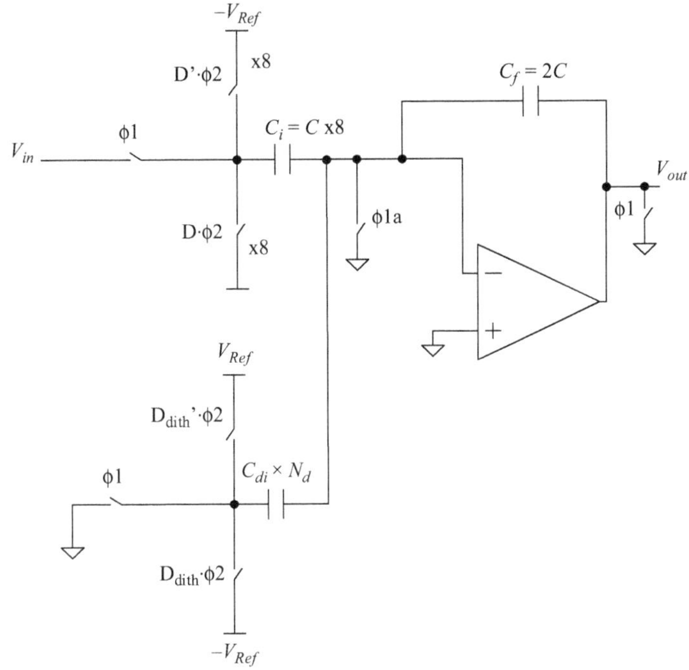

Figure 9.14 A simplified circuit diagram showing the dither injection in the MDAC. © 2014 IEEE. Reprinted, with permission, from Reference 5

randomly generated PN (dither) code. This adds the dither to the main signal as follows:

$$V_{out} = \frac{V_{in}C_t/C_f - \sum_{i=1}^{8} \mathrm{D}_i V_{Ref} C_i/C_f + \sum_{i=1}^{N_d} \mathrm{D_{dith}} V_{Ref} C_{di}/C_f}{1 + (C_t + C_f + C_{dt} + C_p)/(C_f A)} \tag{9.9}$$

where $\mathrm{D_{dith}}$ is the PN (dither) code (± 1), D_i is the DAC thermometer code (± 1), C_{di} is the individual dither capacitance, C_{dt} is the total dither capacitance, C_i is the individual sampling capacitance, C_p is the parasitic capacitance on the summing node, N_d is the number of dither capacitances, and C_t is the total sampling capacitance.

It is important to note that mismatches between the dither capacitances and the main DAC capacitance can result in an error in the gain estimate. The algorithm estimates the gain "seen" by the dither, and has no way of knowing how that relates to the gain that the main DAC signal sees. This mismatch needs to be measured and corrected. Since the capacitor mismatches are relatively constant, it can be corrected once and fused using factory calibration.

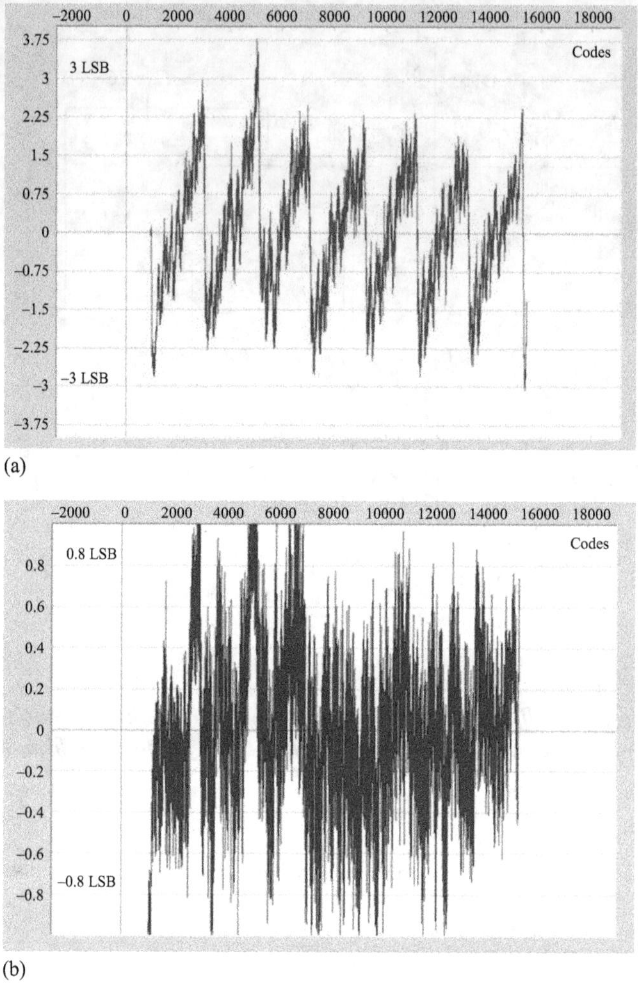

Figure 9.15 An INL of a 14-bit 1GS/s ADC with inter-stage gain error [5]. (a) Before calibration. (b) After calibration

An example INL before and after correlation-based calibration is shown in Figure 9.15 [5]. Before calibration, a clear IGE pattern is visible with INL jumps of about 5 LSBs. After calibration, the INL is fixed to about 1 LSB independent of supply, temperature, and sampling rate [5]. For 14-bit accuracy, a convergence time of a few seconds was needed at 1 GS/s, which translates into millions of samples. Generally, the number of samples required for convergence is proportional to the signal power that needs to be correlated out, and is inversely proportional to the dither power and the square of the allowed convergence error. That is

$$N \propto \frac{V_{\text{signal}}^2}{\varepsilon^2 V_{\text{dither}}^2} \tag{9.10}$$

where N is the number of samples needed for convergence, ε is the convergence error, V_{dither} is the amplitude of the dither signal, and V_{signal} is the amplitude of the signal.

9.1.2.1 Calibration accuracy

The calibration's accuracy and robustness are measured by evaluating the performance with changing conditions such as temperature, supply, aging, sampling rate, and so on. However, an often over-looked measure of the calibration's accuracy is its insensitivity to the input signal's frequency and amplitude. Ideally, a robust background calibration algorithm should be independent of the input signal. This is evaluated by sweeping the input signal's amplitude and frequency to confirm that the calibration is not impacted, and does not require re-convergence [5, 11].

In spite of the effectiveness of the correlation-based calibration algorithms, these techniques rely on injecting a PN calibration signal in the stage's MDAC, but not in its flash. Therefore, this PN signal consumes a portion of the correction range, which reduces the MDAC amplifier's dynamic range and indirectly increases the power consumption of the ADC. It also limits the budget available for comparator offsets and the bandwidth mismatch between the MDAC and flash of the first stage in a SHA-less architecture. To minimize this penalty, it is important to reduce the amplitude of the calibration dither signal [5, 15], which in turn degrades the accuracy and robustness of the calibration. It requires the back-end stages to have higher accuracy than what is required for the input signal. To calibrate the following stage (i.e. stage-2) to this additional accuracy requires even higher accuracy from stage-3, and so on. This represents a vicious cycle that limits the accuracy of the calibration [5].

This dilemma is shown conceptually in Figure 9.16. If the dither amplitude is reduced such that: *V_dither/V_signal* is equal to 1/8, then,

$$\delta/\Delta = \Delta/\Delta' = V\_dither/V\_signal = 1/8 \tag{9.11}$$

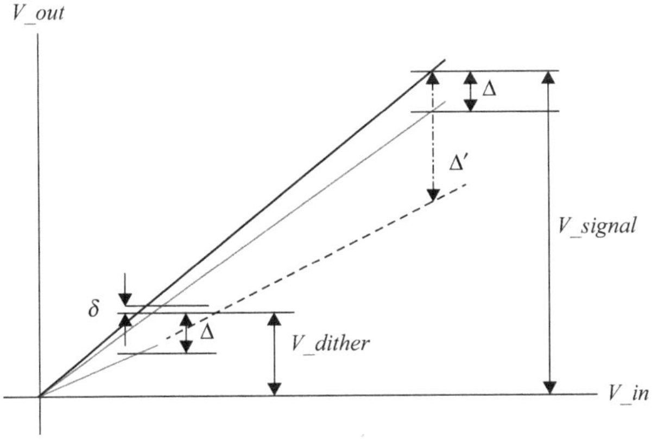

Figure 9.16 A conceptual illustration of the additional accuracy requirement if V_dither is less than V_signal

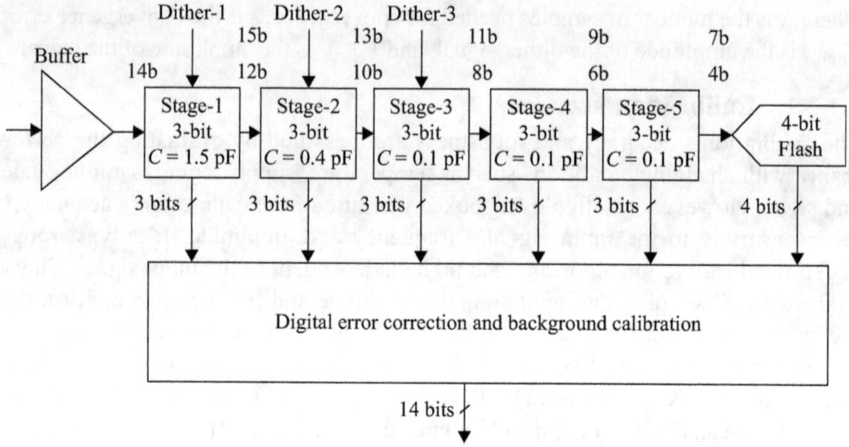

Figure 9.17 The accuracy of each stage, and the required accuracy for the dither processing (top) if V_dither is less than V_signal by a factor of 1/8. The accuracy is increased by a factor of 8 compared to the accuracy in processing the main signal, for example, from 12b to 15b as shown.

where Δ is the acceptable error of a full-scale signal. If the dither signal is processed with that same accuracy in spite of its small amplitude, it will result in a larger error for the full-scale signal of Δ'. Therefore, the required accuracy in processing the dither must be tightened to δ, which is scaled down relative to Δ by the ratio of the dither amplitude to the signal amplitude. For example, if the dither is 1/8th the signal, as given by (9.11), the processing of the dither needs to be more accurate by about 3 bits. This is represented on the stages of the pipeline in Figure 9.17.

It is interesting to spend a few minutes on Figure 9.17. It tells us that in order to calibrate the first stage residue to be 12-bit accurate using a dither signal that is 1/8th the full-scale, we need stage-2 to process that dither with 15-bit accuracy. If we had a stage-2 that is 15-bit accurate, we would not need to calibrate stage-1 to be 12-bit accurate in the first place! This is a clear paradox that needs to be addressed. If this additional accuracy in the back-end stages is not achieved, the estimation of the IGE may be input-dependent. An example of this problem is shown in Figure 9.18, where the IGE changes with the input amplitude by about 9 LSBs. With adequate accuracy in the back end, the change is improved to less than 3 LSBs.

Although this discussion is in the context of correlation-based calibration, the principle applies to all calibration techniques. Whenever we process a calibration signal in the digital domain, we need to ensure that the accuracy is adequate for our purpose. Assuming an "ideal" pipeline back end in analyzing calibration algorithms is a common pitfall that can end up limiting the accuracy and even the viability of the calibration algorithm.

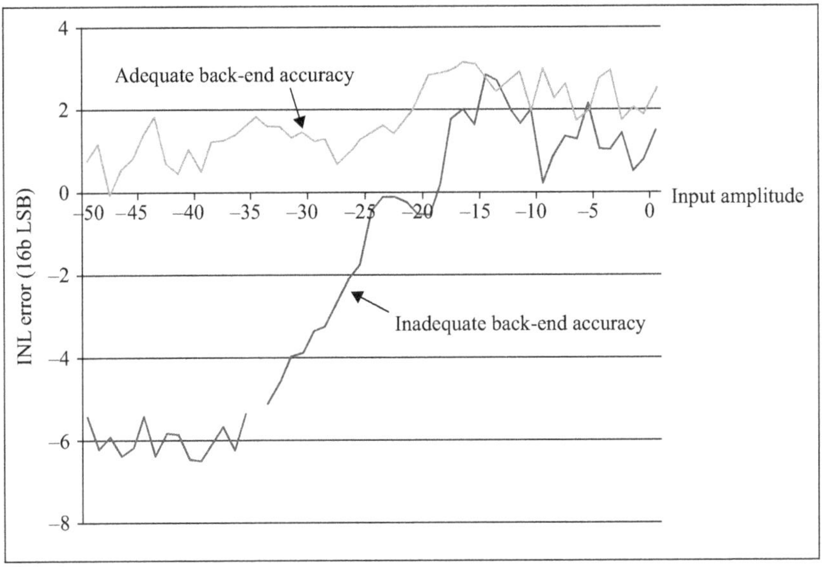

Figure 9.18 A comparison of the INL versus input amplitude for two cases: with and without adequate accuracy in the back end

One way to overcome this limitation is to employ multi-level dithering to enhance the linearity of the following stages. This multi-level dithering can be embedded in the calibration signal itself to perform both the calibration and the dithering [5]. The amplitude and number of dither levels need to be adequate to achieve the additional accuracy required. For example, if the PN calibration signal is 1/8th the input signal, at least 8 levels of dither are required to calibrate the back end. If it is 1/16th the amplitude of the input signal, we would need 16 levels of dither, and so on.

However, a binary number of dither levels leads them to fold on top of each other as they propagate down the pipeline if the inter-stage gains are powers of 2, as shown in Figure 9.19. In order to preserve the number of dither levels as they propagate down the pipeline, an odd number of equally spaced levels is preferred [5]. This arrangement, shown in Figure 9.19, preserves the number of levels for every stage down the pipeline. That is, if we use N equally spaced dither levels, where N is odd, the number of dither levels will be N for every stage down the pipeline.

An example of this implementation is discussed in Reference 5, where 9 levels of dither were used to improve the calibration effectiveness. It is important to note that the dither calibration signal is added to the MDAC only. That makes it effective in dithering the back-end stages, but not necessarily the first few stages of the pipeline. In fact, since it is injected in the MDAC only, it cannot dither its own stage. Moreover, if its amplitude is small, the gain of its stage will not be adequate

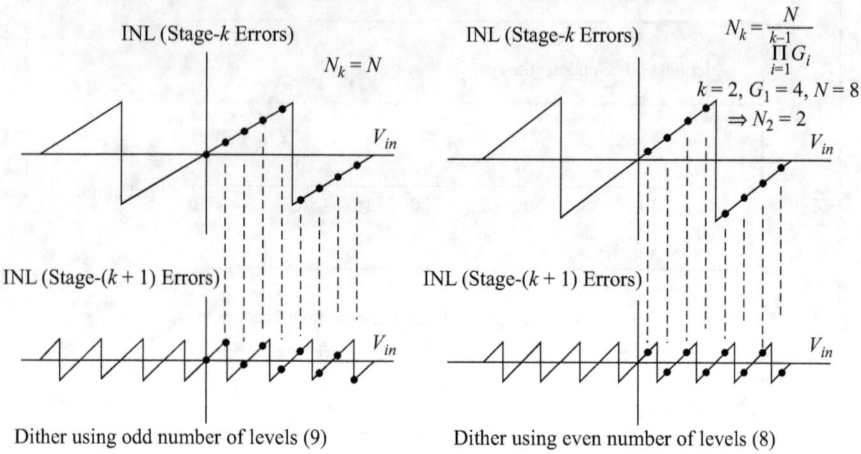

Figure 9.19 Comparison of using a binary number of dither levels versus an odd number of levels. With binary number of levels and stage gain that is also binary, the dither levels fold on each other and their number quickly dwindles as they propagate down the pipeline. With odd number of levels that are properly spaced, the dither levels stay the same and uniformly distributed as they propagate down the pipeline. © 2014 IEEE. Reprinted, with permission, from Reference 5

to amplify it to cover a full sub-range of the next stage. So it may not be effective in dithering the following stage either.

Therefore, it is desirable to add an additional "large" dither signal, to both the MDAC and the flash [39], to dither any residual non-linearity in the first few stages, which cannot be effectively dithered with the small calibration dither signal and to improve the calibration accuracy. It is injected in both the MDAC and the flash to effectively dither the first stage and to avoid using any portion of the correction range. In order to prevent it from consuming any portion of the ADC's dynamic range, an extra comparator is used in the flash as shown in Figure 9.20 and discussed earlier. This additional comparator frees up two halves of a sub-range at the two ends of the ADC's dynamic range, which can be occupied by the large dither without compromising the ADC's dynamic range or its linearity. Since the dither's amplitude is about half a sub-range, it will be effective in dithering the saw-tooth pattern due to the IGE. This is discussed again in more detail later in the "Dither" section of this chapter.

Correlation-based approaches represent an effective approach to calibration. They have relatively low analog overhead, and their digital processing is quite simple. They have reasonable accuracy and robustness. Their main drawback is their long convergence time. They require millions of samples to achieve 14/16-bit accuracy [3, 5]. This can result in long test times and slow response to changes. In addition, the heavy reliance on PN signals may lead to noise and jitter degradation

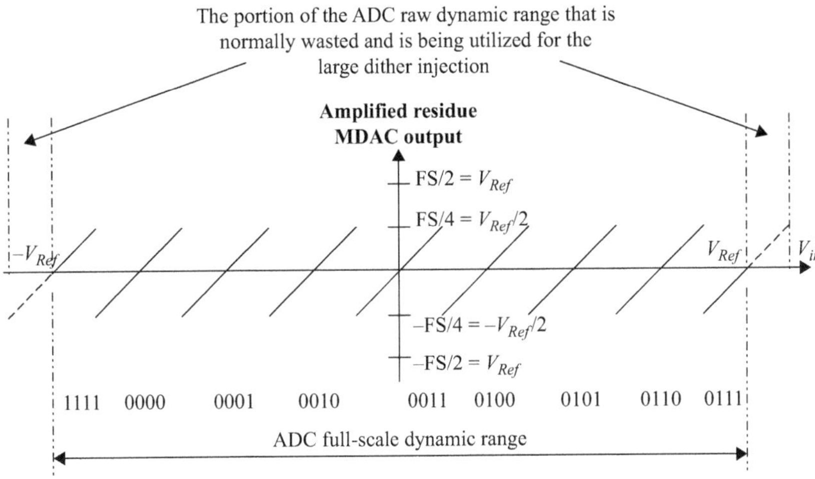

Figure 9.20 A stage-1 residue showing the portion of the dynamic range that is usually not utilized when using an additional comparator in the flash and can be consumed by the large dither signal

due to inadvertent coupling, which can degrade the ADC performance, especially at high input frequencies.

9.1.3 Summing node calibration

The summing-node method is a deterministic calibration approach that relies on measuring the error due to the amplifier's finite open loop gain directly. Examples of this approach are described in References 6 and 7. In a typical switched capacitor MDAC circuit, shown in Figure 9.21, the summing node voltage is equal to $-V_{out}/A$. Analyzing the circuit gives the following expression for the output:

$$V_{out} = \frac{V_{in}C_t/C_f - \sum_{i=1}^{8} D_i V_{Ref} C_i/C_f}{1 + (C_t + C_f + C_p)/(C_f A)} \tag{9.12}$$

where D_i is the DAC code, C_i is the individual sampling/DAC capacitance, C_p is the parasitic capacitance on the summing node, and C_t is the total sampling capacitance. Assuming no capacitor mismatch or DAC errors, the output can be expressed in terms of its ideal value as

$$V_{out} = \frac{V_{out}|_{\text{ideal}}}{1 + (C_t + C_f + C_p)/(C_f A)} \tag{9.13}$$

If we substitute $K = 1/\beta$ in (9.13), where β is the feedback factor, we get

$$V_{out} = \frac{V_{out}|_{\text{ideal}}}{1 + K/A} = V_{out}|_{\text{ideal}} - V_{out}K/A \tag{9.14}$$

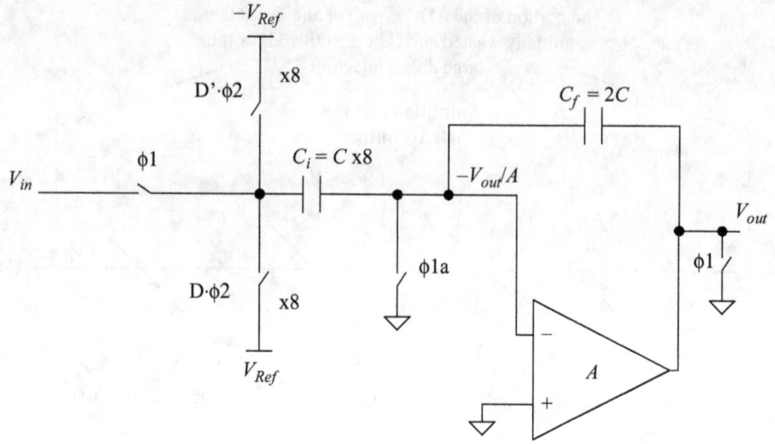

Figure 9.21 A simplified schematic of a switched capacitor MDAC shown as a single-ended circuit for simplicity

which can be rearranged to give

$$V_{out}|_{\text{ideal}} = V_{out} + V_{out}K/A \tag{9.15}$$

It is clear from (9.15) that the error due to the amplifier's finite open loop gain is equal to $V_{out}K/A$, which is proportional to the summing node voltage V_{out}/A. The error can be corrected by sampling the summing node voltage, multiplying it by the constant K, and adding it back to the signal directly in a following stage, or in the digital domain after digitizing using a "shadow pipeline," on a sample-by-sample basis [6]. This is shown conceptually in Figure 9.22.

Alternatively, the sampled summing node voltage can be amplified by gain G_s and digitized using a cheap and slow ADC. The LMS algorithm can be used to estimate the gain error for digital correction [7], as shown in Figure 9.23, which is given by

$$\alpha_{i+1} = \alpha_i - \mu \times D(V_{out1i}) \times [\alpha_i \times D(V_{out1i}) - D(V_{out1i}/A)] \tag{9.16}$$

where α_{i+1} is the estimate of the inverse of the open loop gain ($1/A$), α_i is the previous estimate, μ is the step size of the LMS algorithm, $D(V_{out1})$ is the digital representation of the residue of the stage being calibrated, which is stage-1 in this example, and $D(V_{out1}/A)$ is the digital representation of the summing node voltage. The gain correction is then applied digitally as follows:

$$D(V_{out1\_cal}) = D(V_{out1}) + D(V_{out1}) \times K \times \alpha \tag{9.17}$$

where $D(V_{out1\_cal})$ is the calibrated digital residue, and K is a constant equal to the inverse of the feedback factor.

Alternatively, the error can be processed in the digital domain and the correction done in the analog domain. The steepest descent approach is used to minimize the error, which is proportional to the sampled summing node voltage [7].

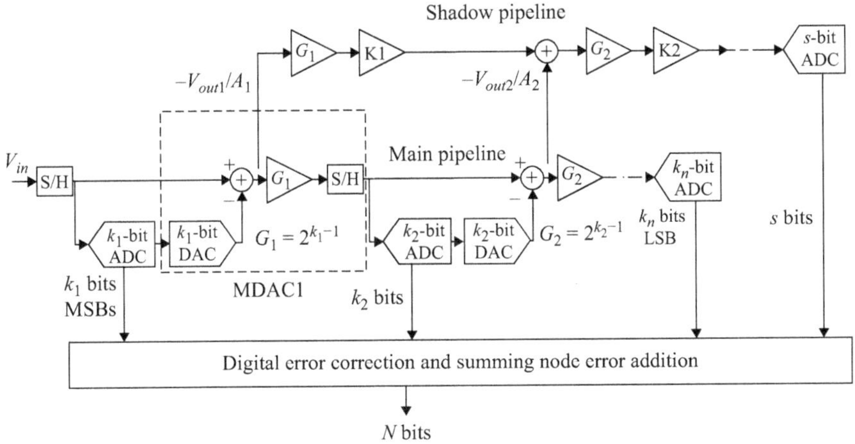

Figure 9.22 A block diagram illustrating the calibration of the amplifier gain error. The summing node voltage is sampled on the stages that are to be calibrated, processed by the shadow pipeline, and the result is added to the digital output with the right polarity on a sample-by-sample basis [6]

An example of the analog adjustment is using programmable positive feedback in the amplifier to enhance the gain. This is shown in Figure 9.24 and is given by

$$\nu_{i+1} = \nu_i - \mu \times D(V_{out1i}) \times [D(V_{out1i}/A)] \tag{9.18}$$

where ν_{i+1} is the new estimate of the analog control parameter, and ν_i is its previous estimate. Convergence occurs when the error, which is the summing node voltage $D(V_{out1}/A)$ is minimized with the desired accuracy. It is important to note that, unlike (9.16), the exact value of the error in (9.18) is irrelevant, as the algorithm attempts to minimize it by feeding back the control parameter ν to the analog domain, regardless of its absolute value.

Equation (9.18) describes an error minimization problem, instead of a system identification one. It converges much faster and requires significantly lower accuracy. However, the correction in the analog domain is possible only if there is a programmable way to control the gain of the amplifier. The gain might be dependent on the sampling rate, temperature, or supply. So the goal of the calibration is to optimize the analog gain such that it is always in its "sweet spot."

An example of an amplifier with programmable open loop gain is shown in Figures 9.25 and 9.26 [7], where positive feedback is used to enhance its gain by providing a negative transconductance that cancels the amplifier's output conductance. This increases the output impedance and the open loop gain, but exact cancellation may require adjustment to some bias voltage or device size that is programmable. On the other hand, if achieving the required gain is not possible in the analog domain, the analog correction approach will not be viable, and digital correction will be necessary.

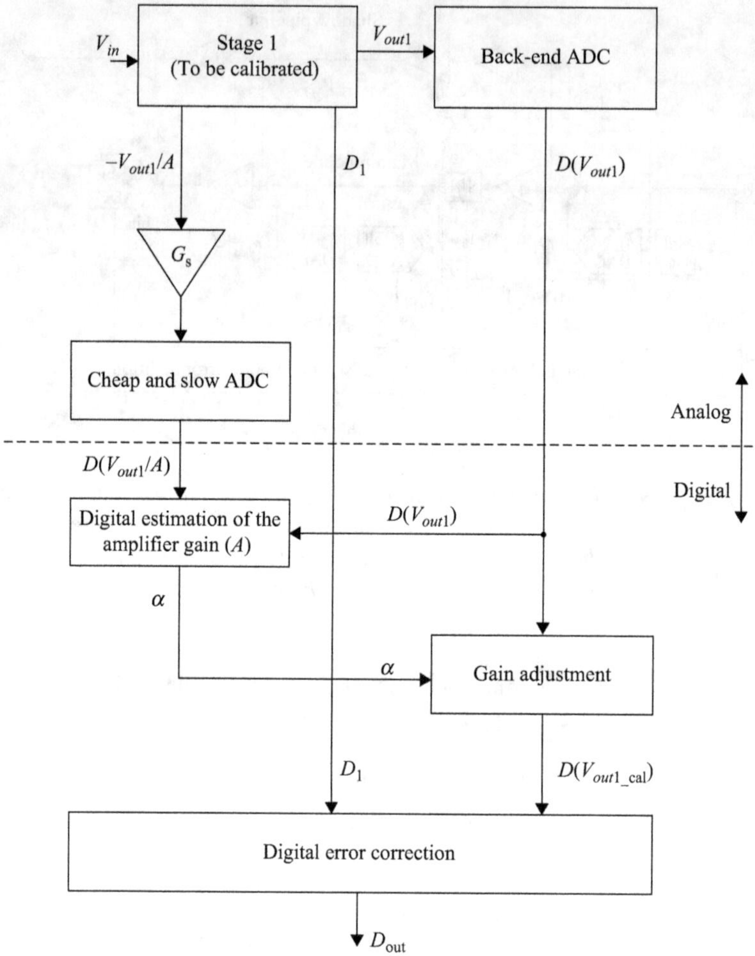

Figure 9.23 A block diagram showing the summing node algorithm with digital correction of the inter-stage gain error. © 2010 IEEE. Reprinted, with permission, from Reference 7

Figure 9.27(a) shows an implementation of the summing node sampling and digitization using a slow and cheap ADC. The timing diagram is shown in Figure 9.27(b). The summing node voltage is sampled on a small capacitor C_{e1} at a much slower rate determined by ϕ2e. Buffering can also be used to reduce the impact of the slow sampling on the summing node of the MDAC. A dummy network can be connected to the summing node in the "off" phases using the clock ϕ2eb to reduce the spurs due to the slow clock of the summing node sampler and quantizer. Once sampled, the summing node voltage is amplified and digitized at a much slower rate for the digital calibration processing.

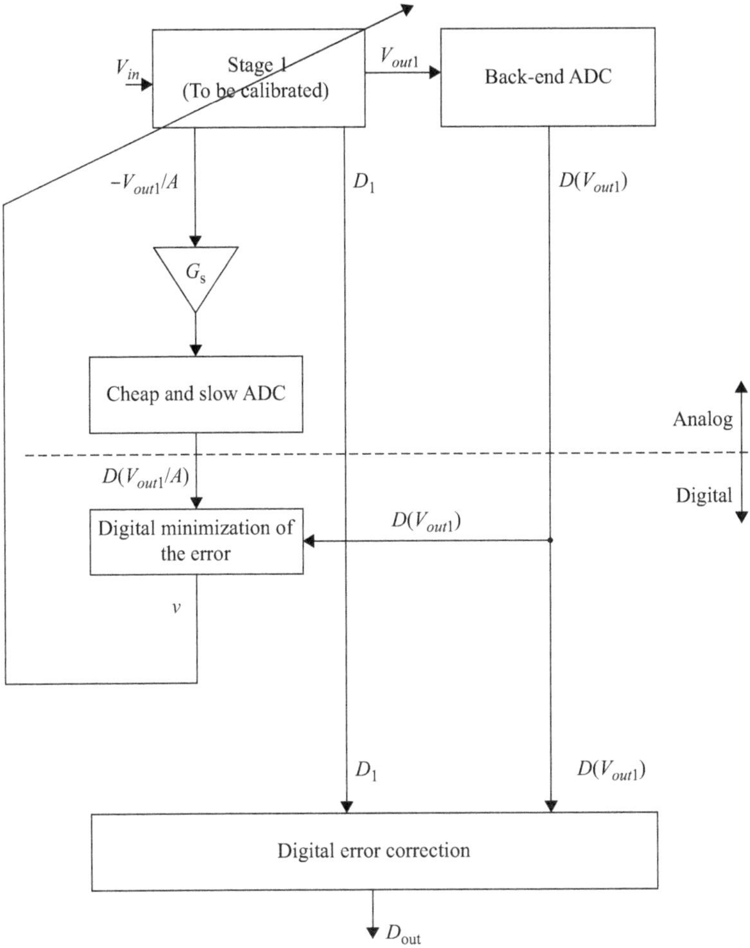

Figure 9.24 A block diagram showing the summing node algorithm with analog correction of the inter-stage gain error. © 2010 IEEE. Reprinted, with permission, from Reference 7

Unlike correlation-based and statistical techniques, the summing node technique measures the error directly in a deterministic fashion. The convergence of this algorithm is much faster and typically requires only a few thousand samples, as opposed to millions of samples for the correlation-based techniques. The convergence speed of the summing node algorithm depends on:

- The required calibration accuracy
- The noise in the calibration path (i.e. the slow ADC)
- The amplitude of the output residue

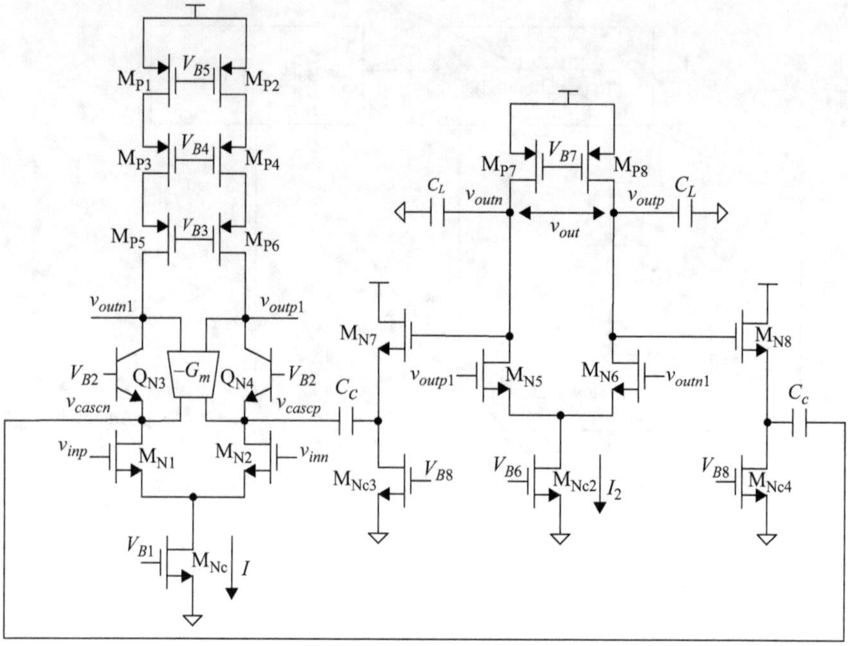

Figure 9.25 An MDAC amplifier employing positive feedback ($-G_m$ block) to enhance the gain. © 2014 IEEE. Reprinted, with permission, from Reference 7

The number of samples N needed for convergence are typically given by

$$N \approx K\left(\frac{V_{Noise\_calpath}}{V_{o1RMS}/A_{needed}}\right)^2 \approx K \times 10^{[(A_{needed\_dB} - SNR_{calpath} - V_{o\_dBFS})/10]} \tag{9.19}$$

where K is a proportionality constant, V_{o1RMS} is the RMS value of the residue signal, V_{o_dBFS} is the output residue amplitude relative to the full scale in dBFS, A_{needed} is the amplifier's open loop gain needed to achieve the desired accuracy, and $V_{Noise\_calpath}$ is the RMS noise voltage of the summing node processing path. An example is shown in Figure 9.28, where about 30,000 samples are needed for convergence with 16-bit accuracy. The noise in the calibration path in this case was about 65 dB and the required gain accuracy was about 110 dB [7]. This was used for a 250 MS/s ADC, where the slow path was operating at 12.5 MS/s. The convergence time was about 2.4 ms, which is orders of magnitude faster than the correlation-based algorithms. Equation (9.19) also shows that the absence of a signal can be problematic and would require the calibration to be frozen.

Since the algorithm measures the error directly, any error in measurement will have a limited second-order effect on the estimation's accuracy. This relaxes the requirement on the accuracy of the error processing. This is in contrast with the correlation-based and statistical approaches, where the calibration processing needs

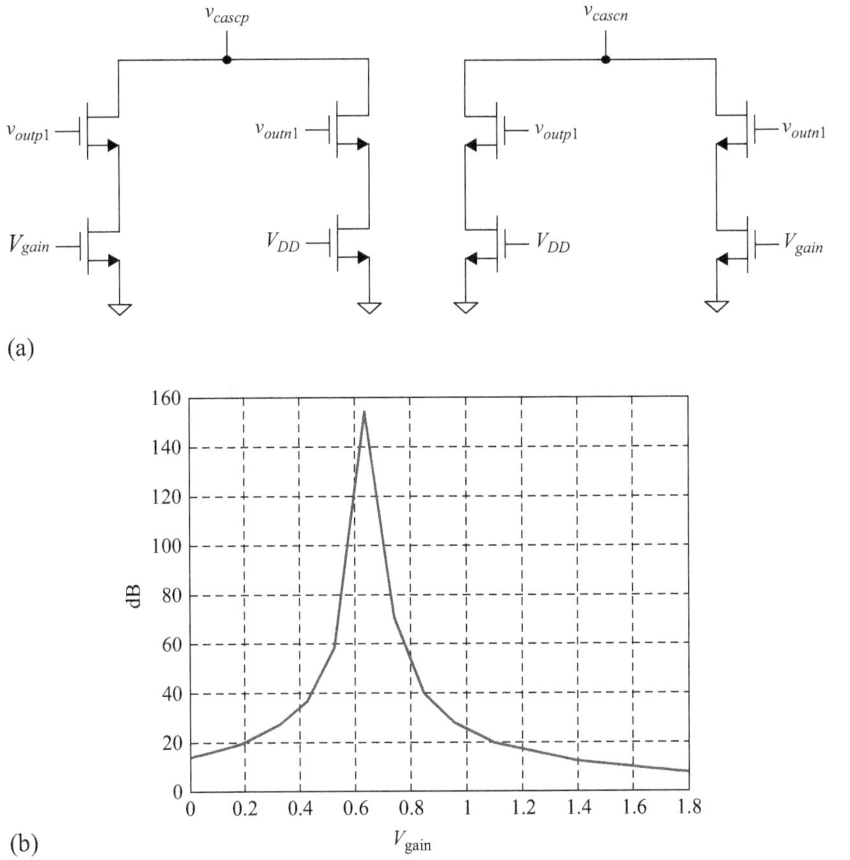

Figure 9.26 (a) An example of a $-G_m$ block in Figure 9.25. (b) The open loop gain of the amplifier as a function of the control voltage V_{gain}. © 2014 IEEE. Reprinted, with permission, from Reference 7

to be as accurate as the overall required accuracy, and sometimes even more as described in the previous section. This relaxation in the required accuracy is demonstrated using the summing node gain correction formula, given by

$$D(V_{out1\_cal}) = D(V_{out1}) + D(V_{out1}) \times K \times \alpha \tag{9.20}$$

For simplicity, we can represent (9.20) as

$$V_{out1\_cal} = V_{out1} + V_{out1}K/A$$

Therefore,

$$\left|\frac{\partial V_{out1\_cal}}{V_{out1\_cal}}\right| = \frac{K}{A} \times \left|\frac{\partial A}{A}\right| \tag{9.21}$$

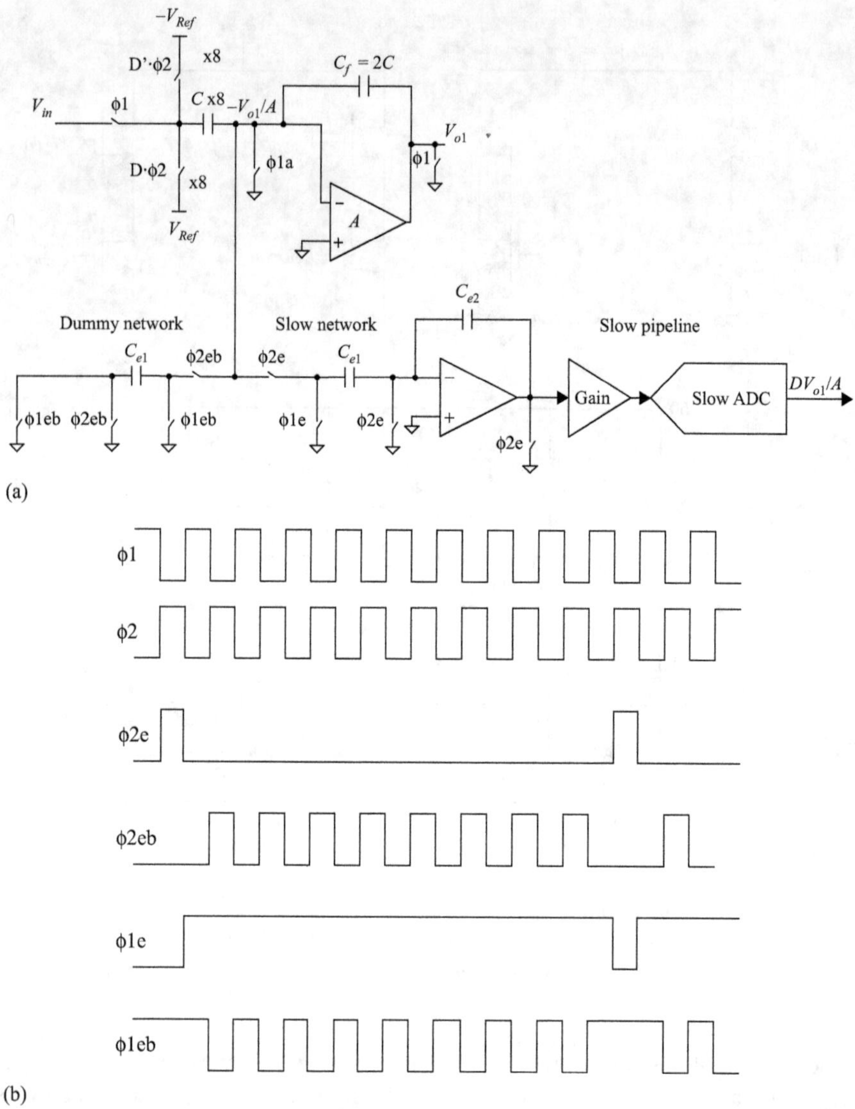

Figure 9.27 (a) A simplified schematic showing the sampling of the summing node voltage and its processing using a slow and cheap network. The dummy network is used to match the slow sampling network in the other phases. (b) Timing diagram showing the sampling of the summing node voltage during $\phi 2e$ and its processing using the slow pipeline during $\phi 1e$. A dummy network is switched in to sample the summing node during the $\phi 2eb$ phases. © 2010 IEEE. Reprinted, with permission, from Reference 7

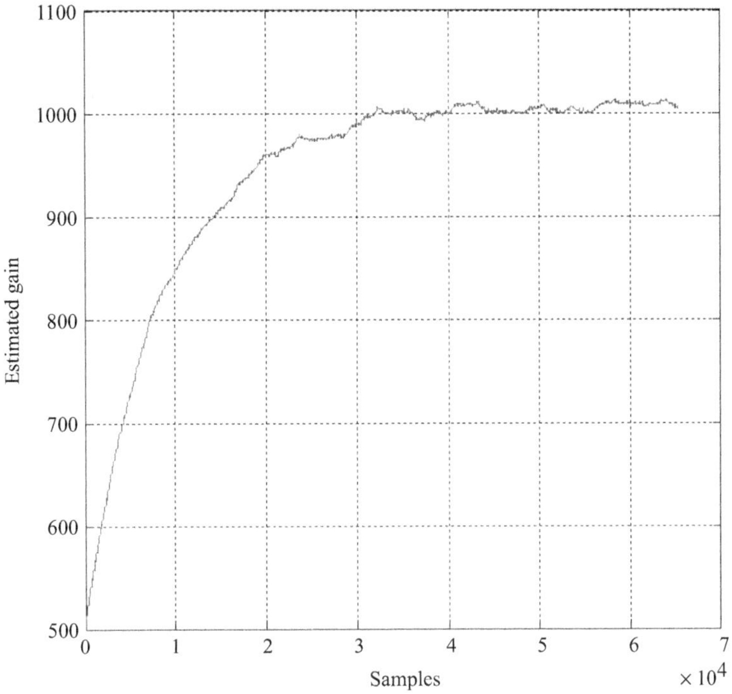

Figure 9.28 An example showing the convergence of the gain to the correct value

that is,

$$\text{The allowed calibration gain error} = \text{required stage accuracy} \times A/K \quad (9.22)$$

where A is the uncalibrated open loop gain and $1/K$ is the feedback factor. For example, if the required stage accuracy is 16 bits, the analog open loop gain A is 80 dB, and the feedback factor is 1/4, then the allowed calibration gain error is 4%, which is significantly less than the required 16-bit accuracy.

On the negative side, the summing node techniques require more analog changes than correlation-based techniques. Digitization of the error requires implementing a slow and cheap ADC. Adding the error to the following stages requires developing analog blocks that perform the error processing and addition. However, as the focus on the amplifier non-idealities increases, with an increasing need to correct the amplifier's non-linearity, deterministic approaches (such as the summing node technique) may become more attractive.

9.1.4 Reference ADC calibration

In this approach, a slow-but-accurate reference ADC is used for calibration, as shown in Figure 9.29. The difference between the output of the reference ADC and

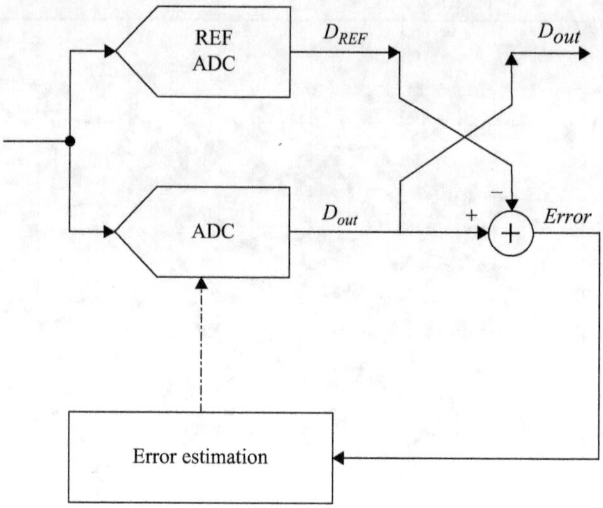

Figure 9.29 A block diagram showing the reference ADC method. The correction can be done in the digital or analog domains

the actual ADC is the error signal, and the LMS algorithm can be used to minimize the error [18, 19], such that

$$G_e[n+1] = G_e[n] - \mu \times V_R[n] \times (V_R[n] \times G_e[n] - V_{REF}[n]) \tag{9.23}$$

where V_R and V_{REF} are the digital representations of the residue voltages of the main and the reference ADCs, respectively, and G_e is the estimate of the inter-stage gain. This can also be represented as follows:

$$G_e[n+1] = G_e[n] - \mu \times D_{REF}[n] \times (D_{out}[n] - D_{REF}[n]) \tag{9.24}$$

where D_{out} and D_{REF} are the outputs of the main and reference ADCs, respectively.

This technique falls under the umbrella of the deterministic approaches. Similar to the summing node calibration techniques, this method has fast convergence time and suffers from significant analog design overhead to develop the reference ADC. Unlike the summing node method, the slow reference ADC needs to be accurate. Its noise can be high, but its linearity must match the required accuracy of the calibrated ADC.

The reference ADC method suffers from limitations due to possible mismatches between the main path and the reference path. Some of these mismatches, such as clock skew and kick-back, can limit the calibration accuracy. In addition, the reference ADC can degrade the performance of the main ADC due to its slow clock, and its loading effect on the input.

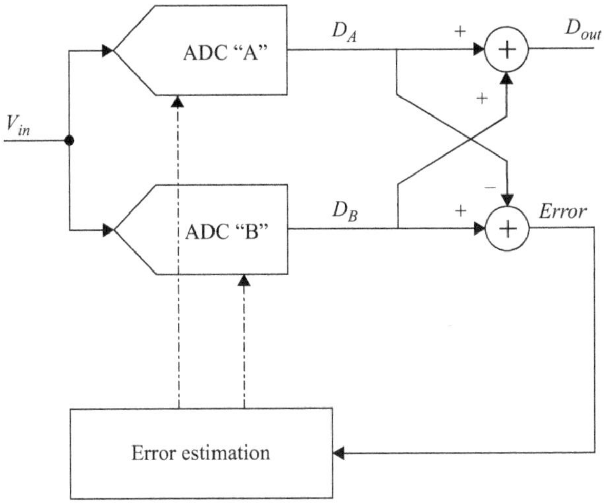

Figure 9.30 A block diagram showing the Split ADC method. The correction can be done in the digital or analog domains. © 2009 IEEE. Reprinted, with permission, from Reference 30

9.1.5 Split-ADC calibration

This technique is similar to the reference ADC method. However, instead of using a slow-but-accurate ADC, the main ADC is divided into two identical halves. The sum (or average) of the two outputs is the overall ADC output. The difference between the two outputs represents the error signal that needs to be minimized by the LMS algorithm [20]. This is shown conceptually in Figure 9.30. In order to ensure that the difference is nulled only when the two ADCs are correctly calibrated, and not just when they are identically wrong, the two halves operate in different residue modes that are randomly selected by forcing the comparators of the sub-ADCs to make different decisions for the same input. The different decision paths taken by the two halves make it unlikely for the error to be minimized when both are equally wrong.

Like the reference ADC approach, this technique is deterministic, has fast convergence time, and requires significant changes on the analog side. In addition, it is sensitive to the offset, gain, and timing mismatches between the two halves, which can lead to calibration errors.

9.1.6 Settling error calibration

An important goal of the digital assistance in high speed pipelined ADCs is to correct the dynamic settling errors, and hence enable higher sampling rates, better accuracy, and lower power consumption compared to what is possible using analog circuit techniques alone. The MDAC amplifier's settling is typically divided into large-signal settling and small-signal settling [21]. The large-signal settling is limited by the slew rate, the DAC reference's large-signal settling, and to some extent the common-mode settling of the amplifier. It is highly non-linear, and results in non-linear distortion in addition to the gain error. The non-linear charge injection can limit the settling linearity as well. On the other hand, small-signal settling is linear and results in IGEs. Therefore, it can be calibrated using the IGE calibration techniques described in the previous sections.

Without loss of generality, if we assume a single pole system, the small-signal settling will be given by

$$v_o = v_{initial}e^{-t_s/\tau} + v_{final}(1 - e^{-t_s/\tau}) \tag{9.25}$$

where t_s is the setting time and τ is the settling time constant, which is given by

$$\tau = \frac{1}{2\pi \times BW_{cl}} = \frac{1}{\beta\omega_u} \tag{9.26}$$

and BW_{cl} is the closed loop bandwidth of the amplifier, β is the feedback factor, and ω_u is the unity gain angular frequency. If the amplifier is reset and starts from zero, the output after a settling time t_s will be given by

$$v_o = v_{final}(1 - e^{-t_s/\tau}) = \left(\frac{V_{in}C_t/C_f - \sum_{i=1}^{8} D_i V_{Refi} C_i/C_f}{1 + 1/\beta A}\right)(1 - e^{-t_s/\tau}) \tag{9.27}$$

If t_s and τ are constant, the output will have a *linear* settling gain error $\delta G/G$ that is given by

$$\delta G/G = e^{-t_s/\tau} \tag{9.28}$$

In the presence of more than one pole, the small-signal settling behavior may be over-damped, under-damped, or critically damped. However, the small-signal settling error will still be linear and cause a gain error, as long as it is within the range of the small-signal settling. In practice, since the large-signal settling time is variable, the initial voltage term $v_{initial}e^{-t_s/\tau}$ in (9.25) cannot be ignored and will degrade the settling error linearity. Moreover, the large-signal settling is highly non-linear, which complicates the calibration further.

Figure 9.31 shows the large- and small-signal settling for two cases A and B. It is clear that in Output A, the large-signal settling is about 50 ps longer than that for Output B. In the former case, the large-signal settling was limited by longer DAC and reference settling time in addition to the slew rate. Up to points 5 and 6, the settling is highly non-linear. The non-linearity decreases substantially until points 2 and 4, where the settling becomes almost entirely linear within the required accuracy. The IGE calibration would be very effective in correcting for settling errors between points 1 and 2, or 3 and 4. Its effectiveness decreases progressively as we move further to the left beyond points 2 and 4 due to the increased non-linearity in the settling errors.

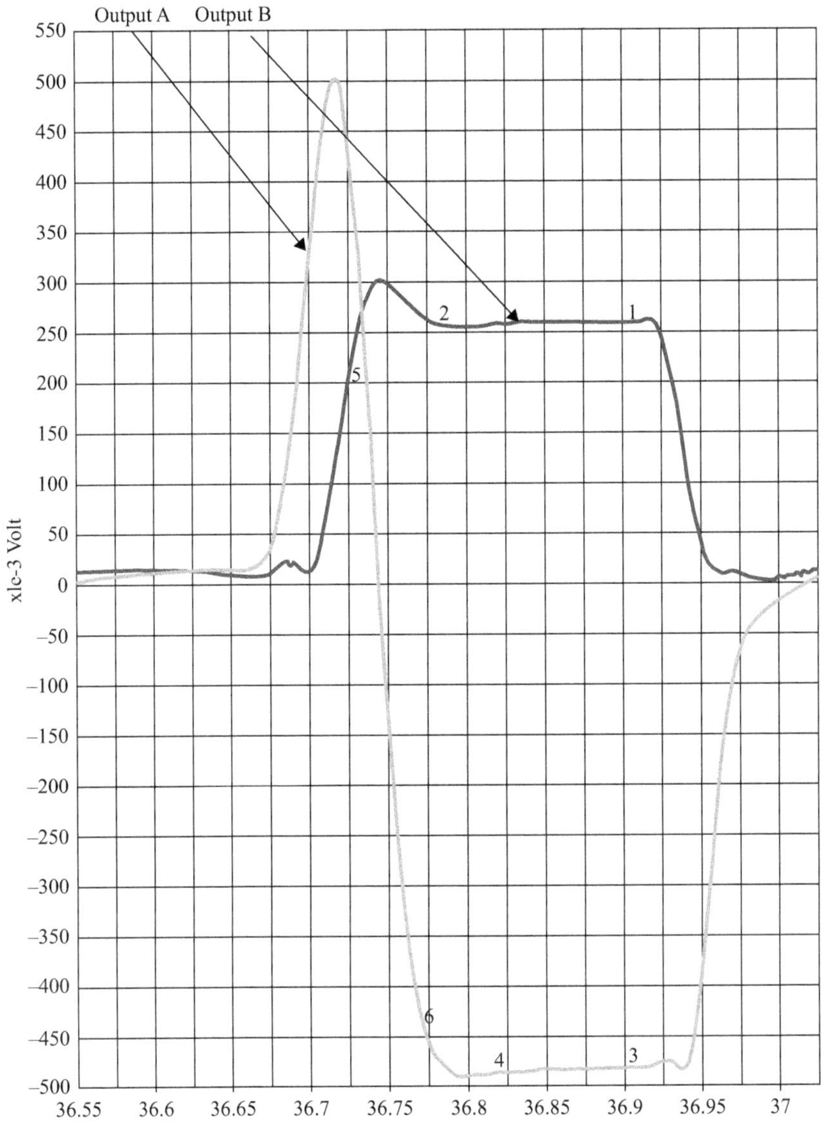

Figure 9.31 A simulation plot of the MDAC output showing the large- and small-signal settling in two cases. A: The curve shows the output going in the wrong direction (due to the DAC delay), then recovering. B: The curve shows a case where the output goes directly to the right direction and hence has a relatively short period of large-signal settling. The large-signal settling time is different by about 40 ps between the two cases

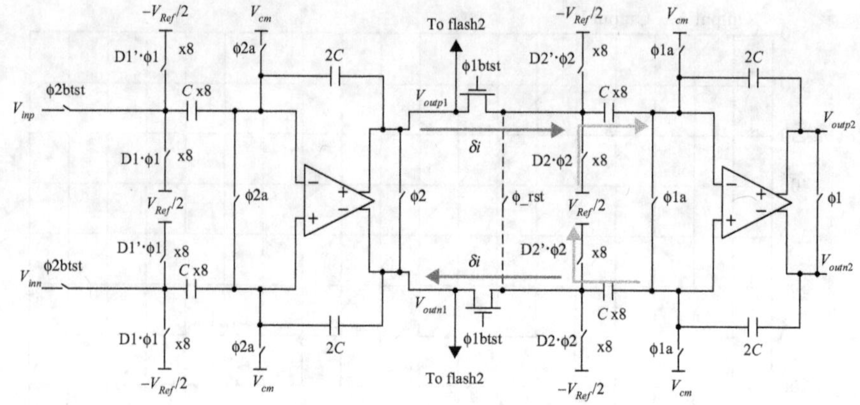

Figure 9.32 A simplified schematic showing the non-linear charge injection from stage-2 onto stage-1. During $\phi 2$, the capacitances of stage-2 are connected to $V_{Ref}/2$ or $-V_{Ref}/2$ based on the DAC code. During $\phi 1$, the capacitances are connected to the output of stage-1. Non-linear charge injection (δi) results from the DAC charge on the capacitance and the charge in the input switch

Another factor that limits the settling linearity is the memory and kick-back errors. This is illustrated in Figure 9.32. When $\phi 2$ is high, the stage-2 DAC is connected to the reference according to its DAC code. When $\phi 1$ is high, the capacitors are connected to the output of the stage-1 amplifier. If the stage-2 capacitors are not reset before being connected to the amplifier output, there will be non-linear charge injection (kick-back) on the stage-1 amplifier because of their initial charge that depends on the previous DAC code. Since that charge is a quantized version of the previous output and hence is highly non-linear, the charge injection will be a non-linear memory error. This indicates that the term $v_{initial}e^{-t_s/\tau}$ in (9.25) is not zero and is highly non-linear. Normally, if the amplifier has adequate time for settling, that non-linear term will dissipate exponentially together with the amplifier's output settling. However, if the settling time is too short, that error will not settle, and may not be corrected by the IGE calibration because of its non-linear nature and its dependence on the previous sample. This indicates that the effective correction of settling errors may sometimes require the correction of the memory and kick-back errors as well [5, 39].

9.1.7 Memory calibration

Memory errors can be caused by dielectric relaxation/absorption in capacitors, incomplete resetting of amplifiers, and incomplete settling from previous charges. It can also be caused by charge injection from the capacitances or switches of the following stages that are not fully settled. The correction of this class of errors can be accomplished using Finite Impulse Response (FIR) filters in the digital domain [12]. The FIR filters will correct the present sample using memory coefficients that apply to the previous samples. Detecting the magnitude of the memory errors can be done using correlation-based approaches by injecting a dither signal in the MDAC, as is done for the IGE calibration, and utilizing the LMS algorithm [13] to detect its memory content, such that

$$Ge_{n+1,k} = Ge_{n,k} - \mu \times V_d[n-k] \times (V_d[n-k] \times Ge_{n,k} - V_R[n]) \tag{9.29}$$

where $V_d[n-k]$ is the kth past dither value and $Ge_{n+1,k}$ is the estimate of the gain coefficient from the kth past sample on the present nth sample. If $k = 0$, then $Ge_{n+1,0}$ is the coefficient for the inter-stage gain and linear settling error. If $k > 0$, then $Ge_{n+1,k}$ is the kth memory coefficient.

The correction and equalization are performed digitally using FIR filters whose M memory taps are $(Ge_{n+1,k})$, as follows:

$$V_{R\_cal}[n] = \sum_{k=0}^{M} V_R[n-k] \times Ge_{n+1,k} \tag{9.30}$$

where V_{R_cal} is the calibrated output residue, and V_R is the uncalibrated residue. We can see from (9.29) and (9.30) that the memory calibration is an extension of the correlation-based IGE calibration, by incorporating the previous samples in addition to the current sample.

9.1.8 Kick-back calibration

A new category of calibrations that has been recently proposed is the kick-back calibration [5]. This calibration corrects for errors due to the non-linear charge injection (kick-back) from the ADC's switched capacitor sampling network onto the ADC's input driver. If the MDAC's input sampling network uses the same capacitors for input sampling and the DAC operation, which is desirable for faster settling and lower noise, a non-linear charge injection occurs due to the DAC charge stored on the capacitors in the previous gain phase. A reset switch controlled by a brief reset pulse is usually used as shown in Figures 9.33 and 9.34. However, the reset pulse consumes a portion of the sampling (acquisition) time, which reduces the time available for the non-linear kick-back coming from other sources to settle. This degrades the distortion as discussed in Chapter 4. In addition, generating such a short pulse increases the power consumption.

This non-linear kick-back can be quite severe in buffer-less ADCs. When an input buffer is used to drive the sampling network, it provides a low output impedance that is capable of driving the sampling capacitance at the desired sampling rate. However, buffers have finite isolation, which can be in the order of only 6–10 dB for source followers. Some of the kick-back propagates to the ADC driver's network whose bandwidth is limited by the driver's impedance and the ADC's input capacitance. This limits the settling accuracy and the resulting performance of the ADC.

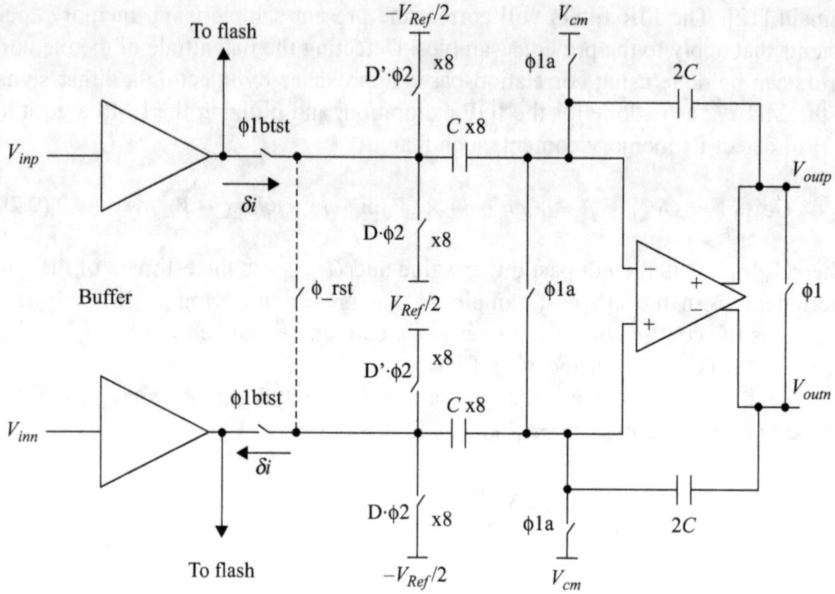

Figure 9.33 A simplified schematics of the first stage MDAC driven by input buffers. The reset switch is used to briefly discharge the sampling capacitances after the gain phase and before the next sampling phase.

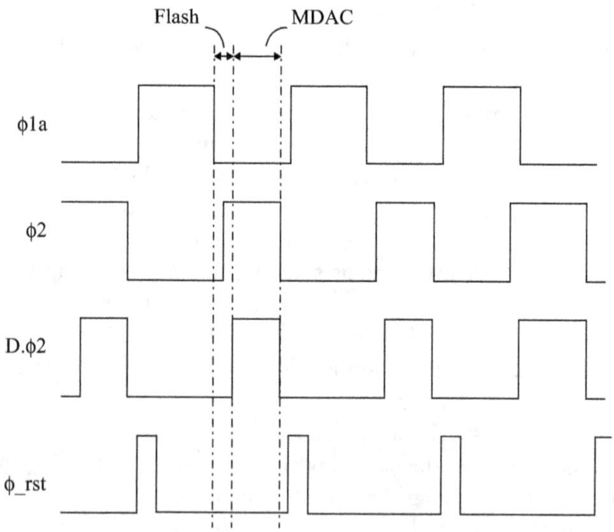

Figure 9.34 A simplified timing diagram showing the MDAC gain phase, the flash time, and input sampling phase. The ϕ_rst clock is used to briefly discharge the capacitances to discharge the previous DAC charge before the next sampling phase. This is done to reduce the non-linear charge injection (kick-back) on the input driver.

Since this non-linear kick-back is proportional to the previous DAC values, the previous codes of the first flash can be used to correct for the resulting distortion as follows [5]:

$$V_{out\_kbcal}[n] = V_o[n] + \sum_{i=1}^{M_{kb}} D_1[n-i] \times Gkb_{n+1,i} \tag{9.31}$$

where D_1 is the digital code of the first stage flash and V_o is the ADC output code. The M_{kb} kick-back coefficients (Gkb_i) are obtained using the LMS algorithm by injecting an uncorrelated PN calibration signal that "kicks" the input during the sample phase. This is shown in Figure 9.35.

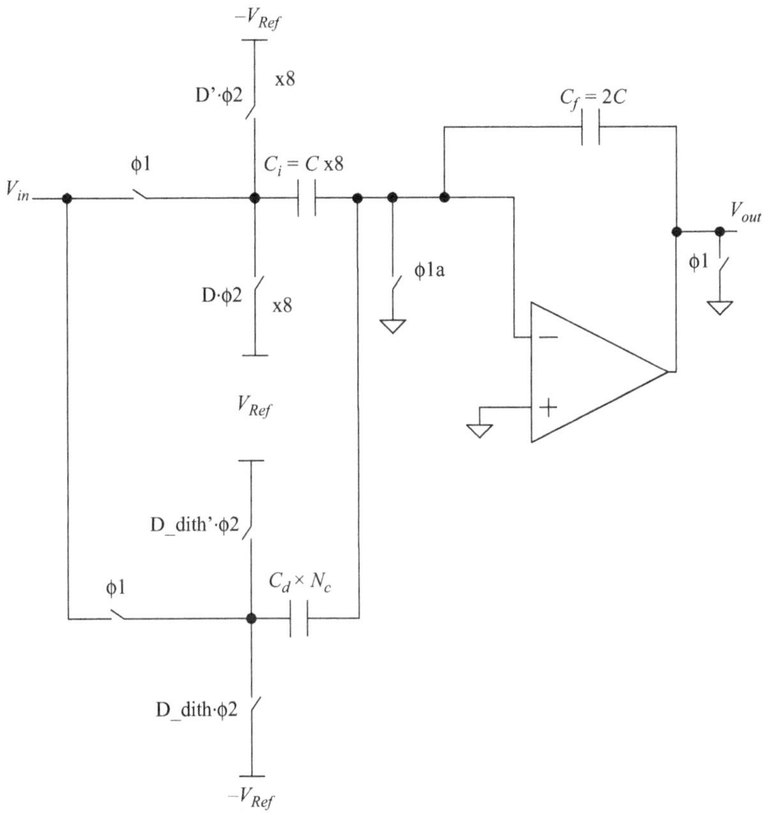

Figure 9.35 A simplified schematic showing the dither injection for the kick-back calibration. During the gain phase, the dither is applied to the dither capacitance. During the sampling phase, the dither capacitance is connected to the input in order to "kick" the input in a manner similar to the sampling capacitances. © 2014 IEEE. Reprinted, with permission, from Reference 5

During the gain/hold phase, the PN signal is stored on additional "kick-back" dither capacitances in a manner similar to the PN signal injection for the inter-stage gain and memory error calibration. In the sampling phase, the kick-back capacitors are connected to the input in parallel with the MDAC sampling capacitors. These capacitances "kick" the input with the stored PN signal charge, which then gets sampled at the end of the sampling phase and digitized with the input signal. The LMS algorithm is used to remove that kick-back PN signal and to estimate the kick-back coefficients using the recursive formula [5]

$$Gkb_{n+1,k} = Gkb_{n,k} - \mu \times Vd[n-k] \times (Vd[n-k] \times Gkb_{n,k} - V_{in}[n]) \quad (9.32)$$

where $Vd[n-k]$ is the kth past dither value and $Gkb_{n+1,k}$ is the estimate of the kick-back coefficient from the kth past sample on the present nth sample.

This calibration is similar to the memory calibration described in the previous section, and is an extension of the correlation-based calibration techniques. Unlike the other calibration methods, this technique involves injecting dither onto the ADC input driver. Since the impedance and bandwidth of the network driving the ADC will affect how it reacts to the injected dither, care must be taken in implementing this technique in order to ensure that adequate number of memory taps are employed.

The kick-back coefficients depend on the process, supply, temperature, sampling rate and the ADC driving network. The kick-back calibration algorithm operates in the background in order to adaptively correct for the changing conditions. An example is shown in Figures 9.36 and 9.37, where the INL improved from 5 LSBs down to less than 1 LSB. Unlike IGEs that manifest themselves as a saw-tooth pattern, kick-back memory errors are usually irregular and depend heavily on the frequency of the input signal.

This technique can be extended to the following pipeline stages to correct for the non-linear charge injection of the DAC charge of one stage onto the previous stage, as discussed earlier and shown in Figure 9.32. In this case, the dither is injected in a way similar to Figure 9.35 and the kick-back error can be corrected as follows

$$V_{R\_cal,i}[n] = V_{R,i}[n] + \sum_{k=0}^{M} D_{i+1}[n-k] \times Gkb_{n+1,k,i} \quad (9.33)$$

where $V_{R\_cal,i}$ is the calibrated output residue of the ith stage, $V_{R,i}$ is the uncalibrated residue of the ith stage, D_{i+1} is the flash code of the following stage, and $Gkb_{n+1,k,i}$ is the kick-back calibration coefficient of the kth previous sample on the present sample obtained using the LMS algorithm as shown in (9.32) when applied to the ith stage.

9.1.9 Residue amplifier non-linearity

In addition to the gain error, the RA may suffer from non-linearity that can limit the performance of the pipelined ADC, even with IGE correction. There are multiple possible causes of the non-linearity, which include the amplifier's slewing,

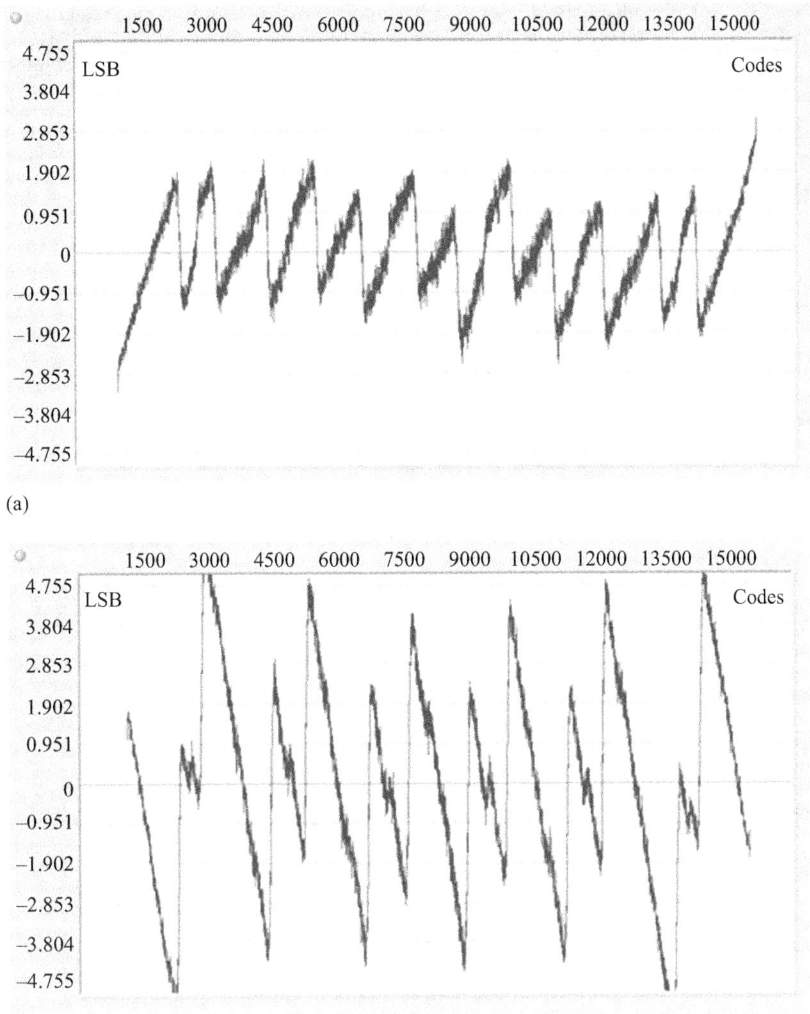

Figure 9.36 INL plots showing the effects of the non-linear kick-back for a 14-bit 1GS/s for an input frequency of (a) 700MHz, (b) 900MHz.

the DAC and reference large-signal settling, the amplifier's common-mode settling, and the non-linear charge injection. It is preferable for the RA to have a low gain that is uniform across its dynamic range, than to have high gain for small amplitudes with a highly compressive non-linear characteristic. The uniformly low gain is more easily correctable with the IGE calibration, and is an example of a calibration-friendly analog design. However, that is not always possible. In addition, we may need to push the speed or performance beyond the linear settling region of

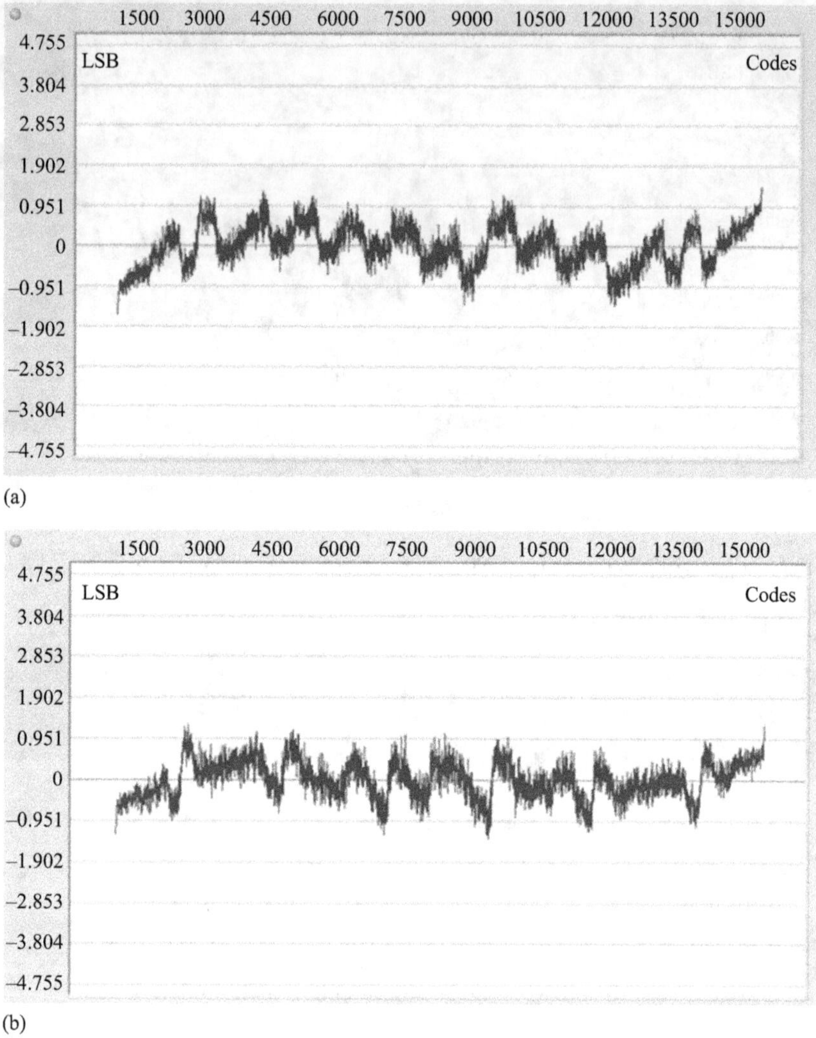

Figure 9.37 INL plots with kick-back calibration for a 14-bit 1GS/s for an input frequency of (a) 700MHz, (b) 900MHz. © 2014 IEEE. Reprinted, with permission, from Reference 5

the amplifier, and hence have to calibrate the amplifier's non-linearity as well. In this section, we discuss several approaches to calibrate the amplifier's non-linearity, which include correlation-based, statistical, and deterministic techniques.

The non-linearity of the inter-stage amplifier can be represented by modifying (9.2) of the ith stage residue V_{oi} as follows

$$V_{oi} = V_{oi}\big|_{\text{ideal}} + \alpha_1 V_{oi} + \alpha_2 V_{oi}^2 + \alpha_3 V_{oi}^3 + \cdots = V_{oi}\big|_{\text{ideal}} + \sum_{k=1}^{m} \alpha_k V_{oi}^k \quad (9.34)$$

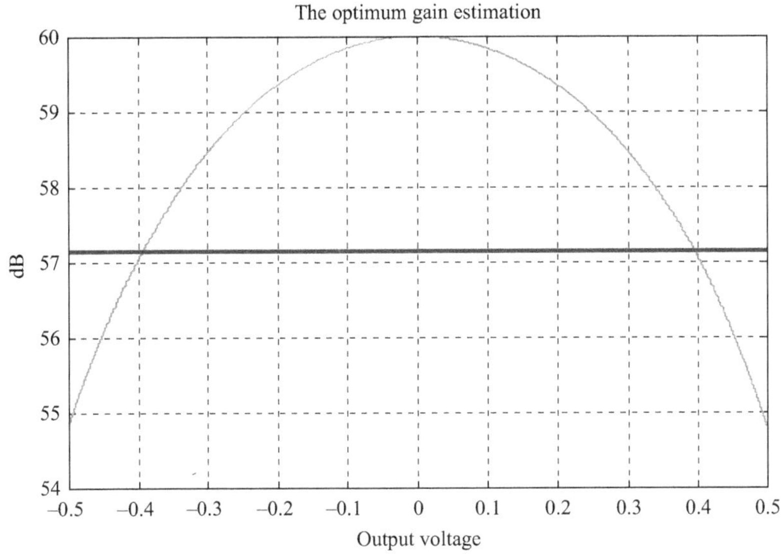

Figure 9.38 A plot of a non-linear amplifier gain with a fixed gain estimate also shown

where α_k is the kth order error term, and m is the highest order non-linearity considered. In the digital domain, these errors can be corrected by applying the opposite non-linearities to the digital residue $D(V_{oi})$ which can be approximated as

$$D(V_{oi})|_{\text{cal}} \approx D(V_{oi}) - \alpha_1 D(V_{oi}) - \alpha_2 D(V_{oi})^2 - \alpha_3 D(V_{oi})^3 - \cdots$$

$$\approx D(V_{oi}) - \sum_{k=1}^{m} \alpha_k D(V_{oi})^k \tag{9.35}$$

where $D(V_{oi})|_{\text{cal}}$ is the digital representation of the ith stage residue, and $D(V_{oi})|_{\text{cal}}$ is the ith stage calibrated digital residue.

In order to illustrate the possible need to calibrate the amplifier's non-linearity in some situations, an example is shown in Figure 9.38 for an amplifier with a compressive gain non-linearity. The gain of the amplifier changes from 60 dB for small signals down to 55 dB near the full-scale. Without any calibration, the SFDR of the resulting pipelined ADC is about 70 dB and the SNDR is limited to 64 dB, as shown in Figure 9.39. Calibrating for the gain error term gives a gain estimate of about 57 dB. The SFDR performance will improve to about 80 dB and the SNDR to about 75 dB, as shown in Figure 9.40. This indicates a substantial improvement, despite the residual non-linearity of the amplifier.

Using piecewise linear approximation of the non-linear gain with two segments, as shown in Figure 9.41, gives an SFDR performance of about 90 dB and SNDR of about 85 dB, as shown in Figure 9.42. Finally, using a third-order polynomial as shown in Figures 9.43 gives an SFDR and SNDR performance of better

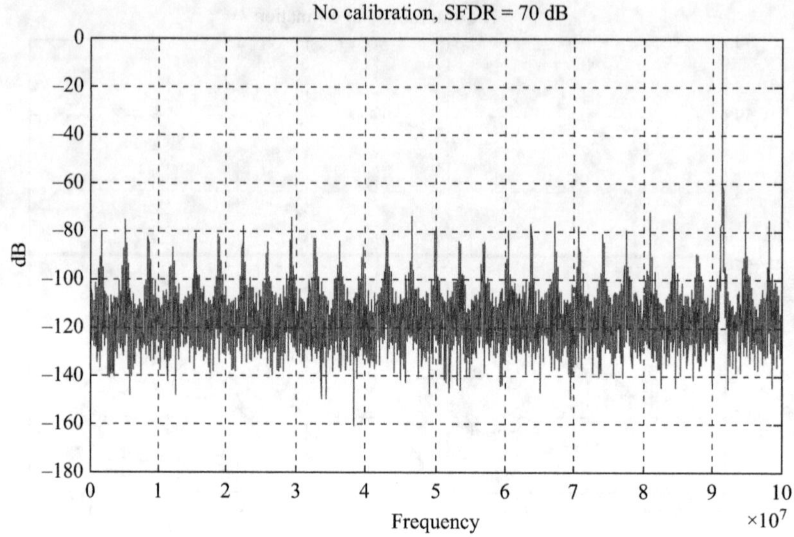

Figure 9.39 *An FFT of the digital output using the MDAC amplifier shown in Figure 9.38 without any correction. SFDR = 70 dB and SNDR = 64 dB*

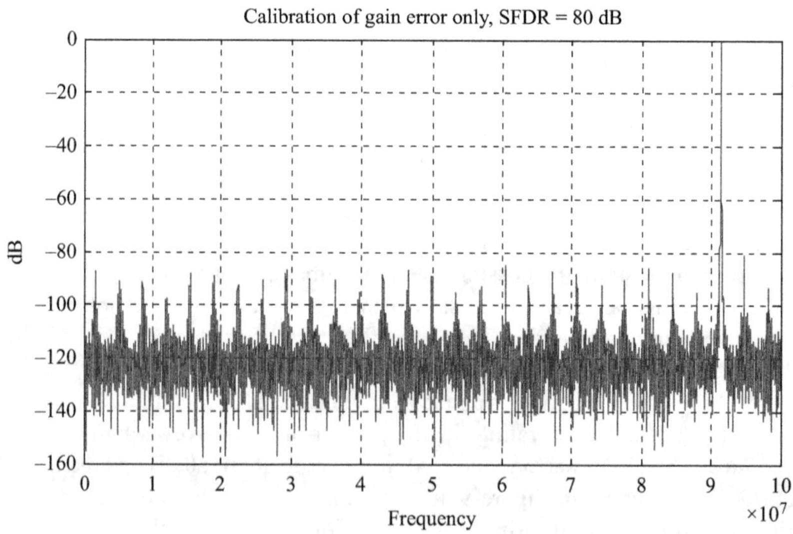

Figure 9.40 *An FFT of the digital output using the MDAC amplifier shown in Figure 9.38 with fixed gain correction using the estimate shown in Figure 9.38. SFDR = 80 dB and SNDR = 75 dB*

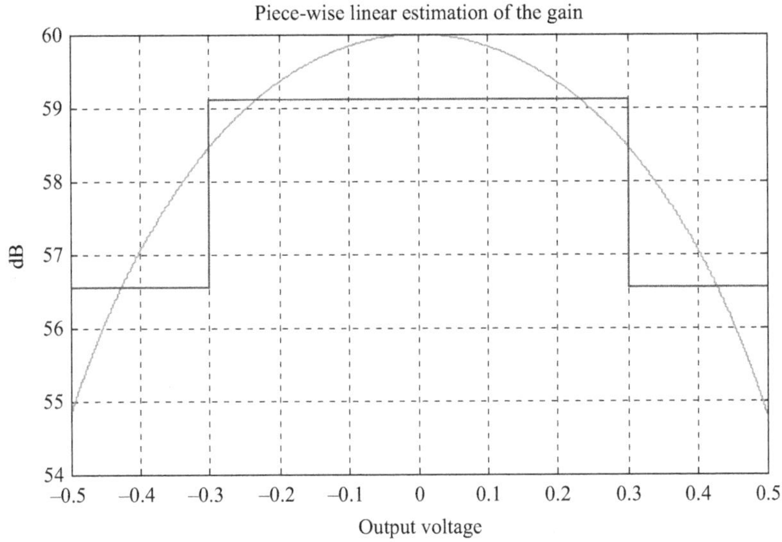

Figure 9.41 *A plot of a non-linear amplifier gain with a two-segment piece-wise linear estimate of the gain also shown*

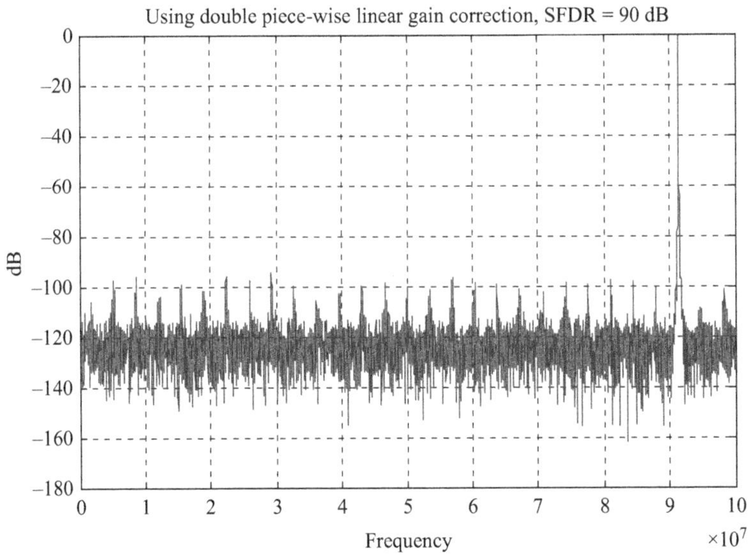

Figure 9.42 *An FFT of the digital output using the MDAC amplifier shown in Figure 9.41 with the piece-wise linear gain correction using the LMS estimates shown in Figure 9.41. SFDR = 95 dB and SNDR = 85 dB*

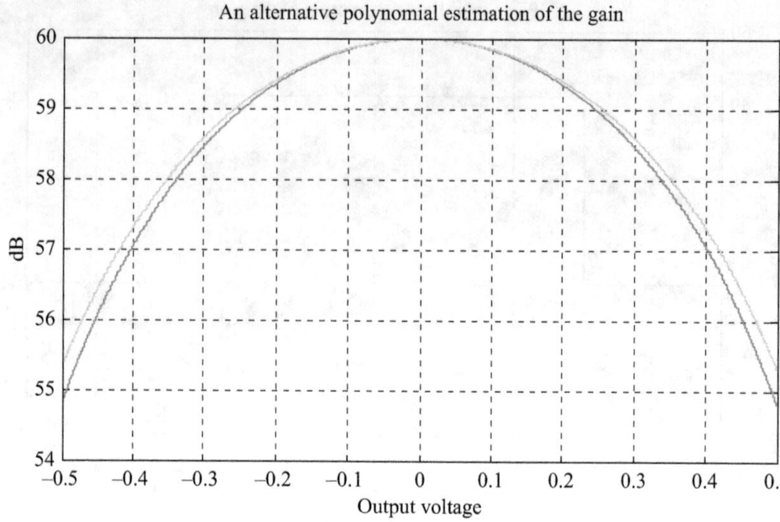

Figure 9.43 A plot of a non-linear amplifier gain with a polynomial estimate of the gain

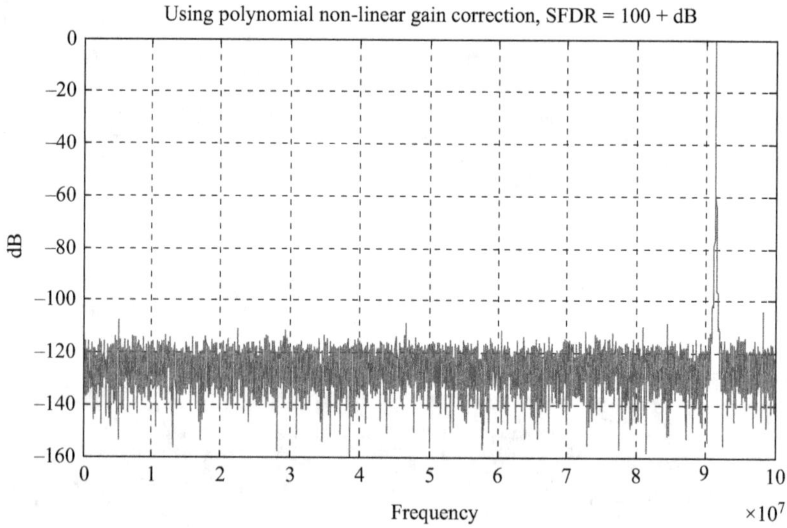

Figure 9.44 An FFT of the digital output using the MDAC amplifier shown in Figures 9.43 with the polynomial correction using the LMS estimate shown in Figures 9.43. SFDR = 110 dB and SNDR = 110 dB

than 110 dB as shown in Figure 9.44. That is, the SNDR will be practically limited by the thermal noise, which is modeled to be quite low in this case. In every case the digital correction represents the *inverse* of the analog error and non-linearity to correct their impact. This example illustrates the importance of correcting the amplifier's non-linearity and the progressive improvement in performance with gradually approaching the amplifier's real characteristics. It also illustrates the effectiveness of the piece-wise linear correction even with a few number of segments. In the following sections, we discuss some techniques to correct for the amplifier non-linearity.

9.1.9.1 Correlation-based method

The correlation-based IGE calibration can be extended to correct for the amplifier's non-linearity. This can be achieved by using multiple uncorrelated PN signals in the stage to be calibrated [9]. In the digital back end, the cross-correlation between the PN signals is estimated and used as a measure of the non-linearity. For example, a third-order non-linearity can be estimated using the cross correlation of three uncorrelated dither signals; a fifth-order non-linearity requires five dither signals, and so on.

An implementation of this calibration approach is discussed in References 8–10, where *m* uncorrelated dither signals are injected into the MDAC. The residue can be represented by

$$V_R[n] = V_R|_{\text{ideal}}[n] + \sum_{k=1}^{m} \alpha_k V_R^k[n] \tag{9.36}$$

where α_1 is the gain error term, α_2 is the second-order distortion term, α_3 is the third-order distortion term, and so on. Ignoring the even order terms, which are minimized by using a differential implementation, the algorithm's goal is to estimate the odd α terms. This is done by estimating the odd γ terms that represent the cross-correlations between the dither signals, such that the first-order term is

$$\gamma_1 = -K_1 \times \mathrm{E}\left[V_{d1}[n]\left(V_R[n] + \sum_{k=1}^{m} V_{dk}[n]\right)\right] \tag{9.37}$$

where E[z] is the estimation (or average value) of z. V_{dk} is the *k*th dither signal, V_R is the residue signal, and K_1 is a constant. The third-order term is expressed as

$$\gamma_3 = -K_3 \times \mathrm{E}\left[V_{d1}[n]V_{d2}[n]V_{d3}[n]\left(V_R[n] + \sum_{k=1}^{m} V_{dk}[n]\right)\right] \tag{9.38}$$

The fifth-order term is

$$\gamma_5 = -K_5 \times \mathrm{E}\left[V_{d1}[n]V_{d2}[n]V_{d3}[n]V_{d4}[n]V_{d5}[n]\left(V_R[n] + \sum_{k=1}^{m} V_{dk}[n]\right)\right] \tag{9.39}$$

where

$$K_i = \frac{A^{-2i}}{i!} \tag{9.40}$$

and A is the amplitude of each calibration dither signal, E[z] is the estimation (or average value) of z, $V_R[n]$ is the residue signal, and $V_{dk}[n]$ is the kth dither signal.

The cross-correlation γ values are not equal to the α values because of the effect of the higher-order terms. It can be shown that [9, 10]

$$\gamma_3 = \alpha_3 + \alpha_5\left[30A^2 + 10Y[n]\right] \tag{9.41}$$

and

$$\gamma_1 = \alpha_1 + \alpha_3\left(13A^2 + 3Y[n]\right) + \alpha_5\left\{241A^4 + 130A^2(Y[n] + 5Z[n])\right\} \tag{9.42}$$

where,

$$Y[n] = \mathrm{E}\left[\left(V_R[n] + \sum_{k=1}^{m} V_{dk}[n]\right)^2\right] \tag{9.43}$$

and

$$Z[n] = \mathrm{E}\left[\left(V_R[n] + \sum_{k=1}^{m} V_{dk}[n]\right)^4\right] \tag{9.44}$$

Therefore, after estimating the γ values, the α values will need to be estimated as shown in Figure 9.45 [9].

In spite of the demonstrated effectiveness of this algorithm, it suffers from long convergence time. The number of samples needed depends on the required

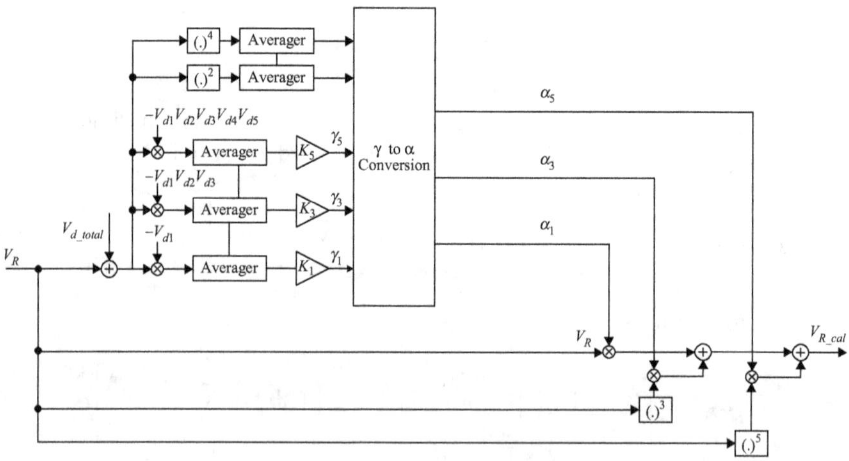

Figure 9.45 A block diagram showing the correlation-based calibration of the MDAC amplifier's IGE and non-linearity. © 2006 IEEE. Reprinted, with permission, from Reference 9

accuracy ε, the amplitude of the signal, the amplitude of the dither, and the number of dither signals, which is related to the order of the non-linearity that needs to be corrected. This is approximated as follows [9]

$$N \propto \frac{V_{signal}^{2k}}{(k!)^2 \varepsilon^2 V_{dither}^{2k}} \tag{9.45}$$

where ε is the allowed convergence error, V_{dither} is the amplitude of each dither signal, V_{signal} is the signal amplitude, and k is the order of the distortion to be corrected. The number of samples required for third-order non-linear correction can be in the billions, which would result in minutes of convergence time for sampling rates below 100 MS/s [9, 10].

9.1.9.2 Statistics-based approach

In statistical approaches, histograms are used to deduce the IGE and the amplifier's non-linearity by statistically estimating the distances between two residue signals or between two sub-ranges of the stage under calibration. This can be done in the factory or foreground using the INL or residue waveforms for a large number of samples in the presence of the input, or by forcing the comparator to switch between two different decisions to measure the corresponding "jump" in the output residue. Extending this approach to correct for the amplifier's non-linearity is straightforward.

Alternatively, non-linear calibration can be employed in the background as proposed in Reference 23. In that work, a PN sequence is used to switch the flash thresholds between two modes and hence create enough variation in the output to estimate the IGE and non-linearity. This is shown in Figure 9.46, where an additional bit is added in the flash ADC to shift the thresholds between the two modes. If the amplifier were linear, the distances h_1 and h_2 would be equal. However, the amplifier's non-linearity causes them to be different, with h_2 typically being less than h_1 in the presence of compression. By measuring the distances h_1 and h_2 over a large number of input samples, estimates of the IGE and non-linearity can be obtained.

The distances h_1 and h_2 are measured by estimating the cumulative histograms of the codes that correspond to the locations of those parameters. Once determined, they can be used to estimate the parameters p_1 and p_2, which are used to perform the gain and non-linear correction, as shown in Figure 9.47.

The estimation is performed using the LMS algorithm. One loop (with step size μ_1) uses an estimate of the parameter h_2 to estimate the gain correction parameter p_1, which is applied to the residue as a gain correction term. The term h_{ideal} is used as a correction factor to account for constant gain errors that need to be zeroed out. The other loop (with step size μ_2) uses estimates of both h_1 and h_2 to estimate the parameter p_2, which is used with look-up tables to correct for the amplifier's non-linearity.

This method requires an input signal with acceptable distributions that can be restrictive to its application. It also requires some analog design overhead, which is

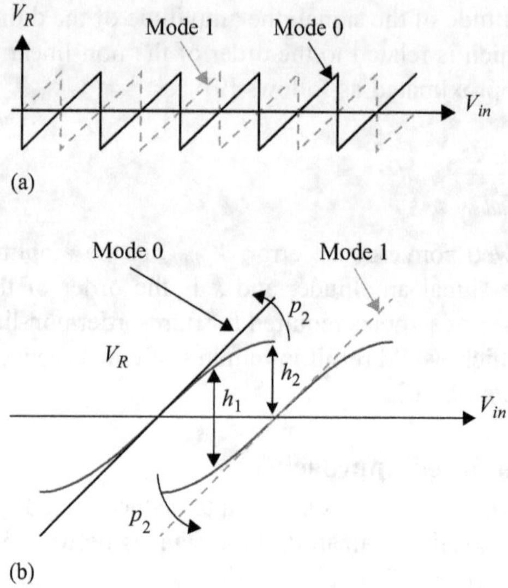

Figure 9.46 A plot of the stage's residue in the two modes of operation. (a) Two modes of operation for a linear residue. (b) Two modes in the presence of non-linearity [22–24]. © 2003 IEEE. Reprinted, with permission, from Reference 23

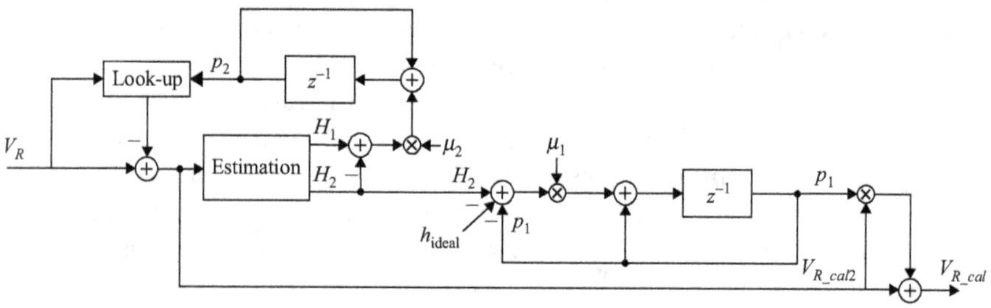

Figure 9.47 A block diagram showing the statistical calibration of the MDAC amplifier's IGE and non-linearity [23, 24]. © 2003 IEEE. Reprinted, with permission, from Reference 23

not substantial. In addition, it requires digital processing and storage that can be significant.

9.1.9.3 Using summing node sampling

A third approach to calibrate the MDAC amplifier's non-linearity is to use a deterministic approach, such as: the summing node algorithm [7]. The summing node voltage is sampled and used to estimate the gain for different ranges of signal amplitudes, using separate buckets for the LMS algorithm. Then, piece-wise linear or polynomial fit can be employed to approximate and correct for the non-linearity.

Unlike correlation-based approaches, this technique has very fast convergence, which is in the order of a few thousand samples, and does not cause dither correlation issues when dividing the samples into buckets.

9.1.10 Coupling calibration

Correlation-based approaches can be extended to correct for coupling errors. This is achieved by injecting an uncorrelated PN signal at the location of the offending signal, and detecting it at the destination. Using the LMS algorithm, the coupling coefficient can be estimated. This is effective in reducing cross-talk, digital coupling, and inter-channel coupling in multi-channel ADCs [25].

9.2 Dither

In spite of the effectiveness of the calibration algorithms, the linearization of the ADCs may still be limited in its accuracy after calibration. *Subtractive dither* can be used to linearize any residual non-linearity by randomizing the non-linear effects and hence spreading them in the noise floor. Subsequently, it is subtracted in the digital back end. Unlike calibration, dithering does not "fix" the errors or improve the SINAD of the ADC. In fact, it sometimes degrades the SINAD because of the imperfect subtraction of the dither signal. However, it can substantially improve the linearity, and hence the SFDR of the ADC. In addition, it can also reduce the variation of the ADC gain with the input amplitude, and improve the accuracy of the background calibration by linearizing the back end, as described earlier.

The dither can be injected on the input signal to propagate down the pipeline and be eventually subtracted in the digital back end. The LMS algorithm can be used to calibrate the dither's subtraction in the background to track changing temperature, supply, and sampling rate, as shown in Figure 9.48.

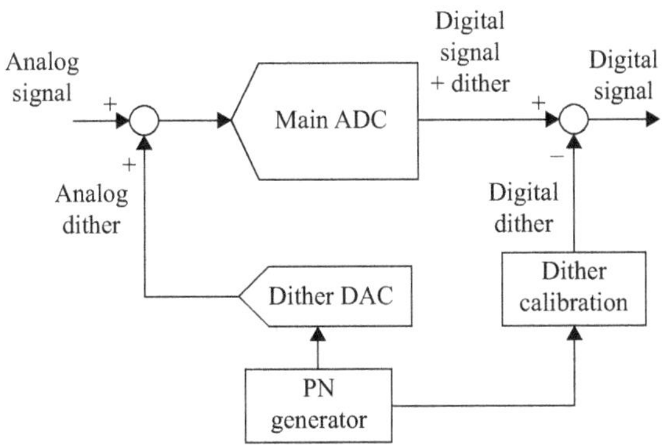

Figure 9.48 A block diagram showing the dither injection and calibration. The dither is used to linearize the ADC

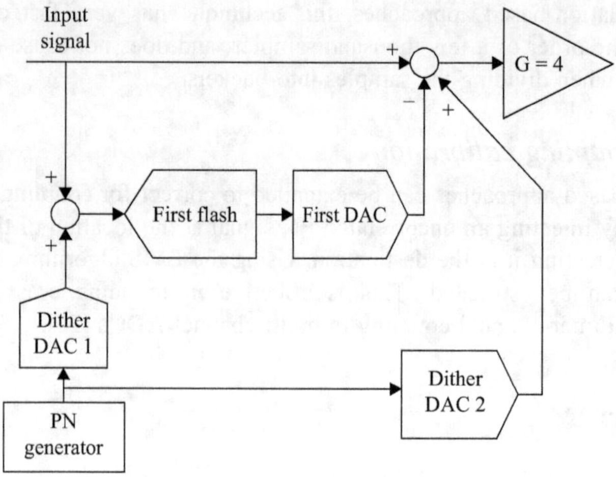

Figure 9.49 A block diagram showing the dither injection in the pipeline stage. The dither is injected in both the MDAC and the flash. It does not use any portion of the correction range

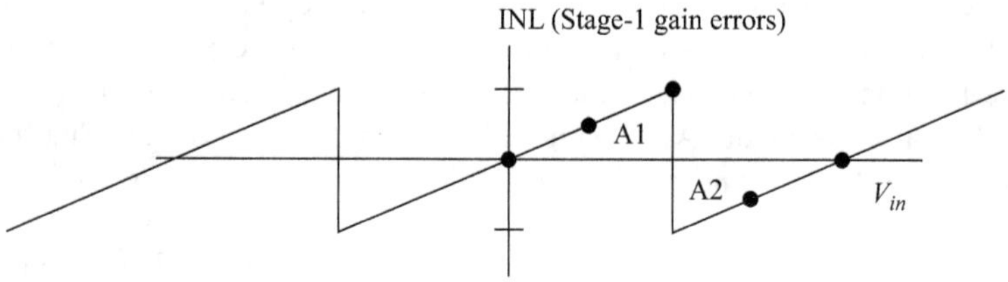

Figure 9.50 A plot of the residue showing the inter-stage gain error and the dither's effectiveness. If A1 is equal to A2, a dither with a magnitude equal to the sub-range size (or multiples of it) is capable of dithering the saw-tooth pattern that results from the inter-stage gain error

In pipelined ADCs, the dither can be injected in both the MDAC and the flash of the first stage, as shown in Figure 9.49. This is equivalent to injecting it on the input of the ADC, which is necessary to effectively dither the IGE errors of the first stage.

It is important to note that a dither signal that is injected on the input of the ADC (i.e. in both the MDAC and flash of the ADC) will "see" the IGE as an INL saw-tooth pattern, as shown in Figure 9.50. Therefore, it will effectively dither that error if the area A1 is equal to A2 *and* if the dither is large enough to span the whole sub-range. However, this dither signal cannot be used as a calibration signal to measure the IGE because it sees the gain of the whole ADC, which is marked by the dotted lines emanating from the origin in Figure 9.51, and not the inter-stage gain.

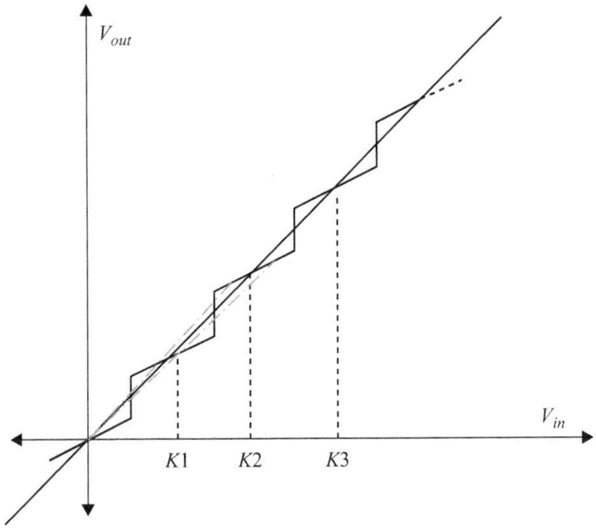

Figure 9.51 *The output of an ADC versus its input in the presence of inter-stage gain errors. The different sub-ranges are denoted by K1, K2, K3, etc. The inter-stage gain error is given by the slope of the segments, the overall gain is given by the slope of the dotted lines from the origin to the operating point*

On the other hand, a dither signal injected in the MDAC *only* (but not in the flash) can be used as a calibration signal to measure the IGE, but cannot effectively dither the IGE of the stage in which it is injected. This dither experiences the inter-stage gain given by the slope of the various sub-ranges in Figure 9.51, but does not "cross" the sub-ranges, and hence cannot dither them. However, it can dither the errors of the back-end stages, as long as it is amplified to be large enough to cover a full sub-range of that back-end stage.

Another observation is that a dither injected in the MDAC only will consume a portion of the correction range, while a dither injected in both the MDAC and flash will not consume (at least ideally) any portion of the correction range. It may consume, however, a portion of the ADC's dynamic range because it is added to the input signal. This can be avoided by using an extra comparator in the first-stage flash, such that the last two sub-ranges on both sides are only half-used when a full-scale input signal is applied, as was shown previously in Figure 9.20 and again in Figure 9.52. The dither can occupy the two remaining half sub-ranges beyond the full-scale range, without compromising the dynamic range or the linearity of the ADC, as long as its peak-to-peak amplitude is equal to or less than a sub-range. Fortunately, this amplitude is usually adequate to effectively dither the IGE of the first stage.

In order to estimate the improvement in SFDR due to dithering, we need to review the SFDR due to a saw-tooth pattern, which is discussed in Chapters 1 and 2.

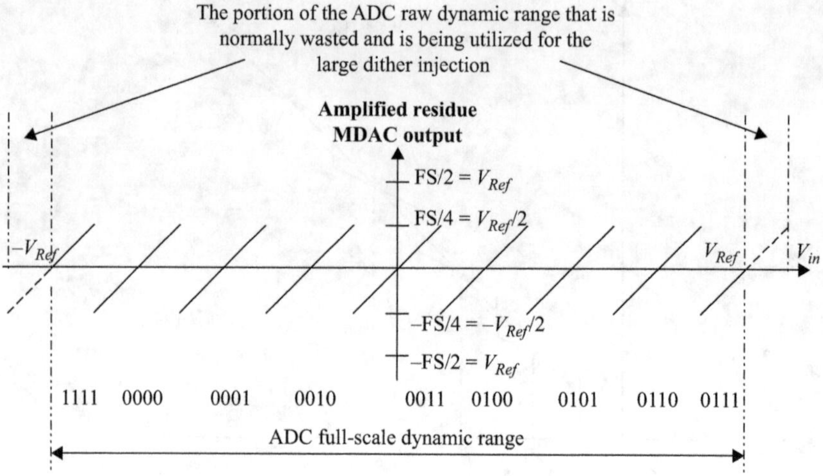

Figure 9.52 A stage-1 residue showing the portion of the dynamic range that is usually not utilized when using an additional comparator in the flash and can be consumed by the large dither signal

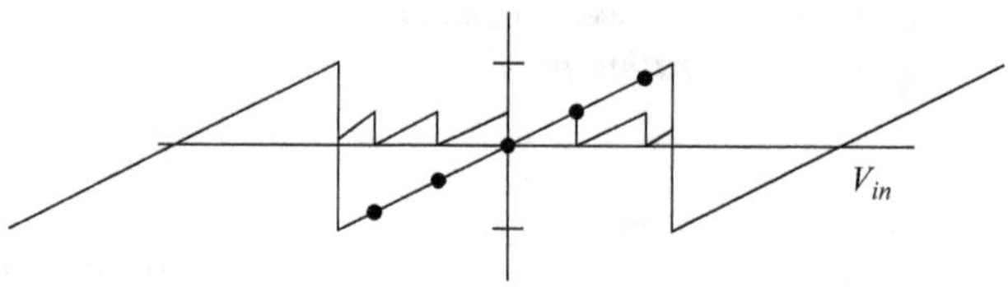

Figure 9.53 A plot of the residue (quantization error) showing the effectiveness of the dither in reducing the spurs due to the quantization noise by increasing the number of segments and reducing the peak magnitude. The total quantization energy is not improved

The saw-tooth error pattern looks like the standard quantization error, whose SFDR for n bits is given by

$$\text{SFDR} = 9n - c \tag{9.46}$$

where c is a constant. Every additional bit, reduces the error energy by 6 dB, and doubles the number of spurs, which results in an additional 3 dB reduction, and hence a total of 9 dB improvement. Since n-bit dithering can be thought of as quantization with an additional n bits, which happen to be random, the SFDR improvement due to dither is expected to follow the same quantization trend. Therefore, dithering a saw-tooth error pattern by n bits of dither (N levels, where $n = \log_2 N$) results in an improvement of about $9n$ dB in SFDR ($30 \log N$) at the same input amplitude. This is shown conceptually in Figure 9.53. In addition, the dither bits act as additional quantization bits in the stage, which effectively increase

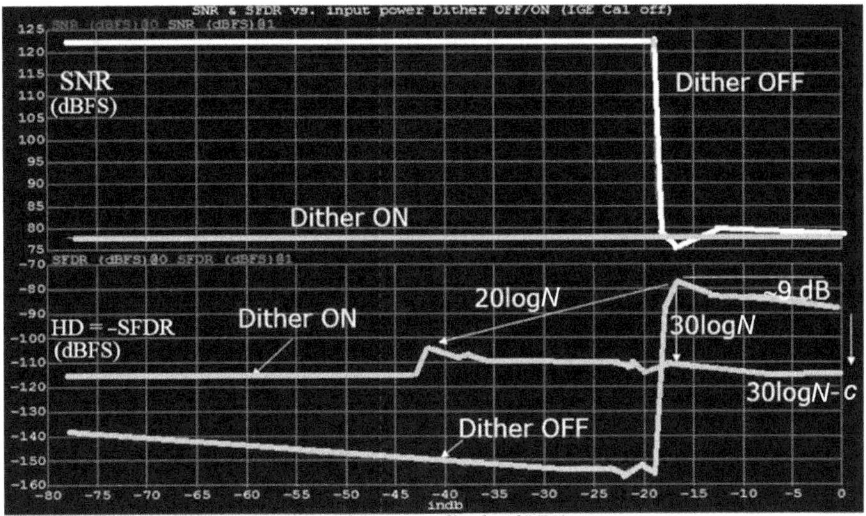

Figure 9.54 A simulation plot showing the effect of dithering with 4 bits (N = 16) on SNR and SFDR for a 3-bit/stage pipeline with IGE in the first stage

the number of sub-ranges and reduce the maximum error amplitude. Therefore, the worst case SFDR, which typically occurs at the boundary of a single sub-range, will occur at a different point that is lower in amplitude by $20 \log N$ and improve by $6n$ ($20 \log N$) in magnitude.

For example, if the pipeline stage is 3 bits, the worst case SFDR normally occurs at -18 dBFS, which is $-6k_1$ ($k_1 = 3$). If this stage is dithered by 16 levels of dither, the SFDR at the same amplitude (i.e. at -18 dBFS) will improve by $30 \log 16$ (i.e. 36 dB better). In addition, the SFDR notch will move by $6n$ ($n = 4$ bits of dithering) from -18 dBFS to -42 dBFS, and its level will improve by $20 \log 16$, which is 24 dB. This is illustrated in Figure 9.54, which shows the SNR and SFDR simulation results with and without dither, in the presence of an IGE in stage-1.

In Figure 9.54, we can also see that the dither does not improve the SNR. In fact, without dither, the SNR improves substantially for input signals that are smaller in amplitude than a single sub-range of stage-1. In the presence of dither, the dithering of the IGE degrades the SNR for very small signals.

So to summarize, if we use N levels of dither, we expect:

- The improvement in SFDR at the same amplitude to be: $30 \log N$
- The improvement in SFDR at the worst case amplitude to be: $20 \log N$
- The amplitude of the worst case SFDR is reduced by: $20 \log N$.

Measurement results are shown in Figures 9.55 and 9.56 for the SFDR and SNR, respectively. This is a 16-bit 250 MS/s with 3-bits in the first stage and 4 bits/stage in the following stages. The results show the impact of using calibration and

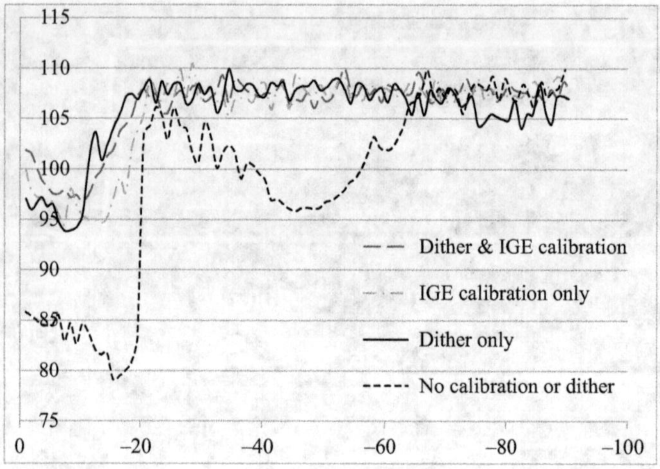

Figure 9.55 A plot of SFDR (in dBFS) versus input amplitude (in dBFS) showing the effect on SFDR due to dither, calibration, and both

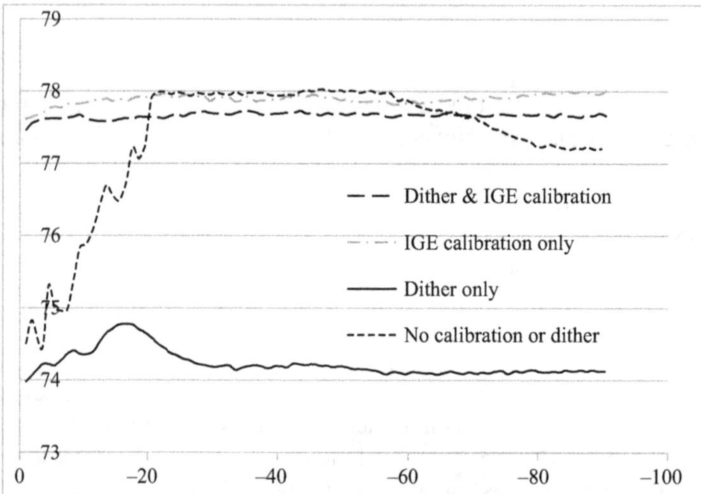

Figure 9.56 A plot of SNR (in dBFS) versus input amplitude (in dBFS) showing the effect on SNR due to dither, calibration, and both

dither. In the baseline curve of Figure 9.55, without calibration or dither, we can see degraded performance due to the inter-stage gain errors. In addition, the notches are clear at the stage-1 sub-range boundary at −18 dBFS ($-6k_1$, $k_1 = 3$) and the stage-2 sub-range boundary at −42 dBFS ($-6k_1 - 6k_2$, $k_1 = 3$, $k_2 = 4$). The SFDR improves substantially with dither or calibration. The best performance is clearly observed with both calibration and dither combined.

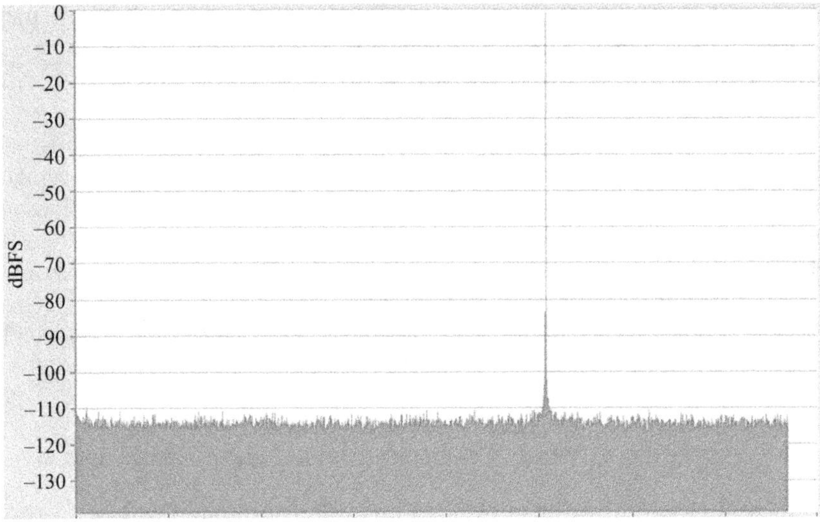

Figure 9.57 An FFT plot of an ADC showing the effectiveness of the dither in delivering an outstandingly clean spectrum with better than 110 dB of SFDR

Similarly, for the SNR curve in Figure 9.56, we see the baseline SNR, and the improvement with calibration. On the other hand, dither alone does not improve the SNR. In fact, it degrades the SNR as expected, and the degradation can be substantial for small signal amplitudes. This is because the calibration fixes the errors thereby improving the SNR, while the dither spreads the errors in the noise floor and hence does not improve the SNR but can actually degrade it in some cases. Once the linearization takes place, the dither is subtracted in the digital back end. Any imperfection in the dither subtraction will result in an SNR degradation.

An example of an output spectrum of an ADC with dither is shown in Figure 9.57. This is an astoundingly excellent linearity which represents the current state of the art and the best published linearity achieved for a high speed ADC in the literature [7].

9.3 Flash sub-ADC calibration

In addition to the MDAC, the calibration of the flash sub-ADC in a pipelined ADC has gained attention. The flash offsets lead to the consumption of a portion of the correction range, which impacts the MDAC's amplifier linearity and power consumption. Calibrating these offsets relaxes the requirements on the MDAC's dynamic range and helps reduce the power consumption. In addition, in SHA-less architectures, the bandwidth and timing mismatches between the MDAC and the flash create errors that also consume a portion of the correction range, and degrade with the input frequency. These mismatches impose an upper limit on the input frequencies that can be sampled by the ADC.

Although these offsets and mismatches can be factory calibrated, it is desirable to develop background calibration techniques that track the variation due to temperature, supply, sampling rate, and aging. Examples of these flash ADC calibrations can be found in References 37–39.

In Reference 39, a background calibration algorithm is described that uses the residue signal to correct the comparator offsets in the background, while averaging out errors due to bandwidth and timing mismatches. This separation of the static offsets from the errors due to phase mismatches is key to achieving accurate offset correction, and to prevent over- or under-correction. The technique is also independent of the input signal characteristics and distribution. It uses the first-stage flash code and the residue of the first stage to generate an error function given by

$$\varepsilon = \mathrm{E}\{[\mathrm{maxL}(V_R)|_{\mathrm{code=x}}] + [\mathrm{minL}(V_R)|_{\mathrm{code=(x+1)}}]\} \tag{9.47}$$

Alternatively, the digital output can be used instead of the residue as follows

$$\varepsilon = \mathrm{E}\{[\mathrm{maxL}(V_{out})|_{\mathrm{code=x}}], [\mathrm{minL}(V_{out})|_{\mathrm{code=(x+1)}}]\} - \mathrm{Ideal\_threshold} \tag{9.48}$$

where ε is the error function used to estimate the offset, V_R is the first stage's digital residue, V_{out} is the ADC digital output, E{z} is the expectation (or average) of all elements of z and E{z,y} is the expectation of all elements of z and y combined. The functions maxL() and minL() are leaky peak and trough detector functions, respectively. A leaky peak detector can be implemented as follows

$$V_p[n+1] = V_p[n] + \alpha(V_R[n] - V_p[n]) \tag{9.49}$$

where V_p is the peak value, V_R is the residue, and the factor α determines the time constant, such that:

$$\alpha = \begin{cases} \text{Large, if } V_R[n] - V_p[n] \text{ is positive => attack time constant is small (fast)} \\ \text{Small, if } V_R[n] - V_p[n] \text{ is negative => decay time constant is large (slow)} \end{cases}$$

On the other hand, the leaky trough detector is implemented as follows

$$V_{tr}[n+1] = V_{tr}[n] + \gamma(V_R[n] - V_{tr}[n]) \tag{9.50}$$

where V_{tr} is the trough value, V_R is the residue, and the factor γ determines the time constant, such that:

$$\gamma = \begin{cases} \text{Large, if } V_R[n] - V_p[n] \text{ is negative => attack time constant is small (fast)} \\ \text{Small, if } V_R[n] - V_p[n] \text{ is positive => decay time constant is large (slow)} \end{cases}$$

These values are accumulated until enough samples are collected to ensure adequate pairing between the two neighboring sub-ranges, and to average out the effects of the phase mismatch and hence eliminate its contribution. This is shown in Figure 9.58 for a 3-bit MDAC that has 8 sub-ranges, or more accurately, 7 sub-ranges plus 2 half sub-ranges. One of the comparators has an offset, which appears as a shift in the threshold value between codes 0110 and 0111. Since this is

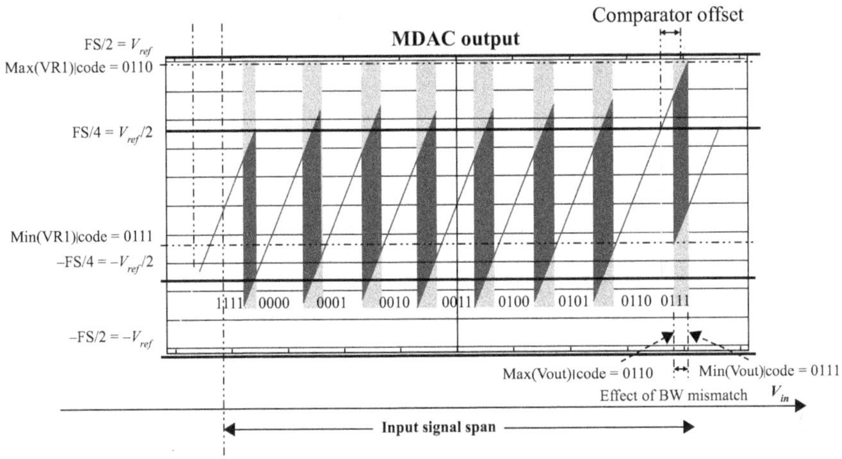

Figure 9.58 A simplified representation of the first-stage residue measured in the lab. The effect of the comparator offset is shown as a shift in one of the sub-ranges. The timing/BW mismatch between the flash and MDAC appears as the shaded areas at the boundaries between the sub-ranges. An example of the parameters used by the algorithm is superimposed on the plot [39]

a SHA-less ADC, the timing and bandwidth mismatches between the flash and the MDAC create errors that consume a portion of the correction range. These mismatch errors depend on the instantaneous amplitude and slope of the input signal. Therefore, they change sign and magnitude with the input signal and appear as shaded regions at the sub-range boundaries in Figure 9.58.

The offset control can then be given by: $V_{th} = V_{th0} + \mu\,\varepsilon$ (9.51)

where μ is a weighting factor that controls the time constant of the algorithm, V_{th} is the comparator's threshold voltage, and ε is the error defined in (9.47) or (9.48). This threshold voltage is fed back to the comparator to control its offset.

It is interesting to note that this flash ADC calibration is facilitated by the pipeline architecture. The MDAC's residue contains the needed information about the comparators' behavior. This can be processed in the digital back end to extract the offset and/or the bandwidth mismatch, and use that to feedback an offset correction parameter to the comparators in the flash, without disrupting their normal operation.

9.4 Calibration of mismatches in interleaved ADCs

In interleaved ADCs, background calibration can be used to estimate and correct inter-channel mismatches. These include offset, gain, and timing mismatches. Offset and gain mismatches are relatively straightforward to estimate and fix in the digital domain. Timing mismatch, on the other hand, can be quite challenging and represents an active area of research. In the following sections, some mismatch calibration techniques are discussed. For simplicity, they will be shown in the context of two-way interleaving. However, the concepts can be applicable to larger numbers of channels.

9.4.1 Offset mismatch calibration

The offset of the ADC can be estimated using a low-pass filter or integrator to filter out the signal. By measuring the offsets of the individual channels, the required DC corrections can be added to their outputs to eliminate the mismatches in the digital domain. This algorithm can be applied in the factory, in the foreground or in the background. This is shown in Figure 9.59 for a two-way interleaved ADC, where the correction is done by removing the offsets from both channels. This technique, however, does not differentiate between the offset and a DC input signal.

Chopping can be employed to preserve the DC input from being removed [26] such that the only energy at DC would be from the ADC offsets. This is shown in Figure 9.60. Since chopping is applied on the analog input, it must be applied again on the digital output to undo its effect on the input signal. This chopping will convert a DC input into a square wave or a random signal depending on the code used for chopping. Therefore, the DC that remains is the DC that is added after the chopping, which is the DC offset.

9.4.2 Gain mismatch calibration

The gain of the ADC can be estimated from its digital output using multiple approaches. One method is to estimate the average power of the output signal for each channel. A mismatch in the average power between the channels is considered

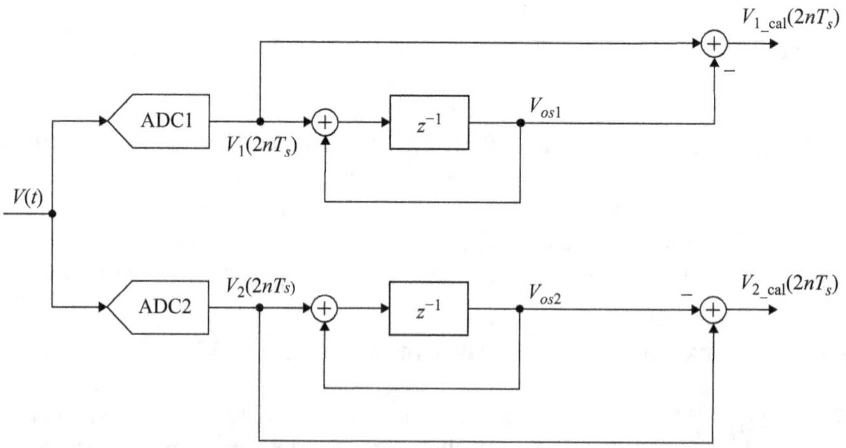

Figure 9.59 A block diagram of an offset mismatch calibration in a two-way interleaved ADC

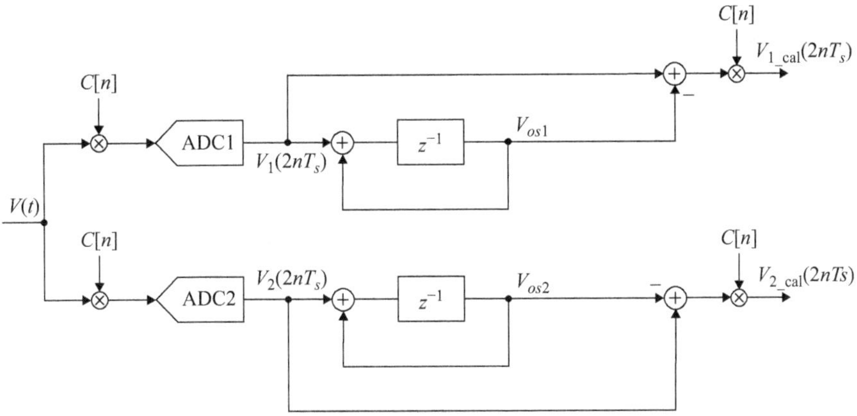

Figure 9.60 A block diagram of an offset mismatch calibration in a two-way interleaved ADC that employs chopping. © 1998, 2002 IEEE. Reprinted, with permission, from References 26 and 27

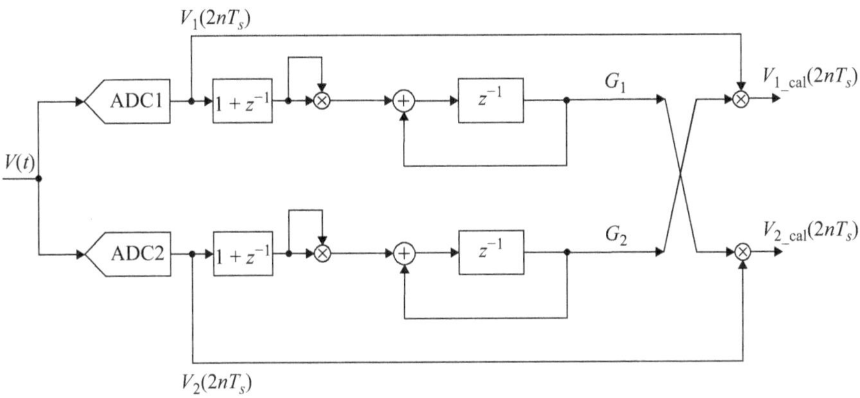

Figure 9.61 A conceptual representation of gain mismatch calibration in a two-way interleaved ADC

a gain mismatch and is corrected using a digital multiplier. This is shown in Figure 9.61. The average power estimation can be done using a squarer or an absolute value estimate of the digital output.

If the input signal's frequency is $f_s/2M$, where M is the number of channels and f_s is the interleaved ADC sampling rate, this will be the Nyquist frequency of each channel. The digital power estimate of this Nyquist signal changes with the sampling instant (phase) even in the absence of mismatches, and hence can confuse the algorithm and lead to misleading results. Therefore, it is important to filter out the energy at the Nyquist frequency before the power estimation. This is performed in Figure 9.61 using the $(1 + z^{-1})$ filter, which has a zero at $z = -1$. This corresponds to a zero at the Nyquist frequency of the individual channels, which acts as a notch filter to remove or attenuate this problematic signal. Once the gain of each channel is estimated, it can be used to normalize the other channels as shown conceptually in Figure 9.61.

Another method for calibrating the gain mismatch involves multiplying the interleaved digital output by $(-1)^n$ [26]. As discussed in Chapter 8, the gain mismatch creates an image of the fundamental signal (f_{in}) that is located at $f_s - f_{in}$ and is in phase with the fundamental. Therefore, this multiplication of the signal by its chopped version produces a DC component that is proportional to the gain mismatch. This can be shown by analyzing the effect of chopping, given by

$$V_{ch}[n] = (-1)^n V[n] \tag{9.52}$$

where $V_{ch}[n]$ is the chopped version of the interleaved signal, and $V[n]$ is the original interleaved signal. In the frequency domain, this gives

$$V_{ch}(e^{j\omega}) = V(e^{j(\omega-\pi)}) \tag{9.53}$$

which in the continuous time domain gives

$$V_{ch}(\omega) = V(\omega - \omega_N) \tag{9.54}$$

where ω_N is the Nyquist angular frequency. Therefore, chopping moves the signal to the location of its image and the correlation between the signal and its chopped version gives an output that is proportional to the gain mismatch ΔG_c. That is,

$$\Delta G_c[n+1] = \Delta G_c[n] + \mu \times V[n] \times V_{ch}[n] \tag{9.55}$$

As shown in Figure 9.62, the interleaved signal $V[n]$ in (9.55) is obtained by adding the outputs from the individual channels after up-sampling and delaying one of the channels by one sample. In addition, a notch filter $(1 + z^{-2})$ at $f_s/4$ is used to filter out the signal at $f_s/4$ that would cause misleading gain estimates as mentioned earlier. Unlike the filtering performed in Figure 9.61, this filtering is performed

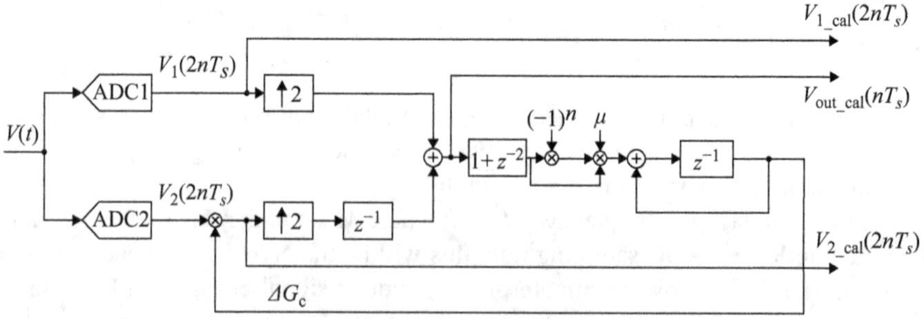

Figure 9.62 A block diagram of gain mismatch calibration in a two-way interleaved ADC [26]. © 2002 IEEE. Reprinted, with permission, from Reference 26

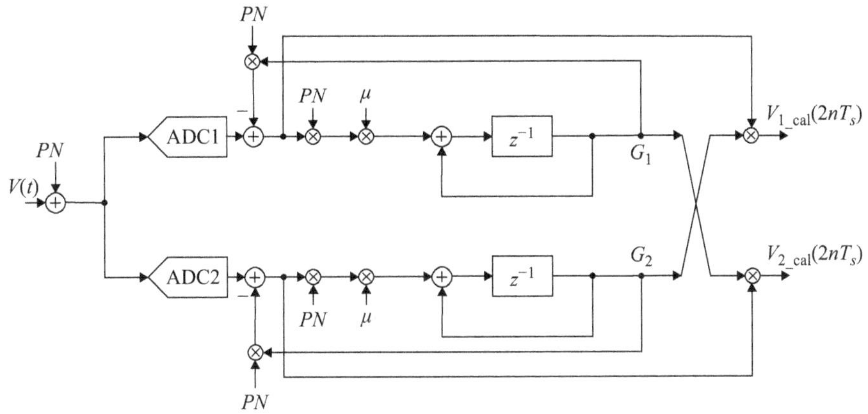

Figure 9.63 A block diagram of a correlation-based gain mismatch calibration in a two-way interleaved ADC using an injected dither signal (PN). © 1998 IEEE. Reprinted, with permission, from Reference 27

after up-sampling. Therefore, the filter needs to be in the form of $(1 + z^{-2})$, instead of the $(1 + z^{-1})$ that was used in Figure 9.61. Finally, the LMS algorithm is used to estimate the gain error and apply it to one of the channels.

These gain estimation techniques rely on the presence of an input signal for the calibration to converge. In the absence of an input signal, the calibration needs to be frozen to prevent wrong convergence. Alternatively, a signal can be injected to ensure activity and continuous calibration at all times. This calibration signal is usually in the form of a pseudo-random signal that is uncorrelated with the input signal. The LMS algorithm can be used to correlate out the signal and estimate the gain of each channel [27]. This is shown in Figure 9.63, and is given by

$$G_i[n+1] = G_i[n] - \mu \times PN[n] \times (PN[n] \times G_i[n] - V_i[n]) \tag{9.56}$$

where G_i is the gain estimate of the ith channel, $PN[n]$ is the injected PN signal, μ is the step size, and $V_i[n]$ is digital of the ith channel.

The gain mismatches can also be estimated using a reference ADC [28, 32, 35, 36], which can be used to correct for multiple kinds of mismatches. Using the LMS algorithm, the gain error of each channel can be eliminated using the formula

$$G_i[n+1] = G_i[n] - \mu \times V_i[n] \times (V_i[n] \times G_i[n] - V_{REF}[n]) \tag{9.57}$$

where G_i is the gain of the ith channel, V_i is the output if the ith channel, μ is the step size, and V_{REF} is the output of the reference ADC. Alternatively, the signs of the signals can be used instead to simplify the digital processing at the expense of convergence time. This is given by

$$G_i[n+1] = G_i[n] - \mu \times \mathrm{sgn}(V_i[n]) \times \mathrm{sgn}(V_i[n] \times G_i[n] - V_{REF}[n]) \tag{9.58}$$

The reference ADC can have high noise, but needs to have a gain that is independent of the input signal amplitude.

9.4.3 Timing mismatch calibration

The background calibration of the timing mismatches in interleaved ADCs is a challenging calibration problem. Since the effect of timing mismatches degrades as the input frequency increases, this problem becomes significantly more challenging for high input frequencies and RF sampling.

Good layout practices are usually employed to improve the timing mismatches. Proximity of the sampling networks of the different channels helps with the matching. In addition, using a common clock whenever possible helps reduce the mismatch errors further. However, with the best effort, the timing mismatch tend to be in the order of a few hundred femto-seconds or worse. This typically results in SINADs that are in the order of 50 dB for an input signal of about 2 GHz. It is important to note that a systematic or fixed timing mismatch can be corrected by applying a fixed adjustment to the sampling clock. However, mismatches that change with temperature, supply, sampling rate, or aging are more problematic as they require algorithms that operate in the background and adaptively track the timing changes. The timing correction can be done in the analog or digital domains.

9.4.3.1 Analog correction

In analog correction, the sampling clock passes through programmable delays that adjust the sampling instant of the different channels to eliminate the timing mismatches. This is shown conceptually in Figure 9.64. The delays can be controlled by switchable small load capacitances in the clock path. Alternatively, the threshold level of the sampling switch, or the level of its clock, can be adjusted in order to achieve finer timing adjustments [39].

9.4.3.2 Digital correction

Digital correction relies on interpolation to deduce the expected values of the samples at the correct time instants in-between the available samples. Digital filtering is typically used to perform the interpolation and correction. This is shown in Figure 9.65 and an example is shown in Figure 9.66 [26]. The discrete-time nature of the digital output creates fundamental limitations on the bandwidth and frequency of the signals whose timing mismatch can be corrected in the digital domain. Increasing the complexity of the interpolator helps, but limitations will

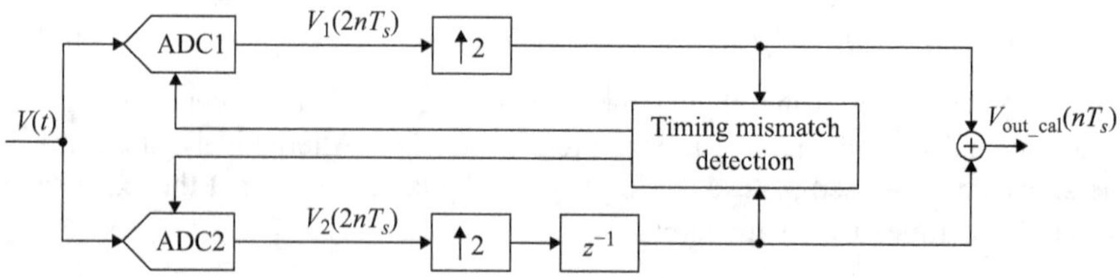

Figure 9.64 A block diagram of a timing mismatch calibration in a two-way interleaved ADC using digital detection and analog correction. © 2011 IEEE. Reprinted, with permission, from Reference 31

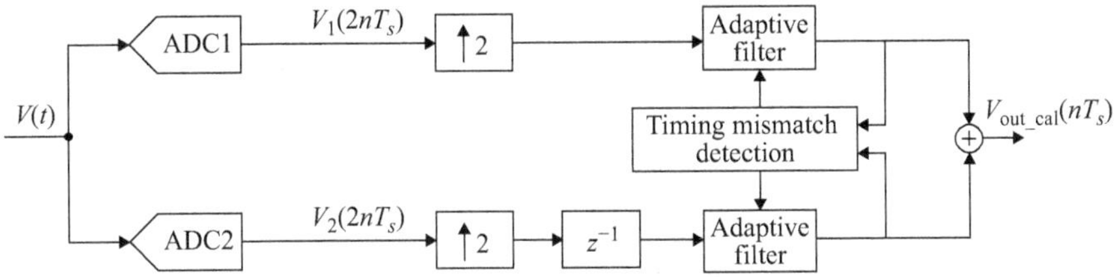

Figure 9.65 A block diagram of a timing mismatch calibration in a two-way interleaved ADC using digital detection and digital correction. © 2011 IEEE. Reprinted, with permission, from Reference 31

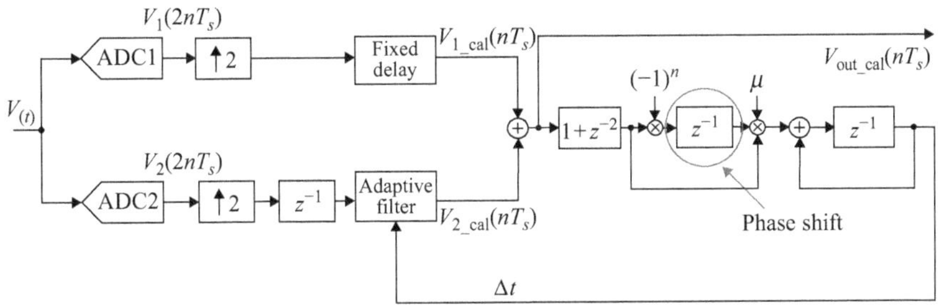

Figure 9.66 A block diagram of timing mismatch calibration in a two-way interleaved ADC using the LMS algorithm and an adaptive filter for correction. © 2002 IEEE. Reprinted, with permission, from Reference 26

still exist. Therefore, analog correction of timing mismatches is often preferred to avoid the complexity and limitations of the digital timing correction [26].

Aside from the correction, the main challenge in the timing mismatch calibration is the background estimation of the mismatch without disrupting the normal operation of the ADC. There are several approaches, which are discussed in the following sections.

9.4.3.3 Estimation using chopping

An example is shown in Figure 9.66, where the correlation between the interleaved signal and a chopped version of it is used to estimate the timing mismatch. This is the same methodology used for the gain mismatch correction of Figure 9.62, where the LMS algorithm is employed to estimate the timing mismatch error. An important difference between calibration of the timing mismatch in Figure 9.66 and the gain mismatch in Figure 9.62 is the phase shift introduced to the chopped signal. This is marked by the circle in Figure 9.66. The timing mismatch causes an image at the same location as the image due to the gain mismatch. However, since the effect of the timing mismatch depends on the slope of the signal, the image is 90° out of phase with the signal. This means that the chopped signal will be orthogonal to the image signal, and hence their product will average to zero. A 90° phase shift is needed for the chopped signal, which requires a Hilbert transform [26]. In this implementation, this is approximated by the delay (z^{-1}), which shifts the phase by 90° near $f_s/4$. However, this $f_s/4$ frequency is the location of the zero created by the notch filter ($1 + z^{-2}$). Moreover, the phase shift created by the delay (z^{-1}) will approach zero at low frequencies and 180° near $f_s/2$. However, near $f_s/4$, some energy will remain and be phase shifted adequately by 90° to give an estimate of the timing mismatch. More elaborate filters can be used to achieve more accurate phase shifting and better detection.

9.4.3.4 Estimation using reference ADC

A reference ADC that samples the input at the same instants as the individual channels can be used to detect the timing mismatch. The LMS algorithm uses the timing error and the signal slope to estimate the timing mismatch. An additional resistance ΔR is added in the input network of the reference ADC. As shown in Figure 9.67 [32], the outputs of the two channels are subtracted to give the error e. The slope D is estimated using one of the channels and the reference channel that has the additional resistance ΔR, which is given by

$$D = V_2 - V_R \tag{9.59}$$

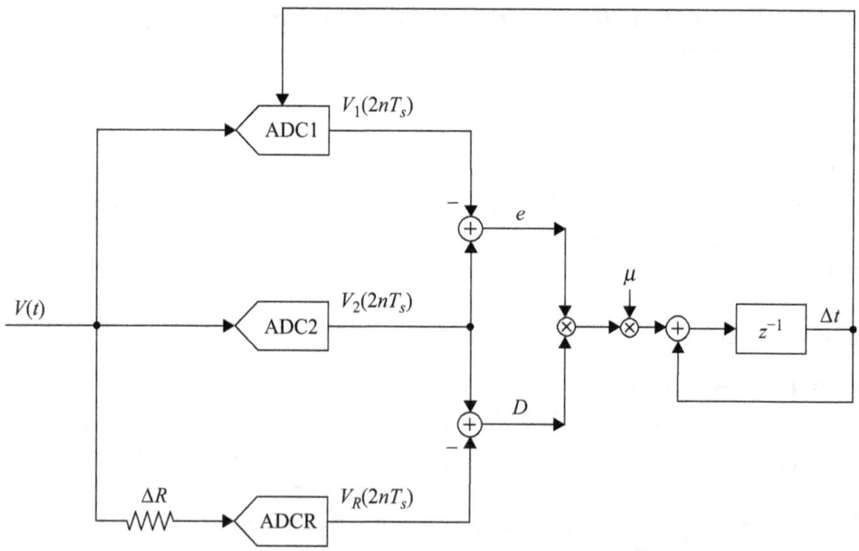

Figure 9.67 A block diagram of a timing mismatch calibration in a two-way interleaved ADC using the LMS algorithm and reference ADC. The correction is done in the analog domain by controlling the timing of one of the ADCs [32]. © 2013 IEEE. Reprinted, with permission, from Reference 32

The error e is given by

$$e = V_1 - V_2 \tag{9.60}$$

The LMS algorithm is implemented by

$$\Delta t_{i+1} = \Delta t_i + \mu \times e \times D \tag{9.61}$$

where Δt is the estimate of the timing mismatch, and μ is the algorithm's step size. The timing mismatch estimate is fed back to one of the ADCs, in order to fix its timing. This feedback loop will converge to a solution where the timing mismatch is minimized.

9.4.3.5 Estimation using cross-correlation

The output of each ADC is cross-correlated with an additional ADC (ADCC) as shown in Figure 9.68 [31]. The feedback loops converge to a point that maximizes the cross-correlations. The reference ADC can be low resolution (even one bit) and can be slow, as long as its sampling instant coincides regularly with each of the ADC channels. If the reference is clocked at the full speed f_s, it will be able to coincide with all the channels, which are running at f_s/M each. However, designing such a fast ADC can be quite challenging. Alternatively, the correlation ADC can

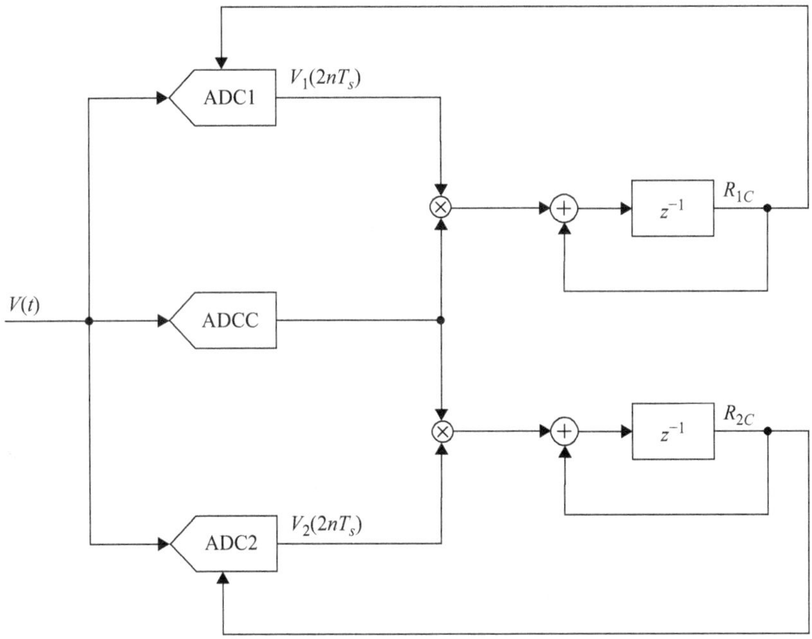

Figure 9.68 A block diagram of a timing mismatch calibration in a two-way interleaved ADC using cross-correlation. The correction is done in the analog domain by controlling the timing of one of the ADCs. © 2011 IEEE. Reprinted, with permission, from Reference 31

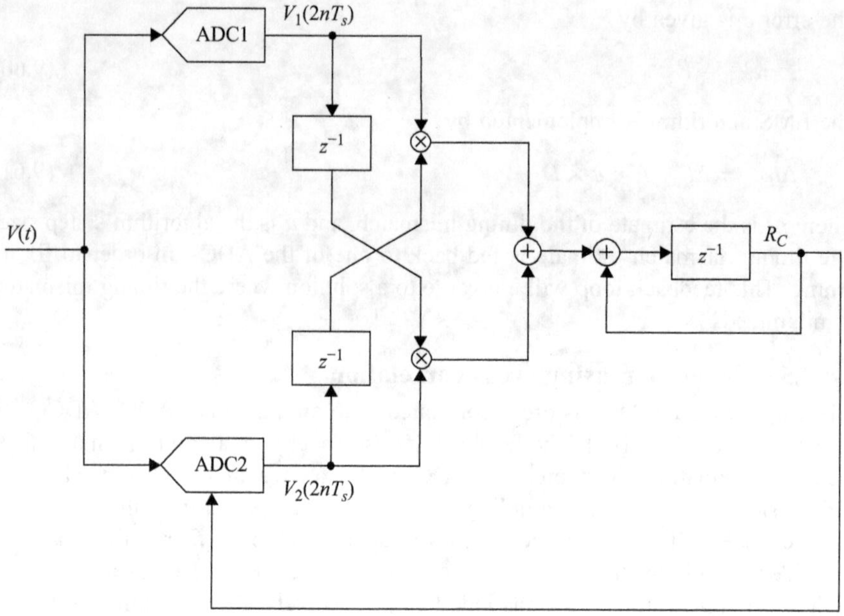

Figure 9.69 A block diagram of a timing mismatch calibration in a two-way interleaved ADC using cross-correlation. The correction is done in the analog domain by controlling the timing of one of the ADCs.

be clocked at a lower rate f_s/K, such that the greatest common divisor between M and K is 1. For example, that condition is satisfied if M is equal to 8 and K is equal to 9. In this case, clocking the correlation ADC at $f_s/9$ would be adequate, as it ensures that its sampling time will cyclically coincide with each of the other channels in a periodic fashion.

An alternative cross-correlation approach that does not require an additional ADC is described is shown in Figure 9.69 [40]. In this approach, the outputs of the two channels are cross-correlated after a delay of one cycle. These delays are necessary to circumvent the fact that the signal and its image are orthogonal, and hence their cross-correlation would have averaged to zero even in the presence of timing mismatch. Therefore, by adding the delays, the loop converges only when the cross-correlation is zero and the mismatch is minimized.

9.4.3.6 Estimation using split ADC

In this approach each of the interleaved ADC cores is broken down into two halves. The differences (errors) between the two halves are used as estimates of the mismatch errors [29, 30]. Shuffling the halves will exercise all combinations, and the LMS algorithm can be used in a manner similar to the reference ADC method. Intuitively, it is clear that an extra half-channel will be needed to allow for all combinations.

As discussed earlier, the split ADC method has analog and digital overhead. However, it has the additional advantage of being compatible with channel randomization, which is discussed in Chapter 8 and later in this chapter.

9.4.4 Other mismatches

In addition to offset, gain, and timing mismatches, interleaved ADCs may suffer from other sources of error, such as bandwidth and non-linearity mismatches. The bandwidth mismatch causes phase mismatch similar to that of the timing mismatch. However, it also causes frequency-dependent gain mismatch that can be very problematic. Examples of techniques to address the bandwidth mismatch are described in References 34 and 35. In practice, designers may prefer to increase the analog bandwidth such that the effect of its mismatch is negligible.

The mismatch in the ADC non-linearity is another source of error that is often left uncorrected. It is usually not a dominant source of error if the harmonic distortion level is low. An example of the research done to calibrate this kind of mismatch using the reference ADC approach is described in Reference 36.

9.4.5 Randomization

In order to improve the spurs due to any residual mismatch errors after calibration, channel randomization can be employed as discussed in Chapter 8. The order of the channel usage is randomized to avoid the periodic pattern that creates the interleaving spurs. This spreads the mismatch spur energy into the noise floor, which improves the linearity and SFDR of the ADC. Unlike calibration, randomization does not fix the mismatches or improve the SNR or SINAD. It merely disperses the spur energy into the noise floor without fixing their cause.

In order to effectively randomize the channels, at least one additional channel is required. This can be intuitively understood if we note that the same channel cannot be used to handle two samples that are spaced less than M samples apart, where M is the number of interleaved channels. This is the speed limitation that necessitated interleaving in the first place. For example, if we have two channels, the randomization cannot be achieved without an additional third channel, because none of the two channels is fast enough to handle two successive sample. This represents a significant overhead.

One of the issues with randomization is that it spreads all of the mismatch spurs' energy in the noise floor, even those that might have existed in benign locations and hence were practically harmless. For example, in two-way interleaving, the spur due to the offset mismatch exists at the Nyquist frequency ($f_s/2$), which may not be a problem for most applications. However, randomization spreads this energy in the noise floor, which makes it imperative to minimize the offset mismatch before randomization. In addition, since the channel shuffling is not completely random, it may create "humps" and "coloring" in the noise floor as discussed in Chapter 8. Another limitation of randomization is the memory effects that can be exacerbated when changing the channel ordering as is done in randomization, thereby causing a degradation in the SINAD and NSD. These memory effects may have to be calibrated before randomization as shown in References 12, 25, and 34.

9.5 Conclusion

In this chapter, we discussed some of the advanced calibration techniques used in pipelined and time-interleaved ADC. These techniques represent the state-of-the-art in digitally assisted ADCs. They enable higher sampling rates, higher performance, lower power consumption, and better integration in fine lithography CMOS processes. It is important to note that the field of digitally assisted converters is an active area of research with developments and breakthroughs occurring at a fast pace. This chapter is meant to present a snapshot of the state of the art with some of the most effective techniques. The common theme is that there are a lot of limitations and active problems to solve. The field of digitally assisted converters is nowhere where it needs to be. So, analog designers should not worry about being obsolete yet.

Problems

1. Using a behavioral modeling language, build a model of a 2-stage pipelined ADC with 3 bits in every stage and a 3-bit back-end flash. The full-scale is 2 V. With 1-bit redundancy, what is the resolution of the whole ADC? What are the inter-stage gain values? Plot an FFT of the digital output using a unity amplitude sine wave with 100 MHz frequency.
   (a) If the inter-stage gain has an error of 1%, plot the FFT and INL of the digital output.
   (b) Apply an IGE correction in the digital domain, and plot the FFT and INL of the digital output.
   (c) Apply an IGE correction using code-dependent addition, and plot the FFT and INL of the digital output.
   (d) Comment on the difference between the plots of parts (b) and (c)?
2. Add a random signal (dither) to the input of the pipeline in previous problem with amplitude equal to −24 dBFS, and subtract it from the digital back end. Plot the FFTs and INLs of the previous problem with the dither added.
3. Model a 4-bit MDAC using SPICE or a behavioral modeling language.
   (a) Investigate the amplifier settling of (9.25) and (9.27), and correcting it using a gain term.
   (b) Force the initial voltage to be non-linear using the formula $v_{initial} = 0.01v_{in}^3 + 0.01v_{in}^5$ or any other non-linear formula. Investigate the effect of the initial condition on the output when the settling time is adequate and inadequate.
4. Given the similarities and differences between the SAR and pipelined ADCs, how do they compare in terms of digital assistance? Discuss how the feedback nature of a SAR ADC compare with the feed-forward pipelined ADC? What kind of errors can be calibrated digitally in a SAR ADC?
5. Discuss if and how the correlation-based calibration algorithm can be applied to a SAR ADC.

6. Simulate the LMS algorithm and apply it to a 1-bit dither signal combined with a sinusoidal signal, in order to accurately subtract the dither from a sinusoidal signal. Investigate the impact of the dither amplitude, the sine wave amplitude, and the step size μ on the results.
7. Simulate the MDAC of Figure 9.21 using a SPICE simulator. Using the summing node voltage to correct the inter-stage gain error of the output.
8. Using a behavioral modeling language, model the IGE calibration method using a reference ADC and the LMS algorithm. Apply it to the pipeline of Problem 1. Could you apply the same method to a SAR ADC with the same resolution?
9. Using a behavioral modeling language, model the IGE calibration method using the split-ADC method and the LMS algorithm. Apply it to the pipeline of Problem 1. Could you apply the same method to a SAR ADC with the same resolution?
10. For the pipeline model of Problem 1, apply the following non-linearity to the inter-stage amplifier: $V_{out} = 0.99\,V_{in} - 0.01\,V_{in}^3$. How does the non-linearity impact the output FFT? Apply gain correction, piece-wise linear correction and polynomial correction, and plot the output FFT in every case.
11. Using a behavioral modeling language, model the effects of offset and gain mismatch in a 2-way interleaved ADC. Model the algorithms described in Figures 9.60 and 9.61 for the offset and gain mismatch calibrations, respectively. Can we extend this approach to 4-way interleaving? If yes, draw the block diagram.
12. Using a behavioral modeling language, model the algorithms described in Figures 9.62 and 9.66 for the gain and timing mismatch calibrations, respectively. Can we extend this approach to 4-way interleaving? If yes, draw the block diagram.
13. Using a behavioral modeling language, model the algorithms described in Figures 9.67 and 9.68 for timing mismatch calibration. Can we extend this approach to 4-way interleaving? If yes, draw the block diagram.
14. What is the magnitude of the inter-stage gain causing the INL in Figure 9.5? What are the most likely causes of the IGE error?
15. What is the magnitude of the inter-stage gain causing the INL in Figure 9.6? What are the most likely causes of the IGE error? What could be causing the breaks within each sub-range? Discuss the possible causes among the following:
    (a) Capacitor mismatch
    (b) Amplifier open loop error
    (c) Reference error
16. What are the possible causes of the DAC errors in the INL of Figure 9.11? What are the magnitudes of the errors on the analog side? Note the symmetry between the two sides and the increase in the magnitude of the errors as we move away from the center.

References

[1] A. N. Karanicolas, H.-S. Lee, and K. L. Barcrania, "A 15-bit 1-Msample/s Digitally Self-Calibrated Pipeline ADC," *IEEE Journal of Solid-State Circuits*, 28, pp. 1207–1215, Dec 1993.

[2] E. Siragusa and I. Galton, "Gain Error Correction Technique for Pipelined Analogue-to-Digital Converters," *Electronic Letters*, 36(7), pp. 617–618, 2000.

[3] E. Siragusa and I. Galton, "A Digitally Enhanced 1.8-V 15-bit 40-MSample/s CMOS Pipelined ADC," *IEEE Journal of Solid-State Circuits*, 39(12), pp. 2126–2138, Dec 2004.

[4] J. Ming and S.H. Lewis, "An 8-bit 80-Msample/s Pipelined Analog-to-Digital Converter with Background Calibration," *IEEE Journal of Solid-State Circuits*, 36(10), pp. 1489–1497, 2001.

[5] A.M.A. Ali, H. Dinc, P. Bhoraskar, *et al.*, "A 14b 1GS/s RF Sampling Pipelined ADC with Background Calibration," *IEEE Journal of Solid-State Circuits*, 49(12), pp. 2857–2867, Dec 2014.

[6] A.M.A. Ali and K. Nagaraj, "Background Calibration of Operational Amplifier Gain Error in Pipelined A/D Converters," *IEEE Transactions on Circuits and Systems II*, 50(9), pp. 631–634, 2003.

[7] A.M.A. Ali, A. Morgan, C. Dillon, *et al.*, "A 16-bit 250-MS/s IF Sampling Pipelined ADC with Background Calibration," *IEEE Journal of Solid-State Circuits*, 45(12), pp. 2602–2612, Dec 2010.

[8] A. Panigada and I. Galton, "A 130mW 100MS/s Pipelined ADC with 69dB SNDR Enabled by Digital Harmonic Distortion Correction," *IEEE ISSCC Digest of Technical Papers*, pp. 162–163, Feb 2009.

[9] A. Panigada and I. Galton, "Digital Background Correction of Harmonic Distortion in Pipelined ADCs," *IEEE Transactions on Circuits and Systems–I: Regular Papers*, 53(9), pp. 1885–1895, Sep 2006.

[10] A. Panigada and I. Galton, "A 130mW 100MS/s Pipelined ADC with 69dB SNDR Enabled by Digital Harmonic Distortion Correction," *IEEE Journal of Solid-0State Circuits*, 44(12), pp. 3314–3328, Dec 2009.

[11] N. Rakuljic and I. Galton, "Suppression of Quantization-Induced Convergence Error in Pipelined ADCs with Harmonic Distortion Correction," *IEEE Transactions on Circuits and Systems I*, 60(3), pp. 593–602, Mar 2013.

[12] A.M.A. Ali, "Methods and structures that reduce memory effects in analog-to-digital converters," US Patent No 6,861,969, Mar 2005.

[13] J.P. Keane, P.J. Hurst, and S.H. Lewis, "Digital Background Calibration for Memory Effects in Pipelined Analog-to-Digital Converters," *IEEE Transactions on Circuits and Systems-I: Regular Papers*, 53(3), pp. 511–525, Mar. 2006.

[14] A.M.A. Ali and A. Morgan, "Calibration methods and structures for pipelined converter systems," US Patent 8,068,045, Nov 2011.

[15] A.M.A. Ali, A.C. Morgan, and S.G. Bardsley, "Correlation-based background calibration of pipelined converters with reduced power penalty," US Patent 7,786,910, Aug 2010.
[16] A.M.A. Ali, "Pipelined converter systems with enhanced accuracy," US Patent 7,271,750, Sep 2007.
[17] A.M.A. Ali, "Method and device for improving convergence time in correlation-based algorithms," US Patent 8,836,558, Sep. 2014.
[18] S. Sonkusale, J. Van der Spiegel, and K. Nagaraj, "True Background Calibration Technique for Pipelined ADC," *Electronic Letters*, pp. 786–788, 36(9), 2000.
[19] Y. Chiu, C.W. Tsang, B. Nikolic, and P.R. Gray, "Least Mean Square Adaptive Digital Background Calibration of Pipelined Analog-to-Digital Converters, *IEEE Transactions on Circuits and Systems-I: Regular Papers*, 51(1), pp. 38–46, Jan 2004.
[20] J.A. McNeill, J. McNeill, M.C.W. Coln, and B.J. Larivee, "Split ADC" Architecture for Deterministic Digital Background Calibration of a 16-bit 1-MS/s ADC, *IEEE Journal of Solid State Circuits*, 40(12), pp. 2437–2445, Dec 2005.
[21] E. Iroaga and B. Murmann, "A 12-Bit 75-MS/s Pipelined ADC Using Incomplete Settling," *IEEE Journal of Solid-State Circuits*, 42(4), pp. 748–756, Apr 2007.
[22] B. Murmann and B.E. Boser, "A 12 b 75 MS/s Pipelined ADC Using Open-Loop Residue Amplification," *IEEE ISSCC Digest of Technical Papers*, 1, pp. 328–497, 2003.
[23] B. Murmann and B.E. Boser, "A 12 b 75 MS/s Pipelined ADC Using Open-Loop Residue Amplification," *IEEE Journal of Solid-State Circuits*, 38(12), pp. 2040–2050, Dec 2003.
[24] B. Murmann and B.E. Boser, "Digital Domain Measurement and Cancellation of Residue Amplifier Nonlinearity in Pipelined ADCs," *IEEE Transactions on Instrumentation and Measurements*, 56(6), pp. 2504–2514, Dec 2007.
[25] A.M.A. Ali and H. Dinc, "Method and device for reducing inter-channel coupling in interleaved and multi-channel ADCs," US Patent 8,471,741, Jun 2013.
[26] S.M. Jamal, D. Fu, N.C.-J. Chang, *et al.*, "A 10-b 120-Msample/s Time-Interleaved Analog-to-Digital Converter with Digital Background Calibration," *IEEE Journal of Solid-State Circuits*, 37(12), pp. 1618–1627, Dec 2002.
[27] D. Fu, K.C. Dyer, S.H. Lewis, and P.J. Hurst, "A Digital Background Calibration Technique for Time-Interleaved Analog-to-Digital Converters," *IEEE Journal of Solid-State Circuits*, 33(12), pp. 1904–1911, Dec 1998.
[28] K. Dyer, D. Fu, S.H. Lewis, and P.J. Hurst, "An Analog Background Calibration Technique for Time-Interleaved Analog-to-Digital Converters," *IEEE Journal of Solid-State Circuits*, 33(12), pp. 1912–1919, Dec 1998.

[29] J.A. McNeill, C. David, M. Coln, and R. Croughwell, "Split ADC" Calibration for All-Digital Correction of Time-Interleaved ADC Errors, *IEEE Transactions on Circuits and Systems-II: Express Briefs*, 56(5), pp. 344–348, May 2009.

[30] J.A. McNeill, M.C.W. Coln, D.R. Brown, and B.J. Larivee, "Digital Background-Calibration Algorithm for "Split ADC" Architecture," *IEEE Transactions on Circuits and Systems-I: Regular Papers*, 56(2), pp. 294–306, Feb 2009.

[31] M. El-Chammas and B. Murmann, "A 12-GS/s 81-mW 5-bit Time-Interleaved Flash ADC with Background Timing Skew Calibration," *IEEE Journal of Solid-State Circuits*, 46 (4), pp. 838–847, Apr 2011.

[32] D. Stepanović and B. Nikolić, "A 2.8 GS/s 44.6 mW Time-Interleaved ADC Achieving 50.9 dB SNDR and 3 dB Effective Resolution Bandwidth of 1.5 GHz in 65 nm CMOS," *IEEE Journal of Solid-State Circuits*, 48(4), pp. 971–982, Apr 2013.

[33] G. Léger, E.J. Peralias, A. Rueda, and J.L. Huertas, "Impact of Random Channel Mismatch on the SNR and SFDR of Time-Interleaved ADCs," *IEEE Transactions on Circuits and Systems-I: Regular Papers*, 51(1), pp. 140–150, Jan 2004.

[34] C.H. Law, P.J. Hurst, and S.H. Lewis, "A Four-Channel Time-Interleaved ADC with Digital Calibration of Interchannel Timing and Memory Errors," *IEEE Journal of Solid-State Circuits*, 45(10), pp. 2091–2103, Oct 2010.

[35] T.-H. Tsai, P.J. Hurst, and S.H. Lewis, "Bandwidth Mismatch and Its Correction in Time-Interleaved Analog-to-Digital Converters," *IEEE Transactions on Circuits and Systems-II: Express Briefs*, 53(10), pp. 1133–1137, Oct 2006.

[36] W. Liu and Y. Chiu, "Time-Interleaved Analog-to-Digital Conversion with Online Adaptive Equalization," *IEEE Transactions on Circuits and Systems-I: Regular Papers*, 59(7), pp. 1384–1395, Jul 2012.

[37] P. Huang, S. Hsien, V. Lu, *et al.*, "SHA-Less Pipelined ADC with *In Situ* Background Clock-Skew Calibration," *IEEE Journal of Solid-State Circuits*, 46(8), pp. 1893–1903, Aug 2011.

[38] M. Brandolini, Y. Shin, K. Raviprakash, *et al.*, "A 5GS/s 150mW 10b SHA-Less Pipelined/SAR Hybrid ADC in 28nm CMOS," *IEEE ISSCC Digest of Technical Papers*, pp. 468–469, Feb 2015.

[39] A.M.A. Ali, H. Dinc, P. Bhoraskar, *et al.*, "A 14-bit 2.5GS/s and 5GS/s RF Sampling ADC with Background Calibration and Dither," *IEEE VLSI Circuits Symposium*, pp. 206–207, 2016.

[40] B. Razavi, "Design Considerations for Interleaved ADCs," *IEEE Journal of Solid-State Circuits*, 48(8), pp. 1806–1817, Aug 2013.

[41] B.D. Sahoo and B. Razavi, "A 12-Bit 200-MHz CMOS ADC," *IEEE Journal of Solid State Circuits*, 44(9), pp. 2366–2380, Sep. 2009.

Chapter 10
Evolution and trends

Over the past two decades, impressive progress has been made in the high speed and high resolution ADC space. The sampling rate and bandwidth have increased at an astounding pace. The resolution and linearity have improved to levels previously unimaginable. The input frequency has skyrocketed to reach the GHz range, and the clock jitter has been reduced to less than 50 fs. This progress has been accomplished while keeping the power consumption roughly the same, if not lower. These advancements were driven in part by the explosive growth in wireless communications and the unquenchable thirst for more bandwidth, higher performance, and lower power to accommodate our smart phones, tablets, laptops, entertainment, and numerous other applications.

In this chapter, we discuss the evolution of the high speed ADC space over the past two decades, while emphasizing trends and future directions. In addition to the overall view of the high speed space, we will focus on the evolution of the high speed and high resolution ADC subspace from the author's experience and perspective [1–4].

10.1 Performance evolution

High speed ADCs are usually defined in terms of resolution (in bits) and speed (in sampling rate). However, as discussed in Chapter 2, there are multiple metrics that are needed to fully describe the ADC's performance. While it is difficult to track the evolution of all the dimensions of performance of the vast high speed ADC space, it may be more insightful to focus on a subset of these metrics to highlight some overarching trends.

In the literature, there are some useful surveys discussing the evolution and trends of data converters [5–9]. There is also an online survey of state of the art developments in various dimensions of performance, which is created and kept up to date by B. Murmann [8]. Some of the plots from this survey are shown in Figures 10.1 and 10.2. Figure 10.1 is a plot of the energy efficiency, in power per sampling rate, as a function of SNDR. The lines corresponding to two figures-of-merit (FOM) are drawn on the plot. The dotted line corresponds to a "Walden FOM" of 5 fJ/step (*FOM*1 in Chapter 2). The solid line corresponds to a Schreier FOM of 175 dB (*FOM*3 in Chapter 2).

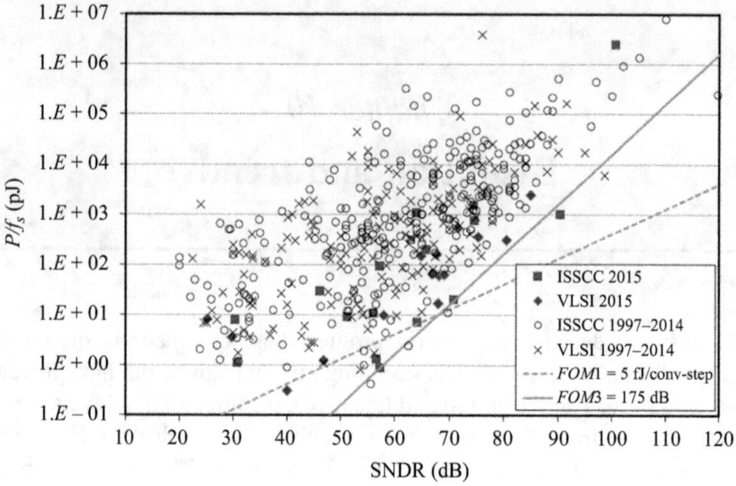

Figure 10.1 A scatter plot of energy efficiency (P/f_s) versus SNDR [8]

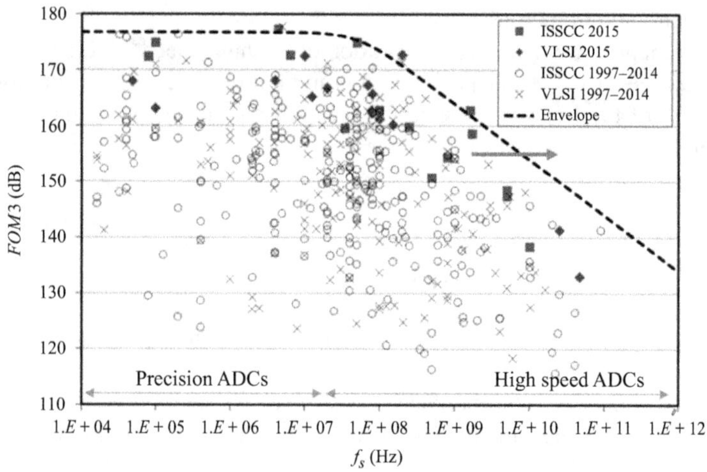

Figure 10.2 A scatter plot of the Schreier Figure-of-Merit (FOM3) versus sampling rate [8]

Figure 10.2 is a plot of the Schreier FOM (*FOM*3) versus sampling rate. High speed ADCs are on the right hand side of the plot (above 10 MS/s), while precision ADCs occupy the left hand side. It is interesting to note the degradation in the FOM for the high speed space compared to the precision space, because of the additional power required to achieve the high speed and the impact of the process technology limitations.

Some of the observations in these surveys are [8, 9]:

- The energy efficiency (P/f_s) improves by a factor of 2 every about 1.6 years.
- The Schreier FOM (*FOM*3) for high speed ADCs has seen an average improvement of about 1.82 dB every year from 1997 to 2011, and about 3 dB every 2 years.
- The term BW $\times$ 2^{ENOB} has remained relatively fixed, which indicates a relatively fixed noise spectral density (NSD).

These trends have been partially due to process and architectural limitations, and partially due to market demands that may drive progress in a certain dimension but not the other. For example, the growth and evolution of the wireless infrastructure market has seen growing demand for higher sampling rates, wider bandwidths, lower power consumption, and higher input frequencies. However, the requirements on the NSD have remained fixed in the range of -151 to -155 dBFS/Hz.

In addition to the whole high speed ADC space, it is interesting to cover the evolution of the subspace of high speed and high resolution ADCs for wireless infrastructure applications. If we start with the resolution and sampling rate, we can see a clear increase in both dimensions over the past 10–15 years. While the state of the art 10 years ago was a 14-bit, 125 MS/s ADC [1], it is currently a 14-bit, 2.5 GS/s non-interleaved ADC, and a 14-bit, 5 GS/s interleaved ADC [4]. Figure 10.3 shows the evolution over the last 8 years with the timeline for each milestone. It shows a 25$\times$ increase in sampling rate of non-interleaved ADCs in 7 years. This represents an exponential growth of doubling the speed approximately every 18 months.

In Figure 10.3, all the non-interleaved ADCs have an NSD of about -153 to -155 dBFS/Hz, and the power is in the range of 1–1.2 W. It is important to note that, as the sampling rate increases, system and ADC designers take advantage of the processing gain to reduce the SINAD, while keeping the NSD the same. On the other hand, if we were to keep the SINAD fixed while increasing the sampling rate, the power consumption would have increased substantially. Every 3 dB increase in SINAD requires doubling the power consumption, as shown in Figure 10.4.

This increase in sampling rate was enabled by faster processes (finer lithography CMOS processes), new architectures, enhanced analog circuit design, and digitally assisted techniques to take advantage of the advances in the process technology (such as device speed and lower parasitics), while minimizing the impact of its drawbacks (such as lower device gain, worse linearity, higher leakage, and smaller dynamic range).

Another important dimension of performance is linearity. In the past decade, the high speed ADC's SFDR has seen a substantial improvement from the 70–80 dB range to the 100 dB range for input frequencies up to 100 MHz [1, 2]. In addition, the input frequency range for the 70–80 dB level of SFDR performance increased substantially from the 10–50 MHz range to the 1–5 GHz range, as shown in Figure 10.5. These SFDR and input frequency improvements were enabled by enhancements in input sampling circuit techniques and process advancements [1–4]. They were driven in wireless base stations by the need to meet the multi-carrier

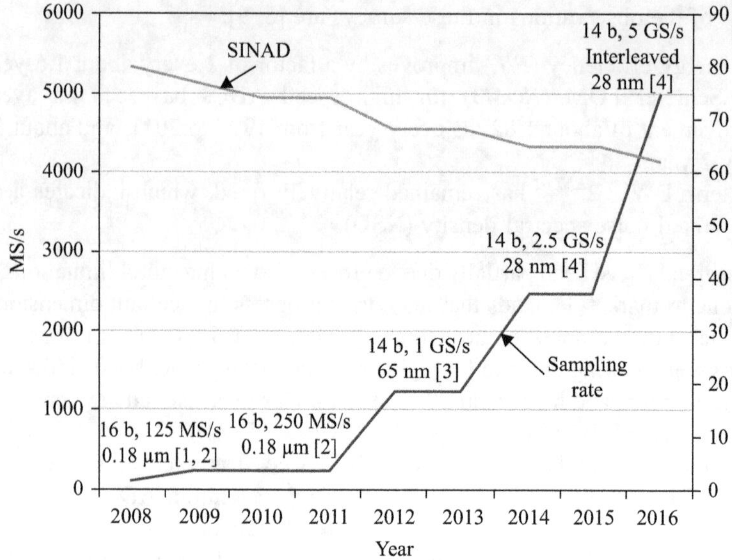

Figure 10.3 Progress of the sampling rate and SINAD versus time for high speed ADCs with resolutions of 14–16 bits and noise spectral density in the range of −153 to −155 dBFS/Hz. The speed increased by 25× in 7 years for the straight non-interleaved pipelined ADCs, and 50× in 8 years for 14-bit interleaved ADCs. The power/core remained essentially the same. As the sampling rate increased, the SINAD decreased to keep the NSD and the power the same

GSM specifications and the desire to achieve RF sampling, and hence move the ADC up the signal chain closer to the antenna. This helps reduce cost and increase flexibility by moving more of the processing to the digital domain.

Power consumption in this ADC space has remained approximately the same at about 1–1.2 W for the same NSD. The last data point, which is the interleaved 5 GS/s ADC, has a 3-dB better NSD and double the power. This indicates that the Schreier Figure-of-Merit (*FOM*3) defined in Chapter 2 remained about the same at around 154–155 dB, for all the data points of the high speed ADCs shown in Figure 10.3, in spite of the increase in sampling rate. This represents a horizontal push in the state-of-the-art line shown in Figure 10.2.

The relation between power consumption and sampling rate is depicted conceptually in Figure 10.6. For a particular process technology (x nm), the power consumption increases linearly with the sampling rate up to a point, beyond which the power starts to increase exponentially. When this point is approached, it becomes inefficient to push the sampling rate higher on the same process. Then, the designer needs to migrate to a finer lithography process (x/2 nm), as shown in the plot, or change the architecture to an interleaved architecture, which enables a theoretical linear increase in power with sampling rate. However, the move to an

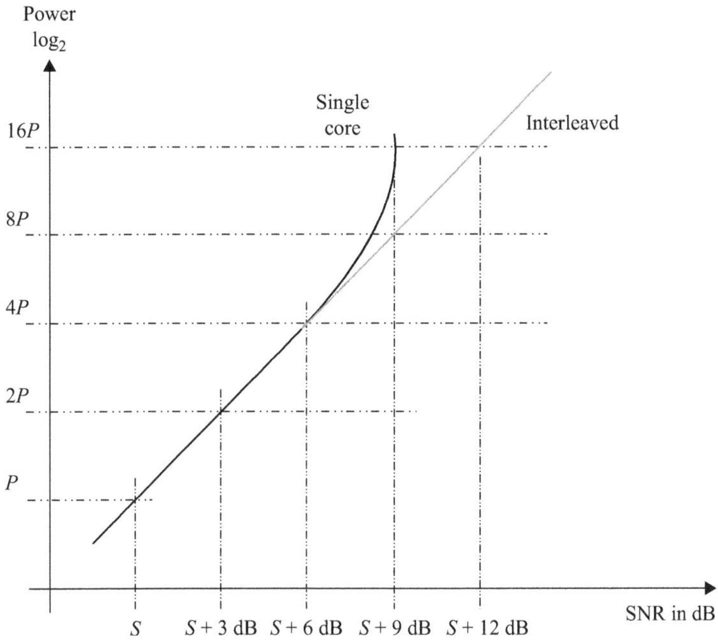

Figure 10.4 Trend curves showing the power consumption versus SNR (or SINAD). Every 3 dB improvement in SINAD requires doubling the power

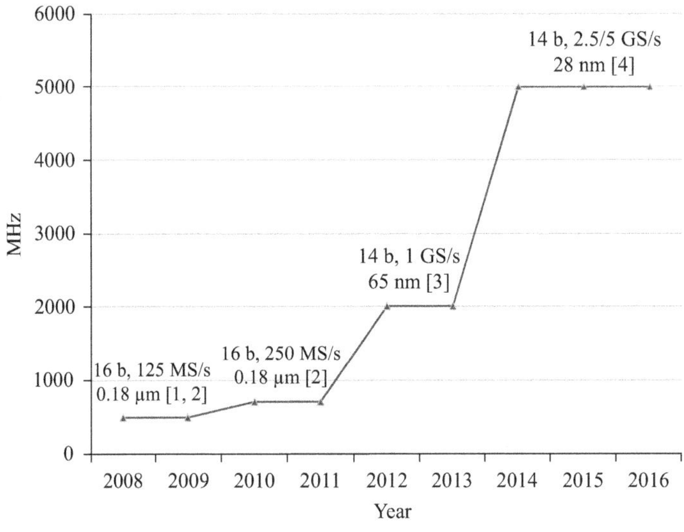

Figure 10.5 Evolution of the maximum input frequency sampling capability (IF/RF sampling)

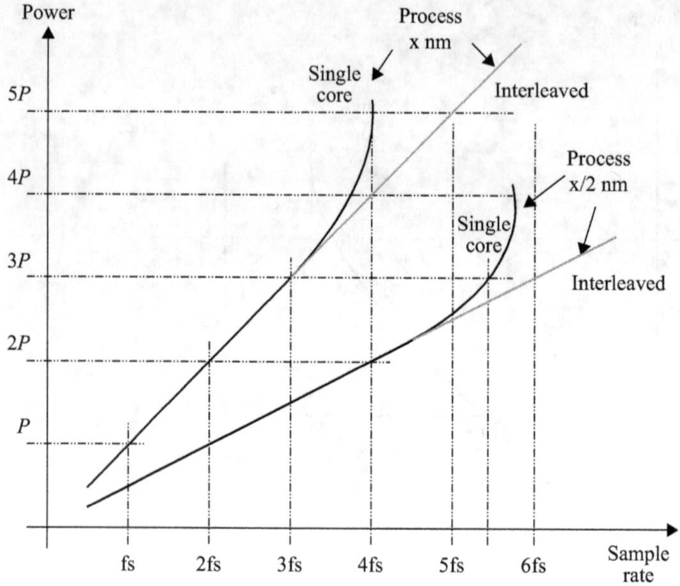

Figure 10.6 Trend curves showing the power consumption versus sampling rate for two hypothetical process nodes x nm and x/2 nm

interleaved architecture adds additional power overhead and creates the dreaded interleaving spurs that need to be addressed or planned around.

Over the past few years, there has been an increased interest in time-interleaved ADCs to push the sampling rate beyond what is possible, or attractive, by the process advancement alone [4, 10]. This has spurred growing interest in researching the interleaving mismatch problems and their calibration, especially timing and bandwidth mismatch calibration. The goal has been to extend the interleaving architecture to the high performance space for resolutions of 10–12 bits and above.

Another trend is the rise in the popularity of SAR ADCs in the high speed space [11]. SAR ADCs are simple, consume low power that scales well with the process lithography, and are compatible with heavy time-interleaving. Tremendous progress has been achieved in pushing the time-interleaved SAR architecture to very high speeds for very low power consumption. Moreover, SAR ADCs have also been incorporated in the pipeline architecture to lower the power consumption of the traditional pipelined ADC, while improving the performance beyond that of the SAR ADC by itself [11–17].

10.2 Process evolution

High speed converter technologists have enjoyed a relatively smooth ride down the CMOS/BiCMOS process curve over the past two decades, which was fraught with doom-and-gloom warnings whenever a new process appeared on the horizon.

This was the case moving from the 0.5 μm process, with its ample 5 V power supply, down to the 0.35 μm process with the then dreaded (yet now lamented) 3.3 V power supply; and further down to the 0.18 μm with its "very tight" 1.8 V supply. Those worries later proved to be non-issues in the face of the creativity and resourcefulness of the analog circuit designers and ADC architects, and with the help of the ever innovative process foundries. Things have changed a bit in the past few years when ADCs moved to the 65 nm and 28 nm processes. While some designers accommodated the ever tighter supplies that shrunk to 1.2 V (65 nm) and 0.9 V (28 nm), others rejected the power penalty of the smaller power supplies on noise-limited ADCs. They preferred to operate portions of the analog signal path on a larger supply (e.g. 1.8 V) where it is most efficient, while using the low supply for the analog and digital blocks that can benefit from the power savings of the lower supply. This dictated the need for creative over-voltage protection circuitry to ensure that the thin oxide devices are not exposed to voltages that exceed their reliability limits. Another consequence of the migration to fine lithography CMOS processes is more reliance on digital assistance, including adaptive background calibrations, and stronger push for higher sampling rates and oversampling to achieve the desired SINAD within the signal band using "Nyquist" ADCs.

The preference to use the larger analog power supply is driven by the need to accommodate a relatively large input signal that is acceptable to the user and can efficiently achieve the desired SINAD and NSD. Lowering the input signal full-scale to fit the smaller supply value can lead to a less efficient design because the SINAD degrades with the input span on a 6 dB/octave curve, while the power saving due to the lower supply in the analog path may follow a 3 dB/octave trend.

For example, if the power supply is cut in half, the analog power would be reduced by 50%. However, if that means reducing the input span by 50%, this would result in a 6 dB loss in SINAD. If the SINAD is limited by the thermal noise, restoring the original SINAD requires increasing the sampling capacitance value by a factor of 4, which increases the current by factor of 4. Therefore, the net power will increase by a factor of 2, which is not an efficient trend.

To demonstrate this trend analytically, let's start with the power S of a sinusoidal input signal with amplitude A, which is given by

$$S = 0.5A^2 \tag{10.1}$$

If we assume that the input amplitude is proportional to the analog power supply value V_{DD}, then

$$S \propto 0.5V_{DD}^2 \tag{10.2}$$

If the noise power N is given by

$$N = kT/C \tag{10.3}$$

then,

$$SINAD \approx \frac{S}{N} \propto \frac{0.5V_{DD}^2}{kT/C} \tag{10.4}$$

that is,

$$SINAD \propto C \times V_{DD}^2 \tag{10.5}$$

The analog power P is proportional to the supply voltage and the capacitance. That is

$$P \propto C \times V_{DD} \tag{10.6}$$

Therefore, from (10.5) and (10.6) the power per SINAD is given by

$$\frac{P}{SINAD} \propto \frac{1}{V_{DD}} \tag{10.7}$$

That is, for the same SINAD in noise-limited ADCs, reducing the supply value can lead to an increase in power consumption if it leads to a proportional reduction in the input full-scale.

In practice, reducing the power supply does not necessarily lead to a proportional reduction in the input full-scale. Moreover, in some cases, it is desirable to reduce the input span for ease of drive regardless of the supply, especially for high input frequencies and high sampling rates. This makes the power consumption trade-off more complicated and dependent on the particular situation. Nevertheless, the cost of reducing the analog power supply has been so steadily increasing that it is hard to find examples where it has led to significant analog power savings in the high speed ADC space.

10.3 Future trends

Given recent advances in process technology, the high speed converter space is expected to continue to evolve, grow, and improve. The limitations of the processes for analog circuit design have dictated using digitally assisted calibration techniques that are clearly here to stay, and will need to continue improving. Now, the designers and architects of high speed converters must be proficient in analog design, mixed-signal design, and signal processing. The most effective calibration techniques are the ones that are inspired by specific needs and limitations on the analog side, and guided by a strong understanding and knowledge of both analog circuits and signal processing. Black box calibration approaches, although useful, tend to be less effective.

In addition to digital assistance, and closely tied to it, is the increased interest in time-interleaved ADCs, which shine in the very high speed and relatively low performance space. Improving the mismatch calibration and architecting the time-interleaved ADCs for application-specific spaces will help make this class of converters more attractive.

Another trend that has been evident in the past decade, and is expected to continue, is the increase in the digital content that is integrated with state-of-the-art ADCs, and the move toward even more integration of both analog and digital functionality. Digital features, such as filtering, decimation, equalization, and spectral shaping, are commonly integrated with the ADC. This creates some challenges for the ADC designers in terms of digital coupling, substrate isolation, power/ground planning, total power consumption, and thermal and packaging issues. The need for integration will only grow more intense as wireless communication starts migrating toward 5G with its massive MIMO architectures. The required performance, bandwidth, and power of converters in this new architecture represent a challenge for the system architects and converter technologists.

As the process technology advances beyond 28 nm toward FinFETs and beyond, there are serious concerns regarding cost, analog friendliness, complexity, and achievable speed improvements. There is also anxiety about whether Moore's law has reached its limit or is slowing down. The answers to these questions have significant ramifications for the high speed converter space, its rate of progress, and its future. However, one thing is certain: the brilliant engineers who have achieved the fascinating progress that we have seen so far in the high speed converter space are capable of pushing this progress forward through innovation and creativity for the foreseeable future.

References

[1] A.M.A. Ali, C. Dillon, R. Sneed, *et al.*, "A 14-bit 125 MS/s IF/RF Sampling Pipelined ADC with 100 dB SFDR and 50 fs Jitter," *IEEE Journal of Solid-State Circuits*, 41(8), pp. 1846–1855, Aug 2006.

[2] A.M.A. Ali, A. Morgan, C. Dillon, *et al.*, "A 16-bit 250-MS/s IF Sampling Pipelined ADC with Background Calibration," *IEEE Journal of Solid-State Circuits*, 45(12), pp. 2602–2612, Dec 2010.

[3] A.M.A. Ali, H. Dinc, P. Bhoraskar, *et al.*, "A 14-bit 1GS/s RF Sampling Pipelined ADC with Background Calibration," *IEEE Journal of Solid-State Circuits*, 49(12), pp. 2857–2867, Dec 2014.

[4] A.M.A. Ali, H. Dinc, P. Bhoraskar, *et al.*, "A 14-bit 2.5GS/s and 5GS/s RF Sampling ADC with Background Calibration and Dither," *IEEE VLSI Circuits Symposium*, pp. 206–207, 2016.

[5] R.H. Walden, "Analog-to-Digital Converter Survey and Analysis," *IEEE Journal of Selected Areas in Communications*, 17(4), pp. 539–550, 1999.

[6] R. Schreier and G. Temes, "Understanding Delta-Sigma Data Converters," IEEE Press, Piscataway, NJ, 2005.

[7] B. Murmann, "A/D Converter Trends: Power Dissipation, Scaling and Digitally Assisted Architectures," *IEEE CICC*, pp. 105–112, 2008.

[8] B. Murmann, "ADC Performance Survey 1997–2015," [Online]. Available at: http://web.stanford.edu/~murmann/adcsurvey.html.

[9] G. Manganaro, "Advanced Data Converters," Cambridge University Press, Cambridge, UK, 2012.

[10] M. Straayer, J. Bales, D. Birdsall, *et al.*, "A 4GS/s Time-Interleaved RF ADC in 65nm CMOS with 4GHz Input Bandwidth," *IEEE ISSCC Digest of Technical Papers*, pp. 464–465, 2016.

[11] R. Kapusta, J. Shen, S. Decker, *et al.*, "A 14b 80MS/s SAR ADC with 73.6dB SNDR in 65nm CMOS," *IEEE ISSCC Digest of Technical Papers*, pp. 472–473, 2013.
[12] C.C. Lee and M.P. Flynn, "A SAR-Assisted Two-Stage Pipeline ADC," *IEEE Journal of Solid State Circuits*, 46(4), pp. 859–869, Apr 2011.
[13] M. Furuta and M. Nozawa, "A 10-bit, 40-MS/s, 1.21 mW Pipelined SAR ADC Using Single-Ended 1.5-bit/cycle Conversion Technique," *IEEE Journal of Solid State Circuits*, 46(6), pp. 1360–1370, Jun 2011.
[14] Y. Zhu, C.-H. Chan, S.-W. Sin, *et al.*, "A 50-fJ 10-b 160-MS/s Pipelined-SAR ADC Decoupled Flip-Around MDAC and Self-Embedded Offset Cancellation," *IEEE Journal of Solid State Circuits*, 47(11), pp. 2614–2626, Nov 2012.
[15] C.-Y. Lin, Y.-H. Wei, and T.-C. Lee, "A 10b 2.6GS/s Time-Interleaved SAR ADC with Background Timing-Skew Calibration," *IEEE ISSCC Digest of Technical Papers*, pp. 468–469, 2016.
[16] J. Wu, A. Chou, T. Li, *et al.*, "A 4GS/s 13b Pipelined ADC with Capacitor and Amplifier Sharing in 16 nm CMOS," *IEEE ISSCC Digest of Technical Papers*, pp. 466–467, 2016.
[17] M. Brandolini, Y. Shin, K. Raviprakash, *et al.*, "A 5GS/s 150mW 10b SHA-Less Pipelined/SAR Hybrid ADC for Direct Sampling Systems in 28nm CMOS," *IEEE Journal of Solid-State Circuits*, 50(12), pp. 2922–2934, Dec 2015.

Index

www.ingramcontent.com/pod-product-compliance
Lightning Source LLC
LaVergne TN
LVHW011902200926
848484LV00002B/504